T0184898

UNITEXT for Physics

For further volumes:
www.springer.com/series/13351

Egidio Landi Degl'Innocenti

Atomic Spectroscopy and Radiative Processes

Translation by Giulio Del Zanna

 Springer

Egidio Landi Degl'Innocenti
Dipartimento di Fisica e Astronomia
Università degli Studi di Firenze
Florence, Italy

ISSN 2198-7882   ISSN 2198-7890 (electronic)
UNITEXT for Physics
ISBN 978-88-470-3905-6   ISBN 978-88-470-2808-1 (eBook)
DOI 10.1007/978-88-470-2808-1
Springer Milan Heidelberg New York Dordrecht London

Translation from the original Italian edition:
Egidio Landi Degl'Innocenti: Spettroscopia atomica e processi radiativi
© Springer-Verlag Italia 2009
© Springer-Verlag Italia 2014
Softcover reprint of the hardcover 1st edition 2014

Printed on acid-free paper

Springer is part of Springer Science+Business Media (www.springer.com)

To Nadine and Vanessa

Preface

This book collects the notes of the undergraduate course I thought at the University of Florence almost uninterruptly for over 25 years. During this time, the program of the course changed. Initially, the topics covered where basically those of atomic spectroscopy. Later, when the course was more specifically addressed to the students of the curriculum in Astrophysics, the program changed with the suppression of some parts and the introduction of new topics, especially those concerning radiative processes and the applications of relativistic quantum electrodynamics.

After a lengthy revision process, the notes, initially handwritten, were collected in 2008 in the Italian edition of this book *Spettroscopia Atomica e Processi Radiativi*, edited by Springer Italia. Thanks to the painful work of translation of a former student of mine, Dr. Giulio Del Zanna, to whom I am deeply indebted, the book can now appear in the present English edition.

The book covers several different topics and it contains enough material for at least two annual courses (at the third- of fourth-year level or at the post-graduate level). By appropriately selecting the topics within this volume, it is possible to follow a number of different didactic paths. For example, if one is interested in teaching a course like "Complements of Electromagnetism and Thermodynamics", the contents of Chaps. 1, 2, 3 and 10 would be appropriate. For a course on "Elementary Atomic Spectroscopy" Chaps. 1, 2, and 6 would suffice, while for a more in-depth course on "Atomic Spectroscopy" Chaps. 4, 5, 7, and 9 could be added. This volume also contains relevant material for other courses such as "Radiative Processes", "Astronomical Spectroscopy", or "Applications of Quantum Electrodynamics".

The diagram below (see Fig. 1) shows the logic connections between the various chapters, as a guideline to choose a learning path. For example, the topics presented in Chap. 15 (second order processes) can only be properly understood if the contents of other chapters are known, in particular Chap. 3 (where the simpler processes are discussed within classical electrodynamics), Chap. 5 (where the Dirac equation is introduced), and Chap. 11 (which discusses the interaction between matter and radiation).

The "introductory" Chaps. 1, 5, 6, and 10 require a basic knowledge of electromagnetism, quantum mechanics, special relativity and statistical thermodynamics,

Fig. 1 This diagram shows
the logic relations between
the various book chapters

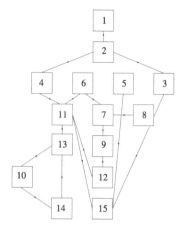

all of which are normally thought within the first three years of the physics degree.

To those readers that do not have solid foundations in the above topics, this volume offers the possibility to learn and understand them better, following the various examples where direct physical applications are provided. This was done on purpose, to avoid a purely theoretical approach, as often found in other works.

The appendices are a significant part of this book. Some of them are dedicated to further develop some of the important results which follow on as natural complements of the topics covered in the text. Others are dedicated to introduce the formalism used throughout the text.

Finally, I would like to thank the lecturers of the *Corso di Laurea in Fisica* of the University of Florence which I attended in the early and mid 1960s. They showed me how understanding physics requires an in-depth quantitative interpretation of the empirical phenomena, without excessive indulgence in the formalism. In particular, my thanks go to Profs. Manlio Mandò, Simone Franchetti and Giuliano Toraldo di Francia, sadly no longer with us, and to Prof. Marco Ademollo. A heartfelt thanks also goes to the generations of students that have followed over the past several years, and whose contribution, in terms of illuminating questioning, has been fundamental to the writing of this book.

Arcetri, Florence, Italy   Egidio Landi Degl'Innocenti
November 20, 2013

Contents

Chapter 1
General Laws of the Electromagnetic Field

Maxwell's equations enclose the main experimental results on electricity and magnetism obtained in the late eighteenth and first half of the nineteenth century by many scientists, in particular Coulomb, Volta, Ørsted, Ampère, Faraday, and Gauss.

The aim of this chapter is to return to these equations and to derive a series of consequences that are particularly relevant for the understanding of electromagnetic phenomena. Specifically, we will see how Maxwell's equations imply the possibility to define two fundamental physical quantities for the electromagnetic field such as the energy density and the momentum density.

We will also see how the solution of Maxwell's equations can be greatly simplified by the introduction of two potentials: the scalar potential and the vector potential, of which we will study the main properties. The last part of the chapter is dedicated to the study of plane electromagnetic waves, a particular solution of Maxwell's equations in vacuum.

1.1 Maxwell's Equations

Maxwell's equations summarise the various empirical laws related to electromagnetic phenomena. In the system of CGS units of Gauss, consistently used throughout this volume, these equations are written in the form[1]

$$\text{div}\,\mathbf{E} = 4\pi\rho, \tag{1.1}$$

$$\text{div}\,\mathbf{B} = 0, \tag{1.2}$$

$$\text{rot}\,\mathbf{B} - \frac{1}{c}\frac{\partial \mathbf{E}}{\partial t} = \frac{4\pi}{c}\mathbf{j}, \tag{1.3}$$

---

[1]The reader who is not familiar with the system of units used here can find in Sect. 16.1 the relations between this system and the Système International (SI), more appropriate for practical applications rather than theoretical ones.

E. Landi Degl'Innocenti, *Atomic Spectroscopy and Radiative Processes*,
UNITEXT for Physics, DOI 10.1007/978-88-470-2808-1_1, © Springer-Verlag Italia 2014

$$\operatorname{rot} \mathbf{E} + \frac{1}{c} \frac{\partial \mathbf{B}}{\partial t} = 0, \tag{1.4}$$

where $\mathbf{E}$ is the electric field, $\mathbf{B}$ is the vector of magnetic induction, ρ is the electric charge density, and $\mathbf{j}$ is the corresponding current density. The first equation is just Coulomb's law for electrical charges in its differential form. The second equation, sometimes called Gilbert's law, is the analogous expression for the so-called magnetic charges, which takes into account that magnetic monopoles do not exist. The third is Ampère's law in its differential form, as modified by Maxwell in order to take into account the displacement current. The fourth is Faraday's induction equation in its differential form.

Written in this form, the equations are very general in the sense that they apply not only to the vacuum but are also valid in the presence of a medium, provided that polarization charges are included in ρ, and magnetic and polarization currents are included in the current density $\mathbf{j}$. More precisely, the electric displacement $\mathbf{D}$ and magnetic $\mathbf{H}$ field vectors are introduced, alongside $\mathbf{E}$ and $\mathbf{B}$, when describing material systems. $\mathbf{E}$ is the electric field vector as measured within a needle-like cavity aligned along the direction of the field itself (to avoid the formation of polarization charges). $\mathbf{B}$ is the magnetic field vector as measured within a cavity of infinitesimal thickness aligned perpendicularly to the lines of force (to avoid the formation of currents).

Maxwell's equations already include the equation of continuity for the electric charge. In fact, taking the divergence of Eq. (1.3) and substituting Eq. (1.1), we obtain

$$-\frac{4\pi}{c} \frac{\partial \rho}{\partial t} = \frac{4\pi}{c} \operatorname{div} \mathbf{j},$$

that is,

$$\frac{\partial \rho}{\partial t} + \operatorname{div} \mathbf{j} = 0,$$

which is the continuity equation, expressing the conservation of the electric charge.

1.2 Energy Transported by the Electric Field

Consider a system of charges and currents subject only to electromagnetic interactions. Denoting by W the mechanical energy per unit volume, we have, recalling that the work of magnetic forces on isolated charges is zero (because the Lorentz force is always perpendicular to the particle velocity),

$$\frac{\partial W}{\partial t} = \mathbf{j} \cdot \mathbf{E}.$$

Getting $\mathbf{j}$ from Eq. (1.3) we obtain

$$\frac{\partial W}{\partial t} = \frac{c}{4\pi} \mathbf{E} \cdot \operatorname{rot} \mathbf{B} - \frac{1}{4\pi} \mathbf{E} \cdot \frac{\partial \mathbf{E}}{\partial t},$$

and subtracting Eq. (1.4) from the right-hand side after taking the scalar product with the vector $c\mathbf{B}/(4\pi)$, we obtain

$$\frac{\partial W}{\partial t} = \frac{c}{4\pi}\mathbf{E}\cdot\mathrm{rot}\,\mathbf{B} - \frac{c}{4\pi}\mathbf{B}\cdot\mathrm{rot}\,\mathbf{E} - \frac{1}{4\pi}\mathbf{E}\cdot\frac{\partial\mathbf{E}}{\partial t} - \frac{1}{4\pi}\mathbf{B}\cdot\frac{\partial\mathbf{B}}{\partial t}.$$

Recalling the vector identity (see Sect. 16.2, Eq. (16.7))

$$\mathrm{div}(\mathbf{E}\times\mathbf{B}) = \mathbf{B}\cdot\mathrm{rot}\,\mathbf{E} - \mathbf{E}\cdot\mathrm{rot}\,\mathbf{B},$$

we can write

$$\frac{\partial}{\partial t}(W+u) + \mathrm{div}\,\mathbf{S} = 0,$$

where

$$u = \frac{1}{8\pi}\left(E^2 + B^2\right), \qquad \mathbf{S} = \frac{c}{4\pi}\mathbf{E}\times\mathbf{B}.$$

This equation, known as Poynting's theorem, represents an energy balance: the electromagnetic field has an energy per unit of volume given by u and it transports energy along the direction of the vector $\mathbf{S}$. The magnitude of this vector is the energy flux density, i.e. the energy flowing per unit of time across the unit area normal to the direction of the vector itself. $\mathbf{S}$ is called Poynting vector.

1.3 Momentum Transported by the Electromagnetic Field

We now define with $\mathbf{Q}$ the momentum per unit volume associated with the same system of charges and currents considered in the previous section. This momentum varies with time following the equation

$$\frac{\partial\mathbf{Q}}{\partial t} = \rho\mathbf{E} + \frac{\mathbf{j}}{c}\times\mathbf{B},$$

where the first term in the right-hand side is due to the electric force, while the second term is due to the Lorentz force. Eliminating ρ and $\mathbf{j}$ using Eq. (1.1) and (1.3), we obtain

$$\frac{\partial\mathbf{Q}}{\partial t} = \frac{1}{4\pi}\mathbf{E}\,\mathrm{div}\,\mathbf{E} + \frac{1}{4\pi}(\mathrm{rot}\,\mathbf{B})\times\mathbf{B} - \frac{1}{4\pi c}\frac{\partial\mathbf{E}}{\partial t}\times\mathbf{B},$$

i.e., recalling the definition of the Poynting vector

$$\frac{\partial\mathbf{Q}}{\partial t} = \frac{1}{4\pi}\mathbf{E}\,\mathrm{div}\,\mathbf{E} + \frac{1}{4\pi}(\mathrm{rot}\,\mathbf{B})\times\mathbf{B} - \frac{1}{c^2}\frac{\partial\mathbf{S}}{\partial t} + \frac{1}{4\pi c}\mathbf{E}\times\frac{\partial\mathbf{B}}{\partial t}.$$

Substituting the $\partial\mathbf{B}/\partial t$ term using Eq. (1.4) and recalling that $\mathrm{div}\,\mathbf{B} = 0$ (Eq. (1.2)), we obtain the symmetric formula

$$\frac{\partial}{\partial t}\left(\mathbf{Q} + \frac{\mathbf{S}}{c^2}\right) = \frac{1}{4\pi}(\mathbf{E}\,\mathrm{div}\,\mathbf{E} + \mathbf{B}\,\mathrm{div}\,\mathbf{B} - \mathbf{E}\times\mathrm{rot}\,\mathbf{E} - \mathbf{B}\times\mathrm{rot}\,\mathbf{B}).$$

Considering the vector identity

$$\text{div}(\mathbf{a}\,\mathbf{a}) - \frac{1}{2}\,\text{grad}\!\left(a^2\right) = \mathbf{a}\,\text{div}\,\mathbf{a} - \mathbf{a} \times \text{rot}\,\mathbf{a},$$

which can be obtained using Eqs. (16.11) and (16.12) in Sect. 16.2, we obtain

$$\frac{\partial}{\partial t}\left(\mathbf{Q} + \frac{\mathbf{S}}{c^2}\right) = \frac{1}{4\pi}\,\text{div}(\mathbf{E}\mathbf{E} + \mathbf{B}\mathbf{B}) - \frac{1}{8\pi}\,\text{grad}\!\left(E^2 + B^2\right).$$

We now introduce the tensor $\mathbf{T}$, named after Maxwell, defined by the expression

$$\mathbf{T} = \frac{1}{4\pi}\,(\mathbf{E}\mathbf{E} + \mathbf{B}\mathbf{B}) - \frac{1}{8\pi}\left(E^2 + B^2\right)\mathbf{U},$$

where $\mathbf{U}$ is the unit tensor ($U_{ij} = \delta_{ij}$). In components, we have

$$T_{ij} = T_{ji} = \frac{1}{4\pi}\,(E_i E_j + B_i B_j) - \frac{1}{8\pi}\left(E^2 + B^2\right)\delta_{ij}.$$

Recalling the tensorial identity (see Eq. (16.13))

$$\text{div}(f\mathbf{T}) = (\text{grad}\,f)\cdot\mathbf{T} + f\,\text{div}\,\mathbf{T},$$

(where f is an arbitrary scalar and $\mathbf{T}$ is an arbitrary tensor), and given that

$$\mathbf{v}\cdot\mathbf{U} = \mathbf{U}\cdot\mathbf{v} = \mathbf{v},$$

(where $\mathbf{v}$ is an arbitrary vector), we finally obtain

$$\frac{\partial}{\partial t}\left(\mathbf{Q} + \frac{\mathbf{S}}{c^2}\right) = \text{div}\,\mathbf{T}.$$

This equation represents a momentum balance and is interpreted as follows. The electromagnetic field possesses a momentum per unit of volume which is given by $\mathbf{S}/c^2$. The electromagnetic field behaves as an elastic material subject to deformations, where the internal forces can be described by a stress tensor. The variation per unit time of the momentum contained within an arbitrary volume is equal to the flux of the tensor $-\mathbf{T}$ across the area that encloses the same volume. If one considers an infinitesimal area dS and its normal unit vector $\mathbf{n}$, the quantity $d\mathbf{F} = -\mathbf{n}\cdot\mathbf{T}\,dS$ represents the momentum flux across dS (defined as positive if directed along $\mathbf{n}$).

1.4 Electromagnetic Potentials

Recalling Eq. (1.2) and the fact that the divergence of the curl of any vector is zero, the magnetic field[2] can be written as a function of a suitable vector potential $\mathbf{A}$

$$\mathbf{B} = \operatorname{rot} \mathbf{A}.$$

Substituting this expression into Eq. (1.4), we obtain

$$\operatorname{rot}\left(\mathbf{E} + \frac{1}{c}\frac{\partial \mathbf{A}}{\partial t}\right) = 0,$$

and recalling that the curl of the gradient of any vector is always zero, we can introduce the scalar potential ϕ by writing

$$\mathbf{E} = -\operatorname{grad}\phi - \frac{1}{c}\frac{\partial \mathbf{A}}{\partial t}.$$

We now substitute these expressions for $\mathbf{B}$ and $\mathbf{E}$ into the non-homogeneous Maxwell equations. From Eq. (1.3) we have

$$\operatorname{rot}(\operatorname{rot}\mathbf{A}) + \frac{1}{c}\frac{\partial}{\partial t}\left(\operatorname{grad}\phi + \frac{1}{c}\frac{\partial \mathbf{A}}{\partial t}\right) = 4\pi\frac{\mathbf{j}}{c},$$

that is, using the vector identity (16.14)

$$\nabla^2 \mathbf{A} - \frac{1}{c^2}\frac{\partial^2 \mathbf{A}}{\partial t^2} - \operatorname{grad}\left(\operatorname{div}\mathbf{A} + \frac{1}{c}\frac{\partial \phi}{\partial t}\right) = -4\pi\frac{\mathbf{j}}{c}.$$

From Eq. (1.1) we also have

$$\nabla^2 \phi + \frac{1}{c}\frac{\partial}{\partial t}\operatorname{div}\mathbf{A} = -4\pi\rho,$$

an equation that can also be written in the form

$$\nabla^2 \phi - \frac{1}{c^2}\frac{\partial^2 \phi}{\partial t^2} + \frac{1}{c}\frac{\partial}{\partial t}\left(\operatorname{div}\mathbf{A} + \frac{1}{c}\frac{\partial \phi}{\partial t}\right) = -4\pi\rho.$$

The system of these last two equations in principle allows one to obtain the potentials $\mathbf{A}$ and ϕ (and therefore the fields $\mathbf{E}$ and $\mathbf{B}$), once ρ, $\mathbf{j}$, and the boundary conditions are known. We will see however that the equations involving $\mathbf{A}$ and ϕ can be simplified using a supplementary condition.

---

[2]The vector $\mathbf{B}$ should be called more appropriately "magnetic induction" instead of "magnetic field", this last name referring to the vector $\mathbf{H}$. However it has now become customary to call $\mathbf{B}$ the magnetic field vector, when no ambiguities are present.

1.5 Gauge Invariance

The potentials $\mathbf{A}$ e ϕ uniquely determine the electric and magnetic fields. However, there exist different $\mathbf{A}$ and ϕ functions that can provide the same $\mathbf{E}$ and $\mathbf{B}$. Indeed, if we substitute

$$\mathbf{A} \to \mathbf{A}' = \mathbf{A} - \operatorname{grad} \chi, \tag{1.5}$$

$$\phi \to \phi' = \phi + \frac{1}{c} \frac{\partial \chi}{\partial t}, \tag{1.6}$$

where χ is an arbitrary function of space and time, we obtain, for the new field $\mathbf{B}'$,

$$\mathbf{B}' = \operatorname{rot}(\mathbf{A} - \operatorname{grad} \chi) = \operatorname{rot} \mathbf{A} = \mathbf{B},$$

and, similarly for the new field $\mathbf{E}'$,

$$\mathbf{E}' = -\operatorname{grad}\left(\phi + \frac{1}{c} \frac{\partial \chi}{\partial t} \right) - \frac{1}{c} \frac{\partial}{\partial t}(\mathbf{A} - \operatorname{grad} \chi) = \mathbf{E}.$$

This property is called *gauge* invariance and can be conveniently used with an appropriate choice of the function χ.

Let us suppose that $\mathbf{A}_0$ and ϕ_0 are the vector and scalar potentials producing the $\mathbf{E}$ and $\mathbf{B}$ fields that we want to determine, given a set of boundary conditions and a distribution of charges and currents. If we consider the gauge transformation

$$\mathbf{A} = \mathbf{A}_0 - \operatorname{grad} \chi,$$

$$\phi = \phi_0 + \frac{1}{c} \frac{\partial \chi}{\partial t},$$

and we choose the function χ so as to satisfy the equation

$$\nabla^2 \chi - \frac{1}{c^2} \frac{\partial^2 \chi}{\partial t^2} = \operatorname{div} \mathbf{A}_0 + \frac{1}{c} \frac{\partial \phi_0}{\partial t},$$

we have

$$\operatorname{div} \mathbf{A} + \frac{1}{c} \frac{\partial \phi}{\partial t} = 0. \tag{1.7}$$

When the potentials are chosen so as to satisfy this equation, it is said that the Lorenz gauge is adopted[3] and the equations for $\mathbf{A}$ and ϕ become

$$\nabla^2 \mathbf{A} - \frac{1}{c^2} \frac{\partial^2 \mathbf{A}}{\partial t^2} = -4\pi \frac{\mathbf{j}}{c}, \tag{1.8}$$

---

[3]The Danish physicist L.V. Lorenz (1829–1891) should not be confused with the more famous Dutch physicist H.A. Lorentz (1853–1928), best known for the transformations and the force that bear his name. Both physicists, almost contemporary, are remembered together for the Lorentz–Lorenz equation which relates the index of refraction of a material with the polarizability coefficient of its molecules.

$$\nabla^2 \phi - \frac{1}{c^2}\frac{\partial^2 \phi}{\partial t^2} = -4\pi\rho. \tag{1.9}$$

These equations can be further simplified with another gauge transformation if there are no free charges. Let us assume that $\mathbf{A}_0$ and ϕ_0 are two potentials that satisfy the Lorenz gauge. If we apply another transformation within the Lorenz gauge, the function χ needs to satisfy the equation

$$\nabla^2 \chi - \frac{1}{c^2}\frac{\partial^2 \chi}{\partial t^2} = 0.$$

On the other hand, in the absence of free electric charges, the potential ϕ_0 satisfies the equation

$$\nabla^2 \phi_0 - \frac{1}{c^2}\frac{\partial^2 \phi_0}{\partial t^2} = 0,$$

which is the same equation satisfied by χ and therefore also by $\partial \chi / \partial t$. If the function χ is chosen so that

$$-\frac{1}{c}\frac{\partial \chi}{\partial t} = \phi_0,$$

it is possible to obtain for the scalar potential

$$\phi = 0, \tag{1.10}$$

so that the Lorenz condition simply becomes

$$\operatorname{div}\mathbf{A} = 0. \tag{1.11}$$

Another particularly interesting gauge is the Coulomb gauge, obtained when the function χ is chosen in order to satisfy

$$\nabla^2 \chi = \operatorname{div}\mathbf{A}_0,$$

where $\mathbf{A}_0$ is, as before, the vector potential that we want to calculate. With this gauge transformation we obtain

$$\operatorname{div}\mathbf{A} = 0,$$

and the resulting equations for $\mathbf{A}$ and ϕ are

$$\nabla^2 \mathbf{A} - \frac{1}{c^2}\frac{\partial^2 \mathbf{A}}{\partial t^2} = -4\pi\frac{\mathbf{j}}{c} + \frac{1}{c}\operatorname{grad}\frac{\partial \phi}{\partial t},$$

$$\nabla^2 \phi = -4\pi\rho.$$

This gauge has the advantage that the scalar potential coincides with the (instantaneous) one of electrostatics, even in the case of non-stationary phenomena involving the presence of charges and currents varying with time.

1.6 Solving Maxwell's Equations in Vacuum

Let us consider a region of space free of charges and currents. As we saw in the previous section, within the Lorenz gauge we can set $\phi = 0$ so that the electromagnetic field is described only by the vector potential, which satisfies the equation

$$\nabla^2 \mathbf{A} - \frac{1}{c^2} \frac{\partial^2 \mathbf{A}}{\partial t^2} = 0,$$

with the supplementary condition of the Lorenz gauge (in vacuum)

$$\operatorname{div} \mathbf{A} = 0.$$

Let us assume that the electromagnetic field is a function of one coordinate, i.e. only depends on the coordinate z of an orthogonal Cartesian reference system. The first equation can then be written

$$\frac{\partial^2 \mathbf{A}}{\partial z^2} - \frac{1}{c^2} \frac{\partial^2 \mathbf{A}}{\partial t^2} = 0.$$

This equation, commonly known as the wave equation, has the general solution

$$\mathbf{A}(z, t) = \mathbf{A}_\mathrm{p}(z - ct) + \mathbf{A}_\mathrm{r}(z + ct),$$

where $\mathbf{A}_\mathrm{p}$ and $\mathbf{A}_\mathrm{r}$ are arbitrary vectors, functions of their arguments. The two terms represent plane waves that are propagating along the positive and negative direction of the z axis with the velocity c. If we consider for example only the first type

$$\mathbf{A}(z, t) = \mathbf{A}_\mathrm{p}(z - ct),$$

we see that, for a fixed time t, the vector potential is constant in any plane $z = $ constant. Furthermore, the value of the vector potential at time t_0 and plane $z = z_0$ changes, within a time $\mathrm{d}t$, by a quantity $\mathrm{d}z$ such that

$$z_0 - ct_0 = (z_0 + \mathrm{d}z) - c(t_0 + \mathrm{d}t),$$

i.e.

$$\mathrm{d}z = c\,\mathrm{d}t.$$

This justifies calling the propagation of the vector potential $\mathbf{A}_\mathrm{p}(z - ct)$ a plane progressive wave. Similarly, we can show that the vector potential $\mathbf{A}_\mathrm{r}(z + ct)$ describes a regressive wave.

We now generalise the above results by assuming that propagation occurs along the unit vector $\mathbf{n}$. We then have

$$\mathbf{A}(\mathbf{r}, t) = \mathbf{a}(f),$$

where the phase of the wave f is given by

$$f = \mathbf{n} \cdot \mathbf{r} - ct,$$

$\mathbf{r}$ being the position vector measured from the origin of the coordinate system. $\mathbf{a}$ is an arbitrary (vectorial) function of f. Applying Lorenz condition (Eq. (1.11)), we have

$$\mathrm{div}\big[\mathbf{a}(f)\big] = \mathbf{a}'(f) \cdot \mathbf{n} = 0, \tag{1.12}$$

where we have taken into account that

$$\frac{\partial f}{\partial x_i} = \frac{\partial}{\partial x_i}\left(\sum_j n_j x_j\right) = n_i,$$

and where $\mathbf{a}'(f)$ indicates the derivative of the vector $\mathbf{a}(f)$ with respect to its argument, i.e.

$$\mathbf{a}'(f) = \frac{\mathrm{d}\mathbf{a}(f)}{\mathrm{d}f}.$$

By integrating Eq (1.12) with respect to f we obtain

$$\mathbf{a}(f) \cdot \mathbf{n} + \mathbf{a}_0 \cdot \mathbf{n} = 0,$$

where $\mathbf{a}_0$ is a constant vector, i.e. independent of $\mathbf{r}$ and t, that we can equate to zero by means of an appropriate gauge transformation (applying Eq. (1.5) and choosing for χ the function $\mathbf{a}_0 \cdot \mathbf{r}$). We then have

$$\mathbf{a}(f) \cdot \mathbf{n} = 0,$$

which means that the vector potential which describes the plane wave is perpendicular to the direction of propagation.

We now consider the vectors $\mathbf{E}$ and $\mathbf{B}$ associated with the wave. With simple algebra we obtain

$$\mathbf{B} = \mathrm{rot}\big[\mathbf{a}(f)\big] = \mathbf{n} \times \mathbf{a}'(f),$$

$$\mathbf{E} = -\frac{1}{c}\frac{\partial \mathbf{a}(f)}{\partial t} = \mathbf{a}'(f).$$

We see that the vectors $\mathbf{E}$ and $\mathbf{B}$ are both perpendicular to the direction of propagation and that

$$\mathbf{B} = \mathbf{n} \times \mathbf{E},$$

which means that the two vectors are perpendicular to each other and such that their directions form with $\mathbf{n}$ a right-handed coordinate system[4] (following the order $\mathbf{n}, \mathbf{E}, \mathbf{B}$), as shown in Fig. 1.1.

---

[4]The fact that the triad of vectors $(\mathbf{n}, \mathbf{E}, \mathbf{B})$ is right-handed and not left-handed is related with the conventions that have historically been established to identify the signs of the electric charges and

Fig. 1.1 The vectors **n**, **E**
and **B** form, in this order,
a right-handed triad

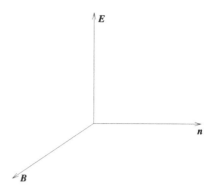

The electric and magnetic energy carried by the wave are equal to each other (because $E^2 = B^2$). The total density of the electromagnetic energy is given by

$$u = \frac{1}{8\pi}\left(E^2 + B^2\right) = \frac{1}{4\pi}E^2,$$

and Poynting vector is

$$\mathbf{S} = \frac{c}{4\pi}\mathbf{E} \times \mathbf{B} = \frac{c}{4\pi}E^2\mathbf{n}.$$

We recall that the magnitude of the Poynting vector is the energy flowing per unit of time across a unit area perpendicular to the unit vector **n**. By comparing the two above equations, it is clear that the energy propagates with speed c.

The momentum density can therefore be written as

$$\mathbf{g} = \frac{\mathbf{S}}{c^2} = \frac{E^2}{4\pi c}\mathbf{n} = \frac{u}{c}\mathbf{n},$$

i.e. the momentum density is directed along the direction of propagation of the wave, and is equal to the energy density divided by c. This is an important classical result that in quantum mechanics becomes $E = cp$, where E is the energy and p the momentum of a photon.

The momentum flux across a unit area perpendicular to the direction of propagation can be calculated with Maxwell's tensor

$$\mathbf{F(n)} = -\mathbf{n} \cdot \mathbf{T}.$$

By substituting the expression for **T** and considering that $\mathbf{E} \cdot \mathbf{n} = \mathbf{B} \cdot \mathbf{n} = 0$, we obtain

$$\mathbf{F(n)} = \frac{1}{4\pi}E^2\mathbf{n}.$$

---

the magnetic masses. We recall that the positive (negative) electric charge is that which is deposited by rubbing a glass (ebonite) rod, while the magnetic mass is positive (negative) if it is attracted by the North (South) terrestrial pole.

Fig. 1.2 A plane
electromagnetic wave falls on
the elementary surface dS.
The momentum absorbed by
the surface is directed along
the unit vector **n**

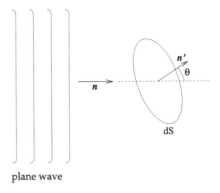

plane wave

More generally, we can calculate the momentum flux across a unit area perpendicular to the direction identified by the unit vector $\mathbf{n}'$. We have

$$\mathbf{F}(\mathbf{n}') = -\mathbf{n}' \cdot \mathbf{T},$$

and substituting

$$\mathbf{F}(\mathbf{n}') = -\frac{1}{4\pi}\left[(\mathbf{n}' \cdot \mathbf{E})\mathbf{E} + (\mathbf{n}' \cdot \mathbf{B})\mathbf{B} - \frac{E^2 + B^2}{2}\mathbf{n}'\right].$$

Recalling that $\mathbf{n}$, $\mathbf{E}$, and $\mathbf{B}$ form an orthogonal triad of vectors, we can write

$$\mathbf{n}' = (\mathbf{n}' \cdot \mathbf{n})\mathbf{n} + \frac{(\mathbf{n}' \cdot \mathbf{E})\mathbf{E}}{E^2} + \frac{(\mathbf{n}' \cdot \mathbf{B})\mathbf{B}}{B^2},$$

from which, given that $E^2 = B^2$, we obtain

$$\mathbf{F}(\mathbf{n}') = \frac{1}{4\pi}E^2(\mathbf{n}' \cdot \mathbf{n})\mathbf{n},$$

i.e. the flux is always directed along the direction of propagation, but contains a projection factor $(\mathbf{n}' \cdot \mathbf{n}) = \cos\theta$, where θ is the angle defined in Fig. 1.2.

1.7 Radiation Pressure

We now consider a plane wave travelling along the direction $\mathbf{n}$ incident on a surface of area dS of a perfectly-absorbing body, oriented as shown in Fig. 1.2. The surface dS absorbs, within the time dt, a momentum $d\mathbf{q}$ given by

$$d\mathbf{q} = \mathbf{F}(\mathbf{n}')\,dS\,dt = \frac{E^2}{4\pi}(\mathbf{n}' \cdot \mathbf{n})\mathbf{n}\,dS\,dt.$$

The projection along the perpendicular to the surface is

$$d\mathbf{q} \cdot \mathbf{n}' = \frac{E^2}{4\pi}(\mathbf{n}' \cdot \mathbf{n})^2 \, dS \, dt = \frac{E^2}{4\pi} \cos^2\theta \, dS \, dt.$$

This momentum transfer is effectively a pressure[5] (radiation pressure) that is

$$P_{\text{rad}} = \frac{d\mathbf{q} \cdot \mathbf{n}'}{dS \, dt} = \frac{E^2}{4\pi} \cos^2\theta = u \cos^2\theta, \qquad (1.13)$$

where u is the energy density.

This result applies only to a perfectly absorbing surface. In the case of a perfectly reflecting surface one must take into account that the amount of momentum transfer is twice, so that the right-hand side of the above equation must be multiplied by a factor of 2.

The pressure exerted by light on a surface was revealed experimentally by Lebedev in 1901 with a delicate experiment which confirmed, within experimental errors, the theoretical value expected from Maxwell's equations. In astrophysics, we now know that radiation pressure has a fundamental importance in the interiors of hot stars and as a mechanism for acceleration of stellar winds. In terms of Astronautics there are some projects whereby a spacecraft would be propelled out of the solar system using the pressure of the solar radiation (cosmic sails).

1.8 Sinusoidal Plane Waves

A particularly important type of plane wave is the sinusoidal plane wave, which is described by a vector potential of the form

$$\mathbf{A}(\mathbf{r}, t) = \mathbf{A}_0 \cos\left[\frac{\omega}{c}(\mathbf{n} \cdot \mathbf{r} - ct) + \varphi\right],$$

where $\mathbf{A}_0$ and φ are two constants known as the amplitude and phase of the wave, ω is the angular frequency, and $\mathbf{n}$ is the unit vector along the direction of propagation. The above expression can also be written in various equivalent forms, introducing e.g. the cyclic frequency (or frequency *tout court*) ν, the period T, the wavelength λ or the wavenumber $\mathbf{k}$. These quantities are related among themselves and the speed of light by the following relations

$$\nu = \frac{\omega}{2\pi} = \frac{1}{T}, \qquad \mathbf{k} = \frac{\omega}{c}\mathbf{n} = k\mathbf{n},$$

---

[5] Actually, for the case shown in figure, we also have a component of the force that is not perpendicular to the surface. These types of components vanish when the radiation incident on the surface is isotropic. In this case the factor $\cos^2\theta$ in Eq. (1.13) is replaced by $1/3$, its average over the solid angle.

$$\lambda = \frac{2\pi}{k}, \qquad \lambda v = c, \qquad \omega = ck.$$

For example, one can write

$$\mathbf{A}(\mathbf{r}, t) = \mathbf{A}_0 \cos\left(\frac{2\pi \mathbf{n} \cdot \mathbf{r}}{\lambda} - 2\pi v t + \varphi\right),$$

or

$$\mathbf{A}(\mathbf{r}, t) = \mathbf{A}_0 \cos(\mathbf{k} \cdot \mathbf{r} - \omega t + \varphi).$$

If for example the last expression is adopted, we obtain for $\mathbf{E}$ and $\mathbf{B}$

$$\mathbf{B}(\mathbf{r}, t) = \mathbf{B}_0 \sin(\mathbf{k} \cdot \mathbf{r} - \omega t + \varphi), \qquad \mathbf{E}(\mathbf{r}, t) = \mathbf{E}_0 \sin(\mathbf{k} \cdot \mathbf{r} - \omega t + \varphi),$$

where

$$\mathbf{B}_0 = -\mathbf{k} \times \mathbf{A}_0, \qquad \mathbf{E}_0 = -\frac{\omega}{c} \mathbf{A}_0.$$

Obviously we still have, at each time t and at each point of space,

$$\mathbf{B}(\mathbf{r}, t) = \mathbf{n} \times \mathbf{E}(\mathbf{r}, t),$$

as we have shown, in more general terms, in the previous section.

The energy and the momentum of a sinusoidal plane wave are functions of space and time of the type $\sin^2(\mathbf{k} \cdot \mathbf{r} - \omega t + \varphi)$. By taking the temporal averages of these quantities, one obtains the corresponding expressions of the previous section, with the terms E^2 and B^2 now substituted with $E_0^2/2$ and $B_0^2/2$, respectively (the average of the square of the sine function over a period is equal to $1/2$).

Sometimes it is convenient to represent a sinusoidal plane wave with a complex exponential, instead of using a real expression, as for example

$$\mathbf{A}(\mathbf{r}, t) = \mathbf{A}_0 \, e^{i(\mathbf{k} \cdot \mathbf{r} - \omega t + \varphi)}, \quad \text{or} \quad \mathbf{A}(\mathbf{r}, t) = \boldsymbol{\mathcal{A}}_0 \, e^{i(\mathbf{k} \cdot \mathbf{r} - \omega t)},$$

where $\boldsymbol{\mathcal{A}}_0$ is a complex vector given by

$$\boldsymbol{\mathcal{A}}_0 = \mathbf{A}_0 \, e^{i\varphi}.$$

By convention, the physical observable (e.g. the electric field) represents the real part of the complex expression. The use of complex exponentials is convenient when performing linear operations, as the real part of the result is equal to the result that is obtained by performing the same linear operations on the real part. However, when non-linear operations are performed (e.g. for the calculation of the energy), one must first take the real part of the complex exponential and then perform the desired operation.

In this regard, we note that if $A(t)$ and $B(t)$ are two complex quantities with the same sinusoidal dependence on time

$$A(t) = \mathcal{A} e^{-i\omega t}, \qquad B(t) = \mathcal{B} e^{-i\omega t},$$

with $\mathcal{A}$ and $\mathcal{B}$ constants, we can write for the time average over a period of the product of their real parts

$$\langle \mathrm{Re}\, A(t)\, \mathrm{Re}\, B(t) \rangle = \frac{1}{2} \mathrm{Re}(\mathcal{A}\mathcal{B}^*) = \frac{1}{2} \mathrm{Re}(\mathcal{A}^*\mathcal{B}).$$

In fact:

$$\mathrm{Re}\, A(t) = \frac{1}{2}\left[\mathcal{A}\,\mathrm{e}^{-\mathrm{i}\omega t} + \mathcal{A}^*\,\mathrm{e}^{\mathrm{i}\omega t}\right], \qquad \mathrm{Re}\, B(t) = \frac{1}{2}\left[\mathcal{B}\,\mathrm{e}^{-\mathrm{i}\omega t} + \mathcal{B}^*\,\mathrm{e}^{\mathrm{i}\omega t}\right],$$

and therefore

$$\langle \mathrm{Re}\, A(t)\, \mathrm{Re}\, B(t) \rangle = \frac{1}{4}\langle \mathcal{A}\mathcal{B}\,\mathrm{e}^{-2\mathrm{i}\omega t} + \mathcal{A}\mathcal{B}^* + \mathcal{A}^*\mathcal{B} + \mathcal{A}^*\mathcal{B}^*\,\mathrm{e}^{2\mathrm{i}\omega t}\rangle.$$

Since $\langle \mathrm{e}^{\pm 2\mathrm{i}\omega t}\rangle = 0$, we have

$$\langle \mathrm{Re}\, A(t)\, \mathrm{Re}\, B(t) \rangle = \frac{1}{4}(\mathcal{A}\mathcal{B}^* + \mathcal{A}^*\mathcal{B}) = \frac{1}{2}\mathrm{Re}(\mathcal{A}\mathcal{B}^*) = \frac{1}{2}\mathrm{Re}(\mathcal{A}^*\mathcal{B}).$$

Chapter 2
Spectrum and Polarisation

Apart from the direction of propagation, the electromagnetic radiation is characterised by two other fundamental properties, typical of waves: the spectrum and the polarisation. These properties are of crucial importance to obtain the physical characteristics of the body that is emitting the radiation that we observe, be it an atom or a star. These properties are encoded in the variation of the electric and magnetic field vectors as a function of time. This chapter deals with the mathematical concepts underlying the definition of the spectrum and the polarisation of the electromagnetic radiation. We will also discuss the physical measurements of these properties as performed with appropriate instruments, such as grating spectrometers and polarimeters, of which we will illustrate the principles of operation.

2.1 Spectrum of the Radiation

The monochromatic plane wave introduced in the previous chapter as the solution of Maxwell's equations is a mathematical abstraction and, as such, can only provide the description of the radiation emitted during a real physical process only as a limiting case. More generally, the electric and magnetic fields associated with the radiation that flows across an infinitesimal area, say fixed in space, are described by appropriate functions of time which govern the spectral and polarimetric characteristics of the radiation itself. However, as we saw in the previous chapter, in the case of a plane wave (not necessarily monochromatic) propagating in vacuum, the magnitudes of the magnetic and electric field vectors ($\mathbf{B}$, $\mathbf{E}$) are the same, and the vectors are perpendicular (both also being perpendicular to the direction of propagation). If one of the two vectors is known, the other is also known. To comprehensively characterise the radiation we can therefore limit ourselves in considering only one of the two, e.g. the electric field, which is described, at the point of observation, by the vector function $\mathbf{E}(t)$.

In order to simplify the notation, we will consider initially a scalar function of time—instead of a vector—of the form $E(t)$. By doing so, we will therefore neglect

E. Landi Degl'Innocenti, *Atomic Spectroscopy and Radiative Processes*,
UNITEXT for Physics, DOI 10.1007/978-88-470-2808-1_2, © Springer-Verlag Italia 2014

all the phenomena that are associated with the direction of the electric field in the plane perpendicular to the direction of propagation. These so-called polarisation phenomena are discussed in later sections within this chapter (Sects. 2.5 and 2.6).

The spectrum of the radiation depends on the temporal variation of the function $E(t)$, and therefore it is not possible to define a spectrum in a precise instant in time. The spectrum of the radiation is in fact defined within a sufficiently long time interval which will be specified later on. If we assume that the function $E(t)$ vanishes for $t \to \pm\infty$, we can define its Fourier transform in the angular frequency space, $\hat{E}(\omega)$, as the complex function given by the integral

$$\hat{E}(\omega) = \frac{1}{2\pi} \int_{-\infty}^{\infty} E(t)e^{i\omega t}\, dt. \tag{2.1}$$

Since $E(t)$ is real, the complex conjugate of the Fourier transform $\hat{E}(\omega)^*$ is such that

$$\hat{E}(\omega)^* = \frac{1}{2\pi} \int_{-\infty}^{\infty} E(t)e^{-i\omega t}\, dt = \hat{E}(-\omega),$$

so that it is always possible to relate the Fourier transform at negative frequencies with the one at positive frequencies. In other words, the Fourier transform at negative frequencies does not contain more information compared to the part at positive frequencies and can, in some sense, be neglected.

The electric field is obtained from the Fourier transform with an anti-transform operation

$$E(t) = \int_{-\infty}^{\infty} \hat{E}(\omega)e^{-i\omega t}\, d\omega.$$

To prove this equation, we first multiply both sides of Eq. (2.1) by $e^{-i\omega t'}$ and then integrate in $d\omega$. We have

$$\int_{-\infty}^{\infty} \hat{E}(\omega)e^{-i\omega t'}\, d\omega = \frac{1}{2\pi} \int_{-\infty}^{\infty} dt\, E(t) \int_{-\infty}^{\infty} e^{-i\omega(t'-t)}\, d\omega.$$

On the other hand, the last integral can be written taking the limit

$$\int_{-\infty}^{\infty} e^{-i\omega(t'-t)}\, d\omega = \lim_{\Omega \to \infty} \int_{-\Omega}^{\Omega} e^{-i\omega(t'-t)}\, d\omega = \lim_{\Omega \to \infty} \frac{2\sin(\Omega\,\Delta t)}{\Delta t}, \tag{2.2}$$

where we have defined $\Delta t = t' - t$. The graph of the function $2\sin(\Omega\,\Delta t)/\Delta t$ is displayed in Fig. 2.1, and shows that the function behaves, for $\Omega \to \infty$, as a Dirac delta function.[1] More precisely

$$\lim_{\Omega \to \infty} \frac{2\sin(\Omega\,\Delta t)}{\Delta t} = 2\pi\,\delta(\Delta t).$$

---

[1] In principle, the Dirac delta is not a function but rather a distribution, i.e. a functional that associates to any real function $f(x)$ a real number $F[f(x)]$. The main properties of the Dirac delta are collected in Sect. 16.3.

Fig. 2.1 Plot of $2\sin(\Omega\,\Delta t)/\Delta t$ as a function of Δt. The function has a maximum at the origin and the first zero occurs for $\Delta t = \pi/\Omega$. At the limit for $\Omega \to \infty$, the central peak increases and becomes narrower around $\Delta t = 0$

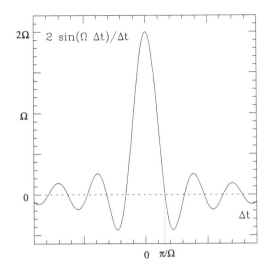

Substituting we obtain

$$\int_{-\infty}^{\infty} \hat{E}(\omega)e^{-i\omega t'}\,d\omega = \frac{1}{2\pi}\int_{-\infty}^{\infty} E(t)2\pi\delta(t'-t)\,dt = E(t'),$$

which is the expression that we wanted to prove.

Let us now consider the energy flux density (energy per unit area and per unit time)[2] transported by the electromagnetic radiation. Recalling the definition of the Poynting vector, the flux density F can be written as

$$F = \frac{c}{4\pi}E^2(t),$$

and the total energy $\mathcal{F}$ crossing the unit area between $t = -\infty$ and $t = +\infty$ is

$$\mathcal{F} = \int_{-\infty}^{\infty} F\,dt = \frac{c}{4\pi}\int_{-\infty}^{\infty} E^2(t)\,dt.$$

Introducing the Fourier transform, we can write

$$\int_{-\infty}^{\infty} E^2(t)\,dt = \int_{-\infty}^{\infty} dt \int_{-\infty}^{\infty} \hat{E}(\omega)e^{-i\omega t}\,d\omega \int_{-\infty}^{\infty} \hat{E}(\omega')^* e^{i\omega't}\,d\omega'.$$

If we evaluate the integral in dt with a procedure similar to the one adopted in Eq. (2.2) we obtain

$$\int_{-\infty}^{\infty} e^{i(\omega'-\omega)t}\,dt = 2\pi\delta(\omega'-\omega),$$

---

[2]The flux density of a given physical quantity (e.g. energy, charge, mass, etc.) is generally defined as the amount of such quantity that flows across the unit area in unit time. However, this definition is not universally accepted and sometimes flux is defined as the quantity which crosses an area (not necessarily an unit area) per unit time. In mathematical physics, the flux of a vector across a closed surface is defined in a different way (recall Gauss theorem), without any reference to the unit of time.

so we obtain the so-called Parseval theorem

$$\int_{-\infty}^{\infty} E^2(t)\,\mathrm{d}t = 2\pi \int_{-\infty}^{\infty} \left| \hat{E}(\omega) \right|^2 \mathrm{d}\omega. \tag{2.3}$$

Substituting into the expression for $\mathcal{F}$ and recalling that $\hat{E}(\omega) = \hat{E}(-\omega)^*$, we then have

$$\mathcal{F} = c \int_0^{\infty} \left| \hat{E}(\omega) \right|^2 \mathrm{d}\omega.$$

Apart from the proportionality factor c, this equation indicates that $|\hat{E}(\omega)|^2$ (the square of the magnitude of the Fourier transform of $E(t)$) is equal to the total energy crossing the unit area (formally between $t = -\infty$ and $t = +\infty$) in the spectral range $\mathrm{d}\omega$. This quantity, $|\hat{E}(\omega)|^2$, is called the spectrum of the electromagnetic radiation.[3] It is evident from the above derivation that this spectrum does not depend on the instantaneous behaviour of the function $E(t)$, but on its variation over a very long time (in principle infinite).

It is obvious that, in practice, any measurement of the spectrum of the radiation can never last for an infinite time, but it is necessarily limited to a time interval $\mathcal{T}$, called the sampling time. This fact obviously prevents one to be able to obtain information on the variability of the function $E(t)$ over periods of the order of or greater than $\mathcal{T}$. This implies that the spectrum of the radiation at frequencies lower than a threshold frequency ω_t (with $\omega_t \simeq 1/\mathcal{T}$) remains undefined. This limitation, however, does not have any practical relevance for the electromagnetic radiation since even if $\mathcal{T}$ is very short, for example one second, the value of ω_t is equal to 1 Hz, and the condition $\omega < \omega_t$ affects a region of the spectrum that is completely irrelevant.

In practice, we often deal with stationary phenomena. Consider for example the radiation emitted by a discharge lamp, or the radiation from a star. It is obvious that in these cases the integral defining the Fourier transform (Eq. (2.1)) is divergent and a particular treatment is then required, involving taking the limit of the integral, in which the sampling time is essential. Let us then consider a sampling time $\mathcal{T}$ sufficiently long to contain all the spectral characteristics of the phenomenon under study. We set artificially to zero the electric field for times t that are outside the sampling time. The Fourier transform, that we now denote by $\hat{E}(\omega, \mathcal{T})$ (to remember the "artificial" dependence on $\mathcal{T}$ that we have introduced), is given by

$$\hat{E}(\omega, \mathcal{T}) = \frac{1}{2\pi} \int_{-\mathcal{T}/2}^{\mathcal{T}/2} E(t) \mathrm{e}^{\mathrm{i}\omega t}\,\mathrm{d}t. \tag{2.4}$$

We now consider the total amount of energy, $\mathcal{F}_{\mathcal{T}}$, crossing the unit area in the time interval between $-\mathcal{T}/2$ and $\mathcal{T}/2$. We have

$$\mathcal{F}_{\mathcal{T}} = \frac{c}{4\pi} \int_{-\mathcal{T}/2}^{\mathcal{T}/2} E^2(t)\,\mathrm{d}t.$$

---

[3]In practice, spectral measurements are usually given in terms of relative units, the function $|\hat{E}(\omega)|^2$ being measured apart from a constant of proportionality. More precisely, we should therefore say that the spectrum of the electromagnetic radiation is proportional to $|\hat{E}(\omega)|^2$.

Starting from this expression, and taking into account that the sampling time can formally be considered as going to ∞, we can follow the above steps leading to the Parseval theorem to obtain a modified expression that is valid for stationary phenomena, that is

$$\mathcal{F}_T = \frac{c}{4\pi} \int_{-T/2}^{T/2} E^2(t)\, \mathrm{d}t = c \int_0^\infty |\hat{E}(\omega, T)|^2 \,\mathrm{d}\omega. \tag{2.5}$$

As we shall see below, when we discuss some specific stationary phenomena, the quantity $|\hat{E}(\omega, T)|^2$ is proportional to T. This enables us to provide a coherent definition of the monochromatic flux, the amount of energy contained in the frequency interval $\mathrm{d}\omega$ that flows across a unit area per unit time. Introducing the symbol F_ω for this quantity, we can write

$$\int_0^\infty F_\omega \,\mathrm{d}\omega = \frac{\mathcal{F}_T}{T} = \frac{c}{T} \int_0^\infty |\hat{E}(\omega, T)|^2 \,\mathrm{d}\omega,$$

from which we obtain

$$F_\omega = \frac{c}{T} |\hat{E}(\omega, T)|^2. \tag{2.6}$$

2.2 Spectra of Some Particular Pulse Shapes

Let us consider some particular cases of the function $E(t)$ and obtain from them the corresponding spectra. If we assume that $E(t)$ is a Gaussian function (case a)

$$E(t) = E_0 e^{-\frac{1}{2}(t/\tau)^2},$$

the Fourier transform is given by (recall Eq. (2.1))

$$\hat{E}(\omega) = \frac{E_0}{2\pi} \int_{-\infty}^\infty e^{-\frac{1}{2}(t/\tau)^2 + \mathrm{i}\omega t} \,\mathrm{d}t.$$

Introducing the reduced variable $x = t/\tau$ and by adding and subtracting the quantity $\omega^2\tau^2/2$ to the exponent we obtain

$$\hat{E}(\omega) = \frac{E_0\tau}{\cdot 2\pi} e^{-\frac{1}{2}\omega^2\tau^2} \int_{-\infty}^\infty e^{-\frac{1}{2}(x - \mathrm{i}\omega\tau)^2} \,\mathrm{d}x,$$

which can be written in terms of the complex variable $z = (x - \mathrm{i}\omega\tau)/\sqrt{2}$ as

$$\hat{E}(\omega) = \frac{E_0\tau}{\sqrt{2\pi}} e^{-\frac{1}{2}\omega^2\tau^2} \int_{\mathcal{L}} e^{-z^2} \,\mathrm{d}z,$$

where $\mathcal{L}$ is a path in the complex plane of the variable z that is parallel to the real axis. The integral is equal to $\sqrt{\pi}$ so we obtain

$$|\hat{E}(\omega)|^2 = \frac{E_0^2\tau^2}{2\pi} e^{-\omega^2\tau^2}.$$

Fig. 2.2 Graph of the
function $E(t)$ and the
corresponding square
magnitude of its Fourier
transform $|\hat{E}(\omega)|^2$ for the
three cases (a), (b), and (c)
discussed in the text. Note the
correspondence between the
characteristic length of time
of the signal τ and the
frequency width of the
spectrum $1/\tau$

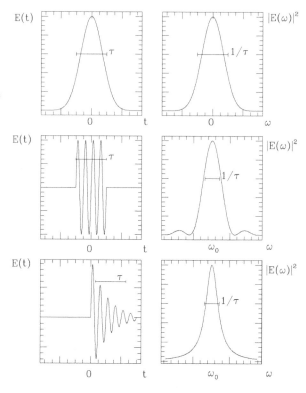

We therefore conclude that the spectrum of a Gaussian is also a Gaussian function.
Similar calculations can be carried out for other functional forms of $E(t)$. For example, (case b)

$$E(t) = \begin{cases} E_0 \sin(\omega_0 t) & \text{for } |t| < \frac{\tau}{2} \\ 0 & \text{for } |t| \geq \frac{\tau}{2}. \end{cases}$$

If τ is a time interval containing a large number of wave periods ($\tau \gg 1/\omega_0$), we
obtain for the spectrum at positive frequencies

$$\left|\hat{E}(\omega)\right|^2 \simeq \frac{E_0^2}{4\pi^2} \frac{\sin^2[(\omega - \omega_0)\tau/2]}{(\omega - \omega_0)^2},$$

with a symmetric contribution, at negative frequencies, centred at the frequency
$-\omega_0$. Finally, if (case c)

$$E(t) = \begin{cases} E_0 \sin(\omega_0 t)e^{-t/\tau} & \text{for } t \geq 0 \\ 0 & \text{for } t < 0, \end{cases}$$

and if τ is again a time interval containing a large number of wave periods ($\tau \gg
1/\omega_0$), we obtain for the spectrum at positive frequencies

$$\left|\hat{E}(\omega)\right|^2 \simeq \frac{E_0^2}{16\pi^2} \frac{1}{(\omega - \omega_0)^2 + (1/\tau)^2},$$

Fig. 2.3 Temporal variation
of the electric field due to the
statistical superposition of
three elementary signals
having the same form

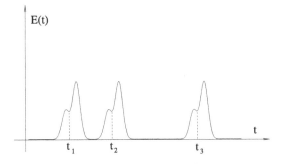

with a symmetric contribution, at negative frequencies, centred at the frequency
$-\omega_0$.

The above functional forms for $E(t)$ and their corresponding spectra are displayed in Fig. 2.2. Considering these cases, it is clear that $\Delta\omega$, the width in frequency of the spectrum of a typical signal, is related to the length of time τ of the signal by

$$\Delta\omega\tau \simeq 1.$$

In other words, the shorter the signal, the wider is the frequency band where the Fourier transform is significantly different from zero, i.e. the wider the spectrum. At the limit, considering a signal having an infinitesimal length of time, such as a Dirac delta, the spectrum becomes constant, i.e. independent of frequency.

2.3 Spectra of Stochastic and Periodic Signals

We now consider a physical situation in which the electric field $E(t)$ is stationary and denote by $\mathcal{T}$ the sampling time. A particular case is that of a stochastic signal, i.e. one in which the field is due to the superposition of a very large number N of elementary signals, all equal to each other in shape, and appearing at random times with frequency $\mathcal{N} = N/\mathcal{T}$. These signals may possibly overlap with each other, as schematically illustrated in Fig. 2.3, and are such that their characteristic time scales are much smaller than the sampling time. Denoting by $t_1, t_2, \ldots, t_N$ the instants of the individual elementary signals, and with $f(t)$ the function of time describing each signal, the electric field is given by the expression

$$E(t) = \sum_{j=1}^{N} f(t - t_j).$$

The Fourier transform of such electric field is

$$\hat{E}(\omega, \mathcal{T}) = \frac{1}{2\pi} \sum_{j=1}^{N} \int_{-\mathcal{T}/2}^{\mathcal{T}/2} f(t - t_j) e^{i\omega t} \, dt.$$

Fig. 2.4 The sum $\mathcal{S}$ of 100 complex numbers each having unit modulus and random phase. As numbers are added, the partial sum moves around the complex plane, starting from the initial point O, following a typical random walk until the last point P is reached

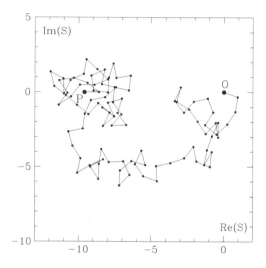

If we make the substitution $t - t_j = \tau$ in the j-th integral and define the Fourier transform of the elementary signal with the usual expression[4]

$$\hat{f}(\omega) = \frac{1}{2\pi} \int_{-\infty}^{\infty} f(\tau) e^{i\omega t},$$

with simple algebra we obtain

$$\hat{E}(\omega, \mathcal{T}) = \hat{f}(\omega)\mathcal{S},$$

where the complex number $\mathcal{S}$ is given by

$$\mathcal{S} = \sum_{j=1}^{N} e^{i\omega t_j}.$$

The Fourier transform of the electric field is therefore given by the product of the Fourier transform of the elementary signal by the sum of N phase factors (N complex numbers each having unit modulus). Given that the times t_j are casually distributed, the phase factors tend to cancel out when considering their sum. The net result is that the modulus of the complex number $\mathcal{S}$ becomes much smaller than N, as shown in Fig. 2.4. This is a situation similar to the one of the Brownian motion, where a particle follows a random walk, moving very slowly from its original location in space.

If we consider the square of the modulus of the transform, we have

$$\left|\hat{E}(\omega, \mathcal{T})\right|^2 = \left|\hat{f}(\omega)\right|^2 |\mathcal{S}|^2 = \left|\hat{f}(\omega)\right|^2 \sum_{j=1}^{N} \sum_{k=1}^{N} e^{i\omega(t_j - t_k)}.$$

---

[4]Since we have assumed that the function $f(t)$ varies over timescales much shorter than $\mathcal{T}$, the integral in dt can be extended between $-\infty$ and ∞ instead of extending it from $-\mathcal{T}/2$ to $\mathcal{T}/2$.

The double sum can be subdivided into two parts. First the N terms with $j = k$ are summed, then the other $N(N - 1)$ ones having $j \neq k$. The first sum results in N, while the second totals to a small contribution, whose ratio with the first tends to zero as N tends to infinity. This is due to the fact that the times t_j and t_k are distributed in a casual way. We therefore obtain, recalling that $N = \mathcal{N}\mathcal{T}$

$$\left|\hat{E}(\omega, \mathcal{T})\right|^2 = \mathcal{N}\mathcal{T}\left|\hat{f}(\omega)\right|^2. \tag{2.7}$$

As we anticipated in Sect. 2.1, the Fourier transform is proportional to the sampling time $\mathcal{T}$. Recalling Eq. (2.6), the monochromatic flux is then

$$F_\omega = c\mathcal{N}\left|\hat{f}(\omega)\right|^2.$$

This expression can be generalised to the case when the electric field can be considered as the incoherent combination of two stochastic signals. Let us assume, for example, that the electric field is the superposition of many elementary signals with the functional forms $f(t)$ and $g(t)$, casually distributed in time with frequencies $\mathcal{N}_f$ and $\mathcal{N}_g$, respectively. If we repeat the above arguments, we obtain for the monochromatic flux

$$F_\omega = c\left(\mathcal{N}_f\left|\hat{f}(\omega)\right|^2 + \mathcal{N}_g\left|\hat{g}(\omega)\right|^2\right).$$

If we assume as before that the temporal distribution is stochastic and the different signals are incoherent, we can generalise the above equation, in the case of a continuous distribution of elementary signals (depending, for example, on a parameter ζ), by writing

$$F_\omega = c\mathcal{N}_{\text{tot}}\langle\left|\hat{f}(\omega)\right|^2\rangle, \tag{2.8}$$

where $\mathcal{N}_{\text{tot}}$ is the total frequency of the elementary signals and $\langle|\hat{f}(\omega)|^2\rangle$ is the average of $|\hat{f}(\omega)|^2$ with respect to the parameter ζ.

We now consider a different case, one where the stationary function $E(t)$ is periodic in time (with period T). According to Fourier's theorem, which is valid for any arbitrary function, the electric field can be expressed by the equation

$$E(t) = \sum_{n=-\infty}^{\infty} \mathcal{E}^{(n)} e^{-in\omega_0 t},$$

where ω_0 is the fundamental angular frequency defined as

$$\omega_0 = \frac{2\pi}{T},$$

and where the $\mathcal{E}^{(n)}$ values (with $\mathcal{E}^{(-n)} = \mathcal{E}^{(n)*}$) are complex quantities, called Fourier components, given by the equation

$$\mathcal{E}^{(n)} = \frac{1}{T}\int_{t_0}^{t_0+T} E(t)e^{2\pi int/T}\,dt = \frac{1}{T}\int_{t_0}^{t_0+T} E(t)e^{in\omega_0 t}\,dt,$$

where t_0 is an arbitrary instant of time.

We now determine the relation between the Fourier transform and the Fourier components $\mathcal{E}^{(n)}$. We start by considering a sampling time $\mathcal{T}$ containing a large number of periods. We have

$$\hat{E}(\omega, \mathcal{T}) = \frac{1}{2\pi} \int_{-\mathcal{T}/2}^{\mathcal{T}/2} E(t) e^{i\omega t} \, dt,$$

that is, substituting the expression for $E(t)$ in terms of the Fourier components

$$\hat{E}(\omega, \mathcal{T}) = \frac{1}{2\pi} \sum_{n=-\infty}^{\infty} \mathcal{E}^{(n)} \int_{-\mathcal{T}/2}^{\mathcal{T}/2} e^{i(\omega - n\omega_0)t} \, dt.$$

Expanding the integral, we have

$$\hat{E}(\omega, \mathcal{T}) = \frac{1}{2\pi} \sum_{n=-\infty}^{\infty} \mathcal{E}^{(n)} \frac{2\sin[(\omega - n\omega_0)\mathcal{T}/2]}{\omega - n\omega_0},$$

and considering that, for $\mathcal{T} \to \infty$,

$$\lim_{\mathcal{T} \to \infty} \frac{2\sin[(\omega - n\omega_0)\mathcal{T}/2]}{\omega - n\omega_0} = 2\pi\delta(\omega - n\omega_0),$$

we obtain

$$\hat{E}(\omega, \mathcal{T}) = \sum_{n=-\infty}^{\infty} \mathcal{E}^{(n)}\delta(\omega - n\omega_0).$$

This expression, although correct, is not appropriate, however, to calculate the square modulus of the Fourier transform as it would produce the square of a Dirac delta function. It would be impossible to give a correct meaning to this function without using mathematical concepts typical of the theory of distributions. To calculate $|\hat{E}(\omega, \mathcal{T})|^2$ we can in any case consider the expression of $\hat{E}(\omega, \mathcal{T})$ as a function of $\mathcal{T}$, evaluate its square modulus, and take its limit for $\mathcal{T} \to \infty$. By doing so we obtain a double sum of the type

$$|\hat{E}(\omega, \mathcal{T})|^2 = \frac{1}{4\pi^2} \sum_{n=-\infty}^{\infty} \sum_{m=-\infty}^{\infty} \mathcal{E}^{(n)}\mathcal{E}^{(m)*}$$
$$\times \frac{2\sin[(\omega - n\omega_0)\mathcal{T}/2]}{\omega - n\omega_0} \frac{2\sin[(\omega - m\omega_0)\mathcal{T}/2]}{\omega - m\omega_0}.$$

For $n \neq m$, in the limit $\mathcal{T} \to \infty$ the two functions of $\mathcal{T}$ in the right-hand side are mutually exclusive, in the sense that where one is non-zero the other is null, and vice versa. The double sum therefore reduces to a single sum by imposing $m = n$. Noting that

$$\lim_{\mathcal{T} \to \infty} \frac{4\sin^2[(\omega - n\omega_0)\mathcal{T}/2]}{(\omega - n\omega_0)^2} = 2\pi\mathcal{T}\delta(\omega - n\omega_0), \tag{2.9}$$

we obtain

$$|\hat{E}(\omega, \mathcal{T})|^2 = \frac{\mathcal{T}}{2\pi} \sum_{n=-\infty}^{\infty} |\mathcal{E}^{(n)}|^2 \delta(\omega - n\omega_0).$$

We have therefore obtained again the result that the Fourier transform is proportional to the sampling time $\mathcal{T}$. Recalling Eq. (2.6) and excluding the zero harmonic (which is equivalent to assume that the temporal average of $E(t)$ is zero), the monochromatic flux of a periodic signal is given by

$$F_\omega = \frac{c}{2\pi} \sum_{n=1}^\infty |\mathcal{E}^{(n)}|^2 \delta(\omega - n\omega_0). \qquad (2.10)$$

The spectrum of a periodic signal is therefore composed of a series of Dirac delta functions centred at the fundamental frequency and at the various harmonics. The weight of each harmonic is proportional to the square of the magnitude of its respective Fourier component.

2.4 Diffraction Grating Spectroscope

The diffraction grating spectroscope can be rightly considered the prototype instrument to measure the spectrum of the radiation, at least for the visible region of the electromagnetic spectrum and the neighboring infrared and ultraviolet regions. In this section we present a simplified treatment of such an instrument, in order to illustrate the connection between the mathematical definition of a spectrum and its practical measurement.

Referring to Fig. 2.5, the radiation coming from a laboratory source (or from e.g. a telescope) is converted into a plane wave by means of a system of lenses and is incident perpendicularly on a plane transmission grating,[5] characterised by the presence of N rules with a (constant) separation between them equal to d, referred to as the grating constant. In the grating plane, the electric field associated with the radiation is described by the function $E(t)$.

The rules on the grating give rise to the phenomenon of diffraction and the expression for the diffracted wave can be computed using the principle of Huygens-Fresnel. For the case of an ideal transmission grating, the principle can be formulated by saying that each section of the grating is the axis of a cylindrical wave whose amplitude, on the emerging side of the grating, is simply given by $E(t)$ in those areas where the wave is transmitted and zero otherwise.

We now consider the wave diffracted by the j-th section of the grating in the direction that forms an angle θ with the direction of the incident wave. The amplitude of such a wave at the distance L from the grating and at the time t is given by an expression of the type

$$E_j^{\mathrm{d}}(t, \theta) = kE\left(t - \frac{L}{c} - j\frac{d\sin\theta}{c}\right),$$

---

[5]The plane transmission grating is the most simple type of diffraction grating. In practice, many other types of gratings can be used (reflection, echelle, saw-tooth, phase-transparency, circular, concave, etc.).

Fig. 2.5 Schematic diagram
of a diffraction grating

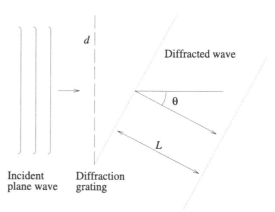

where k is a constant, in general complex,[6] and where the index j numbers the rules,
starting with the first (see Fig. 2.5). Adding the contribution of the diffracted waves
from all the rules of the grating and expressing the electric field of the incident wave
by means of its Fourier integral,[7] we obtain, with simple algebra

$$E^{\mathrm{d}}(t,\theta) = \sum_{j=1}^{N} \int_{-\infty}^{\infty} k \hat{E}(\omega, \mathcal{T}) e^{-i\omega(t-L/c)} e^{ij\omega d \sin\theta/c} \, d\omega.$$

The sum over j can be easily evaluated recalling that for a truncated geometrical
series we have

$$1 + q + q^2 + \cdots + q^{N-1} = \frac{1-q^N}{1-q}.$$

Thus we get

$$E^{\mathrm{d}}(t,\theta) = \int_{-\infty}^{\infty} k \hat{E}(\omega, \mathcal{T}) e^{-i\omega[t-(L+d\sin\theta)/c]} f(\omega,\theta) \, d\omega,$$

where

$$f(\omega,\theta) = \frac{1 - e^{iN\omega d \sin\theta/c}}{1 - e^{i\omega d \sin\theta/c}}.$$

The equation we have obtained for $E^{\mathrm{d}}(t,\theta)$ defines directly the Fourier transform
of the electric field diffracted in the direction θ. With obvious symbols we have

$$\hat{E}^{\mathrm{d}}(\omega,\theta,\mathcal{T}) = k \hat{E}(\omega, \mathcal{T}) e^{i\omega(L+d\sin\theta)/c} f(\omega,\theta).$$

---

[6]For an ideal transmission grating it can be shown that k is given by $a\cos\theta/\lambda$, where a is the size
of the transmission area of each rule and λ is the wavelength (see, e.g., Toraldo di Francia 1958).
In practice, the constant k depends on how the grating is actually built.

[7]We refer here to the case where the radiation has a stationary behaviour, so we need to refer to the
Fourier transform relative to a sampling time $\mathcal{T}$.

In other words, apart from a proportionality factor, the Fourier transform of the electric field diffracted in the direction θ is obtained by multiplying the Fourier transform of the incident field by a phase factor and by the complex function $f(\omega, \theta)$, known as the grating transfer function.

In a common spectroscope, the radiation diffracted by the grating is focused by means of a system of lenses on the detector,[8] which responds with a signal proportional to the square of the incident electric field. At the point on the detector where the radiation coming from the direction θ falls, we have a signal $S(\theta)$ that is given by the expression

$$S(\theta) = K \int_{-T/2}^{T/2} E^{\mathrm{d}}(t, \theta)^2 \, \mathrm{d}t,$$

where K is a constant that depends on the efficiency of the detector (as well as the units in which S is measured), and where T is the time over which the measurement is done (sampling time). Recalling the Parseval theorem in the form of Eq. (2.5) and putting $K' = 4\pi K$, we obtain

$$S(\theta) = K' \int_0^\infty \left| \hat{E}^{\mathrm{d}}(\omega, \theta, T) \right|^2 \mathrm{d}\omega = K' \int_0^\infty k^2 \left| \hat{E}(\omega, T) \right|^2 \left| f(\omega, \theta) \right|^2 \mathrm{d}\omega. \quad (2.11)$$

We now study the behaviour of the $\left| f(\omega, \theta) \right|^2$ function with respect to θ, for ω fixed. From the definition we obtain

$$\left| f(\omega, \theta) \right|^2 = \frac{\sin^2\left(\frac{N\omega d \sin\theta}{2c} \right)}{\sin^2\left(\frac{\omega d \sin\theta}{2c} \right)}.$$

This function has very high maxima for those values of $\sin\theta$ such that the denominator vanishes, i.e. for

$$\sin\theta_m = m \frac{2\pi c}{\omega d}, \quad (2.12)$$

where m is any integer (positive, negative or zero) that characterises the so-called order of the spectrum.[9] In correspondence with such θ values we have, by considering the limit

$$\left| f(\omega, \theta_m) \right|^2 = N^2.$$

Such points are called principal maxima and it can be shown that between any two of them there are $(N - 1)$ points where the function vanishes, corresponding to values of θ such that the numerator is zero but the denominator is not. The first zero adjacent to θ_m is at a distance $\Delta\theta$ such that

$$\frac{\omega d \sin(\theta_m + \Delta\theta)}{2c} = \left(m + \frac{1}{N} \right) \pi,$$

---

[8]Nowadays, the detector is generally a CCD camera or a series of photo-multipliers. Previously, photographic plates were commonly used.

[9]Note that by substituting the angular frequency with the wavelength ($\omega = 2\pi c/\lambda$), Eq. (2.12) can be written in the form $d \sin\theta_m = m\lambda$, which is the equation resulting from the elementary theory of the diffraction grating.

Fig. 2.6 Graph of the square of the magnitude of the grating transfer function. The graph is obtained for $N = 8$. The grating commonly used in the laboratory and in astronomy have much higher values of N. A typical grating for solar observations has $N \simeq 10^5$

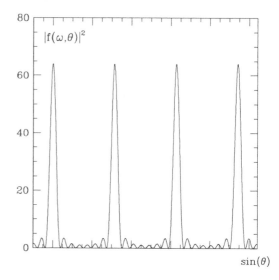

or, by means of a series expansion

$$\Delta\theta = \frac{1}{mN} \tan\theta_m. \tag{2.13}$$

Between the $(N - 1)$ zeroes we have, recalling Rolle's theorem, $(N - 2)$ so-called secondary maxima where the function takes on a value of the order of unity. The graph of the function for $N = 8$ is shown in Fig. 2.6.

As shown in the figure, as N increases the function behaves as a series of delta of Dirac. In first approximation we have, taking into account the height of the peaks and their width

$$\left| f(\omega, \theta) \right|^2 = \sum_m \frac{N \tan\theta_m}{m} \delta(\theta - \theta_m).$$

Alternatively, we can also study the behaviour of the same function with ω for θ fixed. Again we find that the function has very high maxima with value N^2 at the frequencies ω_m given by

$$\omega_m = m \frac{2\pi c}{d \sin\theta}.$$

The first zero next to ω_m is at a distance (in frequency) $\Delta\omega$ such that

$$\frac{(\omega_m + \Delta\omega)d \sin\theta}{2c} = \left(m + \frac{1}{N} \right)\pi,$$

or

$$\Delta\omega = \frac{2\pi c}{Nd \sin\theta}.$$

Again, we have obtained in first approximation a behaviour as a series of Dirac delta (a Dirac comb), given by the equation

$$|f(\omega, \theta)|^2 = \sum_m N \frac{2\pi c}{d \sin \theta} \delta(\omega - \omega_m).$$

Substituting this result in Eq. (2.11) we obtain the following expression for the signal measured by the spectroscope detector in the direction θ

$$S(\theta) = K' k^2 N \frac{2\pi c}{d \sin \theta} \sum_m |E(\omega_m, \mathcal{T})|^2.$$

If we set aside the problems related to the superposition of the spectra of different orders, this formula shows that, if we only consider intervals of θ sufficiently small (so we can neglect the slight dependence on θ contained in the factor $k^2 / \sin \theta$), the signal at the detector effectively provides a measure of the square modulus of the Fourier transform of the electric field of the incident radiation, i.e. of its spectrum. In other words, the spectroscope works as an analog device able to perform the Fourier transform of the incident electric field. Note also that the signal $S(\theta)$ is proportional to the time of the measurement $\mathcal{T}$. This dependence is contained in the square modulus of the Fourier transform (remember the results of Sect. 2.3 and in particular Eq. (2.7)).

The theory of the grating spectroscope presented above also allows us to determine the resolving power of the instrument. Suppose we have an incident radiation composed of two purely monochromatic waves that differ in frequency by a small amount $\delta\omega$. They will be diffracted, at the order m, in two directions forming an angle $\delta\theta$. Differentiating Eq. (2.12) we get

$$\delta\theta = \tan \theta_m \frac{\delta\omega}{\omega}.$$

On the other hand, for the corresponding signals on the detector to be distinct, it is necessary that the angle $\delta\theta$ is greater than the intrinsic width of each signal $\Delta\theta$ given by Eq. (2.13). So we obtain the condition

$$\frac{\delta\omega}{\omega} \geq \frac{1}{mN}.$$

This places a lower limit on the frequency difference $(\delta\omega)_{\min}$ that we must have for the signals to be distinct. Obviously we have

$$\frac{(\delta\omega)_{\min}}{\omega} = \frac{1}{mN}.$$

The resolving power P of an instrument is the ratio between the frequency ω and such minimum difference $(\delta\omega)_{\min}$. The resolving power of a grating spectroscope is therefore given by

$$P = mN.$$

As a practical example, consider a diffraction grating characterised by a number of rules $N = 10^5$ and by a grating constant $d = 1.5$ μm. If the incident radiation covers the entire visible range between 3800 and 7000 Å, the first order spectrum is located between the angles of deflection

$$(\theta_1)_{min} = 14°.7, \qquad (\theta_1)_{max} = 27°.8,$$

the second order spectrum between the angles

$$(\theta_2)_{min} = 30°.4, \qquad (\theta_2)_{max} = 69°.0,$$

while the third order spectrum starts from the angle of deflection

$$(\theta_3)_{min} = 49°.5,$$

and extends until $\theta = 90°$, the angle corresponding to the wavelength of 5000 Å. The spectra of higher orders are not observable. Finally, the resolving power is equal to 10^5 for the first order spectrum, 2×10^5 for the second order spectrum, and 3×10^5 for the third order one. For example, around 5000 Å two spectral lines can be resolved in second order only if they are separated by more than 25 mÅ.

2.5 Polarisation of a Monochromatic Wave

The polarisation phenomena of the electromagnetic radiation are related to the fact that the electric field vector (or the associated magnetic field vector) of a beam of radiation propagating in vacuum can be directed along any direction within the plane perpendicular to the direction of propagation. To describe these phenomena, we begin by considering a plane monochromatic wave of angular frequency ω that propagates along the z direction of a Cartesian reference system that is orthogonal and right-handed (x, y, z). The electric field vector of the wave can be decomposed into two components along the axes x and y. At a given point in space, these components are described by expressions of the type

$$E_x(t) = E_1 \cos(\omega t - \phi_1), \qquad E_y(t) = E_2 \cos(\omega t - \phi_2),$$

where E_1, E_2, ϕ_1 and ϕ_2 are four real quantities. Alternatively, we can use complex quantities and write

$$E_x(t) = \mathrm{Re}\big[\mathcal{E}_1 e^{-i\omega t}\big], \qquad E_y(t) = \mathrm{Re}\big[\mathcal{E}_2 e^{-i\omega t}\big],$$

where

$$\mathcal{E}_1 = E_1 e^{i\phi_1}, \qquad \mathcal{E}_2 = E_2 e^{i\phi_2}.$$

The tip of the electric field vector rotates in the x-y plane along an ellipse which is called the polarisation ellipse. In order to show this fact, let us consider the general equation of an ellipse in terms of its principal axes x' and y'. The equation of the ellipse in parametric form is

$$E_{x'} = E_0 \cos \gamma \cos(\omega t), \qquad E_{y'} = -E_0 \sin \gamma \sin(\omega t),$$

where E_0 is a real and positive quantity and where γ (with $|\gamma| \leq \pi/4$) is a parameter related to the eccentricity of the ellipse ($|\tan \gamma|$ being the ratio of the semi-axes).

Fig. 2.7 The tip of the electric field vector moves over time along the polarisation ellipse

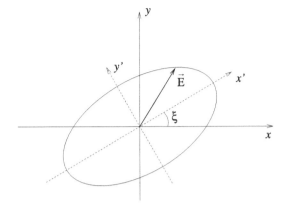

In the previous equation we have assumed that, at time $t = 0$, the electric field is directed along the positive x' axis. Eliminating t, we obtain

$$\frac{E_{x'}^2}{E_0^2 \cos^2 \gamma} + \frac{E_{y'}^2}{E_0^2 \sin^2 \gamma} = 1.$$

The semi-axes of the ellipse are respectively $E_0 |\cos \gamma|$ and $E_0 |\sin \gamma|$; if γ is positive the ellipse is described clockwise, as observed from a person who sees the incoming radiation. The geometry is shown in Fig. 2.7, where the z axis points toward the reader.

If the electric field vector is observed as rotating in a clockwise direction we have positive (or right-handed) elliptical polarisation, otherwise, we have a negative (or left-handed) elliptical polarisation. There are two special cases of elliptical polarisation: if $\gamma = \pm\pi/4$, the ellipse degenerates into a circle and the polarisation is called circular (right and left, respectively). If instead $\gamma = 0$, the ellipse degenerates into a segment, and the polarisation is called linear.

We now relate the geometrical characteristics of the ellipse (amplitude E_0, azimuth of the major semi-axis ξ, and ratio of the semi-axes $\tan \gamma$) to the quantities E_1, E_2, ϕ_1 and ϕ_2, by transforming the components of **E** from the (x', y') to the (x, y) system. Referring to Fig. 2.7, we have

$$E_x = E_{x'} \cos \xi - E_{y'} \sin \xi, \qquad E_y = E_{x'} \sin \xi + E_{y'} \cos \xi,$$

and substituting

$$E_x = E_0 \big[\cos \gamma \cos \xi \cos(\omega t) + \sin \gamma \sin \xi \sin(\omega t) \big],$$
$$E_y = E_0 \big[\cos \gamma \sin \xi \cos(\omega t) - \sin \gamma \cos \xi \sin(\omega t) \big].$$

If we now identify these expressions with the previous ones, we obtain the relations between the E_1, E_2, ϕ_1 and ϕ_2 quantities and the parameters of the ellipse

$$E_1 \cos \phi_1 = E_0 \cos \gamma \cos \xi, \qquad E_1 \sin \phi_1 = E_0 \sin \gamma \sin \xi,$$
$$E_2 \cos \phi_2 = E_0 \cos \gamma \sin \xi, \qquad E_2 \sin \phi_2 = -E_0 \sin \gamma \cos \xi.$$

Taking the square of the four equations we get

$$E_0^2 = E_1^2 + E_2^2;$$

multiplying the first by the fourth and subtracting the product of the second by the third we obtain

$$\sin(2\gamma) = \frac{2E_1 E_2 \sin(\phi_1 - \phi_2)}{E_1^2 + E_2^2};$$

multiplying the first by the third and adding the product of the second by the fourth we have

$$\cos(2\gamma)\sin(2\xi) = \frac{2E_1 E_2 \cos(\phi_1 - \phi_2)}{E_1^2 + E_2^2};$$

subtracting the sum of the squares of the last two from the sum of the squares of the first two we get

$$\cos(2\gamma)\cos(2\xi) = \frac{E_1^2 - E_2^2}{E_1^2 + E_2^2};$$

and finally, dividing the last two equations, we obtain

$$\tan(2\xi) = \frac{2E_1 E_2 \cos(\phi_1 - \phi_2)}{E_1^2 - E_2^2}.$$

These equations allow to obtain the parameters of the ellipse (E_0, γ, ξ) starting from the quantities that describe the electrical oscillation along the axes x and y $(E_1, E_2, \phi_1, \phi_2)$. As shown by the above equations, the parameters of the ellipse depend only on the phase difference $(\phi_1 - \phi_2)$ and not on the absolute phases. The four quantities (bilinear in the components of the electric field) that appear in the previous equations, i.e. $(E_1^2 + E_2^2)$, $(E_1^2 - E_2^2)$, $2E_1 E_2 \cos(\phi_1 - \phi_2)$, and $2E_1 E_2 \sin(\phi_1 - \phi_2)$ are fundamental when characterising the properties of the polarisation of a monochromatic plane wave. The first quantity, $(E_1^2 + E_2^2)$, is proportional to the flux of energy, i.e. to the energy that crosses the unit surface per unit time. We denote this flux with the symbol[10] F_I. We can associate the other quantities with other fluxes that we denote by F_Q, F_U, and F_V. For our monochromatic wave, recalling the definition of the Poynting vector and taking into account that the functions $\sin^2(\omega t)$ and $\cos^2(\omega t)$ are on average equal to $1/2$, averaging over a period we have

$$F_I = \frac{c}{8\pi}\left(E_1^2 + E_2^2\right), \qquad\qquad F_Q = \frac{c}{8\pi}\left(E_1^2 - E_2^2\right),$$
$$F_U = \frac{c}{8\pi} 2E_1 E_2 \cos(\phi_1 - \phi_2), \qquad F_V = \frac{c}{8\pi} 2E_1 E_2 \sin(\phi_1 - \phi_2).$$

It should be noted that the symbols I, Q, U and V, that have been put as indices to the various fluxes, are commonly used to denote different physical quantities,

---

[10]This quantity was denoted with the symbol F in Sect. 2.1, where we neglected the polarisation properties.

related to them by some multiplicative dimensional constants. These quantities are the so-called Stokes parameters, usually introduced in the theory of radiative transfer for polarised radiation. The Stokes parameters have the dimension of the specific intensity of the radiation field and are therefore dimensionally equal to an energy per unit surface, per unit time, per unit spectral interval and per unit solid angle. In the following we will refer to F_I, F_Q, F_U and F_V as the fluxes (of energy) in the four Stokes parameters.

We can relate the geometrical characteristics of the polarisation ellipse to the fluxes in the Stokes parameters by inverting the above equations. We obtain, with easy algebra

$$E_0^2 = \frac{8\pi}{c} F_I, \qquad \sin(2\gamma) = \frac{F_V}{F_I}, \qquad \cos(2\gamma)\cos(2\xi) = \frac{F_Q}{F_I},$$

$$\cos(2\gamma)\sin(2\xi) = \frac{F_U}{F_I}, \qquad \tan(2\xi) = \frac{F_U}{F_Q}.$$

The polarisation ellipse has different forms, depending on the values of the fluxes in the Stokes parameters. Conversely, each polarisation ellipse is characterised by a particular set of the fluxes in the Stokes parameters. In particular, if the ellipse degenerates into a circle ($\gamma = \pm\pi/4$) we have that $F_Q = F_U = 0$, $F_V = \pm F_I$, where the plus sign refers to a circle described in the clockwise direction (for an observer who sees the approaching wave) and the minus sign to a circle described counter-clockwise. In this case we have the so-called pure circular polarisation, positive and negative (or right-handed and left-handed), respectively. If instead the ellipse degenerates into a segment ($\gamma = 0$), we have $F_V = 0$ and $F_Q^2 + F_U^2 = F_I^2$, where the values of F_Q and F_U are related to the angle ξ that the segment forms with the x axis. In this case we have the so-called purely linear polarisation.

The fluxes in the Stokes parameters may also be expressed in terms of the complex amplitudes $\mathcal{E}_1$ and $\mathcal{E}_2$ introduced previously. In fact we have, as is easy to verify

$$F_I = \frac{c}{8\pi}\left(\mathcal{E}_1^*\mathcal{E}_1 + \mathcal{E}_2^*\mathcal{E}_2\right), \qquad F_Q = \frac{c}{8\pi}\left(\mathcal{E}_1^*\mathcal{E}_1 - \mathcal{E}_2^*\mathcal{E}_2\right),$$
$$F_U = \frac{c}{8\pi}\left(\mathcal{E}_1^*\mathcal{E}_2 + \mathcal{E}_2^*\mathcal{E}_1\right), \qquad F_V = \frac{c}{8\pi}\mathrm{i}\left(\mathcal{E}_1^*\mathcal{E}_2 - \mathcal{E}_2^*\mathcal{E}_1\right).$$

$$(2.14)$$

Finally, we note that the fluxes in the Stokes parameters of a monochromatic plane wave are not independent of each other. In fact, squaring the previous expressions, it is easy to verify with some simple algebra that the fluxes in the Stokes parameters are so related

$$F_I^2 = F_Q^2 + F_U^2 + F_V^2. \tag{2.15}$$

This relation is typical for monochromatic waves which, by their nature, are always polarised. In other words, it is impossible to represent a polarisation-free radiation (i.e. with $F_Q = F_U = F_V = 0$) with a monochromatic wave. The previous relation expresses in mathematical terms the fact that only three geometric parameters are sufficient to completely define the ellipse of polarisation. These could be, for example, the amplitude, the ratio between the semi-major and semi-minor axis, and the tilt of the major axis.

2.6 Spectropolarimetric Measurements

The monochromatic plane wave considered in the previous section has polarisation characteristics perfectly defined. It represents, however, a very particular case. In all generality we can assume that the two components of the electric field vector along two directions x and y (perpendicular to the direction of propagation z of a beam of electromagnetic radiation) are described by two arbitrary functions of time, $E_1(t)$ and $E_2(t)$. The spectrum and the characteristics of the polarisation of the radiation depend on the behaviour of these two functions.

We define the Fourier transforms (according to Eq. (2.4)) of the two functions $E_1(t)$ and $E_2(t)$ considering a stationary case and within a sampling time $\mathcal{T}$

$$\hat{E}_1(\omega, \mathcal{T}) = \frac{1}{2\pi} \int_{-\mathcal{T}/2}^{\mathcal{T}/2} E_1(t) e^{i\omega t} \, dt, \qquad \hat{E}_2(\omega, \mathcal{T}) = \frac{1}{2\pi} \int_{-\mathcal{T}/2}^{\mathcal{T}/2} E_2(t) e^{i\omega t} \, dt.$$

We can use these definitions to generalise the various equations we have previously obtained by introducing the monochromatic flux in the single Stokes parameters instead of the monochromatic flux F_ω. Generalising Eq. (2.6), relative to a stationary signal, we obtain

$$
\begin{aligned}
F_\omega^I &= \frac{c}{\mathcal{T}} \left(\hat{E}_1(\omega, \mathcal{T})^* \hat{E}_1(\omega, \mathcal{T}) + \hat{E}_2(\omega, \mathcal{T})^* \hat{E}_2(\omega, \mathcal{T}) \right), \\
F_\omega^Q &= \frac{c}{\mathcal{T}} \left(\hat{E}_1(\omega, \mathcal{T})^* \hat{E}_1(\omega, \mathcal{T}) - \hat{E}_2(\omega, \mathcal{T})^* \hat{E}_2(\omega, \mathcal{T}) \right), \\
F_\omega^U &= \frac{c}{\mathcal{T}} \left(\hat{E}_1(\omega, \mathcal{T})^* \hat{E}_2(\omega, \mathcal{T}) + \hat{E}_2(\omega, \mathcal{T})^* \hat{E}_1(\omega, \mathcal{T}) \right), \\
F_\omega^V &= \frac{c}{\mathcal{T}} i \left(\hat{E}_1(\omega, \mathcal{T})^* \hat{E}_2(\omega, \mathcal{T}) - \hat{E}_2(\omega, \mathcal{T})^* \hat{E}_2(\omega, \mathcal{T}) \right).
\end{aligned}
\tag{2.16}
$$

Generalising Eq. (2.8), relative to a stochastic signal, we obtain

$$
\begin{aligned}
F_\omega^I &= c\mathcal{N}_{\text{tot}} \langle \hat{f}_1(\omega)^* \hat{f}_1(\omega) + \hat{f}_2(\omega)^* \hat{f}_2(\omega) \rangle, \\
F_\omega^Q &= c\mathcal{N}_{\text{tot}} \langle \hat{f}_1(\omega)^* \hat{f}_1(\omega) - \hat{f}_2(\omega)^* \hat{f}_2(\omega) \rangle, \\
F_\omega^U &= c\mathcal{N}_{\text{tot}} \langle \hat{f}_1(\omega)^* \hat{f}_2(\omega) + \hat{f}_2(\omega)^* \hat{f}_1(\omega) \rangle, \\
F_\omega^V &= c\mathcal{N}_{\text{tot}} i \langle \hat{f}_1(\omega)^* \hat{f}_2(\omega) - \hat{f}_2(\omega)^* \hat{f}_1(\omega) \rangle.
\end{aligned}
\tag{2.17}
$$

Finally, generalising Eq. (2.10), relative to a periodic signal, we obtain

$$
\begin{aligned}
F_\omega^I &= \frac{c}{2\pi} \sum_{n=1}^{\infty} \left(\mathcal{E}_1^{(n)*} \mathcal{E}_1^{(n)} + \mathcal{E}_2^{(n)*} \mathcal{E}_2^{(n)} \right) \delta(\omega - n\omega_0), \\
F_\omega^Q &= \frac{c}{2\pi} \sum_{n=1}^{\infty} \left(\mathcal{E}_1^{(n)*} \mathcal{E}_1^{(n)} - \mathcal{E}_2^{(n)*} \mathcal{E}_2^{(n)} \right) \delta(\omega - n\omega_0), \\
F_\omega^U &= \frac{c}{2\pi} \sum_{n=1}^{\infty} \left(\mathcal{E}_1^{(n)*} \mathcal{E}_2^{(n)} + \mathcal{E}_2^{(n)*} \mathcal{E}_1^{(n)} \right) \delta(\omega - n\omega_0), \\
F_\omega^V &= \frac{c}{2\pi} i \sum_{n=1}^{\infty} \left(\mathcal{E}_1^{(n)*} \mathcal{E}_2^{(n)} - \mathcal{E}_2^{(n)*} \mathcal{E}_1^{(n)} \right) \delta(\omega - n\omega_0).
\end{aligned}
\tag{2.18}
$$

The above equations show that the polarimetric characterisation of the electro-magnetic radiation involves the determination of bilinear products of Fourier trans-forms of the type $\hat{E}_i(\omega)^*\hat{E}_j(\omega)$, with $i, j = 1, 2$. In the radio region of the electro-magnetic spectrum, at frequencies less than or of the order of a GHz, this can be done by directly measuring the electric fields $E_1(t)$ and $E_2(t)$ with two antennas arranged along the x and y axes, and obtaining the bilinear expressions with elec-tronic procedures. For the visible and for the neighbouring ultraviolet and infrared regions, this is not possible due to the high frequency of the radiation ($\nu \simeq 10^{15}$ Hz). As we have already said, in these regions of the electromagnetic spectrum the mea-surements of the electric field are carried out with detectors (CCD cameras, photo-multipliers, photographic plates, etc.) that produce a signal proportional to the in-cident energy, i.e. to the square of the electric field. Such measurements therefore only provide information related to integrals of the type

$$\int_{-T/2}^{T/2} \left[E_1(t)^2 + E_2(t)^2\right] dt,$$

where T is the exposure time of the detector. If we limit ourselves in considering only stationary signals, recalling the Parseval theorem (in the form of Eq. (2.5)), and isolating the contribution of the radiation contained within a frequency interval $\Delta\omega$ centred on the frequency ω, the signal at the detector can be expressed in the form

$$S = K\left[\hat{E}_1(\omega, T)^*\hat{E}_1(\omega, T) + \hat{E}_2(\omega, T)^*\hat{E}_2(\omega, T)\right],$$

where K is a dimensional constant that depends on the sensitivity of the detector, the units in which the signal is measured, and the amplitude of the frequency interval considered. This formula shows that, without the use of additional devices, the only measurement that can be done is that of the monochromatic flux F_ω^I.

The polarisation measurements are obtained by placing along the radiation path suitable devices which alter the polarisation characteristics of the incident radiation in a known way. These devices are polarising filters (or simply polarisers) and wave-plates (or retarders). Ideal polarising filters have the property of being completely transparent to the radiation whose electric field vector vibrates along a particular direction (the direction of acceptance or transparency of the filter) and completely opaque to the radiation whose electric field vector vibrates along the direction per-pendicular to it. If we place a filter in such a way that the direction of acceptance makes an angle α with the x axis (see Fig. 2.8 for the conventions used), the detector will respond with a signal of the type

$$S(\alpha) = K\hat{E}_p(\omega, T)^*\hat{E}_p(\omega, T), \tag{2.19}$$

where $\hat{E}_p(\omega, T)$ is the Fourier transform of the component of the electric field along the acceptance axis of the polariser. Referring to Fig. 2.8, such transform can be expressed in the form (obviously, the transforms of the components of a vector are transformed as the components under rotations of the reference system)

$$\hat{E}_p(\omega, T) = \cos\alpha \hat{E}_1(\omega, T) + \sin\alpha \hat{E}_2(\omega, T),$$

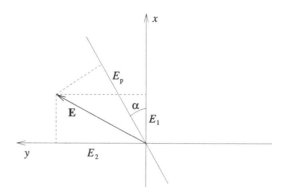

Fig. 2.8 The polarising filter is placed along a direction that forms the angle α with the x axis. The radiation comes from behind the page

so we obtain

$$S(\alpha) = K\left\{\cos^2\alpha\,\hat{E}_1(\omega,\mathcal{T})^*\hat{E}_1(\omega,\mathcal{T}) + \sin^2\alpha\,\hat{E}_2(\omega,\mathcal{T})^*\hat{E}_2(\omega,\mathcal{T})\right.$$
$$\left. + \sin\alpha\cos\alpha\left[\hat{E}_1(\omega,\mathcal{T})^*\hat{E}_2(\omega,\mathcal{T}) + \hat{E}_2(\omega,\mathcal{T})^*\hat{E}_1(\omega,\mathcal{T})\right]\right\}.$$

Inverting Eq. (2.16) and substituting in the previous equation, the signal $S(\alpha)$ can be expressed through the monochromatic fluxes in the Stokes parameters. We obtain

$$S(\alpha) = K'\left[F_\omega^I + \cos(2\alpha)F_\omega^Q + \sin(2\alpha)F_\omega^U\right],$$

where K' is a new constant. This equation shows that the monochromatic fluxes in the Stokes parameters Q and U can be defined by the equations

$$F_\omega^Q = \frac{1}{2K'}\left[S(0°) - S(90°)\right], \qquad F_\omega^U = \frac{1}{2K'}\left[S(45°) - S(135°)\right].$$

The monochromatic flux in the Stokes parameter Q therefore represents the difference between the signal measured by the detector behind a polarising filter oriented under the angle $\alpha = 0°$ (with the direction of acceptance coinciding with the x axis) and the signal measured behind a polarising filter oriented under the angle $\alpha = 90°$ (with the direction of acceptance coinciding with the y axis). The meaning of the other monochromatic flux in the Stokes parameter U is very similar, with the orientation angles of the polariser replaced by $45°$ (direction of acceptance coinciding with the bisector of the x and y axes) and $135°$ (direction of acceptance coinciding with the bisector of the $-x$ and y axes), respectively. Concerning the monochromatic flux in the Stokes parameter I we have

$$F_\omega^I = \frac{1}{2K'}\left[S(\alpha) + S(\alpha + 90°)\right],$$

with α any angle. Alternatively, this same quantity can be measured more simply without interposing any (ideal) polarising filter. From this we see that the monochromatic flux in the Stokes parameter I coincides with the conventional monochromatic flux.

The expressions derived previously show that the monochromatic flux in the Stokes parameter V cannot be measured by placing only a polarising filter. For its measurement it is necessary to introduce a further device, the wave plate, or re-tarder. In general, we can define as an ideal retarder a device that splits the incoming radiation beam into two distinct beams, characterised by different polarisations, in-troduces a phase difference between the two, and then reassembles the two beams into one having a polarisation that differs from that of the input beam. In practice, a retarder can be realised with a birefringent crystal, which is characterised by two axes perpendicular to each other and perpendicular to its optical axis. One of the two axes is the so-called fast axis while the other is the slow axis. The components of the electric field along the fast and slow axes propagate along the optical axis with different indices of refraction, n_f and n_s, respectively, with $n_s > n_f$. This dif-ference between the refractive indices causes within the retarder a phase shift (or retardance) δ. If $\hat{E}_f(\omega)$ and $\hat{E}_s(\omega)$ are the Fourier transforms of the components of the electric field vector along the two axes at the entrance of the retarder, the same transforms at the exit of the retarder, $\hat{E}_f(\omega)'$ and $\hat{E}_s(\omega)'$, are given by (apart from an inessential phase factor)

$$\hat{E}_f(\omega)' = \hat{E}_f(\omega), \qquad \hat{E}_s(\omega)' = \hat{E}_s(\omega)e^{i\delta},$$

where

$$\delta = 2\pi(n_s - n_f)L/\lambda,$$

with L the thickness of the retarder (the wave plate) and $\lambda = 2\pi c/\omega$ the wavelength of the radiation. If $\delta = \pi/2$ we have the so-called quarter-wave plate, while if $\delta = \pi$ we have a half-wave plate, and so on.

It can easily be shown that a quarter-wave plate transforms a circularly polarised beam into a linearly polarised one (with directions that differ by 90° for right or left circular polarisation, respectively). Also, that a half-wave plate rotates by 90° the direction of the linear polarisation (when this direction coincides with the bisector of the angle between the fast and the slow axis). Note also that the retardance of a wave-plate depends strongly on λ, which is the reason why it is difficult to produce the so-called achromatic wave plates (those for which the retardance is independent of the wavelength).

Now suppose we have a quarter-wave plate and we place it along a beam so that the fast axis is directed along the x axis. We then place behind it the polariser, oriented again with an angle α as shown in Fig. 2.8. If $\hat{E}_1(\omega, \mathcal{T})$ and $\hat{E}_2(\omega, \mathcal{T})$ are the Fourier transforms of the components of the electric field along the x and y axes at the entrance of the wave plate, the components at the exit of the wave plate are given by (apart from an inessential phase factor) $\hat{E}_1(\omega, \mathcal{T})$ and $\hat{E}_2(\omega, \mathcal{T})e^{i\pi/2} = i\hat{E}_2(\omega, \mathcal{T})$, respectively. The Fourier transform of the projection of the electric field vector along the axis of acceptance of the polariser is therefore given by

$$\hat{E}_p(\omega, \mathcal{T}) = \cos\alpha\,\hat{E}_1(\omega, \mathcal{T}) + i\sin\alpha\,\hat{E}_2(\omega, \mathcal{T}).$$

The detector will respond with a new signal $T(\alpha)$, still given by the right-hand side of Eq. (2.19). With this expression for $\hat{E}_p(\omega, \mathcal{T})$ we have

$$T(\alpha) = K\{\cos^2\alpha\,\hat{E}_1(\omega, \mathcal{T})^*\hat{E}_1(\omega, \mathcal{T}) + \sin^2\alpha\,\hat{E}_2(\omega, \mathcal{T})^*\hat{E}_2(\omega, \mathcal{T})$$
$$+ i\sin\alpha\cos\alpha[\hat{E}_1(\omega, \mathcal{T})^*\hat{E}_2(\omega, \mathcal{T}) + \hat{E}_2(\omega, \mathcal{T})^*\hat{E}_1(\omega, \mathcal{T})]\},$$

or, in terms of fluxes in the monochromatic Stokes parameters

$$T(\alpha) = K'\left[F_\omega^I + \cos(2\alpha)F_\omega^Q + \sin(2\alpha)F_\omega^V\right].$$

The monochromatic flux in the Stokes parameter V can then be operationally defined by the relation

$$F_\omega^V = \frac{1}{2K'}\left[T\left(45°\right) - T\left(135°\right)\right].$$

In practice, the schematic operations that we have described for the measurement of the fluxes in the Stokes parameters are realised by means of suitable instruments called polarimeters. These instruments are made by one or more wave plates and a polariser at the exit. When both a polarimetric and a spectroscopic analysis of the radiation is required, in general the first precedes the second one, in the sense that the radiation enters the polarimeter before entering the spectroscope. The exit polariser of the polarimeter is generally held in a fixed position because an ordinary diffraction grating is very sensitive to the polarisation of the incoming radiation.

2.7 Properties of the Stokes Parameters

As we have seen in Sect. 2.5, a monochromatic wave has always a well-defined character of polarisation. This ceases to be valid when considering a beam of radiation having a stochastic character. To demonstrate this property, consider the quantity

$$\mathcal{P} = \left(F_\omega^I\right)^2 - \left(F_\omega^Q\right)^2 - \left(F_\omega^U\right)^2 - \left(F_\omega^V\right)^2.$$

Substituting the expressions of Eq. (2.17) relative to a stochastic signal, and developing the calculations we obtain

$$\mathcal{P} = 4c^2\mathcal{N}_{\text{tot}}^2\left[\langle f_1^* f_1\rangle\langle f_2^* f_2\rangle - \langle f_1^* f_2\rangle\langle f_2^* f_1\rangle\right],$$

where we have denoted by f_i ($i = 1, 2$) the Fourier transforms $\hat{f}_i(\omega)$ (to not complicate the notation). We introduce now the complex quantity $\mathcal{A}$ defined by the equation

$$\mathcal{A} = f_2\langle f_1^* f_1\rangle - f_1\langle f_1^* f_2\rangle.$$

The statistical average of its square modulus is, with simple algebra

$$\langle|\mathcal{A}|^2\rangle = \langle f_1^* f_1\rangle\left[\langle f_1^* f_1\rangle\langle f_2^* f_2\rangle - \langle f_1^* f_2\rangle\langle f_2^* f_1\rangle\right],$$

and since the quantities $\langle f_1^* f_1\rangle$ and $\langle|\mathcal{A}|^2\rangle$ are both positive, we can deduce the Cauchy-Schwarz inequality

$$\langle f_1^* f_1\rangle\langle f_2^* f_2\rangle - \langle f_1^* f_2\rangle\langle f_2^* f_1\rangle \geq 0,$$

which, replaced in the expression for $\mathcal{P}$ obtained previously, implies

$$\left(F_\omega^I\right)^2 - \left(F_\omega^Q\right)^2 - \left(F_\omega^U\right)^2 - \left(F_\omega^V\right)^2 \geq 0.$$

Fig. 2.9 A rotation of the
reference system involves a
transformation of the fluxes
in the Q and U Stokes
parameters. The radiation
comes from behind the page

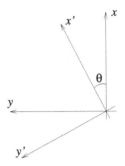

Note that the sign of equality holds only when the quantity $\mathcal{A}$ is null, i.e. when
the ratio of the transforms f_1 and f_2 is such to satisfy the equation

$$\frac{f_1}{f_2} = \frac{\langle f_1^* f_1 \rangle}{\langle f_1^* f_2 \rangle}.$$

In this case the transforms of the two components of the elementary signals along the
x and y axes are characterised by having a constant ratio between the amplitudes and
also a constant phase difference. The case of the monochromatic wave previously
considered is a special case of this situation.

The opposite case is instead that in which the transforms of the two components
are characterised by having equal mean amplitudes

$$\langle f_1^* f_1 \rangle = \langle f_2^* f_2 \rangle,$$

and random phase relation, so that

$$\langle f_1^* f_2 \rangle = \langle f_2^* f_1 \rangle = 0.$$

In this case we have

$$F_\omega^Q = F_\omega^U = F_\omega^V = 0,$$

and we have non-polarised radiation, also called natural radiation.

As we have seen, the fluxes in the Stokes parameters require a reference direction
for their definition. This direction is arbitrary, so it is always necessary to clearly
specify its choice when using these quantities either in theoretical calculations or
in laboratory experiments or astronomical observations. In ground-based night astronomy, for example, it is usual to choose as the reference direction the meridian
across the observed object. In solar physics, or in observation of extended objects,
different reference directions can be chosen, more appropriate to the geometry of
the phenomenon under study. For example, when observing the polarisation at the
solar limb, it is common to choose as the reference direction the tangent to the limb
itself.

As the reference system changes, the Stokes parameters are transformed by sim-
ple linear relations. To show this, we refer to Fig. 2.9 and denote by F_ω^I, F_ω^Q, F_ω^U,
F_ω^V the fluxes in the Stokes parameters relative to the reference direction x. If $(F_\omega^I)'$,
$(F_\omega^Q)'$, $(F_\omega^U)'$, $(F_\omega^V)'$ are the Stokes parameters relative to the direction x', rotated

by an angle θ with respect to x (in the counterclockwise direction looking at the radiation source), the laws of transformation are easily found by noting that

$$f_1' = \cos\theta f_1 + \sin\theta f_2, \qquad f_2' = \cos\theta f_2 - \sin\theta f_1.$$

Substituting in Eq. (2.17) we obtain

$$\left(F_\omega^I\right)' = F_\omega^I, \qquad \left(F_\omega^Q\right)' = \cos(2\theta)F_\omega^Q + \sin(2\theta)F_\omega^U,$$
$$\left(F_\omega^U\right)' = \cos(2\theta)F_\omega^U - \sin(2\theta)F_\omega^Q, \qquad \left(F_\omega^V\right)' = F_\omega^V.$$

As the reference direction changes, only the fluxes in the Stokes parameters relative to the linear polarisation F_ω^Q and F_ω^U change, transforming one in the other according to the previous equations. The other fluxes (in intensity and in circular polarisation) are instead invariant. Note that a rotation of an angle $\theta = \pi$ leaves everything unchanged (the reference direction has no tips).

Finally, we observe that, in practice, the fluxes in the Stokes parameters are often denoted by the symbols I, Q, U, and V, without specifying, in many cases, if they refer to the unity of frequency interval (or angular frequency, or wavelength), or whether they refer to the unit solid angle. Often this does not cause problems because the results of polarimetric measurements are generally expressed in terms of the ratios Q/I, U/I, and V/I, which are independent of any proportionality factors that are implicit in the various definitions. For uniformity in the notation, we simply observe that the I_ν symbol (and the corresponding symbols Q_ν, U_ν and V_ν) should be reserved to express the energy of the radiation having frequency between ν and $\nu + d\nu$ and direction contained within the unit solid angle, flowing, per unit time, through the unit surface perpendicular to the direction of the radiation.

Chapter 3
Radiation from Moving Charges

One of the most important consequences of Maxwell's equations is the emission of electromagnetic radiation by charged particles in accelerated motion. In this chapter we will give a classical description of this phenomenon, emphasizing the general characteristics of the cross sections and the spectral and polarimetric properties of the radiation emitted, for both relativistic and nonrelativistic particles. In particular, we will describe some fundamental physical processes such as Thomson and Rayleigh scattering and give an in-depth discussion of the bremsstrahlung radiation (in nonrelativistic approximation), of the cyclotron radiation and of the synchrotron radiation. The last part of the chapter is dedicated to the study of the radiation due to a large number of particles and its multipole expansion.

3.1 Electromagnetic Potentials Due to Charges and Currents

As we have seen in Chap. 1, the electric field $\mathbf{E}(\mathbf{r}, t)$ and the magnetic field $\mathbf{B}(\mathbf{r}, t)$ at the point of coordinate $\mathbf{r}$ and time t can be obtained in all generality from the electromagnetic potentials $\mathbf{A}(\mathbf{r}, t)$ and $\phi(\mathbf{r}, t)$ by the equations

$$\mathbf{E}(\mathbf{r}, t) = -\operatorname{grad}\phi(\mathbf{r}, t) - \frac{1}{c}\frac{\partial \mathbf{A}(\mathbf{r}, t)}{\partial t}, \qquad \mathbf{B}(\mathbf{r}, t) = \operatorname{rot}\mathbf{A}(\mathbf{r}, t).$$

If the Lorenz gauge is adopted, the electromagnetic potentials satisfy the partial differential equations (Eq. (1.8) and (1.9))

$$\nabla^2 \mathbf{A}(\mathbf{r}, t) - \frac{1}{c^2}\frac{\partial^2}{\partial t^2}\mathbf{A}(\mathbf{r}, t) = -\frac{4\pi}{c}\mathbf{j}(\mathbf{r}, t), \tag{3.1}$$

$$\nabla^2 \phi(\mathbf{r}, t) - \frac{1}{c^2}\frac{\partial^2}{\partial t^2}\phi(\mathbf{r}, t) = -4\pi\rho(\mathbf{r}, t), \tag{3.2}$$

E. Landi Degl'Innocenti, *Atomic Spectroscopy and Radiative Processes*,
UNITEXT for Physics, DOI 10.1007/978-88-470-2808-1_3, © Springer-Verlag Italia 2014

where $\rho(\mathbf{r}, t)$ and $\mathbf{j}(\mathbf{r}, t)$ are the charge and current densities, and the additional condition of Eq. (1.7)

$$\operatorname{div} \mathbf{A}(\mathbf{r}, t) + \frac{1}{c} \frac{\partial}{\partial t} \phi(\mathbf{r}, t) = 0. \tag{3.3}$$

To find the solution of this system of differential equations is convenient to refer to the static case. We consider first the static equation for the scalar potential (Poisson equation)

$$\nabla^2 \phi(\mathbf{r}) = -4\pi \rho(\mathbf{r}), \tag{3.4}$$

and we try to solve it for the particular case

$$\nabla^2 \phi(\mathbf{r}) = -4\pi \delta(\mathbf{r}), \tag{3.5}$$

where $\delta(\mathbf{r})$ is the tridimensional Dirac function defined, in Cartesian coordinates, by $\delta(\mathbf{r}) = \delta(x)\delta(y)\delta(z)$. For obvious reasons in terms of symmetry, the potential ϕ depends only on the magnitude of the vector $\mathbf{r}$. In this case, the Laplacian operator is simply given by (see Eq. (6.7))

$$\nabla^2 = \frac{1}{r^2} \frac{d}{dr} \left(r^2 \frac{d}{dr} \right),$$

which can also be expressed in the more compact form

$$\nabla^2 = \frac{1}{r} \frac{d^2}{dr^2} r. \tag{3.6}$$

The differential equation for ϕ is then, for $r \neq 0$

$$\frac{1}{r} \frac{d^2}{dr^2} \left[r\phi(r) \right] = 0.$$

The most general solution of this equation is of the form

$$r\phi(r) = a + br,$$

with a and b arbitrary constants, so we obtain, for $r \neq 0$,

$$\phi(r) = \frac{a}{r} + b.$$

The constant b defines the value of the potential for $r \to \infty$. Assuming that the potential goes to zero at infinity, such constant is null, so

$$\phi(r) = \frac{a}{r}.$$

To determine the value of the constant a, we recall that the Laplacian operator is given by

$$\nabla^2 = \operatorname{div} \operatorname{grad}.$$

Applying Gauss's theorem to a sphere of arbitrary radius centred at the origin and taking into account that

$$\text{grad}\,\frac{a}{r} = -\frac{a}{r^2}\mathbf{n} = -\frac{a}{r^3}\mathbf{r},$$

where $\mathbf{n}$ is the unit vector along $\mathbf{r}$, we have

$$-4\pi a = -4\pi,$$

or $a = 1$. We have therefore obtained the fundamental result that the solution of the differential equation (3.5) that satisfies the boundary condition of becoming null for $r \to \infty$ is

$$\phi(\mathbf{r}) = \frac{1}{r}.$$

This result can be generalised by considering a translation of the charge. Obviously, the solution of the equation

$$\nabla^2\phi(\mathbf{r}) = -4\pi\delta(\mathbf{r} - \mathbf{r}') \tag{3.7}$$

that satisfies the same boundary condition is

$$\phi(\mathbf{r}) = \frac{1}{|\mathbf{r} - \mathbf{r}'|}.$$

If we finally note that we can always write

$$\rho(\mathbf{r}) = \int \rho(\mathbf{r}')\delta(\mathbf{r} - \mathbf{r}')\,d^3\mathbf{r}',$$

due to the linearity of the Laplacian operator we obtain that the solution of the differential equation (3.4) is the following

$$\phi(\mathbf{r}) = \int \frac{\rho(\mathbf{r}')}{|\mathbf{r} - \mathbf{r}'|}\,d^3\mathbf{r}'.$$

The result we have obtained is very intuitive from a physical point of view and could be anticipated by recalling that a point charge q generates in the space an electric field that derives from a potential of the form $V = q/r$, r being the distance from the charge. The expression given above is just the generalisation of the latter formula to the case of a continuous distribution of charges. Here, we have preferred to give a more formal mathematical proof using a standard method for solving non-homogeneous linear differential equations, known as the method of the Green's function.

We return now to the time-dependent case and begin by solving the equation

$$\nabla^2\phi(\mathbf{r}, t) - \frac{1}{c^2}\frac{\partial}{\partial t^2}\phi(\mathbf{r}, t) = -4\pi f(t)\delta(\mathbf{r}),$$

where $f(t)$ is an arbitrary function of time. As in the previous case, the function ϕ, for obvious reasons in terms of symmetry, can only depend on the magnitude of the vector $\mathbf{r}$ and, in this case, also on time. Expressing the operator ∇^2 using Eq. (3.6), the equation for $\phi(r, t)$ becomes, for $r \neq 0$

$$\frac{1}{r} \frac{\partial^2}{\partial r^2} \left[r\phi(r, t) \right] - \frac{1}{c^2} \frac{\partial^2}{\partial t^2} \phi(r, t) = 0,$$

or

$$\left[\frac{\partial^2}{\partial r^2} - \frac{1}{c^2} \frac{\partial^2}{\partial t^2} \right] \left[r\phi(r, t) \right] = 0.$$

The most general solution of this equation is of the form

$$r\phi(r, t) = g(t \pm r/c),$$

where g is an arbitrary function of its argument, or

$$\phi(r, t) = \frac{g(t \pm r/c)}{r}.$$

In analogy with the stationary case, we impose the condition in the origin applying Gauss's theorem to a sphere of infinitesimal radius. We obtain

$$g(t) = f(t),$$

so the solution is of the form

$$\phi(r, t) = \frac{f(t \pm r/c)}{r}.$$

Of the two solutions that we have obtained only one, the one with the minus sign, has a meaning for the physical problem under consideration. Comparing the result of the time-dependent case with the one obtained in the static case, we see in fact that the potential at a distance r from the origin and at time t has the same expression of the potential of the static case corresponding to the charge $f(t - r/c)$, i.e. to the charge that is at the origin at the so-called retarded time t' defined by

$$t' = t - \frac{r}{c}.$$

This result has a natural explanation related to the fact that electromagnetic signals propagate with speed c. The solution with the plus sign would involve the advanced time $t + r/c$, instead of the retarded one. It has no direct physical interpretation in this problem and should be discarded.

Repeating the arguments for the stationary case, we find that the solution of the differential equation (3.2) is given by the expression

$$\phi(\mathbf{r}, t) = \int \frac{\rho(\mathbf{r}', t')}{|\mathbf{r} - \mathbf{r}'|} \, d^3\mathbf{r}', \tag{3.8}$$

where t' is the retarded time defined by

$$t' = t - \frac{|\mathbf{r} - \mathbf{r}'|}{c}. \qquad (3.9)$$

With similar considerations we obtain for the vector potential, solution of Eq. (3.1), the expression

$$\mathbf{A}(\mathbf{r}, t) = \frac{1}{c} \int \frac{\mathbf{j}(\mathbf{r}', t')}{|\mathbf{r} - \mathbf{r}'|} d^3\mathbf{r}'. \qquad (3.10)$$

Before accepting the solutions given by Eqs. (3.8) and (3.10) for the scalar and vector potentials, it is however necessary to verify that these solutions satisfy the condition imposed by the Lorenz gauge (Eq. (3.3)). We provide in the following such a demonstration. First of all we note that when one is dealing with functions of the type $f(\mathbf{r}, t')$, or $f(\mathbf{r}', t')$ with t' the retarded time, the symbol of the partial derivative becomes ambiguous because the variables $\mathbf{r}$ (or $\mathbf{r}'$) and t' are not independent. In fact, if for example we apply a variation $\delta \mathbf{r}$ to $\mathbf{r}$, we can consider two different types of increments, δf_1 and δf_2, given by

$$\delta f_1 = f(\mathbf{r} + \delta\mathbf{r}, t') - f(\mathbf{r}, t'), \qquad \delta f_2 = f(\mathbf{r} + \delta\mathbf{r}, t' + \delta t') - f(\mathbf{r}, t'),$$

where $\delta t'$ is the variation of t' due to $\delta \mathbf{r}$. We are going to indicate the derivative executed with the increment δf_2 with the usual symbol "∂" of partial derivative. On the other hand, the derivative executed with the increment δf_1 will be indicated with the symbol "δ". The two derivative operations are related by the expression

$$\frac{\partial}{\partial x_k} f(\mathbf{r}, t') = \frac{\delta}{\delta x_k} f(\mathbf{r}, t') + \frac{\partial t'}{\partial x_k} \frac{\partial}{\partial t'} f(\mathbf{r}, t').$$

We also note that the two types of derivatives, ∂ and δ, coincide for functions that depend only on $\mathbf{r}$, and not on time.

We start by evaluating the divergence of the vector $\mathbf{A}(\mathbf{r}, t)$. Since such operation has to be done at t constant, from Eq. (3.10) we have

$$\text{div}\,\mathbf{A}(\mathbf{r}, t) = \sum_k \frac{\partial}{\partial x_k} A_k(\mathbf{r}, t) = \frac{1}{c} \int \sum_k \frac{\partial}{\partial x_k} \left[\frac{j_k(\mathbf{r}', t')}{|\mathbf{r} - \mathbf{r}'|} \right] d^3\mathbf{r}'.$$

To evaluate this integral we take into account the fact that the dependence on x_k is contained both in the denominator and the numerator since the retarded time t' depends on $|\mathbf{r} - \mathbf{r}'|$ and therefore on x_k. If we note that

$$\frac{\partial}{\partial x_k} |\mathbf{r} - \mathbf{r}'| = -\frac{\partial}{\partial x_k'} |\mathbf{r} - \mathbf{r}'|,$$

the expression for div $\mathbf{A}$ can also be written in the form

$$\text{div}\,\mathbf{A}(\mathbf{r}, t) = \frac{1}{c} \int \sum_k \left\{ -\frac{\partial}{\partial x_k'} \left[\frac{j_k(\mathbf{r}', t')}{|\mathbf{r} - \mathbf{r}'|} \right] + \frac{1}{|\mathbf{r} - \mathbf{r}'|} \frac{\delta}{\delta x_k'} j_k(\mathbf{r}', t') \right\} d^3\mathbf{r}'.$$

The first part of the integral can be transformed, using Gauss's theorem, in a surface integral. If we assume that the current density goes to zero at infinity, the integral is null, so

$$\text{div}\,\mathbf{A}(\mathbf{r},t) = \frac{1}{c}\int \frac{1}{|\mathbf{r}-\mathbf{r}'|}\sum_k \frac{\delta}{\delta x_k'} j_k(\mathbf{r}',t')\,d^3\mathbf{r}'.$$

We evaluate now the second term of the Lorenz condition. For Eq. (3.8) we have

$$\frac{1}{c}\frac{\partial}{\partial t}\phi(\mathbf{r},t) = \frac{1}{c}\int \frac{\partial}{\partial t}\frac{\rho(\mathbf{r}',t')}{|\mathbf{r}-\mathbf{r}'|}\,d^3\mathbf{r}'.$$

Taking into account that, once $\mathbf{r}$ and $\mathbf{r}'$ are fixed, we have

$$\frac{\partial}{\partial t}\rho(\mathbf{r}',t') = \frac{\partial}{\partial t'}\rho(\mathbf{r}',t'),$$

we obtain

$$\text{div}\,\mathbf{A}(\mathbf{r},t) + \frac{1}{c}\frac{\partial}{\partial t}\phi(\mathbf{r},t)$$

$$= \frac{1}{c}\int \frac{1}{|\mathbf{r}-\mathbf{r}'|}\left\{\left[\sum_k \frac{\delta}{\delta x_k'} j_k(\mathbf{r}',t')\right] + \frac{\partial}{\partial t'}\rho(\mathbf{r}',t')\right\}d^3\mathbf{r}'. \qquad (3.11)$$

On the other hand, the continuity equation for the charge, written for the point of coordinate $\mathbf{r}'$ and time t', is, with the notations we have introduced,

$$\left[\sum_k \frac{\delta}{\delta x_k'} j_k(\mathbf{r}',t')\right] + \frac{\partial}{\partial t'}\rho(\mathbf{r}',t') = 0,$$

so the term within the curly brackets in Eq. (3.11) is null and the Lorenz condition is verified.

3.2 The Liénard and Wiechart Potentials

In the previous section we have found the expressions of the scalar and vector potentials for an arbitrary distribution of charges and currents. We now apply these expressions to the special case where there is only one point charge e moving in time according to the equation $\mathbf{r}_0(t)$. For the motion of the particle we define the velocity and acceleration vectors according to the usual equations

$$\mathbf{v}(t) = \frac{d}{dt}\mathbf{r}_0(t), \qquad \mathbf{a}(t) = \frac{d}{dt}\mathbf{v}(t) = \frac{d^2}{dt^2}\mathbf{r}_0(t).$$

The charge and current densities due to the point charge can be expressed in terms of the tridimensional Dirac delta. We have

$$\rho(\mathbf{r},t) = e\delta\big[\mathbf{r}-\mathbf{r}_0(t)\big], \qquad \mathbf{j}(\mathbf{r},t) = e\mathbf{v}(t)\delta\big[\mathbf{r}-\mathbf{r}_0(t)\big],$$

so we obtain, from Eqs. (3.8) and (3.10)

$$\phi(\mathbf{r},t) = e \int \frac{\delta[\mathbf{r}' - \mathbf{r}_0(t')]}{|\mathbf{r} - \mathbf{r}'|} \mathrm{d}^3\mathbf{r}', \qquad \mathbf{A}(\mathbf{r},t) = \frac{e}{c} \int \frac{\mathbf{v}(t')\delta[\mathbf{r}' - \mathbf{r}_0(t')]}{|\mathbf{r} - \mathbf{r}'|} \mathrm{d}^3\mathbf{r}',$$

where t' is the retarded time defined in Eq. (3.9). The presence of the delta function allows a simple evaluation of the integrals contained in the previous expressions. We recall that, given two arbitrary functions $f(x)$ and $g(x)$, for the unidimensional Dirac delta function we have (see also Sect. 16.3)

$$\int_{-\infty}^{\infty} f(x)\delta[g(x)]\,\mathrm{d}x = \sum_{i=1}^{N} f(x_i)\frac{1}{|g'(x_i)|},$$

where x_i, with $i = 1, \ldots, N$, are the N solutions of the equation $g(x) = 0$ and where $g'(x)$ is the derivative of the function $g(x)$ with respect to the variable x. For the tridimensional Dirac delta, the previous equation can be generalised into the following expression

$$\int f(\mathbf{r})\delta[\mathbf{g}(\mathbf{r})]\,\mathrm{d}^3\mathbf{r} = \sum_{i=1}^{N} f(\mathbf{r}_i)\frac{1}{|J(\mathbf{r}_i)|},$$

where $\mathbf{r}_i$, with $i = 1, \ldots, N$, are the N solutions of the vector equation $\mathbf{g}(\mathbf{r}) = 0$, and where J is the Jacobian of the transformation $\mathbf{r} = \mathbf{g}(\mathbf{r})$, i.e. the determinant of the Jacobian matrix $J_{kl}(\mathbf{r})$ defined by

$$J_{kl}(\mathbf{r}) = \frac{\partial g_k(\mathbf{r})}{\partial x_l}.$$

Returning to the integrals that we want to calculate, we note that since the particle velocity is necessarily less than c, the equation

$$\mathbf{r}' - \mathbf{r}_0(t') = 0$$

has, with $\mathbf{r}$ and t fixed, only one solution, schematically illustrated in Fig. 3.1. To avoid introducing further notations, we will indicate with $\mathbf{r}'$ and with t' the point and the instant corresponding to such solution. Concerning the Jacobian matrix, we have

$$J_{jk} = \frac{\partial}{\partial x_k'}\{x_j' - [\mathbf{r}_0(t')]_j\} = \delta_{jk} - \frac{\partial}{\partial x_k'}[\mathbf{r}_0(t')]_j.$$

Taking into account the definition of the retarded time (Eq. (3.9)) we have

$$\frac{\partial}{\partial x_k'}[\mathbf{r}_0(t')]_j = \left\{\frac{\partial}{\partial t'}[\mathbf{r}_0(t')]_j\right\}\frac{\partial t'}{\partial x_k'} = -\frac{v_j}{c}\frac{\partial}{\partial x_k'}|\mathbf{r} - \mathbf{r}'|,$$

where v_j is the j-th component of the particle velocity evaluated at the instant t'. To calculate the last derivative, writing

$$\mathbf{R} = \mathbf{r} - \mathbf{r}',$$

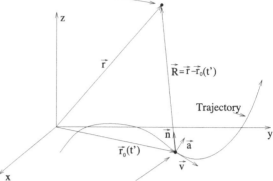

Fig. 3.1 Geometry for the calculation of the electromagnetic field at the point of coordinate **r** at the time t. The vectors **v** and **a** are, respectively, the velocity and acceleration of the particle at the retarded time. The unit vector **n** is directed along the direction defined by the position of the particle at the retarded time and the point where the field is calculated

we have

$$\frac{\partial R_i}{\partial x'_k} = -\delta_{ik},$$

so that

$$\frac{\partial}{\partial x'_k} R^2 = \frac{\partial}{\partial x'_k}\left(\sum_i R_i R_i\right) = -2R_k,$$

and then

$$\frac{\partial}{\partial x'_k} R = \frac{\partial}{\partial x'_k}\sqrt{R^2} = -\frac{R_k}{R} = -n_k,$$

where we have introduced the unit vector **n** to indicate the direction of the vector **R**. Taking into account this result, the Jacobian matrix becomes

$$J_{jk} = \delta_{jk} - \frac{v_j n_k}{c}.$$

We can now calculate the Jacobian. We have

$$J = \det\begin{pmatrix} 1 - \frac{v_x n_x}{c} & -\frac{v_x n_y}{c} & -\frac{v_x n_z}{c} \\ -\frac{v_y n_x}{c} & 1 - \frac{v_y n_y}{c} & -\frac{v_y n_z}{c} \\ -\frac{v_z n_x}{c} & -\frac{v_z n_y}{c} & 1 - \frac{v_z n_z}{c} \end{pmatrix},$$

and, with simple algebra,

$$J = 1 - \frac{\mathbf{v} \cdot \mathbf{n}}{c}.$$

Once we substitute these results into the integrals containing the Dirac delta functions, we get the following expressions for the potentials

$$\phi(\mathbf{r}, t) = \frac{e}{\kappa R}, \qquad \mathbf{A}(\mathbf{r}, t) = \frac{e\mathbf{v}}{c\kappa R}, \tag{3.12}$$

Fig. 3.2 The signals emitted
by a moving source at
successive times, represented
as spherical waves, become
concentrated in the direction
of the velocity and are
rarefied in the opposite
direction. In any case, since
the velocity v is always less
than c, each spherical wave
contains all those emitted at
successive times. This is the
reason why Eq. (3.14) has
only one solution

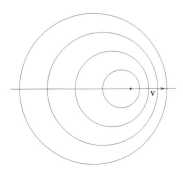

where

$$\kappa = 1 - \frac{\mathbf{v} \cdot \mathbf{n}}{c}, \tag{3.13}$$

and where all the quantities that appear in these equations, i.e. $\mathbf{v}$, R, $\mathbf{n}$, κ need to be evaluated at the retarded time t', solution of the implicit equation

$$t' = t - \frac{|\mathbf{r} - \mathbf{r}_0(t')|}{c}. \tag{3.14}$$

The potentials that we have obtained are called the Liénard and Wiechart potentials. They contain the $1/\kappa$ factor that, as we will see, is of fundamental importance in determining the radiation properties of moving charges. The physical meaning of this factor can be intuitively understood by observing that the signals emitted by a moving source are concentrated in the direction of the velocity, and become rarefied in the opposite direction, as exemplified in Fig. 3.2. Finally, it is interesting to note that, in the static case, the scalar potential of Liénard and Wiechart becomes the electrostatic potential (since $\kappa = 1$), while the vector potential is zero, since $\mathbf{v} = 0$.

3.3 The Electromagnetic Field of a Moving Charge

The electric and magnetic fields at the point of coordinate $\mathbf{r}$ and at time t produced by a point charge (of charge value e) moving along the trajectory $\mathbf{r}_0(t)$ are obtained by applying to the Liénard and Wiechart potentials the general equations defining the scalar and vector potentials, or

$$\mathbf{B}(\mathbf{r}, t) = \operatorname{rot} \mathbf{A}(\mathbf{r}, t), \tag{3.15}$$

$$\mathbf{E}(\mathbf{r}, t) = -\operatorname{grad} \phi(\mathbf{r}, t) - \frac{1}{c} \frac{\partial}{\partial t} \mathbf{A}(\mathbf{r}, t). \tag{3.16}$$

The determination of the fields is therefore reduced to a simple exercise that however involves some mathematical subtleties, requiring a detailed description.

We start by noting that the Liénard and Wiechart potentials contain at the denominator the expression κR and that such expression can be written in the form

$$\kappa R = \left(1 - \frac{\mathbf{v} \cdot \mathbf{n}}{c}\right) R = R - \frac{\mathbf{v} \cdot \mathbf{R}}{c}.$$

We also note that, given a vector of the form

$$\mathbf{X} = \mathbf{x}_a - \mathbf{x}_b,$$

with $\mathbf{x}_a$ and $\mathbf{x}_b$ functions of an arbitrary parameter ζ, we have

$$\frac{\partial}{\partial \zeta}\mathbf{X} = \frac{\partial \mathbf{x}_a}{\partial \zeta} - \frac{\partial \mathbf{x}_b}{\partial \zeta},$$

so that

$$\frac{\partial}{\partial \zeta} X^2 = 2\mathbf{X} \cdot \left[\frac{\partial \mathbf{x}_a}{\partial \zeta} - \frac{\partial \mathbf{x}_b}{\partial \zeta}\right],$$

and then, introducing the unit vector vers($\mathbf{X}$) of the vector $\mathbf{X}$, we have

$$\frac{\partial}{\partial \zeta} X = \frac{\partial}{\partial \zeta}\sqrt{X^2} = \frac{1}{2X}2\mathbf{X} \cdot \left[\frac{\partial \mathbf{x}_a}{\partial \zeta} - \frac{\partial \mathbf{x}_b}{\partial \zeta}\right] = \text{vers}(\mathbf{X}) \cdot \left[\frac{\partial \mathbf{x}_a}{\partial \zeta} - \frac{\partial \mathbf{x}_b}{\partial \zeta}\right].$$

Using this result, we simply obtain the following expressions

$$\frac{\partial R}{\partial t'} = -\mathbf{n} \cdot \mathbf{v}, \qquad \frac{\partial t}{\partial t'} = 1 + \frac{1}{c}\frac{\partial R}{\partial t'} = 1 - \frac{\mathbf{n} \cdot \mathbf{v}}{c} = \kappa,$$

$$\frac{\partial t'}{\partial t} = \frac{1}{\frac{\partial t}{\partial t'}} = \frac{1}{\kappa}, \qquad \text{grad}[\mathbf{r}_0(t')] = -[\text{grad } R]\frac{\mathbf{v}}{c}.$$

For the last equation we have in fact

$$\{\text{grad}[\mathbf{r}_0(t')]\}_{ij} = \frac{\partial}{\partial x_i}[\mathbf{r}_0(t')]_j = \frac{\partial}{\partial t'}[\mathbf{r}_0(t')]_j\frac{\partial t'}{\partial x_i}$$

$$= v_j\frac{\partial}{\partial x_i}\left(t - \frac{R}{c}\right) = -\frac{v_j}{c}\frac{\partial R}{\partial x_i}.$$

On the other hand,

$$\text{grad } R = \text{grad}|\mathbf{r} - \mathbf{r}_0(t')| = \mathbf{n} - \text{grad}[\mathbf{r}_0(t')] \cdot \mathbf{n} = \mathbf{n} + [\text{grad } R]\frac{\mathbf{v} \cdot \mathbf{n}}{c},$$

from which we obtain

$$\text{grad } R = \frac{\mathbf{n}}{\kappa},$$

which allows us to rewrite the equation for the gradient of $\mathbf{r}_0(t')$ in the form

$$\operatorname{grad}\left[\mathbf{r}_0\left(t'\right)\right] = -\frac{\mathbf{n}\mathbf{v}}{\kappa c}.$$

We also have

$$\operatorname{grad}\mathbf{R} = \operatorname{grad}\left[\mathbf{r} - \mathbf{r}_0\left(t'\right)\right] = U + \frac{\mathbf{n}\mathbf{v}}{\kappa c},$$

where U is the unit tensor ($U_{ij} = \delta_{ij}$), and also

$$\operatorname{grad}t' = \operatorname{grad}\left[t - \frac{R}{c}\right] = -\frac{1}{c}\operatorname{grad}R = -\frac{\mathbf{n}}{\kappa c},$$

$$\operatorname{grad}\mathbf{v} = \left(\operatorname{grad}t'\right)\frac{\partial \mathbf{v}}{\partial t'} = -\frac{\mathbf{n}\mathbf{a}}{\kappa c}, \qquad \frac{\partial R}{\partial t} = \frac{\partial R}{\partial t'}\frac{\partial t'}{\partial t} = -\frac{\mathbf{v}\cdot\mathbf{n}}{\kappa},$$

$$\frac{\partial \mathbf{R}}{\partial t} = \frac{\partial \mathbf{R}}{\partial t'}\frac{\partial t'}{\partial t} = -\frac{\mathbf{v}}{\kappa}, \qquad \frac{\partial \mathbf{v}}{\partial t} = \frac{\partial \mathbf{v}}{\partial t'}\frac{\partial t'}{\partial t} = \frac{\mathbf{a}}{\kappa},$$

where $\mathbf{a}$ is the particle acceleration at the retarded time t'. With these equations it is then simple to express both the gradient and the temporal derivative of the product κR. We have

$$\operatorname{grad}(\kappa R) = \operatorname{grad}\left[R - \frac{\mathbf{v}\cdot\mathbf{R}}{c}\right] = \operatorname{grad}R - \frac{1}{c}(\operatorname{grad}\mathbf{v})\cdot\mathbf{R} - \frac{1}{c}(\operatorname{grad}\mathbf{R})\cdot\mathbf{v},$$

i.e.

$$\operatorname{grad}(\kappa R) = \frac{\mathbf{n}}{\kappa} + \frac{1}{c^2\kappa}(\mathbf{a}\cdot\mathbf{R})\mathbf{n} - \frac{\mathbf{v}}{c} - \frac{v^2\mathbf{n}}{c^2\kappa} = \left(1 - \frac{v^2}{c^2}\right)\frac{\mathbf{n}}{\kappa} - \frac{\mathbf{v}}{c} + \frac{1}{c^2\kappa}(\mathbf{a}\cdot\mathbf{R})\mathbf{n},$$

$$\frac{\partial}{\partial t}(\kappa R) = \frac{\partial}{\partial t}\left[R - \frac{\mathbf{v}\cdot\mathbf{R}}{c}\right] = -\frac{\mathbf{n}\cdot\mathbf{v}}{\kappa} - \frac{1}{c\kappa}\mathbf{a}\cdot\mathbf{R} + \frac{v^2}{c\kappa}.$$

We are now able to calculate the electric field. Using Eqs. (3.16) and (3.12) we obtain

$$\mathbf{E}(\mathbf{r}, t) = -e\operatorname{grad}\left(\frac{1}{\kappa R}\right) - \frac{e}{c^2}\frac{\partial}{\partial t}\left(\frac{\mathbf{v}}{\kappa R}\right),$$

or

$$\mathbf{E}(\mathbf{r}, t) = \frac{e}{\kappa^2 R^2}\operatorname{grad}(\kappa R) - \frac{e}{c^2\kappa R}\frac{\partial \mathbf{v}}{\partial t} + \frac{e\mathbf{v}}{c^2\kappa^2 R^2}\frac{\partial}{\partial t}(\kappa R).$$

Substituting the previous expressions, we finally obtain

$$\mathbf{E}(\mathbf{r}, t) = \frac{e}{\kappa^2 R^2}\left[\left(1 - \frac{v^2}{c^2}\right)\frac{\mathbf{n}}{\kappa} - \frac{\mathbf{v}}{c} + \frac{1}{\kappa c^2}(\mathbf{a}\cdot\mathbf{R})\mathbf{n}\right] - \frac{e}{c^2\kappa^2 R}\mathbf{a}$$

$$+ \frac{e\mathbf{v}}{c^2\kappa^2 R^2}\left[-\frac{\mathbf{n}\cdot\mathbf{v}}{\kappa} - \frac{1}{\kappa c}\mathbf{a}\cdot\mathbf{R} + \frac{v^2}{\kappa c}\right].$$

The expression of the electric field contains various terms, some proportional to R^{-2} and others proportional to R^{-1}. If we group the first terms, we obtain the so-called Coulomb term (sometimes also called velocity term) which generalizes to the case of moving charges the usual expression of the Coulomb electrostatic field. When we group the other terms, we get instead the so-called radiation term, sometimes called acceleration term because it is proportional to the acceleration of the charge. Indicating with $[\mathbf{E}(\mathbf{r}, t)]_{\text{Coul}}$ and $[\mathbf{E}(\mathbf{r}, t)]_{\text{rad}}$ the two contributions, by means of simple factorizations we obtain

$$\left[\mathbf{E}(\mathbf{r}, t)\right]_{\text{Coul}} = \frac{e}{\kappa^3 R^2}\left[\left(1 - \frac{v^2}{c^2}\right)\mathbf{n} - \frac{\mathbf{v}}{c}\left(\kappa + \frac{\mathbf{n}\cdot\mathbf{v}}{c} - \frac{v^2}{c^2}\right)\right],$$

or, recalling the expression of κ (Eq. (3.13))

$$\left[\mathbf{E}(\mathbf{r}, t)\right]_{\text{Coul}} = \frac{e}{\kappa^3 R^2}\left[\left(1 - \frac{v^2}{c^2}\right)\left(\mathbf{n} - \frac{\mathbf{v}}{c}\right)\right]. \tag{3.17}$$

Similarly

$$\left[\mathbf{E}(\mathbf{r}, t)\right]_{\text{rad}} = \frac{e}{c^2\kappa^3 R}\left[\left(\mathbf{n} - \frac{\mathbf{v}}{c}\right)(\mathbf{a}\cdot\mathbf{n}) - \kappa\mathbf{a}\right],$$

or, as it is simple to verify,

$$\left[\mathbf{E}(\mathbf{r}, t)\right]_{\text{rad}} = \frac{e}{c^2\kappa^3 R}\mathbf{n}\times\left[\left(\mathbf{n} - \frac{\mathbf{v}}{c}\right)\times\mathbf{a}\right]. \tag{3.18}$$

The expression for the electric field can be put in an alternative form by introducing the typical notations of relativistic mechanics. Defining

$$\boldsymbol{\beta} = \frac{\mathbf{v}}{c}, \qquad \dot{\boldsymbol{\beta}} = \frac{\mathbf{a}}{c},$$

we obtain

$$\mathbf{E}(\mathbf{r}, t) = \frac{e}{\kappa^3 R^2}(1 - \beta^2)(\mathbf{n} - \boldsymbol{\beta}) + \frac{e}{c\kappa^3 R}\mathbf{n}\times\left[(\mathbf{n} - \boldsymbol{\beta})\times\dot{\boldsymbol{\beta}}\right]. \tag{3.19}$$

With similar calculations, we can obtain the expression for the magnetic field. Using Eqs. (3.15) and (3.12) we have

$$\mathbf{B}(\mathbf{r}, t) = \frac{e}{c}\text{rot}\left(\frac{\mathbf{v}}{\kappa R}\right) = \frac{e}{c\kappa R}\text{rot}\,\mathbf{v} + \frac{e}{c}\left[\text{grad}\left(\frac{1}{\kappa R}\right)\right]\times\mathbf{v}.$$

On the other hand, we have

$$\text{rot}\,\mathbf{v} = \left[\text{grad}\,t'\right]\times\mathbf{a} = -\frac{1}{c\kappa}\mathbf{n}\times\mathbf{a},$$

and then, using the above results,

$$\mathbf{B}(\mathbf{r}, t) = -\frac{e}{c^2 \kappa^2 R} \mathbf{n} \times \mathbf{a} - \frac{e}{c \kappa^2 R^2} \left[\left(1 - \frac{v^2}{c^2} \right) \frac{\mathbf{n}}{\kappa} - \frac{\mathbf{v}}{c} + \frac{1}{c^2 \kappa} (\mathbf{a} \cdot \mathbf{R}) \mathbf{n} \right] \times \mathbf{v}.$$

In analogy with the previous transformations for the electric field, we now separate in the right-hand side the terms proportional to R^{-2} from those proportional to R^{-1}. We obtain

$$\mathbf{B}(\mathbf{r}, t) = -\frac{e}{c \kappa^3 R^2} \left(1 - \frac{v^2}{c^2} \right) \mathbf{n} \times \mathbf{v} - \frac{e}{c^2 \kappa^3 R} \left[\kappa \mathbf{n} \times \mathbf{a} + (\mathbf{a} \cdot \mathbf{n}) \mathbf{n} \times \frac{\mathbf{v}}{c} \right].$$

We can rewrite this expression in terms of the vectors $\boldsymbol{\beta}$ and $\dot{\boldsymbol{\beta}}$

$$\mathbf{B}(\mathbf{r}, t) = -\frac{e}{\kappa^3 R^2} (1 - \beta^2) \mathbf{n} \times \boldsymbol{\beta} - \frac{e}{c \kappa^3 R} \left[\mathbf{n} \times \dot{\boldsymbol{\beta}} - (\mathbf{n} \cdot \boldsymbol{\beta}) \mathbf{n} \times \dot{\boldsymbol{\beta}} + (\dot{\boldsymbol{\beta}} \cdot \mathbf{n}) \mathbf{n} \times \boldsymbol{\beta} \right],$$

or, as we can simply verify

$$\mathbf{B}(\mathbf{r}, t) = -\frac{e}{\kappa^3 R^2} (1 - \beta^2) \mathbf{n} \times \boldsymbol{\beta} - \frac{e}{c \kappa^3 R} \left\{ \mathbf{n} \times \dot{\boldsymbol{\beta}} + \mathbf{n} \times \left[\mathbf{n} \times (\boldsymbol{\beta} \times \dot{\boldsymbol{\beta}}) \right] \right\}.$$

Using this expression, together with the one relative to $\mathbf{E}(\mathbf{r}, t)$ (Eq. (3.19)), we easily find the important relation

$$\mathbf{B}(\mathbf{r}, t) = \mathbf{n} \times \mathbf{E}(\mathbf{r}, t), \tag{3.20}$$

which shows that the magnitude of the magnetic field vector is always less or equal than the magnitude of the electric field vector.

Finally, we note that Eqs. (3.19) and (3.20) are very general and, given the relativistic invariance of Maxwell's equations (from which we started), they must be valid in an arbitrary inertial reference system. Considering the nonrelativistic limit to first order in β, it is possible to show that the ordinary laws of electromagnetism valid for stationary phenomena are recovered. The derivation of this property is contained in Sect. 16.4.

3.4 Radiation from a Moving Charge

As we saw in the previous section, the electromagnetic field produced by a moving charge consists of two terms, one inversely proportional to R^2 and the other inversely proportional to R. Obviously, the first term prevails for R tending to zero, while the second term prevails when R tends to infinity. It is interesting to calculate the value of R for which the two terms are of the same order of magnitude. Indicating with R_c such value, we have

$$\frac{e}{\kappa^3 R_c^2} \simeq \frac{ea}{c^2 \kappa^3 R_c},$$

or

$$R_c \simeq \frac{c^2}{a}.$$

Indicating with L the typical dimensions of the region where the charge is moving, and with τ the relative characteristic time, we have

$$a \simeq \frac{L}{\tau^2},$$

so that

$$R_c \simeq \frac{c^2 \tau^2}{L}.$$

On the other hand, as we shall see later, the charge itself radiates at characteristic frequencies $\nu \simeq c/\tau$, so the critical distance R_c can also be written in the form λ^2/L, where λ is the typical wavelength of the radiation emitted by the moving charge. The region where $R \gg R_c$ is called the radiation zone. Within such region, the electromagnetic field is given by Eqs. (3.18) and (3.20) which we rewrite here

$$\mathbf{E}(\mathbf{r}, t) = \frac{e}{c^2 \kappa^3 R} \mathbf{n} \times \left[\left(\mathbf{n} - \frac{\mathbf{v}}{c} \right) \times \mathbf{a} \right], \qquad \mathbf{B}(\mathbf{r}, t) = \mathbf{n} \times \mathbf{E}(\mathbf{r}, t).$$

We recall that the quantities R, κ, $\mathbf{n}$, $\mathbf{v}$, and $\mathbf{a}$ that appear in these equations must be evaluated at the retarded time t'. However, at large distances from the charge ($R \gg L$), the effect of the retarded time on both the distance R and the unit vector $\mathbf{n}$ can be neglected. In this case, both quantities can then be considered as constants.

The first fact to be noted about the electric and magnetic fields in the radiation zone is that they are perpendicular to each other, they are both perpendicular to the unit vector $\mathbf{n}$, and they are equal in magnitude. These are characteristics that we have already encountered in Sect. 1.6 for the plane waves propagating in vacuum. The second fact to be noted concerns the Poynting vector $\mathbf{S}$, given by

$$\mathbf{S} = \frac{c}{4\pi} \mathbf{E} \times \mathbf{B} = \frac{c}{4\pi} \mathbf{E} \times (\mathbf{n} \times \mathbf{E}) = \frac{c}{4\pi} E^2 \mathbf{n}.$$

The Poynting vector at point P is along the direction connecting the charge with the point P.

We now consider the nonrelativistic case where $v \ll c$. In this case, the electric field, at the lowest order in v/c, is given by

$$\mathbf{E}(\mathbf{r}, t) = \frac{e}{c^2 R} \mathbf{n} \times [\mathbf{n} \times \mathbf{a}] = -\frac{e}{c^2 R} [\mathbf{a} - (\mathbf{a} \cdot \mathbf{n})\mathbf{n}].$$

This expression shows that the electric field is perpendicular to $\mathbf{n}$ and lies in the plane defined by $\mathbf{n}$ and $\mathbf{a}$ (the particle acceleration at the retarded time). Therefore, the component of the electric field along a polarisation unit vector $\mathbf{e}_i$ ($i = 1, 2$)

Fig. 3.3 Radiation diagram (or antenna diagram) of a nonrelativistic particle. The acceleration of the particle is directed along the vertical axis. The power emitted along a direction that forms the angle θ with the acceleration is proportional to the segment drawn in the figure

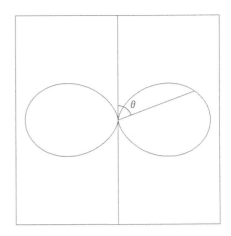

perpendicular to **n** is given by

$$\mathbf{E}(\mathbf{r}, t) \cdot \mathbf{e}_i = -\frac{e}{c^2 R} \mathbf{a} \cdot \mathbf{e}_i. \tag{3.21}$$

Furthermore, indicating with θ the angle between the direction of the acceleration (at the retarded time) and the **n** direction, we have that the Poynting vector at a distance R from the charge is

$$\mathbf{S} = \frac{e^2 a^2 \sin^2 \theta}{4\pi c^3 R^2} \mathbf{n}.$$

This equation shows that the power emitted by a moving charge depends on the direction as $\sin^2 \theta$. If we draw on a graph, for any direction, a segment proportional to the power emitted along the same direction, we obtain a diagram called radiation diagram (also known as radiation pattern or antenna diagram). Such diagram for an accelerated charge (nonrelativistic) is shown in Fig. 3.3. The same equation for **S** can be used to find the total power emitted from the charge. Calculating the flux W of the Poynting vector through a sphere of radius R, and taking into account that

$$\frac{1}{4\pi} \oint \sin^2 \theta \, d\Omega = \frac{2}{3},$$

we obtain the so-called Larmor equation

$$W = \frac{2e^2 a^2}{3c^3}. \tag{3.22}$$

Returning to the general case, it is interesting to note the presence of an important physical phenomenon related to the $1/\kappa^3$ factor we have found in the previous expressions for the radiation field. If we consider for simplicity the case of a charge with acceleration parallel to velocity, the Poynting vector (at a distance R from the

Fig. 3.4 Radiation diagram
of a relativistic particle
having $\beta = 0.8$. Both the
acceleration and the velocity
of the particle are directed
along the vertical axis, the
latter in the direction towards
the top. The power emitted
along a direction that forms
the angle θ with the velocity
(and the acceleration, parallel
or antiparallel to it) is
proportional to the segment
drawn in the figure

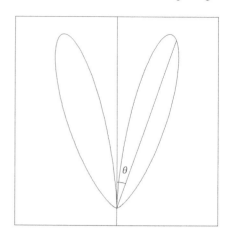

charge and in the direction of the unit vector **n**) is given by

$$\mathbf{S} = \frac{e^2 a^2 \sin^2 \theta}{4\pi c^3 \kappa^6 R^2} \mathbf{n} = \frac{e^2 a^2 \sin^2 \theta}{4\pi c^3 (1 - \beta \cos \theta)^6 R^2} \mathbf{n},$$

where $\beta = v/c$ and where θ is the angle between **n** and the direction of the velocity. For non-zero values of β, the radiation pattern, shown in Fig. 3.4, is profoundly different from the nonrelativistic case. The radiation concentrates in the forward direction (relative to the motion of the particle). This is a typical relativistic effect, known as beaming effect. It becomes more pronounced as the particle velocity approaches the speed of light. The angular width of the cone where the radiation is concentrated can be estimated by noting that, for $\beta \simeq 1$, the factor $1/\kappa$ has a very peaked maximum for $\theta = 0$.

Putting, around $\theta = 0$, $\cos \theta \simeq 1 - \theta^2/2$, we have

$$\frac{1}{\kappa} \simeq \frac{1}{1 - \beta + \beta \theta^2/2} \simeq \frac{1}{1 - \beta + \theta^2/2}.$$

This expression can be rewritten in the form

$$\frac{1}{\kappa} \simeq \frac{2}{\theta^2 + \theta_0^2},$$

where we have introduced the angle θ_0 given by

$$\theta_0 = \sqrt{2(1 - \beta)}.$$

Recalling the expression for the relativistic factor γ (Lorentz factor) given by

$$\gamma = \frac{1}{\sqrt{1 - \beta^2}},$$

Fig. 3.5 Radiation diagram of a relativistic particle having $\beta = 0.8$. The acceleration of the particle is directed along the vertical axis, while the velocity is directed along the horizontal axis, from *left* to *right*. The power emitted along a direction belonging to the plane containing the velocity and the acceleration ($\phi = 0$) and forming the angle θ with the acceleration is proportional to the segment drawn in the figure

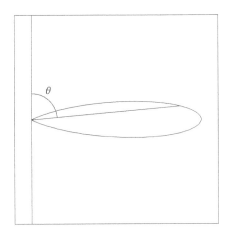

we have, for the ultra-relativistic case

$$\gamma = \frac{1}{\sqrt{(1-\beta)(1+\beta)}} \simeq \frac{1}{\sqrt{2(1-\beta)}},$$

so that we obtain

$$\theta_0 \simeq \frac{1}{\gamma}.$$

The value of θ_0 is precisely the angular width of the cone in which the radiation is concentrated. As the energy of the particle increases (hence the γ factor), the amplitude of the cone becomes increasingly smaller.

When the acceleration is perpendicular to the velocity, the calculation of the Poynting vector is more complicated since cylindrical symmetry is lost. Introducing a coordinate system where the acceleration is directed along the z axis and the velocity along the x axis, and indicating with θ and ϕ the polar coordinates of the direction $\mathbf{n}$, we obtain for the Poynting vector along such direction

$$\mathbf{S} = \frac{e^2 a^2}{4\pi c^3 (1 - \beta \sin\theta \cos\phi)^6 R^2} \left[(1 - \beta \sin\theta \cos\phi)^2 - \left(1 - \beta^2\right) \cos^2\theta\right]\mathbf{n}.$$

The corresponding radiation diagram, relative to the plane containing the velocity and acceleration vectors ($\phi = 0$), is shown in Fig. 3.5.

Finally, in the general case where the acceleration is neither parallel nor perpendicular to the velocity, the expression of the Poynting vector becomes even more complicated. It is reported in Sect. 16.5, where we also demonstrate the generalisation of the Larmor equation for the power emitted by a relativistic charge, that is

$$W = \frac{2e^2}{3c^3} \left(\gamma^6 a_\parallel^2 + \gamma^4 a_\perp^2\right), \tag{3.23}$$

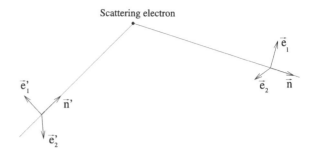

Fig. 3.6 Geometry of Thomson scattering in the general case. The choice of the unit vectors of polarisation is arbitrary

where $a_\parallel$ and $a_\perp$ are the components of the acceleration in the directions parallel and perpendicular to the velocity. It is simple to verify that, in the nonrelativistic case in which $\beta \ll 1$, $\gamma \simeq 1$, being $a_\parallel^2 + a_\perp^2 = a^2$, we obtain the usual expression (3.22) of the Larmor equation.

3.5 Thomson Scattering

Consider a free electron of charge $e = -e_0$ ($e_0 = 4.803 \times 10^{-10}$ ues), and suppose that the electron is subject to the action of an electromagnetic polarised wave of frequency ω propagating along the direction $\mathbf{n}'$. We define a pair of unit vectors $\mathbf{e}_1'$ and $\mathbf{e}_2'$ such that they form with $\mathbf{n}'$ a right-handed triad of unit vectors as shown in Fig. 3.6. We also define with $\mathcal{E}_1'$ and $\mathcal{E}_2'$ the complex components of the electric field vector of the wave along these unit vectors. The motion of the electron, which we assume nonrelativistic, is described by the equation

$$\mathbf{a}(t) = -\frac{e_0}{m}\mathbf{E}'(t),$$

where m is the electron mass and where $\mathbf{E}'(t)$, the electric field of the incoming wave, is given by

$$\mathbf{E}'(t) = \mathrm{Re}(\mathcal{E}'\mathrm{e}^{-\mathrm{i}\omega t}) = \mathrm{Re}[(\mathcal{E}_1'\mathbf{e}_1' + \mathcal{E}_2'\mathbf{e}_2')\mathrm{e}^{-\mathrm{i}\omega t}].$$

We are interested in determining the expression of the electric field emitted from the electron in the radiation zone along the direction defined by the unit vector $\mathbf{n}$ of Fig. 3.6. Introducing the two unit vectors $\mathbf{e}_1$ and $\mathbf{e}_2$ (such to form with $\mathbf{n}$ a right-handed triad) and using complex notations, the components of the acceleration along such unit vectors are given by

$$\mathcal{A}_1 = -\frac{e_0}{m}\mathbf{e}_1 \cdot (\mathcal{E}_1'\mathbf{e}_1' + \mathcal{E}_2'\mathbf{e}_2'), \qquad \mathcal{A}_2 = -\frac{e_0}{m}\mathbf{e}_2 \cdot (\mathcal{E}_1'\mathbf{e}_1' + \mathcal{E}_2'\mathbf{e}_2'), \qquad (3.24)$$

where the complex vector $\mathcal{A}$ is implicitly defined by the equation

$$\mathbf{a}(t) = \mathrm{Re}(\mathcal{A}\mathrm{e}^{-\mathrm{i}\omega t}).$$

Fig. 3.7 Particular case of the geometry of Thomson scattering. The polarisation unit vectors $\mathbf{e}_1$ and $\mathbf{e}_1'$ are perpendicular to the plane containing the directions of the incident and scattered radiation (scattering plane), while $\mathbf{e}_2$ and $\mathbf{e}_2'$ lie in the same plane

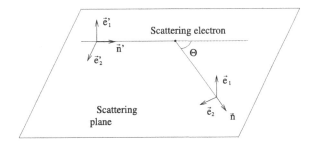

Taking into account Eq. (3.21), the components $\mathcal{E}_1$ and $\mathcal{E}_2$ of the radiation field at the distance R and in the direction $\mathbf{n}$ are given in matrix form by the expression

$$\begin{pmatrix} \mathcal{E}_1 \\ \mathcal{E}_2 \end{pmatrix} = -\frac{r_c}{R} \begin{pmatrix} \mathbf{e}_1 \cdot \mathbf{e}_1' & \mathbf{e}_1 \cdot \mathbf{e}_2' \\ \mathbf{e}_2 \cdot \mathbf{e}_1' & \mathbf{e}_2 \cdot \mathbf{e}_2' \end{pmatrix} \begin{pmatrix} \mathcal{E}_1' \\ \mathcal{E}_2' \end{pmatrix} e^{i\Phi}, \tag{3.25}$$

where $\Phi = \omega R/c$ is an inessential phase factor introduced by the delay, and where r_c, the so-called classical radius of the electron, is defined by

$$r_c = \frac{e_0^2}{mc^2} = 2.818 \times 10^{-13} \text{ cm}. \tag{3.26}$$

The expression that we have obtained is the law of Thomson scattering in terms of electric fields. It contains all the properties of the scattered radiation (spectral, directional, and polarimetric), with in addition the general properties related to the cross section. From a spectral point of view, the electron simply oscillates at the same frequency ω of the incident radiation. The scattered radiation is then, as we say, coherent, i.e. the spectrum is of the type of a Dirac delta centred at the same frequency ω. To analyse the results for the radiation diagram and polarisation it is convenient to adequately choose the unit vectors of polarisation. With the choice sketched in Fig. 3.7, the 2×2 matrix that appears in Eq. (3.25) is greatly simplified and is

$$\begin{pmatrix} 1 & 0 \\ 0 & \cos\Theta \end{pmatrix},$$

where Θ is the angle of scattering.

We can now move on to describe the scattering process in terms of fluxes in the Stokes parameters. For this, we need to recall the expressions of Chap. 2, in particular Eq. (2.14), which relates the fluxes in the Stokes parameters with the components of the electric field. Denoting by $(F_I', F_Q', F_U', F_V')^{\dagger}$ the Stokes vector of the incident radiation and with $(F_I, F_Q, F_U, F_V)^{\dagger}$ that of the scattered radiation, with simple algebra we obtain the following matrix equation

$$
\begin{pmatrix} F_I \\ F_Q \\ F_U \\ F_V \end{pmatrix} = \frac{1}{2} \frac{r_c^2}{R^2} \begin{pmatrix} 1 + \cos^2 \Theta & \sin^2 \Theta & 0 & 0 \\ \sin^2 \Theta & 1 + \cos^2 \Theta & 0 & 0 \\ 0 & 0 & 2 \cos \Theta & 0 \\ 0 & 0 & 0 & 2 \cos \Theta \end{pmatrix} \begin{pmatrix} F_I' \\ F_Q' \\ F_U' \\ F_V' \end{pmatrix} . \quad (3.27)
$$

In particular, for the scattering at the angle Θ of an unpolarised ray we obtain, for the non-zero fluxes

$$
F_I(\Theta) = \frac{r_c^2}{2R^2} \left(1 + \cos^2 \Theta \right) F_I', \qquad F_Q(\Theta) = \frac{r_c^2}{2R^2} \sin^2 \Theta F_I'.
$$

These equations show that the radiation is mostly forward (or backward) scattered relative to the direction of the incident radiation. The ratio $\mathcal{R}$ between the radiation scattered in the Θ direction and the forward scattered one is given by the equation

$$
\mathcal{R} = \frac{1 + \cos^2 \Theta}{2},
$$

and varies between 1 and $1/2$. In addition, the scattered radiation is linearly polarised and the fraction of polarisation is given by

$$
\frac{F_Q(\Theta)}{F_I(\Theta)} = \frac{\sin^2 \Theta}{1 + \cos^2 \Theta},
$$

which implies that the linear polarisation is always positive, i.e. perpendicular to the scattering plane, and that the radiation is 100 % polarised for $\Theta = 90°$. Furthermore, integrating the intensity of the scattered radiation on a sphere of radius R we obtain (recall that the average of $\cos^2 \Theta$ over the solid angle is $1/3$)

$$
W = R^2 \oint F_I(\theta) \, d\Omega = \sigma_T F_I',
$$

where σ_T, the so-called Thomson cross-section, is given by

$$
\sigma_T = \frac{8\pi}{3} r_c^2 = \frac{8\pi e_0^4}{3m^2 c^4} = 6.652 \times 10^{-25} \text{ cm}^2.
$$

3.6 Rayleigh Scattering

Rayleigh scattering is very similar to Thomson scattering, with the only difference that the electron, instead of being free, is bound to an atom or a molecule. From the point of view of classical physics, the bound electron can be described by means of a simple model, due to Lorentz. The model assumes that the action on the electron of the cloud of positive charges present in the atom can be represented as an elastic restoring force of the form $\mathbf{F} = -k\mathbf{x}$, where k is a constant and $\mathbf{x}$ is the position of the electron relative to the barycentre of the positive charges. The law of motion of the bound electron under the action of an electric field of frequency ω is then

$$\frac{d^2\mathbf{x}}{dt^2} = -\omega_0^2\mathbf{x} - \frac{e_0}{m}\mathbf{E}'(t),$$

where $\omega_0 = \sqrt{k/m}$, and where

$$\mathbf{E}'(t) = \text{Re}\left(\mathcal{E}'e^{-i\omega t}\right).$$

The differential equation can be easily solved by searching for a stationary solution of the type $\mathbf{x}(t) = \mathbf{x}_0 e^{-i\omega t}$. Once the solution is found, we then determine the acceleration by differentiating twice with respect to time. The result for the components of the (complex) acceleration along the two unit vectors $\mathbf{e}_1$ and $\mathbf{e}_2$ of Fig. 3.6 is the following

$$\mathcal{A}_1 = -\frac{e_0}{m}\frac{\omega^2}{\omega^2 - \omega_0^2}\mathbf{e}_1 \cdot \left(\mathcal{E}_1'\mathbf{e}_1' + \mathcal{E}_2'\mathbf{e}_2'\right), \qquad \mathcal{A}_2 = -\frac{e_0}{m}\frac{\omega^2}{\omega^2 - \omega_0^2}\mathbf{e}_2 \cdot \left(\mathcal{E}_1'\mathbf{e}_1' + \mathcal{E}_2'\mathbf{e}_2'\right).$$

This expression is very similar to the one we obtained previously for Thomson scattering (Eq. (3.24)), and, as expected, reduces to the previous one for $\omega_0 = 0$ (the case of the free electron). Repeating the same arguments as those developed in the previous section, we obtain exactly the same results with the only difference that the Thomson cross section σ_T must be replaced with the Rayleigh cross section σ_R defined by

$$\sigma_R = \frac{\omega^4}{(\omega^2 - \omega_0^2)^2}\sigma_T = \frac{\omega^4}{(\omega^2 - \omega_0^2)^2}\frac{8\pi}{3}r_c^2.$$

An important aspect of the Rayleigh scattering is the fact that, for $\omega_0 \gg \omega$, the cross section is proportional to ω^4. As this is a good approximation for the visible radiation scattered by nitrogen and oxygen molecules, the most abundant constituents of Earth's atmosphere, and as the light in the sky is just scattered sunlight from such molecules, it follows that the sky is blue.[1] For the same reason, the Sun appears red at sunrise and sunset. The same phenomenon also holds for the moon and the stars.

3.7 Bremsstrahlung

The radiation emitted when a high-speed charged particle (typically an electron) is deflected by passing in the Coulomb field generated by another particle (usually an

---

[1] The blue radiation (4000 Å) is scattered with a cross section approximately 10 times larger than the red radiation (7000 Å).

Fig. 3.8 Geometry for the
bremsstrahlung calculation

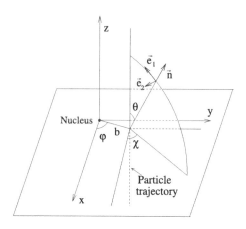

atomic nucleus) is called Bremsstrahlung (a German word meaning "braking radiation" or "deceleration radiation"). In this process the charge is accelerated and is
thus losing energy by radiation. Bremsstrahlung is a fundamental physical process in
astrophysical plasmas. It also has broad technological and laboratory applications.
The X-ray radiation that comes from sources commonly used both in research lab-
oratories and hospitals (as a diagnostic and therapeutic tool) is just bremsstrahlung,
generally produced by accelerating electrons through high potential differences and
making them collide against a metal plate of an element with high Z.

Although the detailed study of the braking radiation requires a quantum treatment, here we approach it from a classical point of view, emphasizing only at the
end of the discussion the influence that quantum-mechanical phenomena can have
on the results. To fix ideas, we consider a unidirectional beam of nonrelativistic
electrons colliding with heavy nuclei (i.e. with mass much greater than that of the
electrons). We then examine the characteristics (spectral, geometric, and polarimetric) of the radiation emitted in this process. With reference to Fig. 3.8 let v be the
velocity of the electron passing close to a nucleus of charge Ze_0. We denote by m
the mass of the electron and by b the impact parameter and assume that the electron is fast enough so we can neglect the deviation of its trajectory with respect to a
straight line. This is well verified if the following inequality holds

$$\frac{1}{2}mv^2 \gg \frac{Ze_0^2}{b},$$

which implies

$$b \gg b_{\text{min}},$$

where

$$b_{\text{min}} = \frac{2Ze_0^2}{mv^2}. \tag{3.28}$$

Under the straight trajectory approximation, and using the coordinate system
(x, y, z) of Fig. 3.8, the position of the electron as a function of time is given by

the equation

$$\mathbf{r}_0(t) = b\cos\varphi\mathbf{i} + b\sin\varphi\mathbf{j} + vt\mathbf{k},$$

where $\mathbf{i}$, $\mathbf{j}$, and $\mathbf{k}$ are three unit vectors directed, respectively, as the x, y, and z axes, φ is the angle that defines (together with the impact parameter) the geometry of the collision, and t is the time measured from the instant when the electrons goes through the x–y plane. The acceleration of the electron can easily be calculated by taking into account the Coulomb force exerted by the nucleus. We have

$$\mathbf{a}(t) = \frac{-Ze_0^2\mathbf{r}_0(t)}{mr_0^3(t)} = -\frac{Ze_0^2}{m(b^2 + v^2t^2)^{3/2}}(b\cos\varphi\mathbf{i} + b\sin\varphi\mathbf{j} + vt\mathbf{k}).$$

We now consider the radiation emitted at large distances along the direction of the unit vector $\mathbf{n}$ and we introduce, as in Fig. 3.8, two unit vectors $\mathbf{e}_1$ and $\mathbf{e}_2$ with the usual convention where $(\mathbf{e}_1, \mathbf{e}_2, \mathbf{n})$ is a right-handed triad. If θ and χ are the polar and azimuthal angles that specify the direction $\mathbf{n}$, we have

$$\mathbf{n} = \sin\theta\cos\chi\mathbf{i} + \sin\theta\sin\chi\mathbf{j} + \cos\theta\mathbf{k},$$

$$\mathbf{e}_1 = -\cos\theta\cos\chi\mathbf{i} - \cos\theta\sin\chi\mathbf{j} + \sin\theta\mathbf{k}, \qquad \mathbf{e}_2 = \sin\chi\mathbf{i} - \cos\chi\mathbf{j}.$$

Applying the nonrelativistic equation (3.21), the components of the electric field at distance R from the nucleus along the direction $\mathbf{n}$ are given by

$$E_1(t + R/c) = \frac{Ze_0^3}{mc^2R(b^2 + v^2t^2)^{3/2}}\left[b\cos\theta\cos(\varphi - \chi) - vt\sin\theta\right],$$

$$E_2(t + R/c) = \frac{Ze_0^3}{mc^2R(b^2 + v^2t^2)^{3/2}}b\sin(\varphi - \chi).$$

Unlike the cases discussed in the previous sections, now the electric field does not have a sinusoidal variation with time. To obtain the spectral and polarimetric properties of the radiation field, we need to consider the Fourier transforms of the two components of the electric field vector. This leads to the need to evaluate integrals of the form

$$\int_{-\infty}^{\infty} \frac{\cos(\omega t)}{(b^2 + v^2t^2)^{3/2}}\,dt, \qquad \int_{-\infty}^{\infty} \frac{t\sin(\omega t)}{(b^2 + v^2t^2)^{3/2}}\,dt,$$

that, with the substitution $\tan x = vt/b$, can be cast in terms of the functions $F(z)$ and $G(z)$ defined by[2]

$$F(z) = \int_0^{\pi/2} \cos(z\tan x)\cos x\,dx, \qquad G(z) = \int_0^{\pi/2} \sin(z\tan x)\sin x\,dx.$$

---

[2] The functions $F(z)$ and $G(z)$ can be related to the modified Bessel functions of second kind, $K_n(z)$. We have $F(z) = zK_1(z)$, $G(z) = zK_0(z)$.

Fig. 3.9 Graph of the
functions $F^2(z)$ and $G^2(z)$

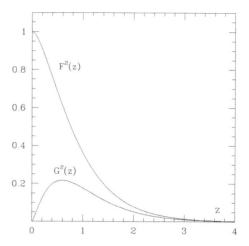

Using these functions (their square is shown in the graph in Fig. 3.9), we obtain for
the Fourier transforms

$$\hat{f}_1(\omega) = \frac{1}{2\pi} \int_{-\infty}^{\infty} E_1(t) e^{i\omega t}\, dt = \frac{Ze_0^3}{\pi mc^2 bvR}$$
$$\times \left[\cos\theta \cos(\varphi - \chi) F(z) - i \sin\theta G(z)\right] e^{i\Phi}, \tag{3.29}$$

$$\hat{f}_2(\omega) = \frac{1}{2\pi} \int_{-\infty}^{\infty} E_2(t) e^{i\omega t}\, dt = \frac{Ze_0^3}{\pi mc^2 bvR} \sin(\varphi - \chi) F(z) e^{i\Phi},$$

where $\Phi = \omega R/c$ is an inessential phase (due to the retarded time) and where

$$z = \frac{\omega b}{v}.$$

We can now evaluate the monochromatic fluxes in the Stokes parameters of the
radiation emitted along the direction $\mathbf{n}$. For this we need to recall the considerations
developed in Sect. 2.3. The expressions that we are going to obtain are the Fourier
transforms of the components, along the unit vectors $\mathbf{e}_1$ and $\mathbf{e}_2$, of the impulses of
the electric field emitted by a single electron moving in proximity of a nucleus with
given values of the impact parameter b and of the angle ϕ. We now think of the physical situation in which there is a uniform flow of electrons, all having velocity v. We
are obviously in the presence of a stochastic phenomenon, described at the microscopic level by a situation described in Fig. 2.3, where the monochromatic fluxes
are given by Eq. (2.17). Denoting by N_e the number density of the electrons of the
beam, in the unit of time we have a number of collisions with impact parameter be-
tween b and $b + db$ and angle between φ and $\varphi + d\varphi$ given by $dN_{coll} = N_e vb\, db\, d\varphi$,
so we obtain

$$F_\omega^I(\mathbf{n}) = cvN_e \int_0^{2\pi} d\varphi \int_0^{\infty} \left[\hat{f}_1(\omega)^* \hat{f}_1(\omega) + \hat{f}_2(\omega)^* \hat{f}_2(\omega)\right] b\, db,$$

$$F_\omega^Q(\mathbf{n}) = cvN_e \int_0^{2\pi} d\varphi \int_0^\infty \left[\hat{f}_1(\omega)^* \hat{f}_1(\omega) - \hat{f}_2(\omega)^* \hat{f}_2(\omega)\right] b\, db,$$

$$F_\omega^U(\mathbf{n}) = cvN_e \int_0^{2\pi} d\varphi \int_0^\infty \left[\hat{f}_1(\omega)^* \hat{f}_2(\omega) + \hat{f}_2(\omega)^* \hat{f}_1(\omega)\right] b\, db,$$

$$F_\omega^V(\mathbf{n}) = cvN_e \int_0^{2\pi} d\varphi \int_0^\infty i\left[\hat{f}_1(\omega)^* \hat{f}_2(\omega) - \hat{f}_2(\omega)^* \hat{f}_1(\omega)\right] b\, db.$$

Substituting the values of the Fourier transforms given by Eq. (3.29), and performing the integration over φ, the fluxes in the Stokes parameters U and V vanish. For the remaining two we have

$$F_\omega^I(\mathbf{n}) = C \int_0^\infty \left[(1 + \cos^2\theta)F^2(z) + 2\sin^2\theta\, G^2(z)\right]\frac{db}{b},$$

$$F_\omega^Q(\mathbf{n}) = -C \sin^2\theta \int_0^\infty \left[F^2(z) - 2G^2(z)\right]\frac{db}{b},$$

where we have put

$$C = \frac{Z^2 e_0^6 N_e}{\pi m^2 c^3 v R^2}.$$

However, we must take into account the fact that the integrals appearing in the above equations are divergent because, for $b \to 0$, the integrands go to infinity as b^{-1}. This divergence is due to the approximation of the straight trajectory that we introduced at the beginning of the calculation. Given that the approximation is not justified for $b < b_{min}$, with b_{min} defined in Eq. (3.28), the divergence can be avoided by changing the first limit of integration from 0 to b_{min}. In this way, we obtain an approximate expression that can only be improved by more complex calculations. So, if we then define the two quantities

$$\mathcal{F} = \int_{b_{min}}^\infty \frac{1}{b} F^2\left(\frac{\omega b}{v}\right) db, \qquad \mathcal{G} = \int_{b_{min}}^\infty \frac{1}{b} G^2\left(\frac{\omega b}{v}\right) db,$$

we obtain

$$F_\omega^I(\mathbf{n}) = C\left[(1 + \cos^2\theta)\mathcal{F} + 2\sin^2\theta\, \mathcal{G}\right], \qquad F_\omega^Q(\mathbf{n}) = -C\sin^2\theta(\mathcal{F} - 2\mathcal{G}).$$

The inequality $\mathcal{F} \gg \mathcal{G}$ is always well verified, as we can infer from the expression of the integrals defining $\mathcal{F}$ and $\mathcal{G}$ and the behaviour at each angular frequency ω (which contributes substantially to the emissivity) of $F^2(z)$ and $G^2(z)$ of Fig. 3.9. This implies that, with good approximation, the radiation diagram has a dependency on θ of the form $(1 + \cos^2\theta)$. This shows that the radiation emitted towards the "poles", i.e. along the electron beam, is twice more intense than that emitted in the "equatorial plane". As regards polarisation, the radiation is linearly polarised with a percentage value, practically independent of the frequency, given by

$$\frac{F_\omega^Q(\mathbf{n})}{F_\omega^I(\mathbf{n})} = -\frac{\sin^2\theta}{1 + \cos^2\theta}.$$

This equation shows that the radiation emitted in the plane perpendicular to the velocity of the colliding particles ($\theta = 90°$) is 100 % linearly polarised. The polarisation direction is contained in the same plane (remember the definition of the two unit vectors of polarisation in Fig. 3.8). Changing the direction of emission, the polarisation keeps the same characteristics (linear and directed perpendicularly to the velocity) but the percentage of polarisation lowers as $\sin^2\theta$ decreases. In particular, the radiation emitted along the direction of the velocity (or in the opposite direction) is not polarised.

Finally, the above equations can also be used to provide an order of magnitude estimate of the total monochromatic power (i.e. integrated over the whole solid angle) of the braking radiation. We perform for this a rather crude approximation on the behaviour of the function $F^2(z)$, assuming that it is equal to 1 in the range of z between 0 and 1 and that it is null for $z > 1$. This implies, for the quantity $\mathcal{F}$,

$$\mathcal{F} = \int_{b_{\min}}^{b_{\max}} \frac{1}{b}\, db = \ln\left(\frac{b_{\max}}{b_{\min}}\right),$$

where

$$b_{\max} = \frac{v}{\omega}.$$

If we also assume that $\mathcal{G} = 0$, integrating the flux $F_\omega^I(\mathbf{n})$ on the sphere of radius R, and substituting the values of $b_{\min}$ and $b_{\max}$, we obtain

$$W_\omega = \frac{16}{3}\frac{Z^2 e_0^6 N_e}{m^2 c^3 v} \ln\left(\frac{mv^3}{2Ze_0^2\omega}\right),$$

or a nearly flat spectral behaviour, since the dependence on ω is only contained in the logarithm.

A more in-depth analysis of the same problem,[3] accomplished without using the straight trajectory approximation, shows that the formula we derived is correct, aside from some slight changes. In the limit of low frequencies ($\omega \to 0$), the factor $\frac{1}{2}$ in the argument of the logarithm must be replaced by the factor $2e^{-\gamma}$, where γ is the Euler-Mascheroni constant defined by

$$\gamma = \lim_{n\to\infty}\left[-\ln n + \sum_{k=1}^{n-1}\frac{1}{k}\right] \simeq 0.57721.$$

---

[3] See Landau and Lifchitz (1966).

Conversely, in the limit of high frequencies ($\omega \to \infty$), the equation must be modified by multiplying the right-hand side by the factor $\pi/\sqrt{3}$ and omitting the logarithm, so to have

$$W_\omega = \frac{16\pi}{3\sqrt{3}} \frac{Z^2 e_0^6 N_e}{m^2 c^3 v}.$$

This formula shows that in the limit of high frequencies the power W_ω is independent of ω. This implies that for the braking radiation we obtain, so to say, a sort of ultraviolet catastrophe similar to that of the classical theory of black body radiation. In fact, defining the total radiated power by the equation

$$W = \int_0^\infty W_\omega \, d\omega,$$

we get a divergent integral. The reason for this fact is because we have completely neglected quantum effects, the most important consequence of which is the appearance of a threshold value for the angular frequency of the emitted photons, ω_{max}, given by

$$\hbar \omega_{max} = \frac{1}{2} m v^2.$$

To obtain an order of magnitude estimate for W, we neglect the minor dependence on the logarithm and we integrate in $d\omega$ between 0 and ω_{max}. We obtain

$$W \simeq \frac{Z^2 e_0^6 N_e}{m^2 c^3 v} \omega_{max} = \frac{Z^2 e_0^6 N_e v}{2 m c^3 \hbar}.$$

This equation allows the introduction of a suitable cross section for the braking radiation, which we indicate with σ_f. The value of the cross section is obtained by dividing the total irradiated power W for the energy flux F_e of the colliding electrons, defined by

$$F_e = N_e v \frac{1}{2} m v^2.$$

With simple algebra we find

$$\sigma_f = \frac{W}{F_e} = Z^2 \alpha \frac{1}{\beta^2} r_c^2,$$

where α is the fine-structure constant ($\alpha = e_0^2/(\hbar c)$), $\beta = v/c$, and r_c is the classical radius of the electron defined in Eq. (3.26). As we can see, the cross section depends on the square of the charge number Z. For this reason, in technical devices that are used for the production of X-rays, the target for the accelerated electrons is constituted by plates of metals with high Z (typically lead). The above formula also shows that it has an inverse quadratic dependence on the velocity of the electrons and, at the limit for $\beta \to 0$ a divergence is obtained. This derives from the approximations that we have introduced, in particular from having assumed that the electron interacts only with the nucleus and not also with the electron cloud present around it.

3.8 Cyclotron Radiation

A nonrelativistic electron moving in a region of space permeated by a magnetic field is subject to the Lorentz force. The motion of the electron is described by the equation

$$\frac{d^2\mathbf{x}}{dt^2} = -\frac{e_0}{mc}\frac{d\mathbf{x}}{dt} \times \mathbf{B},$$

where $\mathbf{B}$ is the magnetic field which we assume uniform and constant in time. In a coordinate system (x, y, z) with the z axis directed along the field, the most general solution of this equation is the following

$$x = x_0 + A\cos(\omega_c t + \phi), \qquad y = y_0 + A\sin(\omega_c t + \phi), \qquad z = z_0 + v_{\|}t,$$

where the quantity

$$\omega_c = \frac{e_0 B}{mc} \tag{3.30}$$

is the so-called cyclotron frequency, and where x_0, y_0, z_0, A, ϕ, and $v_{\|}$ are six constants of integration. The electron has a helical motion, i.e. a circular motion of radius A in the plane $x-y$, superimposed on a uniform motion along the z axis. The circular motion has velocity $v_{\perp} = A\omega_c$ and the motion is counterclockwise if one observes it from the positive tip of the z axis. By appropriately choosing the orientation of the x and y axes, the origin of coordinate system, and the origin of time, the equation of motion can be written in the following "minimal form"

$$x = \frac{v_{\perp}}{\omega_c}\cos(\omega_c t), \qquad y = \frac{v_{\perp}}{\omega_c}\sin(\omega_c t), \qquad z = v_{\|}t.$$

From the equation of motion we can calculate with easy steps the acceleration vector and its components along the unit vectors $\mathbf{e}_1$ and $\mathbf{e}_2$ relative to the direction $\mathbf{n}$, as defined in Fig. 3.10. Without loss of generality, the direction $\mathbf{n}$ can be chosen as belonging to the plane $x-z$ ($\chi = 0$), given the cylindrical symmetry of the problem. We obtain

$$a_1 = v_{\perp}\omega_c\cos\theta\cos(\omega_c t), \qquad a_2 = v_{\perp}\omega_c\sin(\omega_c t),$$

and applying the nonrelativistic equation (3.21), the components of the electric field in a point at large distance R along the direction $\mathbf{n}$ are expressed by the equations

$$E_1(t + R/c) = \frac{e_0 v_{\perp}\omega_c}{c^2 R}\cos\theta\cos(\omega_c t),$$

$$E_2(t + R/c) = \frac{e_0 v_{\perp}\omega_c}{c^2 R}\sin(\omega_c t).$$

Fig. 3.10 Geometry for the calculation of the cyclotron radiation

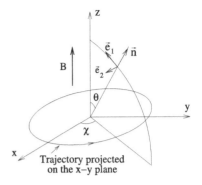

These equations show that the components of the electric field vector are periodic with period $T = 2\pi/\omega_c$. To evaluate the fluxes in the Stokes parameters we then need to use Eq. (2.18) which, in turn, involves the calculation of the Fourier components. On the other hand, since the field has a sinusoidal behaviour with time, all Fourier components are zero except those relative to the fundamental harmonic. Recalling the definition, we have

$$\mathcal{E}_1^{(1)} = \frac{1}{T} \int_0^T E_1(t) e^{i\omega_c t} \, dt, \qquad \mathcal{E}_2^{(1)} = \frac{1}{T} \int_0^T E_2(t) e^{i\omega_c t} \, dt,$$

from which we obtain, with simple algebra,

$$\mathcal{E}_1^{(1)} = \frac{e_0 v_\perp \omega_c}{2c^2 R} \cos\theta e^{i\Phi}, \qquad \mathcal{E}_2^{(1)} = i \frac{e_0 v_\perp \omega_c}{2c^2 R} e^{i\Phi},$$

where $\Phi = \omega_c R/c$ is an inessential phase introduced by the retarded time. The fluxes in the Stokes parameters are then expressed, in matrix form, by the following equation

$$\begin{pmatrix} F_\omega^I \\ F_\omega^Q \\ F_\omega^U \\ F_\omega^V \end{pmatrix} = \frac{e_0^2 v_\perp^2 \omega_c^2}{8\pi c^3 R^2} \begin{pmatrix} 1 + \cos^2\theta \\ -\sin^2\theta \\ 0 \\ -2\cos\theta \end{pmatrix} \delta(\omega - \omega_c).$$

This equation shows that the cyclotron radiation is elliptically polarised. In particular, the radiation emitted along the magnetic field direction ($\theta = 0$ or $\theta = \pi$) is circularly polarised and the one emitted in the plane perpendicular to the magnetic field ($\theta = \pi/2$) is linearly polarised (the polarisation direction being perpendicular to magnetic field). The radiation diagram is of the form $(1 + \cos^2\theta)$, which means that the intensity emitted "at the poles" is twice that emitted "in the equatorial plane". Finally, the total power emitted W can be determined with a double integration of the flux F_ω^I over a sphere of radius R and over frequency. Taking into account that the average over the solid angle of the factor $(1 + \cos^2\theta)$ is $\frac{4}{3}$, we obtain

$$W = \frac{2e_0^2 v_\perp^2 \omega_c^2}{3c^3},$$

or, recalling the definition of the cyclotron frequency (Eq. (3.30))

$$W = \frac{2e_0^4 v_\perp^2 B^2}{3m^2 c^5} = \frac{2}{3} r_c^2 \beta_\perp v_\perp B^2,$$

where $\beta_\perp = v_\perp/c$ and where r_c is the classical radius of the electron. By means of this equation, we can introduce a cross section. Recalling that the density of the magnetic energy is $B^2/(8\pi)$, the flux of energy swept by the electron in its accelerated motion (i.e. non considering the uniform linear motion along the magnetic field) is $v_\perp B^2/(8\pi)$. This amount of energy is transformed by the electron in irradiated energy with a cross section given by

$$\sigma_c = \frac{16\pi}{3} \beta_\perp r_c^2.$$

Recalling the result for the Thomson cross section σ_T, we have

$$\sigma_c = 2\beta_\perp \sigma_T.$$

3.9 Synchrotron Radiation

When we consider the motion of a relativistic electron in a magnetic field, the physical characteristics of the motion are the same as in the nonrelativistic case described in the previous section, with the exception that the frequency decreases. Instead of the cyclotron frequency ω_c, we now have the synchrotron frequency ω_s which depends on the velocity of the particle since

$$\omega_s = \frac{e_0 B}{\gamma mc} = \omega_c \sqrt{1 - \beta^2},$$

where $\beta = v/c$ and where γ is the Lorentz factor.

The analysis of the characteristics of the synchrotron radiation is based on the relativistic formulae presented previously (see in particular Eq. (3.18)). Two factors have a fundamental role in the description of this phenomenon. One is the κ factor, and the other is the effect due to the retarded time, which in this case does not just introduce a simple phase factor in the expressions of the Fourier transforms of the radiation field. A thorough analysis is very complex. In this volume, we simply consider the particular case in which the electron has a circular, rather than helicoidal, trajectory.[4] Referring to Fig. 3.10, we can assume, without loss of generality, that

---

[4]For a more in-depth discussion see e.g. Rybicki and Lightman (1979).

the position of the electron at the retarded time t' is given by

$$\mathbf{x}(t') = \frac{c\beta}{\omega_s}\left[\cos(\omega_s t')\mathbf{i} + \sin(\omega_s t')\mathbf{j}\right].$$

The corresponding expressions for the velocity and the acceleration are obviously given by

$$\mathbf{v}(t') = c\beta\left[-\sin(\omega_s t')\mathbf{i} + \cos(\omega_s t')\mathbf{j}\right],$$

$$\mathbf{a}(t') = -c\beta\omega_s\left[\cos(\omega_s t')\mathbf{i} + \sin(\omega_s t')\mathbf{j}\right].$$

Given the cylindrical symmetry of the problem, we consider the radiation emitted along a direction in the x–z plane (cf. Fig. 3.10). Defining

$$\mathbf{n} = \sin\theta\,\mathbf{i} + \cos\theta\,\mathbf{k},$$

we have, with simple algebra

$$\mathbf{n} - \boldsymbol{\beta} = \left[\sin\theta + \beta\sin(\omega_s t')\right]\mathbf{i} - \beta\cos(\omega_s t')\mathbf{j} + \cos\theta\,\mathbf{k},$$

$$(\mathbf{n} - \boldsymbol{\beta}) \times \mathbf{a} = c\beta\omega_s\left[\cos\theta\sin(\omega_s t')\mathbf{i} - \cos\theta\cos(\omega_s t')\mathbf{j}\right.$$
$$\left. - \left[\sin\theta\sin(\omega_s t') + \beta\right]\mathbf{k}\right],$$

$$\mathbf{n} \times \left[(\mathbf{n} - \boldsymbol{\beta}) \times \mathbf{a}\right] = c\beta\omega_s\left[\cos^2\theta\cos(\omega_s t')\mathbf{i} + \left[\sin(\omega_s t') + \beta\sin\theta\right]\mathbf{j}\right.$$
$$\left. - \sin\theta\cos\theta\cos(\omega_s t')\mathbf{k}\right].$$

The projection of this latter vector on the unit vectors

$$\mathbf{e}_1 = -\cos\theta\,\mathbf{i} + \sin\theta\,\mathbf{k}, \qquad \mathbf{e}_2 = -\mathbf{j},$$

is

$$\mathbf{e}_1 \cdot \left\{\mathbf{n} \times \left[(\mathbf{n} - \boldsymbol{\beta}) \times \mathbf{a}\right]\right\} = -c\beta\omega_s\cos\theta\cos(\omega_s t'),$$

$$\mathbf{e}_2 \cdot \left\{\mathbf{n} \times \left[(\mathbf{n} - \boldsymbol{\beta}) \times \mathbf{a}\right]\right\} = -c\beta\omega_s\left[\sin(\omega_s t') + \beta\sin\theta\right].$$

We are now able to calculate the components of the electric field at time t and distance R from the charge along the direction $\mathbf{n}$. From Eq. (3.18) we obtain

$$E_1(t) = \frac{e_0\beta\omega_s}{cR\kappa^3(t')}\cos\theta\cos(\omega_s t'),$$

$$E_2(t) = \frac{e_0\beta\omega_s}{cR\kappa^3(t')}\left[\sin(\omega_s t') + \beta\sin\theta\right],$$

where

$$\kappa(t') = 1 - \frac{\mathbf{v}(t')\cdot\mathbf{n}}{c} = 1 + \beta\sin\theta\sin(\omega_s t'),$$

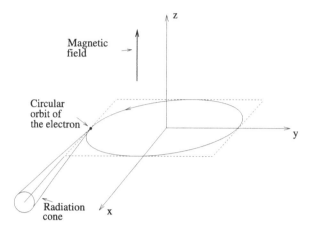

Fig. 3.11 The lighthouse effect. The relativistic particle rotates on the circle emitting most of the radiation within the *cone* drawn in the figure. The case shown here refers to the instant when the emitted radiation is along the x axis. As time goes by, the cone sweeps all the directions in the x–y plane

and where the retarded time t' is related to the time t by

$$t' = t - t_0 + \frac{\beta}{\omega_s} \sin\theta \cos(\omega_s t'),$$

t_0 being the time needed for the light to travel the distance between the centre of the orbit and the point where we calculate the components of the electric field.

As we have seen previously, the κ^{-3} factor becomes very important for relativistic particles, when β approaches unity. In this case, since the particle rotates in a circular orbit, the beaming effect transforms into the "lighthouse effect", as schematically shown in Fig. 3.11. In practice, an observer sees a periodic radiation signal (with period $2\pi/\omega_s$) extremely concentrated in time. Since the signal is also spatially concentrated around $\theta \simeq \pi/2$, the previous equations for E_1 and E_2 show that the radiation is linearly polarised in the direction perpendicular to the z axis, i.e. perpendicular to the magnetic field.

We now examine in more detail the spectropolarimetric characteristics of the emitted radiation. We need to calculate the Fourier components of the E_1 and E_2 quantities, i.e. the values

$$\mathcal{E}_1^{(n)} = \frac{1}{T} \int_0^T E_1(t) e^{in\omega_s t}\, dt, \qquad \mathcal{E}_2^{(n)} = \frac{1}{T} \int_0^T E_2(t) e^{in\omega_s t}\, dt,$$

where $T = 2\pi/\omega_s$. Substituting the previous expressions, we have

$$\mathcal{E}_1^{(n)} = \frac{e_0 \beta \omega_s^2}{2\pi c R} \cos\theta \int_0^{2\pi/\omega_s} \frac{\cos(\omega_s t')}{\kappa^3(t')} e^{in\omega_s t}\, dt,$$

$$\mathcal{E}_2^{(n)} = \frac{e_0 \beta \omega_s^2}{2\pi c R} \int_0^{2\pi/\omega_s} \frac{\sin(\omega_s t') + \beta \sin\theta}{\kappa^3(t')} e^{in\omega_s t} \, dt.$$

To evaluate the integrals, we note that

$$\frac{dt}{dt'} = 1 + \beta \sin\theta \sin(\omega_s t') = \kappa(t'), \qquad \frac{dt'}{dt} = \frac{1}{\kappa(t')},$$

$$\frac{d}{dt}\left[\frac{1}{\kappa(t')}\right] = \frac{dt'}{dt}\frac{d}{dt'}\left[\frac{1}{\kappa(t')}\right] = -\frac{1}{\kappa^3(t')}\beta\omega_s \sin\theta \cos(\omega_s t'),$$

$$\frac{d}{dt}\left[\frac{\cos(\omega_s t')}{\kappa(t')}\right] = \frac{dt'}{dt}\frac{d}{dt'}\left[\frac{\cos(\omega_s t')}{\kappa(t')}\right]$$

$$= -\frac{\omega_s}{\kappa(t')}\left[\frac{\beta \sin\theta \cos^2(\omega_s t')}{\kappa^2(t')} + \frac{\sin(\omega_s t')}{\kappa(t')}\right]$$

$$= -\frac{\omega_s}{\kappa^3(t')}\left[\sin(\omega_s t') + \beta \sin\theta\right].$$

Substituting in the expressions of the Fourier components, we obtain

$$\mathcal{E}_1^{(n)} = -\frac{e_0 \omega_s}{2\pi c R} \cot\theta \int_0^{2\pi/\omega_s} \frac{d}{dt}\left[\frac{1}{\kappa(t')}\right] e^{in\omega_s t} \, dt,$$

$$\mathcal{E}_2^{(n)} = -\frac{e_0 \beta \omega_s}{2\pi c R} \int_0^{2\pi/\omega_s} \frac{d}{dt}\left[\frac{\cos(\omega_s t')}{\kappa(t')}\right] e^{in\omega_s t} \, dt.$$

We now integrate by parts both integrals. The finite factor vanishes because of the periodic conditions. We then perform a change of variable in the integral that is left, by replacing the variable t with t'. Since $dt = \kappa(t') \, dt'$, we have, recalling the relation between t and t',

$$\mathcal{E}_1^{(n)} = \frac{in e_0 \omega_s^2}{2\pi c R} \cot\theta \int_0^{2\pi/\omega_s} e^{in\omega_s(t_0+t')} e^{-in\beta \sin\theta \cos(\omega_s t')} \, dt',$$

$$\mathcal{E}_2^{(n)} = \frac{in e_0 \beta \omega_s^2}{2\pi c R} \int_0^{2\pi/\omega_s} \cos(\omega_s t') e^{in\omega_s(t_0+t')} e^{-in\beta \sin\theta \cos(\omega_s t')} \, dt'.$$

The phase factor originating from the time t_0 is present in both components and can be eliminated without loss of generality. If we perform in both integrals the change of variable $\varphi = \omega_s t'$ and expand the first exponential we have

$$\mathcal{E}_1^{(n)} = \frac{in e_0 \omega_s}{2\pi c R} \cot\theta \int_0^{2\pi} [\cos(n\varphi) + i\sin(n\varphi)] e^{-in\beta \sin\theta \cos\varphi} \, d\varphi,$$

$$\mathcal{E}_2^{(n)} = \frac{in e_0 \beta \omega_s}{2\pi c R} \int_0^{2\pi} \cos\varphi [\cos(n\varphi) + i\sin(n\varphi)] e^{-in\beta \sin\theta \cos\varphi} \, d\varphi.$$

The integrals that appear in these expressions can be simplified noting that the contribution of the term in $\sin(n\varphi)$ is null. The contribution from the term in $\cos(n\varphi)$

can be written in terms of special functions. In fact, the Bessel functions of integer order are defined as the solutions of the differential equation

$$x^2 \frac{d^2 J_n}{dx^2} + x \frac{dJ_n}{dx} + (x^2 - n^2)J_n = 0,$$

and it can be shown that

$$J_n(x) = \frac{i^n}{2\pi} \int_0^{2\pi} \cos(n\varphi) e^{-ix\cos\varphi} \, d\varphi,$$

from which we get by differentiation

$$J'_n(x) = \frac{d}{dx} J_n(x) = -\frac{i^{n+1}}{2\pi} \int_0^{2\pi} \cos\varphi \cos(n\varphi) e^{-ix\cos\varphi} \, d\varphi.$$

Using these relations, it is possible to write the Fourier components in the form

$$\mathcal{E}_1^{(n)} = \frac{1}{i^{n-1}} \frac{ne_0\omega_s}{cR} \cot\theta J_n(n\beta\sin\theta),$$

$$\mathcal{E}_2^{(n)} = -\frac{1}{i^n} \frac{ne_0\beta\omega_s}{cR} J'_n(n\beta\sin\theta).$$

Recalling Eq. (2.18), we can finally write the fluxes in the Stokes parameters by means of the expression

$$\begin{pmatrix} F_\omega^I \\ F_\omega^Q \\ F_\omega^U \\ F_\omega^V \end{pmatrix} = \frac{e_0^2\omega_s^2}{2\pi c R^2} \sum_{n=1}^{\infty} n^2 \begin{pmatrix} \cot^2\theta J_n^2(z) + \beta^2 J_n'^2(z) \\ \cot^2\theta J_n^2(z) - \beta^2 J_n'^2(z) \\ 0 \\ -2\beta\cot\theta J_n(z)J'_n(z) \end{pmatrix} \delta(\omega - n\omega_s),$$

where

$$z = n\beta\sin\theta.$$

It is interesting to consider the limit of these expressions for $\beta \to 0$. Taking into account the expansion in power series

$$J_n(z) = \left(\frac{z}{2}\right)^n \left[\frac{1}{n!} - \frac{z^2}{4(n+1)!} + \cdots\right],$$

at the first order in z we obtain that all the Bessel functions of integer order with $n \geq 2$ are null. For $J_1(z)$ we have

$$J_1(z) = \frac{z}{2}, \qquad J'_1(z) = \frac{1}{2},$$

and we find again the formulae of the cyclotron radiation.

A fundamental characteristic of the spectrum of the synchrotron radiation is that all the harmonics of the frequency ω_s provide a contribution. It is then interesting to find out for which harmonic we have maximum emission. An analysis based on the asymptotic expansions of the Bessel functions[5] shows that for values of β close to 1 the maximum emission (integrated over the solid angle) is for the harmonic characterised by the index n_{max} given by

$$n_{max} \simeq \left(1 - \beta^2\right)^{-3/2} = \gamma^3.$$

For example, for $\beta = 0.99$ we have $n_{max} \simeq 350$. The frequency corresponding to the maximum, ω_{max}, is therefore given by

$$\omega_{max} = \gamma^3 \omega_s = \gamma^2 \omega_c.$$

For ultra-relativistic electrons, the spectrum of the synchrotron radiation therefore becomes a "quasi-continuum" emission. This is true even if the electrons all have the same energy and move along circular orbits (and not helicoidal), as we have assumed here for our limited discussion.

Finally, we evaluate the total power emitted over all solid angles by using the Larmor equation generalised to the relativistic case (Eq. (3.23)). Taking into account that the acceleration is perpendicular to the velocity and that its magnitude is $c\beta\omega_s$, we have

$$W = \frac{2e_0^2}{3c}\gamma^4\beta^2\omega_s^2 = \frac{2e_0^4}{3m^2c^3}\gamma^2\beta^2 B^2,$$

or, in alternative form

$$W = \frac{2}{3}\gamma^2\beta^2 B^2 cr_c^2,$$

where r_c is the classical radius of the electron. We can use the latter equation to introduce a suitable cross section, as we have already done for the cyclotron radiation. Recalling that the density of the magnetic energy is $B^2/(8\pi)$, the flux of energy swept by the electron in its motion is $c\beta B^2/(8\pi)$. This amount of energy is transformed by the electron in irradiated energy with a cross section given by

$$\sigma_s = \frac{16\pi}{3}\gamma^2\beta r_c^2,$$

or, in terms of the Thomson cross section

$$\sigma_s = 2\gamma^2\beta\sigma_T.$$

---

[5]See Landau and Lifchitz (1966).

3.10 Multipolar Expansion in the Radiation Zone

In the previous sections we have discussed the properties of the electromagnetic radiation irradiated by a single charged particle in arbitrary motion. We now consider the case of N charges in motion, instead of a single one. Obviously, for the linearity of Maxwell's equations, the expressions of the electric and magnetic field in the radiation zone can be generalised by simple addition of the fields produced by each charge. We have

$$\mathbf{E}(\mathbf{r}, t) = \sum_{i=1}^{N} \mathbf{E}_i(\mathbf{r}, t), \qquad \mathbf{B}(\mathbf{r}, t) = \sum_{i=1}^{N} \mathbf{B}_i(\mathbf{r}, t) = \sum_{i=1}^{N} \mathbf{n}_i \times \mathbf{E}_i(\mathbf{r}, t),$$

where, recalling Eq. (3.18)

$$\mathbf{E}_i(\mathbf{r}, t) = \frac{e_i}{c^2 \kappa_i^3 R_i} \mathbf{n}_i \times \left[\left(\mathbf{n}_i - \frac{\mathbf{v}_i}{c} \right) \times \mathbf{a}_i \right].$$

In this expression, all the geometrical and dynamical quantities relative to the i-th particle, $\mathbf{n}_i$, κ_i, R_i, $\mathbf{v}_i$, and $\mathbf{a}_i$ must be evaluated at the retarded time of the particle t_i' defined by

$$t_i' = t - \frac{|\mathbf{r} - \mathbf{r}_i(t_i')|}{c},$$

where $\mathbf{r}_i(t)$ is the trajectory of the i-th particle. In general, each particle has a different retarded time.

We now consider the simplified case where all the particles are located within a region of dimension L much less than the distance R from the point where the fields are evaluated ($L \ll R$). In this case, we can assume that the unit vector $\mathbf{n}$ and the distance R are the same for all the particles. With these assumptions, the electric field can be written (neglecting terms of the order of L/R) as

$$\mathbf{E}(\mathbf{r}, t) = \frac{1}{c^2 R} \sum_{i=1}^{N} \frac{e_i}{\kappa_i^3} \mathbf{n} \times \left[\left(\mathbf{n} - \frac{\mathbf{v}_i}{c} \right) \times \mathbf{a}_i \right].$$

We also assume that the charges move with nonrelativistic velocities (i.e. $v_i \ll c$ with $i = 1, \ldots, N$), so all the κ_i are equal to 1 and the second term within the parenthesis in the right-hand side can be neglected. We therefore obtain, for the electric field at the zeroth order in the v/c ratio

$$\mathbf{E}_0(\mathbf{r}, t) = \frac{1}{c^2 R} \mathbf{n} \times \left[\mathbf{n} \times \sum_{i=1}^{N} e_i \mathbf{a}_i \right].$$

We now introduce a further hypothesis, that is we neglect the dependence on the retarded time of the acceleration of the i-th particle. We then consider a point of coordinate $\mathbf{r}_c$ (the "central point" of the charges, for instance their barycentre) and measure the spatial coordinates of the single particles with respect to this point, putting

$$\mathbf{s}_i = \mathbf{r}_i - \mathbf{r}_c.$$

With this definition, given that $|\mathbf{s}_i| \ll |\mathbf{r} - \mathbf{r}_c| \simeq R$, the retarded time t'_i is

$$t'_i = t'_c + \frac{\mathbf{s}_i \cdot \mathbf{n}}{c},$$

where t'_c is the retarded time relative to the central point. The assumption that the dependence on t'_i can be neglected is justified if the typical time τ of the variations in the motion of the particles is much greater than L/c (where L indicates the typical dimension of the spatial region where the particles are located). On the other hand, the characteristic wavelength λ of the radiation emitted by the particles is of the order of $c\tau$, so we must have

$$\lambda \simeq c\tau \gg L.$$

When this condition is verified, having defined the electric dipole moment of the charges $\mathbf{D}$ by the equation

$$\mathbf{D} = \sum_{i=1}^{N} e_i \mathbf{s}_i, \tag{3.31}$$

we have

$$\mathbf{E}_0(\mathbf{r}, t) = \frac{1}{c^2 R} \mathbf{n} \times (\mathbf{n} \times \ddot{\mathbf{D}}),$$

where we have adopted the "dot convention" to indicate the derivative with respect to time and where the second derivative of the electric dipole moment has to be evaluated at the retarded time t'_c.

The radiation described by this equation is called electric dipole radiation. Its radiation diagram is in all respects equal to that one of a nonrelativistic single particle described in Sect. 3.4 and Fig. 3.3. In strict analogy with the previous result, we have that the total emitted power is

$$W_{\text{d.e.}} = \frac{2}{3c^3} [\ddot{\mathbf{D}}]^2.$$

In some cases, the vector $\ddot{\mathbf{D}}$ could be null. In these cases, to determine the radiation properties of the system of charges, it is necessary to consider the contribution of the electric field at the first order in v/c, which we have previously neglected. With reference to the general equation for the electric field, we need to consider three terms originating from: (a) the $\mathbf{v}_i/c$ factor in parenthesis; (b) the multiplicative factor κ_i^{-3}; (c) the correction of the term of order zero in $\mathbf{a}_i$ due to the effect of the retarded time. With simple considerations we therefore obtain for the contribution to the electric field due to the first order expansion in v/c

$$\mathbf{E}_1(\mathbf{r}, t) = \frac{1}{c^3 R} \sum_{i=1}^{N} e_i \mathbf{n} \times \left\{ -(\mathbf{v}_i \times \mathbf{a}_i) + 3(\mathbf{v}_i \cdot \mathbf{n})(\mathbf{n} \times \mathbf{a}_i) + (\mathbf{s}_i \cdot \mathbf{n})(\mathbf{n} \times \dot{\mathbf{a}}_i) \right\},$$

where all the quantities are evaluated at the "central" retarded time t'_c. This expression can be written in another form noting that given any vector $\mathbf{w}$ we have

$$\mathbf{n} \times \mathbf{w} = -\mathbf{n} \times \left[\mathbf{n} \times [\mathbf{n} \times \mathbf{w}] \right],$$

so that, modifying the first term in curly brackets using this equation, we can write

$$\mathbf{E}_1(\mathbf{r}, t) = \frac{1}{c^3 R} \mathbf{n} \times [\mathbf{n} \times \mathcal{A}],$$

where

$$\mathcal{A} = \sum_{i=1}^{N} e_i \left\{ \mathbf{n} \times (\mathbf{v}_i \times \mathbf{a}_i) + 3(\mathbf{v}_i \cdot \mathbf{n})\mathbf{a}_i + (\mathbf{s}_i \cdot \mathbf{n})\dot{\mathbf{a}}_i \right\},$$

that is, developing the double vector product

$$\mathcal{A} = \sum_{i=1}^{N} e_i \left\{ (\mathbf{a}_i \cdot \mathbf{n})\mathbf{v}_i + 2(\mathbf{v}_i \cdot \mathbf{n})\mathbf{a}_i + (\mathbf{s}_i \cdot \mathbf{n})\dot{\mathbf{a}}_i \right\}.$$

We now modify the form of this equation putting

$$\mathbf{v}_i = \dot{\mathbf{s}}_i, \qquad \mathbf{a}_i = \ddot{\mathbf{s}}_i, \qquad \dot{\mathbf{a}}_i = \dddot{\mathbf{s}}_i.$$

With these definitions we have

$$\mathcal{A} = \sum_{i=1}^{N} e_i \left\{ (\ddot{\mathbf{s}}_i \cdot \mathbf{n})\dot{\mathbf{s}}_i + 2(\dot{\mathbf{s}}_i \cdot \mathbf{n})\ddot{\mathbf{s}}_i + (\mathbf{s}_i \cdot \mathbf{n})\dddot{\mathbf{s}}_i \right\}.$$

Let us introduce the magnetic dipole moment of the system with the expression

$$\mathbf{M} = \frac{1}{2c} \sum_{i=1}^{N} e_i \mathbf{s}_i \times \mathbf{v}_i = \frac{1}{2c} \sum_{i=1}^{N} e_i \mathbf{s}_i \times \dot{\mathbf{s}}_i. \tag{3.32}$$

Differentiating twice with respect to time we get

$$\ddot{\mathbf{M}} = \frac{1}{2c} \sum_{i=1}^{N} e_i \{ \dot{\mathbf{s}}_i \times \ddot{\mathbf{s}}_i + \mathbf{s}_i \times \dddot{\mathbf{s}}_i \},$$

and taking its vector product with the unit vector $\mathbf{n}$,

$$\ddot{\mathbf{M}} \times \mathbf{n} = \frac{1}{2c} \sum_{i=1}^{N} e_i \left\{ (\dot{\mathbf{s}}_i \cdot \mathbf{n})\ddot{\mathbf{s}}_i - (\ddot{\mathbf{s}}_i \cdot \mathbf{n})\dot{\mathbf{s}}_i + (\mathbf{s}_i \cdot \mathbf{n})\dddot{\mathbf{s}}_i - (\dddot{\mathbf{s}}_i \cdot \mathbf{n})\mathbf{s}_i \right\}.$$

From this equation, recalling the expression of the vector $\mathcal{A}$, we obtain

$$\mathcal{A} - c\ddot{\mathbf{M}} \times \mathbf{n} = \frac{1}{2} \sum_{i=1}^{N} \left\{ 3(\dot{\mathbf{s}}_i \cdot \mathbf{n})\dot{\mathbf{s}}_i + 3(\dot{\mathbf{s}}_i \cdot \mathbf{n})\ddot{\mathbf{s}}_i + (\dddot{\mathbf{s}}_i \cdot \mathbf{n})\mathbf{s}_i + (\mathbf{s}_i \cdot \mathbf{n})\dddot{\mathbf{s}}_i \right\}.$$

We now define the symmetric tensor $\mathcal{Q}$ using the following dyadic product

$$\mathcal{Q} = \sum_{i=1}^{N} e_i \mathbf{s}_i \mathbf{s}_i. \tag{3.33}$$

As we will see in more detail below, this tensor is closely related to the tensor of the electric quadrupole moment as defined in electrostatics. Differentiating it three times with respect to time and taking the scalar product with the unit vector $\mathbf{n}$ (either to the right or to left since the tensor is symmetric) we easily obtain

$$\mathbf{n} \cdot \dddot{\mathcal{Q}} = \dddot{\mathcal{Q}} \cdot \mathbf{n} = \sum_{i=1}^{N} e_i \left\{ (\dddot{\mathbf{s}}_i \cdot \mathbf{n})\mathbf{s}_i + 3(\ddot{\mathbf{s}}_i \cdot \mathbf{n})\dot{\mathbf{s}}_i + 3(\dot{\mathbf{s}}_i \cdot \mathbf{n})\ddot{\mathbf{s}}_i + (\mathbf{s}_i \cdot \mathbf{n})\dddot{\mathbf{s}}_i \right\},$$

so that finally the vector $\mathcal{A}$ is expressed, in terms of the magnetic dipole moment and the tensor $\mathcal{Q}$, by the equation

$$\mathcal{A} = c\ddot{\mathbf{M}} \times \mathbf{n} + \frac{1}{2}\mathbf{n} \cdot \dddot{\mathcal{Q}}.$$

Substituting this equation in the expression for the electric field $E_1(\mathbf{r}, t)$ and introducing the vector $\mathbf{Q}$ by the equation

$$\mathbf{Q} = \mathbf{n} \cdot \mathcal{Q},$$

we obtain

$$E_1(\mathbf{r}, t) = \frac{1}{c^2 R} \left[\mathbf{n} \times \ddot{\mathbf{M}} + \frac{1}{2c}\mathbf{n} \times (\mathbf{n} \times \dddot{\mathbf{Q}}) \right].$$

It is interesting to note that, when calculating the field $E_1(\mathbf{r}, t)$, the tensor $\mathcal{Q}$ can be substituted by the tensor of the electric quadrupole moment, generally defined in electrostatic by the expression

$$\mathcal{Q} = \sum_{i=1}^{N} e_i \left(\mathbf{s}_i \mathbf{s}_i - \frac{1}{3}s_i^2 \mathbf{U} \right),$$

where $\mathbf{U}$ is the unit tensor. Obviously we have

$$\mathcal{Q} = \mathcal{Q} - \frac{1}{3} \sum_i e_i s_i^2 \mathbf{U},$$

so the two tensors $\mathcal{Q}$ and $\mathcal{Q}$ differ by a quantity that is proportional to the unit tensor. If we adopt the expression for $\mathcal{Q}$ instead of $\mathcal{Q}$ to define the vector $\mathbf{Q}$, we obtain another vector which differs from $\mathbf{Q}$ by a quantity that is proportional to $\mathbf{n}$. This quantity does not contribute to the electric field $\mathbf{E}_1(\mathbf{r}, t)$ because of the double vector product. Hence, to calculate this field, the above definition of $\mathbf{Q}$ can be substituted with the equivalent definition

$$\mathbf{Q} = \mathbf{n} \cdot \mathcal{Q}.$$

3.11 Radiation Diagram for the Multipolar Components

As we have seen in the previous section, the contribution of the electric field in the radiation zone due to the corrections of the first order in v/c can, under a number of assumptions, be decomposed into the sum of a term due to magnetic dipole radiation and a term due to electric quadrupole radiation, or

$$\mathbf{E}_1(\mathbf{r}, t) = \mathbf{E}_{\text{d.m.}}(\mathbf{r}, t) + \mathbf{E}_{\text{q.e.}}(\mathbf{r}, t),$$

where

$$\mathbf{E}_{\text{d.m.}} = \frac{1}{c^2 R} \mathbf{n} \times \ddot{\mathbf{M}}, \qquad \mathbf{E}_{\text{q.e.}} = \frac{1}{2c^3 R} \mathbf{n} \times (\mathbf{n} \times \dddot{\mathbf{Q}}).$$

We will now determine the radiation diagrams relative to the two different types of radiation. If there is only magnetic dipole radiation, evaluating the Poynting vector with the usual expression valid in the radiation zone, i.e.

$$\mathbf{S}(\mathbf{r}, t) = \frac{c}{4\pi} E^2(\mathbf{r}, t) \mathbf{n},$$

we get

$$\mathbf{S}(\mathbf{r}, t) = \frac{\sin^2 \theta [\ddot{\mathbf{M}}]^2}{4\pi c^3 R^2} \mathbf{n},$$

where θ is the angle between the two vectors $\mathbf{n}$ and $\ddot{\mathbf{M}}$. The radiation diagram is in all respects analogous to that of the electric dipole radiation shown in Fig. 3.3. The only difference between these two cases is due to the fact that in the electric dipole radiation there is a double vector product, $\mathbf{n} \times (\mathbf{n} \times \ddot{\mathbf{D}})$, while in the case of the magnetic dipole we have a single vector product, $\mathbf{n} \times \ddot{\mathbf{M}}$.

This difference has consequences only on the polarisation characteristics of the two types of emission. In the first case the electric field vector lies in the plane containing the direction of propagation $\mathbf{n}$ and the $\ddot{\mathbf{D}}$ vector, while in the second case the electric field vector is directed perpendicularly to the plane containing the direction of propagation and the $\dot{\mathbf{M}}$ vector. For the total power we obtain a formula totally analogous to that of the electric dipole emission, i.e.

$$W_{\text{d.m.}} = \frac{2}{3c^3}[\ddot{\mathbf{M}}]^2.$$

Let us now analyse the case of the quadrupole emission assuming that only such radiation exists. The Poynting vector is given by

$$\mathbf{S}(\mathbf{r},t) = \frac{1}{16\pi c^5 R^2}(\mathbf{n} \times \dddot{\mathbf{Q}})^2 \mathbf{n},$$

or, recalling the definition of the vector $\mathbf{Q}$

$$\mathbf{S}(\mathbf{r},t) = \frac{1}{16\pi c^5 R^2}\left[\mathbf{n} \times (\mathbf{n} \cdot \dddot{\mathscr{Q}})\right]^2 \mathbf{n},$$

where the tensor $\dddot{\mathscr{Q}}$ could alternatively be substituted by the tensor $\dddot{\mathcal{Q}}$. In what follows we adopt the latter definition, noting that the trace of the tensor $\mathcal{Q}$ is null[6] and that such property also applies to $\dddot{\mathcal{Q}}$. Recall that we can always find a suitable reference system (x, y, z) in which a symmetric Cartesian tensor such as $\mathcal{Q}$ turns out to be diagonal. We can therefore write, in such reference system

$$\dddot{\mathcal{Q}} = A\mathbf{i}\mathbf{i} + B\mathbf{j}\mathbf{j} + C\mathbf{k}\mathbf{k},$$

where

$$A = \dddot{\mathcal{Q}}_{xx}, \qquad B = \dddot{\mathcal{Q}}_{yy}, \qquad C = \dddot{\mathcal{Q}}_{zz},$$

with

$$A + B + C = 0.$$

If we identify in this system the arbitrary direction $\mathbf{n}$ with the polar angles θ and ϕ, i.e. we put

$$\mathbf{n} = \sin\theta \cos\phi\,\mathbf{i} + \sin\theta \sin\phi\,\mathbf{j} + \cos\theta\,\mathbf{k},$$

we obtain, with simple algebra

$$\mathbf{n} \times (\mathbf{n} \cdot \dddot{\mathcal{Q}}) = \sin\theta \cos\theta \sin\phi(C - B)\mathbf{i} + \sin\theta \cos\theta \cos\phi(A - C)\mathbf{j}$$
$$+ \sin^2\theta \sin\phi \cos\phi(B - A)\mathbf{k},$$

and the Poynting vector is

$$\mathbf{S}(\mathbf{r},t) = \frac{1}{16\pi c^5 R^2}\Big[\sin^2\theta \cos^2\theta \sin^2\phi(B - C)^2 + \sin^2\theta \cos^2\theta \cos^2\phi(C - A)^2$$
$$+ \sin^4\theta \sin^2\phi \cos^2\phi(A - B)^2\Big]\mathbf{n}.$$

---

[6]This property simplifies the derivation of the following equations.

Fig. 3.12 Radiation diagram
for the quadrupolar emission,
in the simple case when
$A = B$. The diagram has
rotational symmetry around
the vertical axis

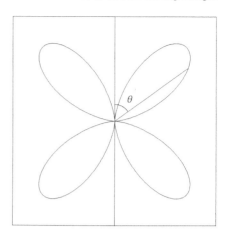

This formula expresses the radiation diagram for the electric quadrupole emission. This diagram is relatively complicated and in general lacks of symmetry. Only in the simplified case where $A = B$ (which implies $C = -2A$), we obtain the following expression, having rotational symmetry around the z axis, shown in Fig. 3.12

$$\mathbf{S}(\mathbf{r}, t) = \frac{9}{16\pi c^5 R^2} \sin^2 \theta \cos^2 \theta A^2 \mathbf{n}.$$

Returning to the general case, the expression for the total radiated power over the entire solid angle can easily be determined by observing that the average over the solid angle of the three functions

$$\sin^2 \theta \cos^2 \theta \sin^2 \phi, \qquad \sin^2 \theta \cos^2 \theta \cos^2 \phi, \qquad \sin^4 \theta \sin^2 \phi \cos^2 \phi$$

is $\frac{1}{15}$. We therefore obtain

$$W_{\text{q.e.}} = \frac{1}{60c^5}\left[(B - C)^2 + (C - A)^2 + (A - B)^2\right],$$

and, recalling that $A + B + C = 0$, so that

$$A^2 + B^2 + C^2 = -2(AB + BC + CA),$$

we have

$$W_{\text{q.e.}} = \frac{1}{20c^5}\left(A^2 + B^2 + C^2\right) = \frac{1}{20c^5}\left[(\dddot{Q}_{xx})^2 + (\dddot{Q}_{yy})^2 + (\dddot{Q}_{zz})^2\right].$$

Finally we note that, given an arbitrary tensor of rank two T_{ij}, we can define the square of its magnitude $\mathbf{T}^2$ by the equation

$$\mathbf{T}^2 = \sum_{ij} T_{ij}^2,$$

and it can be shown that such quantity is invariant under rotations of the reference system. In particular, in the reference system in which the tensor is diagonal, we have

$$\mathbf{T}^2 = T_{xx}^2 + T_{yy}^2 + T_{zz}^2.$$

The expression for the power of quadrupole radiation can then be written, in an arbitrary system, in the form

$$W_{\text{q.e.}} = \frac{1}{20c^5} \sum_{ij} (\dddot{Q}_{ij})^2. \tag{3.34}$$

We finally remark that the expressions for the radiation of electromagnetic waves we have obtained in the last two sections can be "translated" in a heuristic way to describe the radiation of gravitational waves. Section 16.6 is dedicated to this topic.

Chapter 4
Quantisation of the Electromagnetic Field

In its interaction with matter, the electromagnetic radiation has a characteristic behaviour by which the absorption and emission processes occur in the form of quanta
of energy commonly referred to as photons. These phenomena cannot be described
on the basis of the classical theory developed in the previous chapter. They need a
specific treatment capable of unifying, within a formally consistent theory, the basic
concepts of electromagnetism and quantum mechanics. Such theory, called quantum electrodynamics, was developed starting from 1930 with the contribution of
eminent physicists, including Dirac and Feynman. In this chapter we will provide a
basic introduction to the formalism, so-called of second quantisation, that is today
commonly used in the context of quantum electrodynamics to introduce the concept
of photon. The applications of this formalism to describe the interaction between
matter and radiation and the study of specific physical processes will be introduced
in subsequent chapters (Chaps. 11 and 15).

4.1 Harmonic Oscillator, Operators of Creation and Annihilation

We consider a one-dimensional harmonic oscillator. Its simplest practical realisation
consists of a point-like particle of mass m moving along a straight line and subject
to the action of an elastic restoring force. Defining with x the coordinate measured
along this straight line from the position of equilibrium, the equation of motion is

$$m\ddot{x} = -kx,$$

where k is the constant of the restoring force. It is well known that this equation has
harmonic solutions characterised by the angular frequency

$$\omega = \sqrt{\frac{k}{m}}.$$

E. Landi Degl'Innocenti, *Atomic Spectroscopy and Radiative Processes*,
UNITEXT for Physics, DOI 10.1007/978-88-470-2808-1_4, © Springer-Verlag Italia 2014

In the formalism of analytical mechanics, the physical system is described by the Hamiltonian function $\mathcal{H}$ which represents the total energy, given by

$$\mathcal{H} = \frac{p^2}{2m} + \frac{1}{2}kx^2,$$

where p is the momentum of the particle. Following the conventions of analytical mechanics, we denote by the symbol q the coordinate x. Also, replacing k with ω the Hamiltonian can be rewritten in the form

$$\mathcal{H} = \frac{p^2}{2m} + \frac{1}{2}m\omega^2 q^2.$$

In this equation, p represents the kinetic moment conjugate to the variable q. The equations of motion for q and p are obtained using the well-known Hamilton equations

$$\dot{p} = -\frac{\partial \mathcal{H}}{\partial q} = -m\omega^2 q,$$

$$\dot{q} = \frac{\partial \mathcal{H}}{\partial p} = \frac{p}{m}.$$

We introduce now, in place of the variables q and p, some linear combinations of them of the form

$$a = C(m\omega q + ip),$$
$$a^* = C(m\omega q - ip),$$

where $i = \sqrt{-1}$ is the imaginary unity and where C is a real constant whose value will be specified later. For the equations of motion we have, with simple algebra

$$\dot{a} = C(m\omega\dot{q} + i\dot{p}) = -i\omega a,$$

and, similarly,

$$\dot{a}^* = i\omega a^*.$$

As we can see, with the introduction of the variables a and a^*, the differential equations become decoupled and therefore are more easily solved.

Turning to the quantum description, q and p should be thought of as Hermitian linear operators acting on a suitable Hilbert space. The two operators must also comply with the commutation rule

$$[q, p] = qp - pq = i\hbar,$$

where we have introduced the usual symbol for the commutator between two operators, and where $\hbar$ is Planck's constant divided by 2π ($\hbar = 1.055 \times 10^{-27}$ erg s). With the introduction of the quantities a and a^* (that in the quantum formulation become the operators a and $a^\dagger$), the operators q and p are expressed by the equations

$$q = \frac{1}{2Cm\omega}(a + a^\dagger),$$

$$p = -\frac{i}{2C}(a - a^\dagger),$$

and the Hamiltonian becomes (taking care not to alter the order of the operators)

$$\mathcal{H} = \frac{1}{4mC^2}(a^\dagger a + aa^\dagger).$$

We now evaluate the commutator between the operators a and $a^\dagger$. We have

$$[a, a^\dagger] = C^2[m\omega q + ip, m\omega q - ip] = 2C^2 m\hbar\omega,$$

and we choose the constant C in order to have

$$[a, a^\dagger] = 1. \tag{4.1}$$

For this, it is sufficient to set $C = 1/\sqrt{2m\hbar\omega}$, so that the Hamiltonian becomes

$$\mathcal{H} = \frac{1}{2}\hbar\omega(aa^\dagger + a^\dagger a),$$

and, taking into account the commutation rule between a and $a^\dagger$, we obtain

$$\mathcal{H} = \hbar\omega\left(a^\dagger a + \frac{1}{2}\right). \tag{4.2}$$

The Hamiltonian has been reduced to a very simplified form with the introduction of the operators a and $a^\dagger$ which, for reasons that will be clear later, are named, respectively, operator of annihilation and creation. Now we will find the eigenvalues and eigenvectors of this Hamiltonian neglecting, in a certain way, its origin and using only the commutation rule between the operators a and $a^\dagger$. We start by calculating some commutators

$$[\mathcal{H}, a] = \hbar\omega[a^\dagger a, a] = \hbar\omega[a^\dagger, a]a = -\hbar\omega a,$$
$$[\mathcal{H}, a^\dagger] = \hbar\omega[a^\dagger a, a^\dagger] = \hbar\omega a^\dagger[a, a^\dagger] = \hbar\omega a^\dagger.$$

Suppose now that we know a particular eigenvector $|H\rangle$ of the Hamiltonian and its corresponding eigenvalue H

$$\mathcal{H}|H\rangle = H|H\rangle,$$

and consider the scalar product $\langle H | a^\dagger a | H \rangle$. With simple algebra we obtain

$$\hbar\omega \langle H | a^\dagger a | H \rangle = \left(H - \frac{1}{2}\hbar\omega \right)\langle H | H \rangle.$$

We note that the left-hand side is positive, being the product of the quantity $\hbar\omega$ by the norm of the vector $a | H \rangle$. Since on the other hand also the norm of the vector $|H\rangle$ is positive, we have for the eigenvalue H

$$H \geq \frac{1}{2}\hbar\omega,$$

where the equal sign is verified only when $a | H \rangle = 0$. By applying the Hamiltonian on the vector $a | H \rangle$ we find, using the commutation rules derived previously,

$$\mathcal{H}a|H\rangle = (a\mathcal{H} - \hbar\omega a)|H\rangle = (H - \hbar\omega)a|H\rangle.$$

This equation shows that, if $|H\rangle$ is an eigenvector of the Hamiltonian corresponding to the eigenvalue H, the $a|H\rangle$ vector is then also an eigenvector of the Hamiltonian, corresponding to the eigenvalue $(H - \hbar\omega)$. If we apply this property repeatedly, we find that, in general, the vector $a^n|H\rangle$ (with n integer) is also an eigenvector of the Hamiltonian, corresponding to the eigenvalue $(H - n\hbar\omega)$. In this way, we obtain a sequence of eigenvectors with eigenvalues progressively smaller. This sequence needs to be interrupted because otherwise it would lead to eigenvalues that do not satisfy anymore the condition $H \geq \frac{1}{2}\hbar\omega$. The only way to interrupt the sequence is to have, for a certain integer m,

$$a^m|H\rangle = 0.$$

This means that the vector $a^{m-1}|H\rangle$ is the eigenvector that corresponds to the eigenvalue $\frac{1}{2}\hbar\omega$. Such eigenvector, indicated in what follows with the symbol $|0\rangle$, is such that

$$a|0\rangle = 0, \qquad \mathcal{H}|0\rangle = \frac{1}{2}\hbar\omega|0\rangle.$$

If we now consider the vector $a^\dagger|0\rangle$, we have, recalling the previous commutation rules

$$\mathcal{H}a^\dagger|0\rangle = \left(a^\dagger\mathcal{H} + \hbar\omega a^\dagger \right)|0\rangle = \frac{3}{2}\hbar\omega a^\dagger|0\rangle,$$

and, similarly, for any integer n

$$\mathcal{H}a^{\dagger n}|0\rangle = \left(n + \frac{1}{2} \right)\hbar\omega a^{\dagger n}|0\rangle.$$

In this way we have obtained, starting from the eigenvector $|0\rangle$, a sequence of eigenvectors corresponding to eigenvalues that are progressively larger. The eigenvalues of the Hamiltonian are then given by the quantities $\frac{1}{2}\hbar\omega$, $(1 + \frac{1}{2})\hbar\omega$, ...,

$(n + \frac{1}{2})\hbar\omega, \ldots$. Such eigenvalues have the corresponding eigenvectors $|0\rangle, a^\dagger|0\rangle$, $\ldots, a^{\dagger n}|0\rangle, \ldots$.

The eigenvectors that we have found in this way are however not normalised. To find the norm of the generic eigenvector we need to evaluate the quantity $\langle 0|a^n a^{\dagger n}|0\rangle$ by taking into account the equation (which can be obtained with an induction procedure)

$$[a, a^{\dagger n}] = na^{\dagger n-1}.$$

For the norm of the vector $a^{\dagger n}|0\rangle$ we have

$$\langle 0|a^n a^{\dagger n}|0\rangle = \langle 0|a^{n-1} a a^{\dagger n}|0\rangle = \langle 0|a^{n-1}(a^{\dagger n} a + na^{\dagger n-1})|0\rangle$$
$$= n\langle 0|a^{n-1} a^{\dagger n-1}|0\rangle.$$

By applying this equation successively we finally obtain

$$\langle 0|a^n a^{\dagger n}|0\rangle = n!\langle 0|0\rangle.$$

If we suppose that the eigenvector $|0\rangle$ is normalised, we can obtain the other normalised eigenvectors (which we indicate with $|n\rangle$) by the expression

$$|n\rangle = \frac{1}{\sqrt{n!}} a^{\dagger n}|0\rangle.$$

It is simple to show that the action of the a and $a^\dagger$ operators on the eigenvector $|n\rangle$ is such that

$$a|n\rangle = \sqrt{n}|n-1\rangle,$$
$$a^\dagger|n\rangle = \sqrt{n+1}|n+1\rangle.$$

These equations show the reason why these operators are called of annihilation and creation (the quantum number n is interpreted as an occupation number of hypothetical particles).

4.2 Expansion of the Electromagnetic Field in Fourier Series

Consider the electromagnetic field enclosed in a cubic cavity, the so-called "box", of side L and volume $V = L^3$. In the absence of charges and currents, as seen in Sect. 1.5, we can use the special case of the Lorenz gauge in which the scalar potential $\phi(\mathbf{x}, t)$ is zero (Eq. (1.10)). In this way, the field is described only by the vector potential $\mathbf{A}(\mathbf{x}, t)$ that is subject to the additional condition (not invariant relativistically) of Eq. (1.11)

$$\text{div}\,\mathbf{A}(\mathbf{x}, t) = 0,$$

and that satisfies the wave equation

$$\nabla^2 \mathbf{A}(\mathbf{x}, t) - \frac{1}{c^2} \frac{\partial^2}{\partial t^2} \mathbf{A}(\mathbf{x}, t) = 0.$$

We now fix an arbitrary time t and expand the vector potential in Fourier series of the variable $\mathbf{x}$. Requiring that it satisfies the so-called conditions of periodicity, i.e.

$$\mathbf{A}(x, y, z, t) = \mathbf{A}(x + m_x L, y + m_y L, z + m_z L),$$

with m_x, m_y, m_z arbitrary integers (positive, negative, or null), we get

$$\mathbf{A}(\mathbf{x}, t) = \sum_{\mathbf{k}} \mathbf{C}_{\mathbf{k}}(t) e^{i\mathbf{k} \cdot \mathbf{x}},$$

with the sum extended to all the $\mathbf{k}$ of the form

$$\mathbf{k} = \left(n_x \frac{2\pi}{L}, n_y \frac{2\pi}{L}, n_z \frac{2\pi}{L} \right),$$

with n_x, n_y, n_z arbitrary integers (positive, negative, or null). Being $\mathbf{A}(\mathbf{x}, t)$ a real function, the complex vector $\mathbf{C}_{\mathbf{k}}(t)$ satisfies the conjugation property

$$\mathbf{C}_{\mathbf{k}}(t)^* = \mathbf{C}_{-\mathbf{k}}(t).$$

We now require that the vector potential satisfies the wave equation. We obtain for $\mathbf{C}_{\mathbf{k}}(t)$ the differential equation

$$\frac{d^2 \mathbf{C}_{\mathbf{k}}(t)}{dt^2} = -\omega_{\mathbf{k}}^2 \mathbf{C}_{\mathbf{k}}(t),$$

where we have introduced the angular frequency $\omega_{\mathbf{k}}$ (relative to the wave vector $\mathbf{k}$) defined by the expression

$$\omega_{\mathbf{k}} = ck,$$

k being the magnitude of the vector $\mathbf{k}$. The differential equation can easily be solved and gives

$$\mathbf{C}_{\mathbf{k}}(t) = \mathbf{C}_{\mathbf{k}}^{(-)} e^{-i\omega_{\mathbf{k}} t} + \mathbf{C}_{\mathbf{k}}^{(+)} e^{i\omega_{\mathbf{k}} t},$$

with $\mathbf{C}_{\mathbf{k}}^{(-)}$ and $\mathbf{C}_{\mathbf{k}}^{(+)}$ constants. On substituting we obtain

$$\mathbf{A}(\mathbf{x}, t) = \sum_{\mathbf{k}} \mathbf{C}_{\mathbf{k}}^{(-)} e^{i(\mathbf{k} \cdot \mathbf{x} - \omega_{\mathbf{k}} t)} + \sum_{\mathbf{k}} \mathbf{C}_{\mathbf{k}}^{(+)} e^{i(\mathbf{k} \cdot \mathbf{x} + \omega_{\mathbf{k}} t)},$$

with

$$\mathbf{C}_{\mathbf{k}}^{(-)*} = \mathbf{C}_{-\mathbf{k}}^{(+)}.$$

The vector potential has been decomposed in this way into progressive and regressive waves. Taking into account that $\omega_{\mathbf{k}} = \omega_{-\mathbf{k}}$, the exponential appearing in the regressive wave can be transformed by the equation

$$e^{i(\mathbf{k}\cdot\mathbf{x}+\omega_{\mathbf{k}}t)} = e^{-i(-\mathbf{k}\cdot\mathbf{x}-\omega_{-\mathbf{k}}t)},$$

and changing in the second sum the index $\mathbf{k}$ in $-\mathbf{k}$ we get

$$\mathbf{A}(\mathbf{x},t) = \sum_{\mathbf{k}} \mathbf{C}_{\mathbf{k}}^{(-)} e^{i(\mathbf{k}\cdot\mathbf{x}-\omega_{\mathbf{k}}t)} + \sum_{\mathbf{k}} \mathbf{C}_{\mathbf{k}}^{(-)*} e^{-i(\mathbf{k}\cdot\mathbf{x}-\omega_{\mathbf{k}}t)},$$

which clearly shows that the function $A(\mathbf{x},t)$ is real.

We finally impose the gauge supplementary condition ($\operatorname{div} A(\mathbf{x},t) = 0$). We obtain, for each value of $\mathbf{k}$, the transversality condition

$$\mathbf{k} \cdot \mathbf{C}_{\mathbf{k}}^{(-)} = 0.$$

We can satisfy this condition in the following way. We define for each wave vector $\mathbf{k}$ two polarisation unit vectors (in general complex) $\mathbf{e}_{\mathbf{k}\lambda}$ ($\lambda = 1, 2$) that are both perpendicular to $\mathbf{k}$ and to each other, i.e. such that they satisfy the relations

$$\mathbf{e}_{\mathbf{k}\lambda} \cdot \mathbf{k} = 0,$$

$$\mathbf{e}_{\mathbf{k}\lambda} \cdot \mathbf{e}_{\mathbf{k}\lambda'}^{*} = \delta_{\lambda\lambda'}.$$

Having introduced these unit vectors we can now satisfy the transversality condition by writing

$$\mathbf{C}_{\mathbf{k}}^{(-)} e^{-i\omega_{\mathbf{k}}t} = \sum_{\lambda} c_{\mathbf{k}\lambda}(t) \mathbf{e}_{\mathbf{k}\lambda},$$

where $c_{\mathbf{k}\lambda}(t)$ is an oscillating function that satisfies the differential equation

$$\frac{d}{dt} c_{\mathbf{k}\lambda}(t) = -i\omega_{\mathbf{k}} c_{\mathbf{k}\lambda}(t).$$

We therefore obtain for the vector potential the final expression

$$\mathbf{A}(\mathbf{x},t) = \sum_{\mathbf{k}\lambda} \left[c_{\mathbf{k}\lambda}(t) \mathbf{e}_{\mathbf{k}\lambda} e^{i\mathbf{k}\cdot\mathbf{x}} + c_{\mathbf{k}\lambda}^{*}(t) \mathbf{e}_{\mathbf{k}\lambda}^{*} e^{-i\mathbf{k}\cdot\mathbf{x}} \right],$$

from which we can obtain the expressions for the electric and magnetic field vectors

$$\mathbf{E}(\mathbf{x},t) = -\frac{1}{c}\frac{\partial}{\partial t}\mathbf{A}(\mathbf{x},t) = \frac{i}{c}\sum_{\mathbf{k}\lambda} \omega_{\mathbf{k}}\left[c_{\mathbf{k}\lambda}(t)\mathbf{e}_{\mathbf{k}\lambda}e^{i\mathbf{k}\cdot\mathbf{x}} - c_{\mathbf{k}\lambda}^{*}(t)\mathbf{e}_{\mathbf{k}\lambda}^{*}e^{-i\mathbf{k}\cdot\mathbf{x}} \right],$$

$$\mathbf{B}(\mathbf{x},t) = \operatorname{rot}\mathbf{A}(\mathbf{x},t) = i\sum_{\mathbf{k}\lambda}\mathbf{k}\times\left[c_{\mathbf{k}\lambda}(t)\mathbf{e}_{\mathbf{k}\lambda}e^{i\mathbf{k}\cdot\mathbf{x}} - c_{\mathbf{k}\lambda}^{*}(t)\mathbf{e}_{\mathbf{k}\lambda}^{*}e^{-i\mathbf{k}\cdot\mathbf{x}} \right].$$

The sums appearing in these expressions are extended to all $\mathbf{k}$ values such that the periodicity conditions are satisfied, and once $\mathbf{k}$ is fixed, to the two possible states of polarisation. Each $(\mathbf{k}, \lambda)$ pair defines a so-called mode of the radiation field within the cavity. We now want to find the number of modes for which the magnitude of the wave vector is between k and $k + \mathrm{d}k$ and the direction of the wave vector is within the solid angle $\mathrm{d}\Omega$. We note that in the wavenumber space the tips of the possible vectors $\mathbf{k}$ are located on a cubic grid with side $2\pi/L$. Thus we obtain that this number is

$$\mathrm{d}N = 2\left(\frac{2\pi}{L}\right)^{-3} k^2 \, \mathrm{d}k \, \mathrm{d}\Omega = \frac{\mathcal{V}}{4\pi^3} k^2 \, \mathrm{d}k \, \mathrm{d}\Omega, \tag{4.3}$$

where $\mathcal{V}$ is the volume of the cavity and where the additional factor of 2 has been included to take into account the two possible polarisation states.

4.3 The Quantum Analog

The expansion in Fourier series obtained in the previous section is based on purely classical considerations. The quantum analog is obtained in Schrödinger representation by interpreting the classical variables $c_{\mathbf{k}\lambda}(t)$ and $c_{\mathbf{k}\lambda}^*(t)$ (namely the time-dependent coefficients that appear in the expansion in Fourier series of the vector potential and the fields) as quantum operators $c_{\mathbf{k}\lambda}$ and $c_{\mathbf{k}\lambda}^\dagger$ acting on a suitable Hilbert space. These operators are independent of time. In this way, also the classical quantities $\mathbf{A}(\mathbf{x}, t)$, $\mathbf{E}(\mathbf{x}, t)$ and $\mathbf{B}(\mathbf{x}, t)$ become operators independent of time. For the operator $\mathbf{A}(\mathbf{x})$ we have, for example,

$$\mathbf{A}(\mathbf{x}) = \sum_{\mathbf{k}\lambda}\left[c_{\mathbf{k}\lambda}\mathbf{e}_{\mathbf{k}\lambda}e^{i\mathbf{k}\cdot\mathbf{x}} + c_{\mathbf{k}\lambda}^\dagger\mathbf{e}_{\mathbf{k}\lambda}^*e^{-i\mathbf{k}\cdot\mathbf{x}}\right],$$

with similar expressions for the operators $\mathbf{E}(\mathbf{x})$ and $\mathbf{B}(\mathbf{x})$.

We must now define the operator that corresponds to the Hamiltonian of the radiation field. To do this, we rely on the correspondence principle and write, recalling the expression for the energy density of the field

$$\mathcal{H} = \int_{\mathcal{V}} u(\mathbf{x}) \, \mathrm{d}^3\mathbf{x} = \frac{1}{8\pi}\int_{\mathcal{V}}\left[\mathbf{E}^2(\mathbf{x}) + \mathbf{B}^2(\mathbf{x})\right]\mathrm{d}^3\mathbf{x},$$

where $\mathbf{E}(\mathbf{x})$ and $\mathbf{B}(\mathbf{x})$ are the electric and magnetic field operators, respectively. On substituting the operator expressions for $\mathbf{E}(\mathbf{x})$ and $\mathbf{B}(\mathbf{x})$ we obtain an expression containing a double sum over the indices $(\mathbf{k}, \lambda)$ and $(\mathbf{k}', \lambda')$. Taking into account the periodicity conditions, we then note that the only terms producing a non-zero contribution to the integral are those which contain exponential products of the type

$$e^{i\mathbf{k}\cdot\mathbf{x}}e^{-i\mathbf{k}\cdot\mathbf{x}},$$

for which the integration in d^3x is simply $\mathcal{V}$. For each of the electric ($\mathcal{H}_{el.}$) and magnetic ($\mathcal{H}_{ma.}$) contributions we obtain four distinct terms

$$
\begin{aligned}
\mathcal{H}_{el.} &= \frac{1}{8\pi} \int_{\mathcal{V}} \mathbf{E}^2(\mathbf{x}) d^3x \\
&= -\frac{\mathcal{V}}{8\pi c^2} \sum_{\mathbf{k}\lambda\lambda'} \omega_{\mathbf{k}}^2 \big[c_{\mathbf{k}\lambda} c_{-\mathbf{k}\lambda'} (\mathbf{e}_{\mathbf{k}\lambda} \cdot \mathbf{e}_{-\mathbf{k}\lambda'}) + c_{\mathbf{k}\lambda}^\dagger c_{-\mathbf{k}\lambda'}^\dagger (\mathbf{e}_{\mathbf{k}\lambda}^* \cdot \mathbf{e}_{-\mathbf{k}\lambda'}^*) \\
&\qquad - c_{\mathbf{k}\lambda} c_{\mathbf{k}\lambda'}^\dagger (\mathbf{e}_{\mathbf{k}\lambda} \cdot \mathbf{e}_{\mathbf{k}\lambda'}^*) - c_{\mathbf{k}\lambda}^\dagger c_{\mathbf{k}\lambda'} (\mathbf{e}_{\mathbf{k}\lambda}^* \cdot \mathbf{e}_{\mathbf{k}\lambda'}) \big], \\
\mathcal{H}_{ma.} &= \frac{1}{8\pi} \int_{\mathcal{V}} \mathbf{B}^2(\mathbf{x}) d^3x \\
&= -\frac{\mathcal{V}}{8\pi} \sum_{\mathbf{k}\lambda\lambda'} \big[c_{\mathbf{k}\lambda} c_{-\mathbf{k}\lambda'} (\mathbf{k} \times \mathbf{e}_{\mathbf{k}\lambda}) \cdot (-\mathbf{k} \times \mathbf{e}_{-\mathbf{k}\lambda'}) \\
&\qquad + c_{\mathbf{k}\lambda}^\dagger c_{-\mathbf{k}\lambda'}^\dagger (\mathbf{k} \times \mathbf{e}_{\mathbf{k}\lambda}^*) \cdot (-\mathbf{k} \times \mathbf{e}_{-\mathbf{k}\lambda'}^*) - c_{\mathbf{k}\lambda} c_{\mathbf{k}\lambda'}^\dagger (\mathbf{k} \times \mathbf{e}_{\mathbf{k}\lambda}) \cdot (\mathbf{k} \times \mathbf{e}_{\mathbf{k}\lambda'}^*) \\
&\qquad - c_{\mathbf{k}\lambda}^\dagger c_{\mathbf{k}\lambda'} (\mathbf{k} \times \mathbf{e}_{\mathbf{k}\lambda}^*) \cdot (\mathbf{k} \times \mathbf{e}_{\mathbf{k}\lambda'}) \big].
\end{aligned}
$$

We note that, for any arbitrary unit vectors $\mathbf{e}_1$ and $\mathbf{e}_2$ that are perpendicular to the vector $\mathbf{k}$, we have

$$
(\mathbf{k} \times \mathbf{e}_1) \cdot (\mathbf{k} \times \mathbf{e}_2) = k^2 (\mathbf{e}_1 \cdot \mathbf{e}_2),
$$

and given that $k^2 = \omega_{\mathbf{k}}^2/c^2$, we obtain for the first two summands that appear within the square brackets in the right-hand side that the electric and magnetic contributions cancel out. For the last two summands, the electric and magnetic contributions are equal. Finally, taking into account that

$$
\mathbf{e}_{\mathbf{k}\lambda} \cdot \mathbf{e}_{\mathbf{k}\lambda'}^* = \delta_{\lambda\lambda'},
$$

we obtain the expression

$$
\mathcal{H} = \frac{\mathcal{V}}{4\pi c^2} \sum_{\mathbf{k}\lambda} \omega_{\mathbf{k}}^2 \big[c_{\mathbf{k}\lambda} c_{\mathbf{k}\lambda}^\dagger + c_{\mathbf{k}\lambda}^\dagger c_{\mathbf{k}\lambda} \big].
$$

In order to write this Hamiltonian as the sum of Hamiltonians that we have previously discussed, we replace the operators $c_{\mathbf{k}\lambda}$ and $c_{\mathbf{k}\lambda}^\dagger$ with the operators $a_{\mathbf{k}\lambda}$ and $a_{\mathbf{k}\lambda}^\dagger$ defined by

$$
a_{\mathbf{k}\lambda} = \frac{1}{c} \sqrt{\frac{\omega_{\mathbf{k}} \mathcal{V}}{2\pi \hbar}} \, c_{\mathbf{k}\lambda}.
$$

In this way, the Hamiltonian becomes

$$
\mathcal{H} = \sum_{\mathbf{k}\lambda} \frac{1}{2} \hbar \omega_{\mathbf{k}} \big[a_{\mathbf{k}\lambda} a_{\mathbf{k}\lambda}^\dagger + a_{\mathbf{k}\lambda}^\dagger a_{\mathbf{k}\lambda} \big], \tag{4.4}
$$

and the vector potential operator can be written in the form

$$\mathbf{A}(\mathbf{x}) = \sum_{\mathbf{k}\lambda} c \sqrt{\frac{2\pi\hbar}{\omega_{\mathbf{k}}V}} \left[a_{\mathbf{k}\lambda} \mathbf{e}_{\mathbf{k}\lambda} e^{i\mathbf{k}\cdot\mathbf{x}} + a_{\mathbf{k}\lambda}^{\dagger} \mathbf{e}_{\mathbf{k}\lambda}^{*} e^{-i\mathbf{k}\cdot\mathbf{x}} \right]. \tag{4.5}$$

To complete the quantisation of the electromagnetic field we still need to know the commutation rules for the operators $a_{\mathbf{k}\lambda}$ and $a_{\mathbf{k}\lambda}^{\dagger}$. They can be derived from the correspondence principle using the fact that we know the equations of motion that are governing the temporal evolution of the corresponding classical quantities $(c_{\mathbf{k}\lambda}(t) \sim \exp(-i\omega_{\mathbf{k}}t))$. If $|\psi\rangle$ is any state vector of the quantum system, the classical observable corresponding to the operator $a_{\mathbf{k}\lambda}$ is given by the expectation value $\langle\psi|a_{\mathbf{k}\lambda}|\psi\rangle$, and such observable must satisfy the equation

$$\frac{d}{dt}\langle\psi|a_{\mathbf{k}\lambda}|\psi\rangle = -i\omega_{\mathbf{k}}\langle\psi|a_{\mathbf{k}\lambda}|\psi\rangle.$$

Taking into account that the state vector $|\psi\rangle$ satisfies the Schrödinger equation, we have

$$\frac{d}{dt}\langle\psi|a_{\mathbf{k}\lambda}|\psi\rangle = \frac{i}{\hbar}\langle\psi|[\mathcal{H}, a_{\mathbf{k}\lambda}]|\psi\rangle,$$

and, substituting the expression for $\mathcal{H}$,

$$\frac{d}{dt}\langle\psi|a_{\mathbf{k}\lambda}|\psi\rangle = \sum_{\mathbf{k}'\lambda'} \frac{i}{2}\omega_{\mathbf{k}'} \langle\psi|[a_{\mathbf{k}'\lambda'}a_{\mathbf{k}'\lambda'}^{\dagger} + a_{\mathbf{k}'\lambda'}^{\dagger}a_{\mathbf{k}'\lambda'}, a_{\mathbf{k}\lambda}]|\psi\rangle.$$

Identifying the two expressions for the derivative of the observable, we get

$$\omega_{\mathbf{k}}\langle\psi|a_{\mathbf{k}\lambda}|\psi\rangle = -\frac{1}{2}\sum_{\mathbf{k}'\lambda'} \omega_{\mathbf{k}'} \langle\psi|[a_{\mathbf{k}'\lambda'}a_{\mathbf{k}'\lambda'}^{\dagger} + a_{\mathbf{k}'\lambda'}^{\dagger}a_{\mathbf{k}'\lambda'}, a_{\mathbf{k}\lambda}]|\psi\rangle.$$

Since the expression on the right-hand side must contain (as that one on the left-hand side) only quantities that depend on the indices $\mathbf{k}$ and λ, and since this must be verified for any state vector $|\psi\rangle$, we must have

$$[a_{\mathbf{k}\lambda}, a_{\mathbf{k}'\lambda'}] = 0 \quad \text{for } \mathbf{k}' \neq \mathbf{k}, \ \lambda' \neq \lambda, \tag{4.6}$$

$$[a_{\mathbf{k}\lambda}, a_{\mathbf{k}'\lambda'}^{\dagger}] = 0 \quad \text{for } \mathbf{k}' \neq \mathbf{k}, \ \lambda' \neq \lambda, \tag{4.7}$$

which means that the operators a and $a^{\dagger}$ relative to different modes are commuting. The previous equation then becomes

$$\langle\psi|a_{\mathbf{k}\lambda}|\psi\rangle = -\frac{1}{2}\langle\psi|a_{\mathbf{k}\lambda}[a_{\mathbf{k}\lambda}^{\dagger}, a_{\mathbf{k}\lambda}] + [a_{\mathbf{k}\lambda}^{\dagger}, a_{\mathbf{k}\lambda}]a_{\mathbf{k}\lambda}|\psi\rangle.$$

Given that the state vector is arbitrary, this equation can be satisfied only putting

$$[a_{\mathbf{k}\lambda}, a_{\mathbf{k}\lambda}^{\dagger}] = 1, \tag{4.8}$$

and the Hamiltonian assumes the form

$$\mathcal{H} = \sum_{k\lambda} \hbar\omega_k \left(a_{k\lambda}^\dagger a_{k\lambda} + \frac{1}{2} \right). \tag{4.9}$$

By comparing the results contained in Eqs. (4.6)–(4.9) with those in Sect. 4.1 (Eqs. (4.1) and (4.2)), we conclude that the quantum Hamiltonian of the electromagnetic field is equal to the sum of an infinite number of Hamiltonians of harmonic oscillators, independent of each other. Each oscillator is associated with a mode of the radiation field characterised by the wave vector $\mathbf{k}$ (hence by the angular frequency ω_k) and by the unit vector of polarisation $\mathbf{e}_{k\lambda}$. Recalling the results of Sect. 4.1, we have that the energy of each mode can only take one of the values

$$E_{k\lambda} = \hbar\omega_k \left(n + \frac{1}{2} \right),$$

where n is an arbitrary integer greater than or equal to 0. This result can be interpreted by saying that the mode contains a number n of photons all having the energy $\hbar\omega_k$. The number n takes also the name of "occupation number" of the mode. In particular, if $n = 0$, the energy has the value $\frac{1}{2}\hbar\omega_k$, which is called the zero-energy (or vacuum energy) of the mode. In many cases, this energy has a purely formal importance, and can be neglected in most applications. It should be noted that, since the number of modes is infinite, the zero energy of the electromagnetic field is infinite. To interpret consistently the physical reality we are therefore forced to "renormalise" the energy of the vacuum state by imposing that it is zero.

Since the total Hamiltonian is equal to the sum of many independent Hamiltonians, it follows that its eigenvalues are equal to the sum of the eigenvalues of the individual Hamiltonians. The total Hamiltonian eigenvalues are then of the form

$$E = \sum_{k\lambda} \hbar\omega_k \left(n_{k\lambda} + \frac{1}{2} \right),$$

and are identified by the set of integers $n_{k\lambda}$ that specify the number of photons present in each mode. Regarding the identification of the eigenvectors, namely the stationary states of the radiation field, one can use a generalisation of the formalism introduced in Sect. 4.1 and write the eigenvector in the compact form

$$|n_1, n_2, \ldots, n_{k\lambda}, \ldots\rangle = |n_1\rangle|n_2\rangle \cdots |n_{k\lambda}\rangle \cdots,$$

i.e. as the direct product of many kets, each defined in a different Hilbert space. Each operator $a_{k\lambda}$ or $a_{k\lambda}^\dagger$ acts only on the vector $|n_{k\lambda}\rangle$ of the corresponding mode, so we have

$$a_{k\lambda}|n_1, n_2, \ldots, n_{k\lambda}, \ldots\rangle = \sqrt{n_{k\lambda}}|n_1, n_2, \ldots, n_{k\lambda} - 1, \ldots\rangle, \tag{4.10}$$

$$a_{k\lambda}^\dagger|n_1, n_2, \ldots, n_{k\lambda}, \ldots\rangle = \sqrt{n_{k\lambda} + 1}|n_1, n_2, \ldots, n_{k\lambda} + 1, \ldots\rangle, \tag{4.11}$$

which justifies the name given to $a_{\mathbf{k}\lambda}$ and $a_{\mathbf{k}\lambda}^\dagger$ as operators of annihilation and creation of photons.

The formalism introduced in this section is called the formalism of second quantisation. This name is not totally justified for the quantisation of the electromagnetic field but originates by the fact that very similar procedures can be applied to the quantisation of particle fields. If for example we want to introduce the quantisation of a field of nonrelativistic particles of mass m, we start with the Schrödinger equation for the wave function of the particles

$$i\hbar\frac{\partial}{\partial t}\psi(\mathbf{x}, t) = -\frac{\hbar^2}{2m}\nabla^2\psi(\mathbf{x}, t),$$

we next expand the wave function $\psi(\mathbf{x}, t)$ in Fourier series, and then interpret the coefficients of the expansion as quantum operators. In this case it actually makes sense to speak of second quantisation, as the particle field is already described by a quantum mechanical wave equation. In the case of the electromagnetic field, instead, the wave equation for the photons is a classical equation which follows from Maxwell equations.

Finally, it should be noted that the formalism we have introduced here is not covariant, since the expression assumed for the electromagnetic gauge ($\phi = 0$, div $\mathbf{A} = 0$) is not covariant. The quantisation of the electromagnetic field in covariant form is more elegant and certainly more satisfactory from a theoretical point of view,[1] but requires the use of a more complex formalism that is not necessary for many applications, such as those that will be discussed in Chaps. 11 and 15 where we deal with the interaction between matter and radiation.

4.4 Intensity and Photons

The concept of photon introduced in the previous section can be shown to be closely related to the physical quantities traditionally used for the description of radiation phenomena such as the specific intensity and the energy density of the radiation field.

Consider an arbitrary point P and a surface element dS whose normal direction is identified by the unit vector $\mathbf{\Omega}$. We denote by dE_ν the energy of the electromagnetic field within the frequency interval $(\nu, \nu + d\nu)$ and direction within the solid angle $d\Omega$ centred around $\mathbf{\Omega}$, flowing in the infinitesimal time dt through dS (see Fig. 4.1). The dE_ν is obviously proportional to the product $dS\,d\nu\,dt\,d\Omega$ so we have

$$dE_\nu = I_\nu(P, \mathbf{\Omega}, t)\,dS\,d\nu\,dt\,d\Omega.$$

This equation implicitly defines the quantity $I_\nu(P, \mathbf{\Omega}, t)$ which is called the specific intensity (or intensity *tout court*) of the radiation field and that is generally a

---

[1] See for example Bjorken and Drell (1965).

Fig. 4.1 The radiation contained in the solid angle dΩ, centred around Ω, flows through the surface element dS

function not only of frequency, but also of direction and time. If the intensity does not depend on time, the radiation field is called stationary, if it does not depend on direction it is called isotropic and if does not depend on the point P it is called homogeneous.

Instead of the specific intensity I_ν we can also consider the quantity I_λ (which bears the same name) defined in the same way, except for the frequency interval dν that is replaced by the wavelength interval dλ. Similarly, we can also consider the quantity I_ω, also defined by the same relation, with dω (interval of angular frequency) replacing dν (frequency interval). Evidently, for corresponding intervals we have

$$I_\nu \, d\nu = I_\lambda \, d\lambda = I_\omega \, d\omega,$$

and since d$\lambda = \lambda^2 \, d\nu/c$, d$\omega = 2\pi \, d\nu$, we also have

$$I_\lambda = \frac{c}{\lambda^2} I_\nu, \qquad I_\omega = \frac{1}{2\pi} I_\nu. \tag{4.12}$$

Another quantity closely related to the specific intensity is the energy density of the radiation field. Such quantity, denoted by the symbol $u_\nu(P, \Omega, t)$, is defined with respect to a volume element dV centred around the point P. If we denote by d$\mathcal{E}_\nu$ the electromagnetic energy contained within dV, having a frequency between ν and $\nu + d\nu$ and direction contained in dΩ, we have, by definition,

$$d\mathcal{E}_\nu = u_\nu(P, \Omega, t) \, dV \, d\nu \, d\Omega.$$

Since the electromagnetic radiation propagates in vacuum with velocity c, a relation between I_ν and u_ν immediately follows, i.e.

$$I_\nu(P, \Omega, t) = c u_\nu(P, \Omega, t). \tag{4.13}$$

Let us now consider the modes of the radiation field that correspond to the frequency range $(\nu, \nu + d\nu)$ and to the directions contained in the solid angle dΩ. The number of these modes is given by Eq. (4.3), which is, converting the dk in the dν by the relation $k = 2\pi \nu/c$,

$$dN = \frac{2\mathcal{V}\nu^2}{c^3} \, d\nu \, d\Omega, \tag{4.14}$$

where $\mathcal{V}$ is the normalisation volume, which we can identify with the infinitesimal volume dV. If we denote by $n_\nu(P, \Omega, t)$ the number of photons in each mode and since each photon has energy $h\nu$, we get

$$d\mathcal{E}_\nu = n_\nu(P, \boldsymbol{\Omega}, t) h\nu \frac{2\nu^2}{c^3} \, dV \, d\nu \, d\Omega,$$

and recalling the definition of u_ν we obtain the relation

$$u_\nu(P, \boldsymbol{\Omega}, t) = \frac{2h\nu^3}{c^3} n_\nu(P, \boldsymbol{\Omega}, t),$$

or in terms of specific intensity

$$I_\nu(P, \boldsymbol{\Omega}, t) = \frac{2h\nu^3}{c^2} n_\nu(P, \boldsymbol{\Omega}, t). \tag{4.15}$$

It is important to note that a similar expression is also valid classically. In this case, instead of considering the number of photons per mode we have more simply an energy per mode $\epsilon_\nu(P, \Omega, t)$ and we write

$$I_\nu(P, \Omega, t) = \frac{2\nu^2}{c^2} \epsilon_\nu(P, \Omega, t). \tag{4.16}$$

The number of photons per mode is a very important quantity that can be used to characterise in a physical way the light sources. As we shall see in Chap. 10, for a thermal source such number is given by

$$n_\nu = \frac{1}{e^{h\nu/(k_B T)} - 1},$$

where T is the temperature. In the visible at 5000 Å, for example, a normal incandescent lamp with a tungsten filament ($T \simeq 2700$ K) produces a radiation with about 2×10^{-5} photons per mode, while this number is about 10^{-2} for the solar radiation ($T \simeq 5800$ K). The radiation emitted by a laser instead has a higher number of photons per mode, and can reach typical values of the order of 10^7 or even higher.

Chapter 5
Relativistic Wave Equations

In order to provide a quantitative basis to the study of atomic spectra, discussed in the next chapters of this book, it is first necessary to present a thorough study of the relativistic equations for atomic particles, with particular emphasis on the Dirac equation for the electron. In this chapter we will see how it is possible to describe, within quantum mechanics, the dynamical properties of a relativistic particle, either free or moving in a stationary electromagnetic field. Once an appropriate matrix formalism will be introduced, a number of physical consequences will naturally follow. We note that elementary treatises of atomic spectroscopy often introduce such physical consequences (e.g. electron spin, spin-orbit interaction, etc.) in a phenomenological way.

5.1 The Dirac Equation for a Free Particle

The relation between momentum and energy for a free nonrelativistic particle of mass m is given by the equation

$$E = \frac{p^2}{2m},$$

where $\mathbf{p}$ is the momentum, which coincides with the kinetic moment conjugate with the position variable $\mathbf{x}$. The motion of the particle is described in quantum mechanics by the Schrödinger equation

$$i\hbar \frac{\partial}{\partial t} |\psi\rangle = \frac{p^2}{2m} |\psi\rangle,$$

where $\mathbf{p}$ is the operator associated with the classical variable momentum. The operator $\mathbf{p}$ must satisfy the fundamental quantum mechanical commutation rule between two operators associated with two variables canonically conjugated, q and p, namely

$$[q, p] = i\hbar.$$

E. Landi Degl'Innocenti, *Atomic Spectroscopy and Radiative Processes*,
UNITEXT for Physics, DOI 10.1007/978-88-470-2808-1_5, © Springer-Verlag Italia 2014

The explicit expression of the operator $\mathbf{p}$ depends on the particular representation adopted for the state vectors in the Hilbert space. If one chooses the representation in which the vectors are expressed as linear combinations of the eigenvectors of the position operator (wave function representation), the operator $\mathbf{x}$ is a simple multiplier while the operator $\mathbf{p}$ is

$$\mathbf{p} = -i\hbar\,\mathrm{grad}.$$

In such representation the Schrödinger equation assumes the form

$$i\hbar\frac{\partial}{\partial t}\psi(\mathbf{x}, t) = -\frac{\hbar^2}{2m}\nabla^2\psi(\mathbf{x}, t),$$

and, in accordance with the postulates of quantum mechanics, the quantity $|\psi(\mathbf{x}, t)|^2$ represents the probability density of finding the particle in the neighborhood of the point $\mathbf{x}$ at time t.

Turning to relativistic dynamics, the momentum-energy relation becomes

$$E^2 = p^2c^2 + m^2c^4,$$

or

$$E = \sqrt{p^2c^2 + m^2c^4},$$

where m is the rest mass of the particle and c is the speed of light. We could therefore in principle write the Schrödinger equation as

$$i\hbar\frac{\partial}{\partial t}\psi(\mathbf{x}, t) = \sqrt{p^2c^2 + m^2c^4}\,\psi(\mathbf{x}, t),$$

where p^2 is the operator $-\hbar^2\nabla^2$. The square root in the right-hand side does, however, present serious problems as to the meaning to be given to the operator which acts on the wave function. If, for example, one replaces the square root with its power series expansion, one gets powers of all orders in ∇^2 and the equation becomes practically insolvable.

A far simpler equation is instead obtained by translating in quantum mechanical terms the quadratic relation between energy and momentum written above. The resulting equation, known as the Klein-Gordon equation, is

$$\left(\nabla^2 - \frac{1}{c^2}\frac{\partial^2}{\partial t^2}\right)\psi(\mathbf{x}, t) = \frac{m^2c^2}{\hbar^2}\psi(\mathbf{x}, t).$$

However this equation, besides presenting the inconvenience of being a second order differential equation with respect to time (unlike Schrödinger equation that is of the first order), correctly describes only the relativistic particles that do not have spin, such as, for example, mesons.

The problem of determining an equation that could describe relativistic particles with spin, such as the electron, was brilliantly solved by Dirac. In 1928, he proposed a wave equation in which the operators $\partial/\partial x_i$ appeared linearly as the operator $\partial/\partial t$. This is a quite natural approach for a relativistic theory in which the time coordinate is not privileged with respect to the space coordinates. The equation proposed by Dirac, which obviously bears his name, is

$$i\hbar\frac{\partial}{\partial t}\psi(\mathbf{x}, t) = \left[-i\hbar c\left(\alpha_1\frac{\partial}{\partial x_1} + \alpha_2\frac{\partial}{\partial x_2} + \alpha_3\frac{\partial}{\partial x_3}\right) + \beta mc^2\right]\psi(\mathbf{x}, t),$$

and can also be written in the more compact form

$$i\hbar\frac{\partial}{\partial t}\psi(\mathbf{x}, t) = \mathcal{H}_D\psi(\mathbf{x}, t) = \left(c\boldsymbol{\alpha}\cdot\mathbf{p} + \beta mc^2\right)\psi(\mathbf{x}, t), \tag{5.1}$$

where $\mathcal{H}_D$ is the so-called Dirac Hamiltonian and where α_1, α_2, α_3, and β are four dimensionless quantities (operators) that by definition commute with the operators $\partial/\partial t$ and $\partial/\partial x_i$. Such quantities are determined by imposing that the correct relation between energy and momentum is obtained from the Dirac equation. Differentiating with respect to time Eq. (5.1) and substituting the same equation, we obtain

$$-\hbar^2\frac{\partial^2}{\partial t^2}\psi(\mathbf{x}, t) = \left(c\boldsymbol{\alpha}\cdot\mathbf{p} + \beta mc^2\right)\left(c\boldsymbol{\alpha}\cdot\mathbf{p} + \beta mc^2\right)\psi(\mathbf{x}, t),$$

and we must therefore require that

$$\left(c\boldsymbol{\alpha}\cdot\mathbf{p} + \beta mc^2\right)\left(c\boldsymbol{\alpha}\cdot\mathbf{p} + \beta mc^2\right) = p^2c^2 + m^2c^4.$$

Expanding the expression (taking care not to alter the order of the non-commuting operators), we obtain

$$c^2\sum_{ij}\alpha_i\alpha_j p_i p_j + mc^3\sum_i(\alpha_i\beta + \beta\alpha_i)p_i + m^2c^4\beta^2 = c^2\sum_i p_i p_i + m^2c^4,$$

so, to have identity between the left and right hand side we must have

$$\{\alpha_i, \alpha_j\} = 2\delta_{ij}, \qquad \{\alpha_i, \beta\} = 0, \qquad \beta^2 = 1, \tag{5.2}$$

where we have introduced the anticommutator symbol defined by

$$\{A, B\} = AB + BA.$$

The simplest quantities that satisfy this algebra are matrices, of which we now analyze the properties. First of all, since the square of each of the four matrices is equal to unity, the eigenvalues must be equal to ± 1. Moreover, being $\beta^2 = 1$ and $\beta\alpha_i = -\alpha_i\beta$, the trace of each matrix α_i obeys the relation

$$\text{Tr}(\alpha_i) = \text{Tr}(\beta^2\alpha_i) = -\text{Tr}(\beta\alpha_i\beta).$$

On the other hand, given the cyclic property of the trace of a matrix, we also have that

$$\text{Tr}(\alpha_i) = \text{Tr}(\beta^2\alpha_i) = \text{Tr}(\beta\alpha_i\beta).$$

Having obtained that $\text{Tr}(\alpha_i) = -\text{Tr}(\alpha_i)$, we must therefore necessarily have

$$\text{Tr}(\alpha_i) = 0.$$

In a very similar way, it can also be shown that $\text{Tr}(\beta) = 0$. So, since the trace is the sum of the eigenvalues, and since these can only be equal to ± 1 (as we have seen), it follows that the matrices must be of even order. Since it is not possible to build four 2×2 matrices that obey the algebra of Eq. (5.2), the choice falls on 4×4 matrices. So the α_i and β quantities introduced in the Dirac equation can be represented by matrices of the fourth order. They are called the Dirac matrices.

The choice of the Dirac matrices is not unique. One possible representation, which is particularly suitable to treat the nonrelativistic limit, is given in terms of 2×2 matrices by the following expressions

$$\alpha_i = \begin{pmatrix} 0 & \sigma_i \\ \sigma_i & 0 \end{pmatrix}, \qquad \beta = \begin{pmatrix} I & 0 \\ 0 & -I \end{pmatrix}, \tag{5.3}$$

where I is the unit matrix of order 2 and where the σ_i are the Pauli 2×2 matrices

$$\sigma_1 = \begin{pmatrix} 0 & 1 \\ 1 & 0 \end{pmatrix}, \qquad \sigma_2 = \begin{pmatrix} 0 & -i \\ i & 0 \end{pmatrix}, \qquad \sigma_3 = \begin{pmatrix} 1 & 0 \\ 0 & -1 \end{pmatrix}.$$

From their expressions, it can easily be shown that the Pauli matrices satisfy the relations

$$[\sigma_i, \sigma_j] = 2i \sum_k \epsilon_{ijk} \sigma_k,$$

$$\{\sigma_i, \sigma_j\} = 2\delta_{ij},$$

where ϵ_{ijk} is the Ricci (or Levi-Civita) tensor (see Sect. 16.2). The two previous equations can be condensed into a single one

$$\sigma_i \sigma_j = \delta_{ij} + i \sum_k \epsilon_{ijk} \sigma_k.$$

It is relatively simple to verify that the four matrices defined in Eq. (5.3), i.e.

$$\alpha_1 = \begin{pmatrix} 0 & 0 & 0 & 1 \\ 0 & 0 & 1 & 0 \\ 0 & 1 & 0 & 0 \\ 1 & 0 & 0 & 0 \end{pmatrix}, \qquad \alpha_2 = \begin{pmatrix} 0 & 0 & 0 & -i \\ 0 & 0 & i & 0 \\ 0 & -i & 0 & 0 \\ i & 0 & 0 & 0 \end{pmatrix},$$

$$\alpha_3 = \begin{pmatrix} 0 & 0 & 1 & 0 \\ 0 & 0 & 0 & -1 \\ 1 & 0 & 0 & 0 \\ 0 & -1 & 0 & 0 \end{pmatrix}, \qquad \beta = \begin{pmatrix} 1 & 0 & 0 & 0 \\ 0 & 1 & 0 & 0 \\ 0 & 0 & -1 & 0 \\ 0 & 0 & 0 & -1 \end{pmatrix},$$

satisfy the algebra of the Dirac matrices. It is important to stress that an infinite number of representations of the Dirac matrices exist. In fact, given a set of Dirac matrices, any other set obtained via an arbitrary similarity transformation is also a set of Dirac matrices.[1] We finally note that the matrix character of the Dirac equation automatically implies that the wave function is not a scalar anymore, but a four-components quantity, called a spinor.

---

[1] A particularly useful representation for the ultra-relativistic limit is the so-called Majorana representation.

The Dirac equation for the free particle can be solved exactly. We search for a solution of the form

$$\psi(\mathbf{x}, t) = W e^{i(\mathbf{q} \cdot \mathbf{x} - Et)/\hbar},$$

where W is a four-components spinor, independent of $\mathbf{x}$ and t. Substituting in Eq. (5.1), we find that the spinor W must satisfy the equation

$$EW = \left(c\boldsymbol{\alpha} \cdot \mathbf{q} + \beta mc^2\right) W.$$

We can determine the possible values for E by solving the characteristic equation

$$\text{Det} \begin{pmatrix} mc^2 - E & 0 & cq_z & cq_- \\ 0 & mc^2 - E & cq_+ & -cq_z \\ cq_z & cq_- & -mc^2 - E & 0 \\ cq_+ & -cq_z & 0 & -mc^2 - E \end{pmatrix} = 0,$$

where we have set

$$q_+ = q_x + iq_y, \qquad q_- = q_x - iq_y.$$

By solving the determinant, we obtain the equation of fourth-order in E

$$\left(E^2 - c^2 q^2 - m^2 c^4\right)^2 = 0,$$

which has the four solutions

$$E_1 = E_2 = \varepsilon, \qquad E_3 = E_4 = -\varepsilon,$$

where

$$\varepsilon = \sqrt{c^2 q^2 + m^2 c^4}.$$

We have obtained in this way two solutions with positive energy, with the correct relativistic relation between energy and momentum, but also two solutions with negative energy (also having the correct relativistic relation between energy and momentum). The negative energy solutions derive naturally from the Dirac equation, and are a peculiar characteristic of relativistic wave equations (it is in fact possible to show that also the Klein-Gordon equation has solutions with negative energy). Once the eigenvalues are found, one can determine the corresponding eigenvectors. We can easily verify that the four spinors W_1, W_2, W_3, and W_4, given by

$$W_1 = \frac{1}{\sqrt{2\varepsilon(\varepsilon + mc^2)}} \begin{pmatrix} \varepsilon + mc^2 \\ 0 \\ cq_z \\ cq_+ \end{pmatrix}, \qquad W_2 = \frac{1}{\sqrt{2\varepsilon(\varepsilon + mc^2)}} \begin{pmatrix} 0 \\ \varepsilon + mc^2 \\ cq_- \\ -cq_z \end{pmatrix},$$

$$W_3 = \frac{1}{\sqrt{2\varepsilon(\varepsilon + mc^2)}} \begin{pmatrix} -cq_z \\ -cq_+ \\ \varepsilon + mc^2 \\ 0 \end{pmatrix}, \qquad W_4 = \frac{1}{\sqrt{2\varepsilon(\varepsilon + mc^2)}} \begin{pmatrix} -cq_- \\ cq_z \\ 0 \\ \varepsilon + mc^2 \end{pmatrix},$$

are four orthogonal and normalised eigenvectors corresponding, respectively, to the four eigenvalues E_1, E_2, E_3, and E_4. These expressions show that, in the nonrelativistic limit, the last two components of the spinors with positive energy are negligible, compared to the first two. The opposite is true for the spinors with negative energy.

The negative-energy solutions were properly interpreted by Dirac himself, who predicted the existence of antiparticles. The interpretation given by Dirac, here exemplified by the case of the electrons, is the following: we can think that the vacuum state, i.e. the state in which no particles are present, is actually a situation in which all the states with negative energy are filled in accordance with the Pauli principle (the Fermi sea). When an electron that is excited (for example by absorption of a quantum of radiation of energy higher than the threshold value $2mc^2$, equal to about 1 MeV) moves from one negative energy state to a state of positive energy, it leaves in the Fermi sea a "hole". This hole behaves as a particle in all respects similar to the electron, but of opposite charge, called a positron (or positon or positive electron). In other words, one can think that the excitation of an electron from one state of negative energy to one state of positive energy produces the creation of an electron-positron pair, with a consequent annihilation of a quantum of radiation. The positrons were discovered experimentally by Anderson in 1932 with a Wilson cloud chamber experiment. The discovery of the antiproton occurred much later in 1955, thanks to Chamberlain and Segré, when particle accelerators with energies of the order of some GeV became available.

5.2 The Dirac Equation for the Electron in an Electromagnetic Field

The classic Hamiltonian of a charged, relativistic particle moving in an electromagnetic field, is written in the form

$$\mathcal{H} = \sqrt{c^2 \left(\mathbf{p} - \frac{e}{c}\mathbf{A}\right)^2 + m^2 c^4} + e\phi, \tag{5.4}$$

where e is the algebraic value of the charge of the particle, m is its rest mass, $\mathbf{A}$ and ϕ are the electromagnetic potentials, and $\mathbf{p}$ is the kinetic moment conjugate to the particle position, and related to the momentum π by the relation

$$\pi = \mathbf{p} - \frac{e}{c}\mathbf{A}.$$

The expression for the Hamiltonian can be derived in various ways. Here, we simply verify that it produces the correct equations of motion. To do so, we obtain such equations using the Hamilton equations

$$\frac{d\mathbf{x}}{dt} = \mathbf{v} = \frac{\partial \mathcal{H}}{\partial \mathbf{p}} = c^2 \left(\mathbf{p} - \frac{e}{c}\mathbf{A}\right)\left[c^2\left(\mathbf{p} - \frac{e}{c}\mathbf{A}\right)^2 + m^2 c^4\right]^{-1/2},$$

$$\frac{d\mathbf{p}}{dt} = -\frac{\partial \mathcal{H}}{\partial \mathbf{x}} = ce(\text{grad}\,\mathbf{A}) \cdot \left(\mathbf{p} - \frac{e}{c}\mathbf{A}\right)\left[c^2\left(\mathbf{p} - \frac{e}{c}\mathbf{A}\right)^2 + m^2 c^4\right]^{-1/2} - e\,\text{grad}\,\phi.$$

From the second equation, introducing $\mathbf{v}$, we have

$$\frac{d\mathbf{p}}{dt} = \frac{e}{c}(\text{grad}\,\mathbf{A}) \cdot \mathbf{v} - e\,\text{grad}\,\phi.$$

If we now take into account the vector identity (see Eq. (16.12))

$$\mathbf{v} \times \text{rot}\,\mathbf{A} = (\text{grad}\,\mathbf{A}) \cdot \mathbf{v} - \mathbf{v} \cdot \text{grad}\,\mathbf{A},$$

and recall that $\text{rot}\,\mathbf{A} = \mathbf{B}$, we obtain

$$\frac{d\mathbf{p}}{dt} = \frac{e}{c}\mathbf{v} \times \mathbf{B} + \frac{e}{c}\mathbf{v} \cdot \text{grad}\,\mathbf{A} - e\,\text{grad}\,\phi.$$

For the momentum π we then have

$$\frac{d\pi}{dt} = \frac{d}{dt}\left(\mathbf{p} - \frac{e}{c}\mathbf{A}\right) = \frac{e}{c}\mathbf{v} \times \mathbf{B} + \frac{e}{c}\mathbf{v} \cdot \text{grad}\,\mathbf{A} - e\,\text{grad}\,\phi - \frac{e}{c}\frac{d\mathbf{A}}{dt}.$$

The last term in the right-hand side represents the total (or Eulerian) derivative of the vector potential along the trajectory of the particle. It is

$$\frac{d\mathbf{A}}{dt} = \frac{\partial\mathbf{A}}{\partial t} + \mathbf{v} \cdot \text{grad}\,\mathbf{A},$$

so that, recalling the expression for the electric field in terms of the electromagnetic potentials, we have

$$\frac{d\pi}{dt} = e\mathbf{E} + \frac{e}{c}\mathbf{v} \times \mathbf{B}.$$

From the first of Hamilton equations we can also obtain, with simple algebra, that

$$\pi = \frac{m}{\sqrt{1 - v^2/c^2}}\mathbf{v}.$$

The two above equations show that the Hamiltonian is correct, since it produces the well-known relativistic equations for the motion of a charged particle in an electromagnetic field.

Comparing the expression for the Hamiltonian (5.4) with that of the free particle, we can see that, for a charged particle in an electromagnetic field, there is, between the quantities

$$E - e\phi, \qquad \mathbf{p} - \frac{e}{c}\mathbf{A},$$

the same relation that exists, for the free particle, between E and $\mathbf{p}$. This implies that, within relativistic classical mechanics, the description of a charged particle in an electromagnetic field is obtained from that of a free particle by the following formal substitutions in the Hamiltonian

$$\mathbf{p} \to \mathbf{p} - \frac{e}{c}\mathbf{A}, \qquad E \to E - e\phi. \qquad (5.5)$$

This substitution is known as the minimal coupling rule (or principle). For the correspondence principle we can therefore write the Dirac equation for a charged particle in an electromagnetic field applying the minimal coupling rule to Eq. (5.1). We have

$$i\hbar\frac{\partial}{\partial t}\psi(\mathbf{x},t) = \left[c\boldsymbol{\alpha}\cdot\left(\mathbf{p} - \frac{e}{c}\mathbf{A}\right) + \beta mc^2 + e\phi\right]\psi(\mathbf{x},t), \qquad (5.6)$$

where $\boldsymbol{\alpha}$ and β are the 4×4 matrices previously introduced, i.e. the Dirac matrices, whereas $\mathbf{A}$ and ϕ are simply multiplication operators that depend on position and, in general, on time.

We finally note that if we perform a gauge transformation on the electromagnetic potentials, i.e.

$$\mathbf{A}' = \mathbf{A} - \operatorname{grad}\chi, \qquad \phi' = \phi + \frac{1}{c}\frac{\partial\chi}{\partial t},$$

we obtain an identical Dirac equation for the wave function $\psi'(\mathbf{x},t)$ given by

$$\psi'(\mathbf{x},t) = \psi(\mathbf{x},t)e^{-i\delta},$$

where

$$\delta = \frac{e\chi}{\hbar c}.$$

We therefore obtain the result that, for gauge transformations, the wave function is transformed according to a phase factor. The phase is proportional to the function χ, generating the gauge transformation itself. In particular, the square of the absolute value of the wave function is invariant under gauge transformations.

5.3 Nonrelativistic Limit of the Dirac Equation

To find the nonrelativistic limit of the Dirac equation for a charged particle in an electromagnetic field, it is convenient to write the wave function $\psi(\mathbf{x},t)$, a spinor of order 4, in the form

$$\psi(\mathbf{x},t) = \begin{pmatrix} \chi \\ \xi \end{pmatrix} e^{-i(mc^2+\epsilon)t/\hbar}, \qquad (5.7)$$

where χ and ξ are two spinors of order 2, and where the total energy has been written in the form $(mc^2 + \epsilon)$ so that ϵ represents the particle energy less the rest energy mc^2. The nonrelativistic limit is obtained by imposing that $\epsilon \ll mc^2$. At the same time, though, we also need to impose that the other relevant energies, such as the electrostatic energy, are also small when compared to the rest energy.

Substituting the wave function (5.7) in Eq. (5.6) and assuming that the spinors χ and ξ are independent of time, as the electromagnetic potentials, we have

$$\left(mc^2 + \epsilon\right)\begin{pmatrix} \chi \\ \xi \end{pmatrix} = c\boldsymbol{\alpha}\cdot\left(\mathbf{p} - \frac{e}{c}\mathbf{A}\right)\begin{pmatrix} \chi \\ \xi \end{pmatrix} + mc^2\beta\begin{pmatrix} \chi \\ \xi \end{pmatrix} + e\phi\begin{pmatrix} \chi \\ \xi \end{pmatrix},$$

and recalling the explicit expression of the Dirac matrices (Eq. (5.3)), we obtain two coupled equations for the spinors χ and ξ

$$c\boldsymbol{\sigma} \cdot \left(\mathbf{p} - \frac{e}{c}\mathbf{A}\right)\xi = (\epsilon - e\phi)\chi,$$

$$(2mc^2 + \epsilon - e\phi)\xi = c\boldsymbol{\sigma} \cdot \left(\mathbf{p} - \frac{e}{c}\mathbf{A}\right)\chi.$$

We now obtain ξ from the second equation by multiplying both sides by the operator $(2mc^2 + \epsilon - e\phi)^{-1}$. We have

$$\xi = \left(2mc^2 + \epsilon - e\phi\right)^{-1} c\boldsymbol{\sigma} \cdot \left(\mathbf{p} - \frac{e}{c}\mathbf{A}\right)\chi, \tag{5.8}$$

and substituting this expression in the first equation, we get

$$c\boldsymbol{\sigma} \cdot \left(\mathbf{p} - \frac{e}{c}\mathbf{A}\right)\left(2mc^2 + \epsilon - e\phi\right)^{-1} c\boldsymbol{\sigma} \cdot \left(\mathbf{p} - \frac{e}{c}\mathbf{A}\right)\chi = (\epsilon - e\phi)\chi. \tag{5.9}$$

This is an exact equation; the nonrelativistic limit is obtained by assuming that

$$(\epsilon - e\phi) \ll mc^2,$$

and expanding the operator $(2mc^2 + \epsilon - e\phi)^{-1}$ in power series

$$\left(2mc^2 + \epsilon - e\phi\right)^{-1} = \frac{1}{2mc^2}\left[1 - \frac{\epsilon - e\phi}{2mc^2} + \left(\frac{\epsilon - e\phi}{2mc^2}\right)^2 + \cdots\right]. \tag{5.10}$$

The subsequent orders of the nonrelativistic limit of the Dirac equation are obtained by considering the various terms in the square brackets in the right-hand side. If only the first term is considered, the limit at the zero order is obtained. If the first two terms are considered, the limit at the first order is obtained, and so on.

5.4 Zero Order Limit, Pauli Equation

We now consider the nonrelativistic limit at the lowest order, i.e. we neglect all terms except unity in Eq. (5.10). We substitute the result into Eqs. (5.8) and (5.9). The equations for the spinors χ and ξ then are

$$\frac{1}{2m}\left[\boldsymbol{\sigma} \cdot \left(\mathbf{p} - \frac{e}{c}\mathbf{A}\right)\right]^2 \chi = (\epsilon - e\phi)\chi, \qquad \xi = \frac{1}{2mc}\boldsymbol{\sigma} \cdot \left(\mathbf{p} - \frac{e}{c}\mathbf{A}\right)\chi.$$

This last equation shows that, since

$$\left|\mathbf{p} - \frac{e}{c}\mathbf{A}\right| = |\pi| \ll mc$$

for a nonrelativistic particle, we have, for the absolute values of the two spinors

$$|\xi| \ll |\chi|.$$

We consider now again the equation for χ and note that, if $\mathbf{a}$ and $\mathbf{b}$ are any two vectors that commute with the Pauli matrices (but not necessarily between themselves), we have, due to the property of these matrices,

$$(\boldsymbol{\sigma} \cdot \mathbf{a})(\boldsymbol{\sigma} \cdot \mathbf{b}) = \mathbf{a} \cdot \mathbf{b} + i\boldsymbol{\sigma} \cdot \mathbf{a} \times \mathbf{b}.$$

By taking this into account, the equation for χ becomes

$$\left[\frac{1}{2m}\left(\mathbf{p} - \frac{e}{c}\mathbf{A}\right)^2 + \frac{i}{2m}\boldsymbol{\sigma} \cdot \left(\mathbf{p} - \frac{e}{c}\mathbf{A}\right) \times \left(\mathbf{p} - \frac{e}{c}\mathbf{A}\right) \right]\chi = (\epsilon - e\phi)\chi.$$

The second term in square brackets (which would be zero if the two quantities $\mathbf{p}$ and $\mathbf{A}$ were not operators) can be transformed using the commutation rule

$$[p_i, A_j] = -i\hbar \frac{\partial A_j}{\partial x_i}.$$

We obtain, with simple algebra

$$\left(\mathbf{p} - \frac{e}{c}\mathbf{A}\right) \times \left(\mathbf{p} - \frac{e}{c}\mathbf{A}\right) = \frac{ie\hbar}{c}\operatorname{rot}\mathbf{A} = \frac{ie\hbar}{c}\mathbf{B},$$

where $\mathbf{B}$ is the magnetic field vector. Substituting this result we finally obtain the nonrelativistic limit of the Dirac equation at the zero order, or

$$\left[\frac{1}{2m}\left(\mathbf{p} - \frac{e}{c}\mathbf{A}\right)^2 - \frac{e\hbar}{2mc}\boldsymbol{\sigma} \cdot \mathbf{B} + e\phi \right]\chi = \epsilon\chi.$$

This equation is known as the Pauli equation and, apart from the second factor in square brackets, can be obtained directly by applying the principle of minimal coupling (Eq. (5.5)) to the Schrödinger equation for the free particle. The additional term, proportional to $\boldsymbol{\sigma} \cdot \mathbf{B}$, was introduced phenomenologically by Pauli to describe the coupling between the intrinsic magnetic moment of the electron and the magnetic field. Such term contains, in fact, the vector $\boldsymbol{\sigma}$ which, as we shall see later, is proportional to the spin of the electron. It is interesting to note that in the Dirac theory the spin appears in a natural way as a relativistic effect of zero order.

We now consider a special case of the Pauli equation, namely the case of a nonrelativistic particle moving in a constant magnetic field, as well as in an electrostatic potential ϕ. Such magnetic field has an associated vector potential of the form

$$\mathbf{A} = \frac{1}{2}\mathbf{B} \times \mathbf{x},$$

which satisfies the condition $\operatorname{div}\mathbf{A} = 0$. In fact, taking into account the vector identity (16.9), we have

$$\operatorname{rot}\mathbf{A} = -\frac{1}{2}\mathbf{B} \cdot \operatorname{grad}\mathbf{x} + \frac{1}{2}\mathbf{B}\operatorname{div}\mathbf{x} = \frac{1}{2}(-\mathbf{B} + 3\mathbf{B}) = \mathbf{B}.$$

On the other hand we have that

$$\frac{1}{2m}\left(\mathbf{p} - \frac{e}{c}\mathbf{A}\right)^2 = \frac{p^2}{2m} - \frac{e}{2mc}(\mathbf{p} \cdot \mathbf{A} + \mathbf{A} \cdot \mathbf{p}) + \frac{e^2}{2mc^2}A^2.$$

Fig. 5.1 A charged particle moving along a closed circuit encompassing the area a produces, for Ampère's equivalence principle, the magnetic moment $\boldsymbol{\mu}$. The case drawn in the figure corresponds to a positive value of the charge

This quantity can be transformed considering that

$$\mathbf{p} \cdot \mathbf{A} + \mathbf{A} \cdot \mathbf{p} = 2\mathbf{A} \cdot \mathbf{p} - i\hbar \operatorname{div} \mathbf{A} = 2\mathbf{A} \cdot \mathbf{p} = \mathbf{B} \times \mathbf{x} \cdot \mathbf{p} = \mathbf{B} \cdot \mathbf{M},$$

where $\mathbf{M} = \mathbf{x} \times \mathbf{p}$ is the orbital angular momentum of the particle, which we write in the form $\mathbf{M} = \hbar \boldsymbol{\ell}$. Also,

$$A^2 = \frac{1}{4}(\mathbf{B} \times \mathbf{x})^2 = \frac{1}{4}\big[B^2 x^2 - (\mathbf{B} \cdot \mathbf{x})^2\big] = \frac{1}{4} B^2 x_\perp^2,$$

where $x_\perp$ is the component of $\mathbf{x}$ in the plane perpendicular to $\mathbf{B}$. Substituting these intermediate results, the Pauli equation for a charged particle, moving in a uniform magnetic field and in a electric field with potential ϕ, is

$$\left[\frac{p^2}{2m} - \frac{e\hbar}{2mc}(\boldsymbol{\ell} + \boldsymbol{\sigma}) \cdot \mathbf{B} + \frac{e^2}{8mc^2} B^2 x_\perp^2 + e\phi\right]\chi = \epsilon\chi. \tag{5.11}$$

The physical meaning of the various terms appearing in this equation is immediate: $p^2/(2m)$ is the kinetic energy of the particle; $-e\hbar(\boldsymbol{\ell} \cdot \mathbf{B})/(2mc)$ is the interaction energy of the orbital angular momentum with the magnetic field; $-e\hbar(\boldsymbol{\sigma} \cdot \mathbf{B})/(2mc)$ (the term introduced phenomenologically by Pauli) is the energy of the interaction of the intrinsic angular momentum (spin) with the magnetic field; $e\phi$ is the electrostatic energy; and, finally $e^2 B^2 x_\perp^2/(8mc^2)$ is a term that is always positive, and is called diamagnetic term.

The presence of the interaction term between the orbital angular momentum and the magnetic field can also be justified with elementary physical considerations if we suppose that the particle is moving in a central field. In this case, classically, the particle is in fact subject to a periodic motion characterised by a closed orbit (see Fig. 5.1). We can in principle associate an electric current i to the particle and therefore, in accordance with Ampère principle of equivalence, a "classic" magnetic moment $\boldsymbol{\mu}_c$ given by the expression

$$\boldsymbol{\mu}_c = \frac{i}{c}\mathbf{a},$$

where $\mathbf{a}$ is a vector whose magnitude is the area of the orbit, and is directed normally to the plane of the orbit, with the convention of the right-hand rule (rule of the corkscrew). If T is the orbital period, the current is e/T, and

$$\boldsymbol{\mu}_c = \frac{e}{cT}\mathbf{a}.$$

The quantity $\mathbf{a}/T$ is the areolar velocity of the particle, which can be related to the angular momentum. In fact

$$\boldsymbol{\mu}_{\mathrm{c}} = \frac{e}{c}\frac{\mathbf{a}}{T} = \frac{e}{2c}\mathbf{x} \times \mathbf{v} = \frac{e}{2mc}\mathbf{x} \times \mathbf{p} = \frac{e\hbar}{2mc}\boldsymbol{\ell}.$$

If we recall the classic expression for the energy of a dipole in a magnetic field $(-\boldsymbol{\mu}_{\mathrm{c}} \cdot \mathbf{B})$, we have

$$E_{\mathrm{magnetic}} = -\frac{e\hbar}{2mc}\boldsymbol{\ell} \cdot \mathbf{B},$$

that is exactly the term that appears in the Pauli equation.

5.5 First Order Limit

The nonrelativistic limit of the Dirac equation at the first order is obtained by considering the first two terms in the square brackets in Eq. (5.10) and substituting that equation into Eq. (5.9). We obtain

$$\frac{1}{2m}\boldsymbol{\sigma} \cdot \left(\mathbf{p} - \frac{e}{c}\mathbf{A}\right)\left(1 - \frac{\epsilon - e\phi}{2mc^2}\right)\boldsymbol{\sigma} \cdot \left(\mathbf{p} - \frac{e}{c}\mathbf{A}\right)\chi = (\epsilon - e\phi)\chi.$$

We now exchange the order of the first two factors which appear in the left-hand side. To do this we must take into account the commutation rule

$$\left[\boldsymbol{\sigma} \cdot \left(\mathbf{p} - \frac{e}{c}\mathbf{A}\right), 1 - \frac{\epsilon - e\phi}{2mc^2}\right] = \frac{e}{2mc^2}\boldsymbol{\sigma} \cdot [\mathbf{p}, \phi] = -\frac{i\hbar e}{2mc^2}\boldsymbol{\sigma} \cdot \mathrm{grad}\,\phi.$$

Repeating calculations similar to those described in the previous section, the Dirac equation becomes

$$\left(1 - \frac{\epsilon - e\phi}{2mc^2}\right)\left[\frac{1}{2m}\left(\mathbf{p} - \frac{e}{c}\mathbf{A}\right)^2 + \frac{i}{2m}\boldsymbol{\sigma} \cdot \left(\mathbf{p} - \frac{e}{c}\mathbf{A}\right) \times \left(\mathbf{p} - \frac{e}{c}\mathbf{A}\right)\right]\chi$$
$$-\frac{i\hbar e}{4m^2c^2}\left[\mathrm{grad}\,\phi \cdot \left(\mathbf{p} - \frac{e}{c}\mathbf{A}\right) + i\boldsymbol{\sigma} \cdot \mathrm{grad}\,\phi \times \left(\mathbf{p} - \frac{e}{c}\mathbf{A}\right)\right]\chi = (\epsilon - e\phi)\chi.$$

We now further expand the calculations in the simplified case of a purely electrostatic field, for which we can assume $\mathbf{A} = 0$. Taking into account that $\mathbf{p} \times \mathbf{p} = 0$, the above equation becomes

$$\left(1 - \frac{\epsilon - e\phi}{2mc^2}\right)\frac{p^2}{2m}\chi - \frac{i\hbar e}{4m^2c^2}[\mathrm{grad}\,\phi \cdot \mathbf{p} + i\boldsymbol{\sigma} \cdot \mathrm{grad}\,\phi \times \mathbf{p}]\chi = (\epsilon - e\phi)\chi.$$

Since at the zero order we have

$$\frac{p^2}{2m}\chi = (\epsilon - e\phi)\chi,$$

the term

$$\frac{\epsilon - e\phi}{2mc^2}\frac{p^2}{2m}\chi,$$

that is already a first-order correction, can be substituted (neglecting higher orders) with any of the two expressions

$$\frac{p^4}{8m^3c^2}\chi, \quad \text{or} \quad \frac{(\epsilon - e\phi)^2}{2mc^2}\chi. \tag{5.12}$$

Furthermore, supposing that the potential in which the particle moves is a central potential ($\phi = \phi(r)$), and taking into account that

$$\text{grad}\,\phi = \left(\frac{\partial\phi}{\partial r}\right)\text{vers}(\mathbf{r}) = \frac{1}{r}\left(\frac{\partial\phi}{\partial r}\right)\mathbf{r},$$

we can apply the transformations

$$\text{grad}\,\phi \cdot \mathbf{p} = -i\hbar\frac{\partial\phi}{\partial r}\frac{\partial}{\partial r},$$

$$\text{grad}\,\phi \times \mathbf{p} = \frac{1}{r}\frac{\partial\phi}{\partial r}\mathbf{r}\times\mathbf{p} = \frac{1}{r}\frac{\partial\phi}{\partial r}\hbar\boldsymbol{\ell}.$$

Substituting, we obtain the final equation for the nonrelativistic limit at the first order, in the simplified case of a purely electrostatic potential having spherical symmetry

$$\left[\frac{p^2}{2m} + e\phi - \frac{p^4}{8m^3c^2} - \frac{e\hbar^2}{4m^2c^2}\frac{\partial\phi}{\partial r}\frac{\partial}{\partial r} + \frac{e\hbar^2}{4m^2c^2}\frac{1}{r}\frac{\partial\phi}{\partial r}\boldsymbol{\sigma}\cdot\boldsymbol{\ell}\right]\chi = \epsilon\chi. \tag{5.13}$$

The first two terms represent the usual kinetic energy and the zero-order electrostatic energy. We can now try to provide a physical interpretation to the other terms that appear in the equation. The term proportional to p^4 is a relativistic correction to the kinetic energy. We have in fact, with obvious notations

$$E_{\text{kin}} = \sqrt{c^2p^2 + m^2c^4} - mc^2 = mc^2\left(\sqrt{1 + \frac{p^2}{m^2c^2}} - 1\right) = \frac{p^2}{2m} - \frac{p^4}{8m^3c^2} + \cdots.$$

The term proportional to $\boldsymbol{\sigma}\cdot\boldsymbol{\ell}$ describes the so-called spin-orbit interaction and was introduced empirically in the Schrödinger equation even before it was given a formally correct explanation through the Dirac equation. The physical interpretation of the spin-orbit interaction is to be found in the coupling between the magnetic dipole $\boldsymbol{\mu}_{\text{c}}$ associated with the angular momentum of the particle and the magnetic dipole $\boldsymbol{\mu}_{\text{s}}$ associated with the intrinsic angular momentum (spin) of the particle. If we assume that the two dipoles are parallel, one has for the interaction energy

$$E_{\text{dipole}-\text{dipole}} = \frac{1}{r^3}\boldsymbol{\mu}_{\text{c}}\cdot\boldsymbol{\mu}_{\text{s}}, \tag{5.14}$$

where r is the distance between the two dipoles, i.e. the distance of the particle from the origin of the central potential in which it is moving. On the other hand, if we assume, as suggested by the results obtained in the previous section, that

$$\boldsymbol{\mu}_{\text{c}} = \frac{e\hbar}{2mc}\boldsymbol{\ell}, \qquad \boldsymbol{\mu}_{\text{s}} = \frac{e\hbar}{2mc}\boldsymbol{\sigma},$$

we have

$$E_{\text{dipole-dipole}} = \frac{e^2\hbar^2}{4m^2c^2} \frac{1}{r^3} \boldsymbol{\sigma} \cdot \boldsymbol{\ell}.$$

This quantity coincides exactly with that appearing in the Dirac equation if one chooses for ϕ the Coulomb potential due to a charge $-e$ (opposite to that of the particle), as in the case of the hydrogen atom. In this case we have in fact $\phi = -e/r$, and the term in the Dirac equation becomes

$$\frac{e\hbar^2}{4m^2c^2} \frac{1}{r} \frac{\partial\phi}{\partial r} \boldsymbol{\sigma} \cdot \boldsymbol{\ell} = \frac{e^2\hbar^2}{4m^2c^2} \frac{1}{r^3} \boldsymbol{\sigma} \cdot \boldsymbol{\ell} = E_{\text{dipole-dipole}}.$$

The additional term that appears in the Dirac equation, i.e. the term proportional to $\text{grad}\,\phi \cdot \mathbf{p}$, is known as the Darwin term and has no classical analogue. From the point of view of the calculation, its expectation value on an arbitrary wave function can be transformed in the form of a contact term. If ψ is the wave function (supposed here scalar and real), the expectation value of the Darwin term (written in its general form, without supposing to have a spherically symmetric potential) is

$$E_{\text{Darwin}} = \frac{e\hbar^2}{4m^2c^2} \int \psi \,\text{grad}\,\phi \cdot \text{grad}\,\psi \,dV.$$

This integral can be transformed taking into account that

$$\psi \,\text{grad}\,\phi \cdot \text{grad}\,\psi = \frac{1}{2} \text{grad}\,\psi^2 \cdot \text{grad}\,\phi = \frac{1}{2}\left[\text{div}\left(\psi^2 \,\text{grad}\,\phi\right) - \psi^2 \nabla^2\phi\right].$$

If we assume that the wave function and the electric field become null at infinity, for Gauss's theorem the first term in square bracket gives a zero contribution to the integral and we have

$$E_{\text{Darwin}} = -\frac{e\hbar^2}{8m^2c^2} \int \psi^2 \nabla^2\phi \,dV.$$

This expression shows that, for the calculation of its expectation value, the contributions to the Darwin term are only the points where there are electric charges that generate the potential ϕ. For example, if we assume that ϕ is due to a single charge q located at the point $\mathbf{x}_0$, recalling that

$$\nabla^2\phi = -4\pi q\delta(\mathbf{x} - \mathbf{x}_0),$$

we have

$$E_{\text{Darwin}} = \frac{4\pi eq\hbar^2}{8m^2c^2} \psi^2(\mathbf{x}_0). \tag{5.15}$$

5.6 The Dirac Equation Describes a Particle of Spin 1/2

In the previous sections we have already introduced the concept of spin, though in a way that was not completely accurate. We now want to prove in a rigorous way that a particle described by the Dirac equation has an intrinsic angular momentum equal to $1/2$ in units of $\hbar$. To do this, we note that if an observable is a constant of motion, it must have the property that its associated operator commutes with the Hamiltonian. Since the total angular momentum $\mathbf{j}$ of a free particle is preserved in time, we must have

$$[\mathcal{H}, \mathbf{j}] = 0.$$

We now consider the commutator of the Dirac Hamiltonian with the orbital angular momentum of the particle $\boldsymbol{\ell} = \mathbf{x} \times \mathbf{p}/\hbar$. Recalling Eq. (5.1), we have

$$[\mathcal{H}, \boldsymbol{\ell}] = \frac{1}{\hbar}[c\boldsymbol{\alpha} \cdot \mathbf{p} + \beta mc^2, \mathbf{x} \times \mathbf{p}],$$

and for the i-th component of the commutator we get

$$[\mathcal{H}, \ell_i] = \frac{1}{\hbar} \sum_{kl} \epsilon_{ikl} \left[\sum_j c\alpha_j p_j + \beta mc^2, x_k p_l \right]$$

$$= \frac{c}{\hbar} \sum_{jkl} \epsilon_{ikl} \alpha_j [p_j, x_k] p_l = -ic \sum_{kl} \epsilon_{ikl} \alpha_k p_l.$$

We then have, in intrinsic form

$$[\mathcal{H}, \boldsymbol{\ell}] = -ic\boldsymbol{\alpha} \times \mathbf{p}.$$

We can deduce that the orbital angular momentum of the particle is not a constant of motion since it does not commute with the Hamiltonian.

Let us now consider the matrices of order four, $\boldsymbol{\tau}$, defined by

$$\boldsymbol{\tau} = -\frac{i}{2} \boldsymbol{\alpha} \times \boldsymbol{\alpha}, \tag{5.16}$$

where α_i are the Dirac matrices. Recalling their definition in terms of 2×2 matrices (Eq. (5.3)), we have

$$\tau_i = -\frac{i}{2} \sum_{jk} \epsilon_{ijk} \begin{pmatrix} 0 & \sigma_j \\ \sigma_j & 0 \end{pmatrix} \begin{pmatrix} 0 & \sigma_k \\ \sigma_k & 0 \end{pmatrix} = -\frac{i}{2} \sum_{jk} \epsilon_{ijk} \begin{pmatrix} \sigma_j\sigma_k & 0 \\ 0 & \sigma_j\sigma_k \end{pmatrix}.$$

On the other hand, for the property of the Pauli matrices the following relation holds

$$\sum_{jk} \epsilon_{ijk} \sigma_j \sigma_k = 2i\sigma_i,$$

so we get

$$\tau_i = \begin{pmatrix} \sigma_i & 0 \\ 0 & \sigma_i \end{pmatrix}.$$

The matrices $\boldsymbol{\tau}$ are therefore a generalisation to order 4 of the 2×2 Pauli matrices. It is simple to show that they satisfy the same relations, i.e.

$$[\tau_i, \tau_j] = 2i \sum_k \epsilon_{ijk} \tau_k, \qquad \{\tau_i, \tau_j\} = 2\delta_{ij}, \qquad \tau_i \tau_j = \delta_{ij} + i \sum_k \epsilon_{ijk} \tau_k.$$

Considering the commutator $[\mathcal{H}, \boldsymbol{\tau}]$ we have

$$[\mathcal{H}, \tau_i] = -\frac{i}{2} \sum_{kl} \epsilon_{ikl} \left[\sum_j c\alpha_j p_j + \beta mc^2, \alpha_k \alpha_l \right]$$

$$= -\frac{i}{2} \sum_{kl} \epsilon_{ikl} \left(\sum_j cp_j [\alpha_j, \alpha_k \alpha_l] + mc^2 [\beta, \alpha_k \alpha_l] \right).$$

Evaluating the two commutators appearing in the right-hand side we get

$$[\alpha_j, \alpha_k \alpha_l] = 2\delta_{jk} \alpha_l - 2\delta_{jl} \alpha_k, \qquad [\beta, \alpha_k \alpha_l] = 0,$$

so, with simple algebra we finally obtain

$$[\mathcal{H}, \boldsymbol{\tau}] = 2ic\boldsymbol{\alpha} \times \mathbf{p}.$$

Introducing the vector $\mathbf{j}$ through the relation

$$\mathbf{j} = \boldsymbol{\ell} + \frac{1}{2}\boldsymbol{\tau},$$

we obtain

$$[\mathcal{H}, \mathbf{j}] = 0.$$

The vector $\mathbf{j}$ is a constant of motion and is therefore natural to identify it with the total angular momentum vector. The previous relation shows that it is obtained by summing the orbital angular momentum with a vector, that we indicate with $\mathbf{s}$, and that represents the intrinsic angular momentum (spin) of the particle

$$\mathbf{s} = \frac{1}{2}\boldsymbol{\tau}.$$

The vector $\mathbf{s}$ satisfies the relations (immediately derived from those of the vector $\boldsymbol{\tau}$)

$$[s_i, s_j] = i \sum_k \epsilon_{ijk} s_k, \qquad \{s_i, s_j\} = \frac{1}{2}\delta_{ij}, \qquad s_i s_j = \frac{1}{4}\delta_{ij} + \frac{i}{2} \sum_k \epsilon_{ijk} s_k.$$

These properties show that the vector $\mathbf{s}$ satisfies all the requirements to be considered, quite rightly, an angular momentum (see Sect. 7.9). Moreover, being

$$s^2 = \sum_i s_i s_i = \frac{3}{4},$$

we obtain that the intrinsic angular momentum is $\frac{1}{2}$ (recall that, within the angular momentum theory, the eigenvalue of the operator J^2 is $J(J+1)$).

In the nonrelativistic limit of the Dirac equation discussed in the previous sections, and in particular at the zero order, we have seen that the main contribution to the interaction between a charge particle and a magnetic field is described by a term in the Hamiltonian given by

$$\mathcal{H}_{\text{magnetic}} = -\frac{e\hbar}{2mc}(\boldsymbol{\ell} + \boldsymbol{\sigma}) \cdot \mathbf{B}.$$

This expression is consistent with the idea that a magnetic dipole $\boldsymbol{\mu}_\ell$ is associated to the orbital angular momentum, and a magnetic moment $\boldsymbol{\mu}_s$ is associated to the spin. In the case of the electron, for which $e = -e_0$, with $e_0 = 4.803 \times 10^{-10}$ esu, the dipoles are given by

$$\boldsymbol{\mu}_\ell = -\mu_0 \boldsymbol{\ell}, \qquad \boldsymbol{\mu}_s = -\mu_0 \boldsymbol{\sigma},$$

where we have introduced the quantity μ_0, known as the Bohr magneton, defined by[2]

$$\mu_0 = \frac{e_0 \hbar}{2mc} = 9.274 \times 10^{-21} \text{ erg G}^{-1}. \tag{5.17}$$

The momentum $\boldsymbol{\mu}_\ell$ has already been interpreted in classical terms (see Sect. 5.4). Concerning $\boldsymbol{\mu}_s$, we note that, since $\boldsymbol{\sigma}$ is the representation of order 2 of the matrix $\boldsymbol{\tau}$ (of order 4), and being $\boldsymbol{\tau} = 2\mathbf{s}$, we can rewrite it as

$$\boldsymbol{\mu}_s = -2\mu_0 \mathbf{s}.$$

The ratio between the magnetic moment (in units of Bohr magnetons) and the corresponding angular momentum (in units of $\hbar$) is called the gyromagnetic ratio. So we have obtained that the gyromagnetic ratio of the electron is -1 for the orbital angular momentum and is -2 for the spin. This anomalous behavior of the magnetic moment associated with the spin was assumed in the phenomenological theory of the electron given by Pauli. The fact that this results in a natural way from the Dirac equation is one of the great successes of this theory.

5.7 Solution of the Dirac Equation in a Magnetic Field

The Dirac equation can be solved exactly in a number of particular cases. As an application to the theory developed in this chapter, we now seek for the solution relative to a charge moving in a constant magnetic field. Such field can be described in terms of the electromagnetic potentials

$$\phi = 0, \qquad \mathbf{A} = \frac{1}{2}\mathbf{B} \times \mathbf{x},$$

and the Dirac equation becomes

$$i\hbar \frac{\partial}{\partial t} \psi(\mathbf{x}, t) = H_B \psi(\mathbf{x}, t),$$

where H_B, the Dirac Hamiltonian in the case of a constant magnetic field, is given by

$$H_B = c\boldsymbol{\alpha} \cdot \left(\mathbf{p} - \frac{e}{c}\mathbf{A} \right) + \beta mc^2.$$

---

[2]Recall that in c.g.s. units, the unit of measure for the magnetic induction is the gauss, shortened with "G".

To solve this equation, we start by noting that the operator H_B^2 has a much simpler expression than the operator H_B. We have, in fact

$$H_B^2 = \left[c\boldsymbol{\alpha} \cdot \left(\mathbf{p} - \frac{e}{c}\mathbf{A} \right) + \beta mc^2 \right] \left[c\boldsymbol{\alpha} \cdot \left(\mathbf{p} - \frac{e}{c}\mathbf{A} \right) + \beta mc^2 \right].$$

The expression in the right-hand side can be expanded taking into account the algebra of the matrices α and β (Eq. (5.2)). Recalling the definition of the matrices τ (Eq. (5.16)) and the commutation rule

$$[p_i, A_j] = -i\hbar \frac{\partial A_j}{\partial x_i},$$

we find, with some transformations,

$$H_B^2 = c^2 \left(\mathbf{p} - \frac{e}{c}\mathbf{A} \right)^2 - e\hbar c \boldsymbol{\tau} \cdot \mathbf{B} + m^2 c^4.$$

Now suppose we have solved the eigenvalue equation for the operator H_B^2 and denote by ϵ^2 the eigenvalues (which must necessarily be real and positive) and by Φ the eigenvectors. By definition we have

$$\left(H_B^2 - \epsilon^2 \right)\Phi = 0,$$

and the equation can be written in the two alternative forms

$$(H_B - \epsilon)(H_B + \epsilon)\Phi = 0, \qquad (H_B + \epsilon)(H_B - \epsilon)\Phi = 0.$$

These two equations show that the two wave functions

$$\Psi_+ = (H_B + \epsilon)\Phi, \qquad \Psi_- = (H_B - \epsilon)\Phi,$$

are eigenfunctions of the Hamiltonian H_B, corresponding to the eigenvalues ϵ and $-\epsilon$, respectively. By means of this algorithm it is then possible to relate the solution of the Dirac equation to the solution of the eigenvalue equation for the operator H_B^2.

Taking into account the explicit expression for the vector potential, and introducing a system of Cartesian coordinates (x, y, z) with the z axis directed along the direction of the magnetic field, the operator H_B^2 takes the form

$$H_B^2 = c^2 \left(p_x + \frac{eB}{2c}y \right)^2 + c^2 \left(p_y - \frac{eB}{2c}x \right)^2 + c^2 p_z^2 - e\hbar c B\tau_3 + m^2 c^4.$$

The spinorial character of the operator is included only in the diagonal matrix τ_3, explicitly given by

$$\tau_3 = \begin{pmatrix} 1 & 0 & 0 & 0 \\ 0 & -1 & 0 & 0 \\ 0 & 0 & 1 & 0 \\ 0 & 0 & 0 & -1 \end{pmatrix}.$$

Furthermore, all of the commutators that can be constructed from any two out of the five terms of the operator H_B^2 are null, with the exception of the commutator between the first two terms, which is different from zero. In order to transform the operator H_B^2 in a sum of commuting operators, we perform a change of variable introducing the operators

$$a = \gamma\left(p_y - \frac{eB}{2c}x\right) - i\delta\left(p_x + \frac{eB}{2c}y\right),$$

$$a^\dagger = \gamma\left(p_y - \frac{eB}{2c}x\right) + i\delta\left(p_x + \frac{eB}{2c}y\right),$$

where γ and δ are two real constants to be carefully chosen. With these transformations, we obtain the expression for H_B^2

$$H_B^2 = \frac{c^2}{4\gamma^2}\left(a^2 + a^{\dagger 2} + aa^\dagger + a^\dagger a\right) - \frac{c^2}{4\delta^2}\left(a^2 + a^{\dagger 2} - aa^\dagger - a^\dagger a\right)$$
$$+ c^2 p_z^2 - e\hbar c B\tau_3 + m^2 c^4,$$

and, for the commutator between the operators a and $a^\dagger$,

$$[a, a^\dagger] = \frac{2eB\hbar}{c}\gamma\delta.$$

If we now impose that

$$\gamma^2 = \delta^2, \qquad \gamma\delta = \frac{c}{2eB\hbar},$$

the operator H_B^2 becomes the sum of commuting operators for which the eigenvalue problem has already been solved. With simple algebra, we have

$$H_B^2 = 2|e|\hbar c B\left(a^\dagger a + \frac{1 \mp \tau_3}{2}\right) + c^2 p_z^2 + m^2 c^4,$$

where $|e|$ is the absolute value of the charge of the particle and where the sign in front of τ_3 is minus for positive charges and plus for negative charges. Recalling the results about the harmonic oscillator of Sect. 4.1, and taking into account that the eigenvalues of τ_3 are ± 1, the eigenvalues of the operator H_B^2 are of the form

$$\epsilon^2 = m^2 c^4 + c^2 q_z^2 + 2|e|\hbar c Bn,$$

where q_z is the (continuous) eigenvalue of the operator p_z and where n is an arbitrary integer that is positive or null. Finally, for the eigenvalues of the Hamiltonian H_B, we have

$$\epsilon = \pm\sqrt{m^2 c^4 + c^2 q_z^2 + 2|e|\hbar c Bn} \quad (n = 0, 1, 2, \ldots).$$

The energy levels that we have found are called Landau levels. They are characterised by a continuous parameter q_z (the component of the momentum of the particle along the direction of the magnetic field) and by an integer index, n (that instead characterises the motion in the plane perpendicular to the magnetic field). As in the case of the Dirac equation for the free particle, there are levels with positive energy and levels with negative energy.

The equation that expresses the eigenvalues of the energy can also be written in a more meaningful form

$$\epsilon = \pm mc^2 \sqrt{1 + \frac{q_z^2}{m^2 c^2} + 2\frac{B}{B_q} n},$$

where we have introduced the "quantum magnetic field" B_q defined by

$$B_q = \frac{m^2 c^3}{|e|\hbar},$$

that, in the case of electrons, is 4.414×10^{13} G. In the particular case where $q_z \ll mc$ and $B \ll B_q$, the square root can be expanded in a power series and at the lowest order we obtain

$$\epsilon = \pm \left(mc^2 + \frac{q_z^2}{2m} + \hbar \omega_c n \right),$$

where the quantity ω_c, defined by

$$\omega_c = \frac{|e|B}{mc},$$

is the cyclotron frequency that we have already encountered in Chap. 3 (Eq. (3.30)).

Chapter 6
Atoms with a Single Valence Electron

The spectroscopic analysis of the radiation emitted by atomic and molecular sub-
stances in the most varied conditions of pressure and temperature has been carried
on for a long time, starting from the discovery of absorption lines in the solar spectrum by Fraunhofer in 1817. These studies have resulted, over the years, in the
emergence of a new discipline of experimental and theoretical physics that is called
spectroscopy. This discipline had a fundamental historical importance for the understanding of the atomic structure, although it was not until the advent of quantum
mechanics that a rigorous interpretation of observations in the laboratory and astrophysical plasmas could be given. In this volume we present the basic concepts of
spectroscopy using a modern approach, starting by giving a description (at increasing levels of sophistication) of the simpler spectra, namely those related to atoms
containing only one valence electron (hydrogenic atoms, alkali metals and related
isoelectronic sequences). The complications introduced by the presence of more
valence electrons are described in later chapters.

6.1 Hydrogen Atom, Bohr Theory

The spectrum of the hydrogen atom is the simplest and the first one for which an
adequate theoretical interpretation was obtained. The regularities appearing in the
observed wavelengths of the hydrogen lines of the visible spectrum were quantitatively formulated by Balmer, a Swiss secondary school teacher, in 1866. Balmer
discovered that the wavelengths of these lines could be accurately expressed by the
empirical formula

$$\lambda = \lambda_B \frac{n^2}{n^2 - 4},$$

where λ_B is a constant (equivalent to about 3647 Å), and n is an integer ($n = 3, 4,$
5, etc.). Today, this expression is normally written in the form

$$\bar{\nu} = R_H \left(\frac{1}{2^2} - \frac{1}{n^2} \right) \quad (n > 2),$$

E. Landi Degl'Innocenti, *Atomic Spectroscopy and Radiative Processes*,
UNITEXT for Physics, DOI 10.1007/978-88-470-2808-1_6, © Springer-Verlag Italia 2014

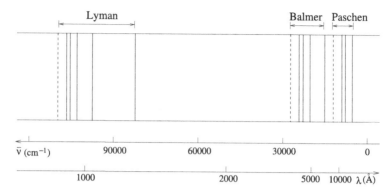

Fig. 6.1 Spectrum of the hydrogen atom, from the ultraviolet to the infrared. The first spectral lines of the Lyman, Balmer, and Paschen series are indicated by vertical lines. The limits of the series are indicated by dashed lines. The horizontal scale is linear in the wavenumber

where $\bar{\nu}(=1/\lambda)$ is the wavenumber of the spectral line and R_H $(=4/\lambda_B)$ is the so-called Rydberg constant. If we substitute the integer 2 in the previous formula with other integers, we obtain other series of lines. The complete spectrum of the hydrogen atom (cf. Fig. 6.1 for a schematic diagram) can be obtained with the formula

$$\bar{\nu} = R_H \left(\frac{1}{m^2} - \frac{1}{n^2} \right) \quad (n > m). \tag{6.1}$$

The $m = 1$ series falls in the ultraviolet and is called the Lyman series. The $m = 2$ series falls in the visible and is obviously called the Balmer series. The other ones fall in the infrared at progressively longer wavelengths and are called the Paschen ($m = 3$), Brackett ($m = 4$), Pfund ($m = 5$), and Humphreys ($m = 6$) series.

The expression for $\bar{\nu}$ can be rewritten as

$$\bar{\nu} = T_m - T_n,$$

where

$$T_k = \frac{R_H}{k^2} \quad \text{(with } k \text{ integer)}. \tag{6.2}$$

This means that the wavenumber of any hydrogen spectral line can be obtained from the difference between two "spectroscopic terms" of the form R_H/k^2. It turns out that this property is very general, in that it applies to the spectra of all the other elements. Experiments have in fact verified that all the wavenumbers of the multitude of spectral lines of a given atom (or ion) can always be obtained from differences of a much smaller number of terms (although, in general, these terms cannot be expressed by simple expressions such as Eq. (6.2)). This fact was historically quite important, and is called the Rydberg-Ritz principle. It states that *all the spectral lines of an element (in any ionisation state) can be obtained from the difference between any two spectroscopic terms, characteristic of the element (or ion). The number of terms is much smaller than the number of possible spectral lines.*

The first theoretical interpretation of the hydrogen spectrum was given by Bohr with a relatively simple model, where he combined ad-hoc quantisation hypotheses with the known laws of classical mechanics. From the modern perspective of quantum mechanics, Bohr's theory is obsolete, however it constitutes a very good introduction to atomic physics and is therefore worth describing it here.

The basic starting point of Bohr's theory is the so-called planetary model of the atom, as it emerged from the pioneering experiments carried out by Rutherford. He showed that an atom is composed of an extremely small (almost point-like) positively-charged nucleus and a cloud of electrons orbiting around it. Bohr considered the simplest atom, i.e. the hydrogenic one, where a single electron is orbiting a central nucleus of charge Ze_0 (e_0 is the absolute value of the electron electric charge and Z is an integer). Bohr started with the following assumptions: (a) of the infinite orbits that, according to classical physics, an electron can describe in its motion around the nucleus, only few ones, verifying some suitable quantization rules, are allowed; in stark contrast to the theory of classical electromagnetism, the electron does not irradiate as it moves along these orbits notwithstanding its accelerated motion. (b) A "quantum" of radiation is either emitted or absorbed following a "transition" of the electron between two allowed orbits, at a frequency

$$\nu = \frac{\Delta E}{h},$$

where ΔE is the difference between the energies of the two orbits, and h is Planck's constant ($h = 6.626 \times 10^{-27}$ erg s). It is interesting to notice that the Rydberg-Ritz principle is implicitly contained in this second hypothesis, the spectroscopic terms being given by the energies of the orbits (apart from a $1/(ch)$ factor).

If we assume for simplicity that the electron follows a circular orbit, from Newton's second law we obtain

$$\frac{Ze_0^2}{r^2} = \frac{mv^2}{r}, \tag{6.3}$$

where r is the radius of the orbit, v is the electron velocity and m its mass. Bohr added to this classic expression the quantisation condition[1] that the angular momentum of the electron must be an integer multiple of the constant $\hbar = h/2\pi$

$$mvr = n\hbar \quad (n = 1, 2, 3, \ldots).$$

By eliminating the velocity in the two equations we obtain for the radius of the orbit characterized by the "quantum number" n

$$r_n = \frac{\hbar^2}{me_0^2} \frac{n^2}{Z}.$$

---

[1] In reality, the original treatment that Bohr introduced is different, being based on the correspondence principle. However, the present is a modern adaptation that does not alter the spirit of Bohr's reasoning.

The quantity $\hbar^2/(me_0^2)$, has the dimension of a length and is called the radius of the first Bohr orbit. It is usually indicated as a_0 and is equivalent to 0.529×10^{-8} cm, i.e. 0.529 Å. The previous expression is therefore usually written as

$$r_n = a_0 \frac{n^2}{Z}.$$

The energy of the electron on the orbit of radius r is

$$E = \frac{1}{2}mv^2 - \frac{Ze_0^2}{r} = -\frac{Ze_0^2}{2r},$$

where we have used Eq. (6.3) and we have assumed a zero electrostatic energy at infinity. Substituting the r_n value, we obtain that the energy of the electron on the orbit characterized by the quantum number n is

$$E_n = -\frac{e_0^2}{2a_0} \frac{Z^2}{n^2} = -\frac{me_0^4}{2\hbar^2} \frac{Z^2}{n^2}.$$

The velocity v_n of the electron on the n-th orbit can be obtained from the quantisation rule on the angular momentum. We have

$$v_n = \frac{e_0^2}{\hbar} \frac{Z}{n},$$

and, introducing the dimensionless quantity α, defined as the fine-structure constant,

$$\alpha = \frac{e_0^2}{\hbar c} = \frac{1}{137.036} = 7.29735 \times 10^{-3},$$

we obtain

$$v_n = \alpha c \frac{Z}{n},$$

which shows that for the hydrogen atom ($Z = 1$) we should expect relativistic corrections of the order of α^2. We can obtain the orbital period of the electron using the expressions for r_n and v_n. We have

$$T_n = \frac{2\pi r_n}{v_n} = \frac{2\pi \hbar^3}{me_0^4} \frac{n^3}{Z^2},$$

i.e. we obtain the analog of the third Kepler law

$$\frac{r_n^3}{T_n^2} = \frac{Ze_0^2}{4\pi^2 m}, \quad \text{independent of } n.$$

Finally, using the definition of the fine-structure constant, the energy E_n can also be written as

$$E_n = -mc^2 \frac{\alpha^2 Z^2}{2n^2}.$$

These results assume that the nucleus has infinite mass. To take into account its finite mass, we need to substitute the electron mass m with the reduced mass m_r

$$m_r = \frac{mM_n}{m + M_n},$$

where M_n is the mass of the nucleus. Considering that we always have $m \ll M_n$ we can expand in series the previous equation, obtaining

$$m_r \simeq m\left(1 - \frac{m}{M_n}\right).$$

This shows that the correction for the reduced mass is of the order of 0.05 % for the hydrogen atom. The formal proof of the reduced mass correction is based on the two-body theorem that we are going to recall here.

Given two bodies of mass m_1 and m_2, if the first exerts the force $\mathbf{F}$ on the second, according to Newton's third law, the second exerts the force $-\mathbf{F}$ on the first, so that, using Newton's second law, the motion of the respective centres of mass follows

$$m_2\ddot{\mathbf{x}}_2 = \mathbf{F}, \qquad m_1\ddot{\mathbf{x}}_1 = -\mathbf{F}.$$

For the relative motion, described by the unit vector $\mathbf{x} = \mathbf{x}_2 - \mathbf{x}_1$, we therefore have

$$\ddot{\mathbf{x}} = \ddot{\mathbf{x}}_2 - \ddot{\mathbf{x}}_1 = \left(\frac{1}{m_2} + \frac{1}{m_1}\right)\mathbf{F},$$

that is

$$m_r\ddot{\mathbf{x}} = \mathbf{F},$$

where

$$m_r = \frac{m_1 m_2}{m_1 + m_2}.$$

This last equation shows that the motion of the second body (in our case the electron) relative the first one (the nucleus) is the same as the absolute motion of a body experiencing the same force and of mass equal to the reduced mass. Considering the correction due to the reduced mass, the energy of the hydrogenic atom in the n-th orbit is

$$E_n = -\frac{m_r e_0^4}{2\hbar^2} \frac{Z^2}{n^2}. \tag{6.4}$$

According to Bohr's second assumption, the wavenumber of the quantum emitted in the transition between the n-th and the m-th orbits ($n > m$) is

$$\bar{\nu} = \frac{E_n - E_m}{hc} = RZ^2\left(\frac{1}{m^2} - \frac{1}{n^2}\right),$$

where R, the Rydberg's constant for the hydrogenic atom, is given by

$$R = \frac{m_r e_0^4}{4\pi c\hbar^3},$$

and in particular, for the hydrogen atom,

$$R_H = \frac{mM_p}{m + M_p} \frac{e_0^4}{4\pi c\hbar^3},$$

M_p being the proton mass.

As can be seen, Bohr's model leads to an expression for the wavenumbers of the spectral lines of the hydrogen atom which coincides with that observed (Eq. (6.1)). More quantitatively, we can compare the numerical values, theoretical and experimental, obtained for R_H. By substituting the values of the atomic constants, good agreement is obtained. This fact was historically one of the most convincing proofs of Bohr's theory.[2]

As we have noted, the formulae derived in this section are applicable not only to the spectrum of the hydrogen atom but also to the spectra of hydrogenic (or hydrogen-like) atoms, i.e. atoms consisting of a single electron orbiting around a nucleus with charge Ze_0, with $Z > 1$. Such spectra are those of singly-ionised helium, He^+ ($Z = 2$, spectrum of He II); of lithium two times ionised: Li^{++} ($Z = 3$, spectrum of Li III), of beryllium three times ionised, etc.[3] The spectra of hydrogen-like atoms are entirely similar to the spectrum of hydrogen, with the difference of a $1/Z$ scale factor in the size of the orbits and of a Z^2 factor in the energies, i.e. in the wavenumbers or frequencies (in addition, there is a further difference, of the order of a fraction of $1/1000$, due to the effect of the reduced mass on the Rydberg constant). The Balmer series of ionised helium, for example, is found in the ultraviolet rather than in the visible.

The quantisation rules introduced by Bohr only apply to bound orbits, i.e. to those (elliptical, in classical physics) corresponding to negative energies. The orbits that correspond to positive energies (hyperbolic, in classical physics) are thus all "possible". In addition to transitions between orbits of negative energy, transitions between orbits having positive and negative energy are also possible. Also, transitions between orbits having positive energy, are possible. In the first case, the wavenumber of the quantum of energy is given by

$$\bar{\nu} = \frac{\epsilon}{hc} + \frac{R}{n^2},$$

where n is the quantum number of the orbit having negative energy, and ϵ is the kinetic energy of the electron at infinity on the hyperbolic orbit. ϵ can have any positive or zero value. Therefore, the series of lines becomes a continuous spectrum at wavenumbers longer than the limit (R/n^2). For the hydrogen spectrum, for example, we have the Lyman continuum for $\lambda < 912$ Å, the Balmer continuum for $\lambda < 3647$ Å, etc. This type of transition corresponds, in absorption, to the expulsion of an electron from the atom (photoelectric effect or photoionisation), while in

---

[2]It must be said that at the time when Bohr published his results, the physical constants were known with poor precision and the mere coincidence between the theoretical and experimental value for R_H was not by itself sufficient evidence to convince the scientific community of the validity of his model. Today the situation has changed radically and the Rydberg constant is one of the best known physical constants with a large number of significant digits ($R_H = 1.0967758341 \times 10^5$ cm^{-1}).

[3]The spectrum of an element n-times ionised is indicated by the symbol of the element followed by the number ($n + 1$) written in Roman numerals. For example, the spectrum of Na I is the spectrum of neutral sodium. The spectrum of C IV is the spectrum of carbon three times ionised. That of Fe XV is the spectrum of iron ionised 14 times, and so on.

Fig. 6.2 Grotrian diagram of the hydrogen atom. The levels of the continuum are represented by the *shaded area* that, in principle, extends indefinitely

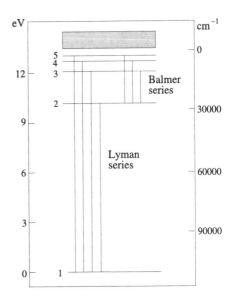

emission it corresponds to the reverse process, electronic recombination. The observation of the continuum limit associated with the ground level is very important, as it provides directly the ionisation potential of the atom (or ion). If we indicate with $\bar{\nu}_\infty$ the wavenumber corresponding to the limit of this series, the ionisation potential (in eV) can be obtained simply by the formula

$$\chi\,(\mathrm{eV}) = 1.2398 \times 10^{-4}\,\bar{\nu}_\infty\,\left(\mathrm{cm}^{-1}\right).$$

For the hydrogen atom, for example, given that $\bar{\nu}_\infty = R_\mathrm{H}$ we obtain a ionisation potential of 13.598 eV.

A transition between two orbits having positive energy is characterized by the wavenumber

$$\bar{\nu} = \frac{\epsilon - \epsilon'}{hc},$$

where ϵ and ϵ' are the kinetic energies at infinity of the electron on the initial and final hyperbolic orbits. This transition corresponds, in emission, to *Bremsstrahlung* (which means "braking radiation" in German), and in absorption to inverse *Bremsstrahlung*.[4]

A particularly useful graphical representation of spectroscopic terms and spectral lines is obtained by plotting in a diagram the energy levels by means of horizontal lines on a vertical scale of energy (or wave number). Figure 6.2 gives an example of such a diagram (called Grotrian diagram) for the hydrogen atom. The energy of a level can be obtained by reading the scale on the left, and the zero value is conventionally assigned to the ground level. The wavenumber of the corresponding

---

[4]See Sect. 3.7 for the classic theory of *Bremsstrahlung*.

Fig. 6.3 System of spherical
coordinates (r, θ, ϕ) and
corresponding unit vectors $\mathbf{e}_r$,
$\mathbf{e}_\theta$, $\mathbf{e}_\phi$

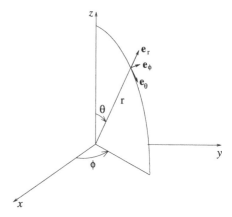

term can be read instead on the right scale, where, again by convention, the zero
corresponds to the ionisation limit. Any spectral line is represented by a vertical line
that connects two energy levels. The length of the line, measured on the right-hand
scale, gives directly the wavenumber of the spectral line and thus its wavelength.

Finally, we note that Bohr's theory, deduced for circular orbits, was later generalised by Sommerfeld to the case of elliptical orbits. We will not discuss such a
theory here, given that, nowadays, it is interesting almost exclusively from an historical point of view.

6.2 Schrödinger's Equation in Spherical Coordinates

The atomic structure, i.e. the energies of the atomic levels and their corresponding spectra are nowadays determined by solving the time-independent Schrödinger
equation. For a particle of mass m moving in a force field where a potential can be
defined, the equation is the following

$$\mathcal{H}|\psi\rangle = \left[\frac{p^2}{2m} + V\right]|\psi\rangle = E|\psi\rangle,$$

where V is the potential energy of the particle. In the wave function representation,
where the operator $\mathbf{p}$ is $-i\hbar\,\mathrm{grad}$, the equation, for a time-independent potential,
becomes

$$\left[-\frac{\hbar^2}{2m}\nabla^2 + V(\mathbf{x})\right]\psi(\mathbf{x}) = E\psi(\mathbf{x}).$$

If the field has spherical symmetry, the solution is simplified by introducing the
spherical coordinates r, θ, ϕ, defined implicitly by the equations (see Fig. 6.3)

$$x = r\sin\theta\cos\phi, \qquad y = r\sin\theta\sin\phi, \qquad z = r\cos\theta,$$

where (x, y, z) is a right-handed orthogonal Cartesian system.

We now need to express the Laplace operator ∇^2 in spherical coordinates. Given that

$$\nabla^2 = \text{div grad},$$

we have to write in spherical coordinates both the gradient and the divergence operators. We first introduce the three unit vectors $\mathbf{e}_r$, $\mathbf{e}_\theta$ and $\mathbf{e}_\phi$ which form a right-handed triad, as shown in Fig. 6.3. These vectors can be expressed as a linear combination of the three unit vectors $\mathbf{i}, \mathbf{j}, \mathbf{k}$, aligned along the x, y, z axes, by the equations

$$\mathbf{e}_r = \sin\theta\cos\phi\,\mathbf{i} + \sin\theta\sin\phi\,\mathbf{j} + \cos\theta\,\mathbf{k},$$

$$\mathbf{e}_\theta = \cos\theta\cos\phi\,\mathbf{i} + \cos\theta\sin\phi\,\mathbf{j} - \sin\theta\,\mathbf{k},$$

$$\mathbf{e}_\phi = -\sin\phi\,\mathbf{i} + \cos\phi\,\mathbf{j}.$$

Through them, we can write the infinitesimal distance dP between two points having spherical coordinates (r, θ, ϕ) and $(r + dr, \theta + d\theta, \phi + d\phi)$ as

$$dP = \mathbf{e}_r\,dr + \mathbf{e}_\theta r\,d\theta + \mathbf{e}_\phi r\sin\theta\,d\phi.$$

Given an arbitrary scalar function $f(\mathbf{r})$, we have, from the definition of the gradient, that

$$df = \text{grad } f \cdot dP = (\text{grad } f)_r\,dr + (\text{grad } f)_\theta r\,d\theta + (\text{grad } f)_\phi r\sin\theta\,d\phi,$$

where we have indicated with the symbols $(\text{grad } f)_{r,\theta,\phi}$ the three components of the gradient of the function f along the three unit vectors $\mathbf{e}_r, \mathbf{e}_\theta, \mathbf{e}_\phi$. On the other hand, given that we also have

$$df = \frac{\partial f}{\partial r}\,dr + \frac{\partial f}{\partial \theta}\,d\theta + \frac{\partial f}{\partial \phi}\,d\phi,$$

by comparing these last two equations, and since f is an arbitrary function, we obtain the expressions for the spherical components of the gradient

$$\text{grad}_r = \frac{\partial}{\partial r}, \qquad \text{grad}_\theta = \frac{1}{r}\frac{\partial}{\partial \theta}, \qquad \text{grad}_\phi = \frac{1}{r\sin\theta}\frac{\partial}{\partial \phi}. \tag{6.5}$$

The expression for the divergence operator can be obtained in a similar way. Considering the flux of an arbitrary vector $\mathbf{v}$ through the surface of the infinitesimal volume of Fig. 6.4, we have, for Gauss's theorem

$$(\text{div } \mathbf{v})r^2\sin\theta\,dr\,d\theta\,d\phi = \frac{\partial}{\partial r}\left(v_r r^2\sin\theta\right)dr\,d\theta\,d\phi$$
$$+ \frac{\partial}{\partial \theta}(v_\theta r\sin\theta)\,dr\,d\theta\,d\phi + \frac{\partial}{\partial \phi}(v_\phi r)\,dr\,d\theta\,d\phi,$$

from which we obtain, as it can be easily proved,

$$\text{div } \mathbf{v} = \frac{1}{r^2}\frac{\partial}{\partial r}\left(r^2 v_r\right) + \frac{1}{r\sin\theta}\frac{\partial}{\partial \theta}(\sin\theta\,v_\theta) + \frac{1}{r\sin\theta}\frac{\partial}{\partial \phi}v_\phi. \tag{6.6}$$

Fig. 6.4 The figure shows
the infinitesimal volume used
to find the expression for the
divergence operator in
spherical coordinates

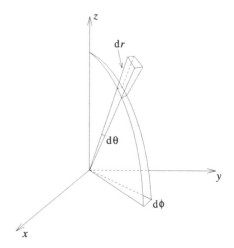

Recalling that $\nabla^2 = \text{div grad}$, together with Eq. (6.5), we then obtain the expression for the Laplace operator of a scalar function in spherical coordinates

$$\nabla^2 = \frac{1}{r^2}\frac{\partial}{\partial r}\left(r^2\frac{\partial}{\partial r}\right) + \frac{1}{r^2 \sin\theta}\frac{\partial}{\partial\theta}\left(\sin\theta\frac{\partial}{\partial\theta}\right) + \frac{1}{r^2 \sin^2\theta}\frac{\partial^2}{\partial\phi^2}. \qquad (6.7)$$

Using this expression, the time-independent Schrödinger equation for a particle moving in a central potential is

$$\mathcal{H}\psi(r,\theta,\phi) = E\psi(r,\theta,\phi),$$

where the Hamiltonian, in spherical coordinates, is

$$\mathcal{H} = -\frac{\hbar^2}{2m}\left[\frac{1}{r^2}\frac{\partial}{\partial r}\left(r^2\frac{\partial}{\partial r}\right) + \frac{1}{r^2 \sin\theta}\frac{\partial}{\partial\theta}\left(\sin\theta\frac{\partial}{\partial\theta}\right) + \frac{1}{r^2 \sin^2\theta}\frac{\partial^2}{\partial\phi^2}\right] + V(r).$$

Before discussing the solution of this equation, we need to also find the expressions, in spherical coordinates, for the operators associated with the orbital angular momentum. Recalling the definition

$$\boldsymbol{\ell} = \frac{1}{\hbar}\mathbf{x}\times\mathbf{p} = -\mathrm{i}\mathbf{x}\times\text{grad}, \qquad (6.8)$$

we have

$$\boldsymbol{\ell} = -\mathrm{i}r\mathbf{e}_r \times \left(\mathbf{e}_r\frac{\partial}{\partial r} + \mathbf{e}_\theta\frac{1}{r}\frac{\partial}{\partial\theta} + \mathbf{e}_\phi\frac{1}{r\sin\theta}\frac{\partial}{\partial\phi}\right),$$

from which we obtain, given that $\mathbf{e}_r \times \mathbf{e}_\theta = \mathbf{e}_\phi$, $\mathbf{e}_r \times \mathbf{e}_\phi = -\mathbf{e}_\theta$,

$$\boldsymbol{\ell} = -\mathrm{i}\left(\mathbf{e}_\phi\frac{\partial}{\partial\theta} - \mathbf{e}_\theta\frac{1}{\sin\theta}\frac{\partial}{\partial\phi}\right).$$

This equation allows us to obtain the three components of the orbital angular momentum along the x, y, z axes (cf. Fig. 6.3). Taking into account the relations between the triads $(\mathbf{e}_r, \mathbf{e}_\theta, \mathbf{e}_\phi)$ and $(\mathbf{i}, \mathbf{j}, \mathbf{k})$, we have

$$\ell_x = \mathbf{i} \cdot \boldsymbol{\ell} = i\left(\sin\phi \frac{\partial}{\partial\theta} + \cot\theta \cos\phi \frac{\partial}{\partial\phi} \right),$$

$$\ell_y = \mathbf{j} \cdot \boldsymbol{\ell} = i\left(-\cos\phi \frac{\partial}{\partial\theta} + \cot\theta \sin\phi \frac{\partial}{\partial\phi} \right),$$

$$\ell_z = \mathbf{k} \cdot \boldsymbol{\ell} = -i\frac{\partial}{\partial\phi}.$$

For ℓ^2 we have

$$\ell^2 = \boldsymbol{\ell} \cdot \boldsymbol{\ell} = -\left(\mathbf{e}_\phi \frac{\partial}{\partial\theta} - \mathbf{e}_\theta \frac{1}{\sin\theta} \frac{\partial}{\partial\phi} \right) \cdot \left(\mathbf{e}_\phi \frac{\partial}{\partial\theta} - \mathbf{e}_\theta \frac{1}{\sin\theta} \frac{\partial}{\partial\phi} \right).$$

To calculate this quantity we need first to find the derivatives of the unit vectors with respect to the spherical coordinates. With simple geometrical considerations we find

$$\frac{\partial}{\partial r}\mathbf{e}_r = 0, \qquad \frac{\partial}{\partial\theta}\mathbf{e}_r = \mathbf{e}_\theta, \qquad \frac{\partial}{\partial\phi}\mathbf{e}_r = \sin\theta\,\mathbf{e}_\phi,$$

$$\frac{\partial}{\partial r}\mathbf{e}_\theta = 0, \qquad \frac{\partial}{\partial\theta}\mathbf{e}_\theta = -\mathbf{e}_r, \qquad \frac{\partial}{\partial\phi}\mathbf{e}_\theta = \cos\theta\,\mathbf{e}_\phi,$$

$$\frac{\partial}{\partial r}\mathbf{e}_\phi = 0, \qquad \frac{\partial}{\partial\theta}\mathbf{e}_\phi = 0, \qquad \frac{\partial}{\partial\phi}\mathbf{e}_\phi = -\sin\theta\,\mathbf{e}_r - \cos\theta\,\mathbf{e}_\theta.$$

We obtain

$$\ell^2 = -\frac{\partial^2}{\partial\theta^2} - \cot\theta \frac{\partial}{\partial\theta} - \frac{1}{\sin^2\theta} \frac{\partial^2}{\partial\phi^2},$$

that is

$$\ell^2 = -\frac{1}{\sin\theta} \frac{\partial}{\partial\theta}\left(\sin\theta \frac{\partial}{\partial\theta} \right) - \frac{1}{\sin^2\theta} \frac{\partial^2}{\partial\phi^2}.$$

Using the previous expression for $\boldsymbol{\ell}$, we can find the commutation relations among the components of the angular momentum. We have

$$\boldsymbol{\ell} \times \boldsymbol{\ell} = -\left(\mathbf{e}_\phi \frac{\partial}{\partial\theta} - \mathbf{e}_\theta \frac{1}{\sin\theta} \frac{\partial}{\partial\phi} \right) \times \left(\mathbf{e}_\phi \frac{\partial}{\partial\theta} - \mathbf{e}_\theta \frac{1}{\sin\theta} \frac{\partial}{\partial\phi} \right),$$

from which we obtain, as it can be easily proved

$$\boldsymbol{\ell} \times \boldsymbol{\ell} = i\boldsymbol{\ell}.$$

This is the fundamental relation containing the commutation relations among the components of the total angular momentum. From this relation we obtain directly

$$[\ell^2, \boldsymbol{\ell}] = 0,$$

i.e. the square of the orbital angular momentum commutes with each of its components. With the introduction of the operator ℓ^2, the Schrödinger equation can be written as

$$\left[-\frac{\hbar^2}{2mr^2} \frac{\partial}{\partial r}\left(r^2 \frac{\partial}{\partial r} \right) + \frac{\hbar^2}{2mr^2}\ell^2 + V(r) \right] \psi(r,\theta,\phi) = E\psi(r,\theta,\phi). \qquad (6.9)$$

Considering that ℓ^2 operates only on the variables θ and ϕ, and that ℓ_z operates only on the variable ϕ, we can search for a solution of the Schrödinger equation that is at the same time an eigenfunction of the three commuting operators $\mathcal{H}$, ℓ^2 and ℓ_z. In order to obtain this, we start by determining the eigenfunctions of the operator ℓ_z, namely the functions Φ_μ such that

$$-i\frac{\partial}{\partial\phi}\Phi_\mu = \mu\Phi_\mu.$$

By integrating this equation we obtain, apart from an arbitrary multiplicative function of the variables r and θ,

$$\Phi_\mu(\phi) = e^{i\mu\phi}.$$

The function $\Phi_\mu(\phi)$ must be single-valued, which happens if we have

$$\mu = m,$$

with m an integer (positive, negative or null). We therefore obtain that the eigenvalues of the ℓ_z operator are the integers m and the corresponding eigenfunctions are of the type

$$\Phi_m(\phi) = e^{im\phi}.$$

Let us now determine the common set of eigenfunctions of the operators ℓ^2 and ℓ_z. To do this we seek functions of the form $\Theta_\lambda(\theta)\Phi_m(\phi)$ such as to satisfy the equation

$$\left[-\frac{1}{\sin\theta}\frac{\partial}{\partial\theta}\left(\sin\theta\frac{\partial}{\partial\theta}\right) - \frac{1}{\sin^2\theta}\frac{\partial^2}{\partial\phi^2}\right]\Theta_\lambda\Phi_m = \lambda\Theta_\lambda\Phi_m.$$

Substituting the expression for Φ_m and performing the change of variable defined by

$$x = \cos\theta,$$

we obtain for the function Θ_λ the differential equation

$$\left(1-x^2\right)\frac{d^2\Theta_\lambda}{dx^2} - 2x\frac{d\Theta_\lambda}{dx} + \left(\lambda - \frac{m^2}{1-x^2}\right)\Theta_\lambda = 0.$$

To solve this equation, we substitute

$$\Theta_\lambda(x) = \left(1-x^2\right)^{|m|/2}f_\lambda(x)$$

and performing the derivatives, we obtain for f_λ the differential equation

$$\left(1-x^2\right)f_\lambda''(x) - 2x\left(1+|m|\right)f_\lambda'(x) + \left(\lambda - |m| - m^2\right)f_\lambda(x) = 0.$$

Finally, we seek a solution for the function f_λ in the form of a power series

$$f_\lambda(x) = \sum_k c_k x^k.$$

By performing the derivatives and substituting, we obtain a recurrence relation between the coefficients c_k of the form

$$c_{k+2} = \frac{(k+|m|)(k+|m|+1) - \lambda}{(k+2)(k+1)}c_k.$$

If the series is not truncated we obtain, for $k \to \infty$, that the c_{k+2}/c_k ratio tends to 1. The series is therefore divergent for $x = \pm 1$. In order to have a finite function, the series needs to be truncated, which implies that the eigenfunction λ satisfies the expression

$$\lambda = l(l+1),$$

with l integer and with

$$l \geq |m|.$$

Note that the maximum degree of the polynomial, k_{max}, is

$$k_{max} = l - |m|,$$

and that the degree of the polynomial is even or odd depending on whether k_{max} is even or odd.

These functions are, except for a multiplicative constant, well known in mathematical physics. They are called Legendre functions (for $m = 0$) and associated Legendre functions of the first kind (for m arbitrary) and are usually denoted, respectively, by the symbols $P_l(x)$ and $P_l^{|m|}(x)$. It is possible to show that the associated Legendre functions satisfy the orthogonality conditions

$$\int_{-1}^{1} P_l^{|m|}(x) P_{l'}^{|m|}(x)\, dx = 0, \quad \text{if } l \neq l'.$$

In summary, we have now found that the common set of eigenfunctions of the operators ℓ^2 and ℓ_z are characterized by two integer quantum numbers m and l, and are of the form

$$P_l^{|m|}(\cos\theta)\, e^{im\phi}.$$

By multiplying these functions by an appropriate factor, we obtain the so-called spherical harmonics, $Y_{lm}(\theta, \phi)$. The factor is chosen so that the functions are normalised to unity over the solid angle and in such a way that they satisfy additional properties of the angular momentum (see Eq. (6.10), which involves the so-called *shift* operators $\ell_\pm$). The definition of the spherical harmonics is the following

$$Y_{lm}(\theta, \phi) = \sqrt{\frac{2l+1}{4\pi} \frac{(l-|m|)!}{(l+|m|)!}} (-1)^{(m+|m|)/2} P_l^{|m|}(\cos\theta) e^{im\phi},$$

and their fundamental properties are summarised in the following equations

$$\ell^2 Y_{lm}(\theta, \phi) = \left[-\frac{1}{\sin\theta} \frac{\partial}{\partial\theta}\left(\sin\theta \frac{\partial}{\partial\theta}\right) - \frac{1}{\sin^2\theta} \frac{\partial^2}{\partial\phi^2}\right] Y_{lm}(\theta, \phi)$$

$$= l(l+1) Y_{lm}(\theta, \phi),$$

$$\ell_z Y_{lm}(\theta, \phi) = -i\frac{\partial}{\partial\phi} Y_{lm}(\theta, \phi) = m Y_{lm}(\theta, \phi),$$

$$\ell_\pm Y_{lm}(\theta, \phi) = (\ell_x \pm i\ell_y) Y_{lm}(\theta, \phi) = \pm e^{\pm i\phi}\left(\frac{\partial}{\partial\theta} \pm i\cot\theta \frac{\partial}{\partial\phi}\right) Y_{lm}(\theta, \phi)$$

$$= \sqrt{(l \pm m + 1)(l \mp m)} Y_{lm\pm 1}(\theta, \phi),$$

$$(6.10)$$

$$Y_{lm}^*(\theta,\phi) = (-1)^m Y_{l-m}(\theta,\phi), \qquad Y_{lm}(\pi-\theta,\phi+\pi) = (-1)^l Y_{lm}(\theta,\phi), \quad (6.11)$$

$$\int_0^{2\pi} d\phi \int_0^\pi d\theta \sin\theta\, Y_{lm}^*(\theta,\phi) Y_{l'm'}(\theta,\phi) = \delta_{ll'}\delta_{mm'}. \quad (6.12)$$

The explicit expressions of the simplest spherical harmonics are the following

$$Y_{00} = \sqrt{\frac{1}{4\pi}}, \qquad Y_{10} = \sqrt{\frac{3}{4\pi}}\cos\theta, \qquad Y_{1\pm1} = \mp\sqrt{\frac{3}{8\pi}}\sin\theta\, e^{\pm i\phi}. \quad (6.13)$$

Returning to the Schrödinger equation in the form (6.9), we seek a solution of the type

$$\psi(r,\theta,\phi) = R(r)Y_{lm}(\theta,\phi) = \frac{1}{r}P(r)Y_{lm}(\theta,\phi),$$

where $R(r)$ is the "ordinary" radial function, $P(r)$ is the so-called reduced radial function, and both functions depend, in general, on the quantum number l. The reduced radial function must satisfy the boundary condition

$$P(0) = 0,$$

so that the wave function ψ is finite at the origin. By substituting, we obtain for the function $P(r)$ the so-called radial Schrödinger equation

$$-\frac{\hbar^2}{2m}\frac{d^2}{dr^2}P(r) + V_{\text{eff}}(r)P(r) = EP(r), \quad (6.14)$$

where the effective potential energy $V_{\text{eff}}(r)$ is given by

$$V_{\text{eff}}(r) = V(r) + \frac{\hbar^2 l(l+1)}{2mr^2}.$$

Equation (6.14) is in all respects similar to that one for the one-dimensional motion of the particle, with the only difference that a centrifugal potential term needs to be added to the potential energy. This term, which vanishes for $l = 0$, has the effect of keeping the particle away from the origin. Its importance increases quadratically with increasing orbital angular momentum.

It has to be remarked that the presence of the centrifugal potential is not a special characteristic of quantum mechanics but a similar potential is present in classical physics when studying the motion of a particle in a central potential. In a central potential, the angular momentum $\mathbf{M}$ is constant, and it is convenient to introduce a system of polar coordinates (r,ϕ) in the plane of the orbit (defined as the plane perpendicular to the angular momentum vector). In these coordinates, the conservation of angular momentum implies

$$mr^2\dot\phi = M,$$

with M constant. On the other hand, for the theorem of conservation of mechanical energy, we have

$$\frac{1}{2}mv^2 + V(r) = \frac{1}{2}m\dot r^2 + \frac{1}{2}mr^2\dot\phi^2 + V(r) = E,$$

with E constant. Substituting for $\dot{\phi}$ the value obtained from the conservation of angular momentum, we have

$$\frac{1}{2}m\dot{r}^2 + V(r) + \frac{M^2}{2mr^2} = E,$$

that is in fact the equation describing the unidimensional motion of a particle in an "effective" potential containing the additional term due to the centrifugal potential. The quantum mechanical version of this equation is indeed the radial Schrödinger equation (Eq. (6.14)).

6.3 Hydrogen Atom, Quantum Theory

We now apply the findings of the previous section to the hydrogenic atom. The potential energy of the electron is given by $-Ze_0^2/r$, hence the radial Schrödinger equation (Eq. (6.14)) is

$$-\frac{\hbar^2}{2m_r}\frac{\mathrm{d}^2}{\mathrm{d}r^2}P(r) + \left[-\frac{Ze_0^2}{r} + \frac{\hbar^2 l(l+1)}{2m_r r^2}\right]P(r) = EP(r),$$

where we have introduced the reduced mass m_r since the two-body theorem can be directly generalised to quantum mechanics. To solve this equation it is convenient to introduce dimensionless variables. Recalling the results of Bohr's theory, we introduce the parameters ξ and ϵ with

$$r = \xi\frac{a_0}{Z}, \qquad E = -\epsilon\frac{Z^2 e_0^2}{2a_0},$$

where the radius of the first Bohr orbit now includes the reduced mass instead of the electron mass:

$$a_0 = \frac{\hbar^2}{m_r e_0^2}.$$

By performing the substitution, we obtain the differential equation

$$\frac{\mathrm{d}^2}{\mathrm{d}\xi^2}P(\xi) + \left[\frac{2}{\xi} - \frac{l(l+1)}{\xi^2} - \epsilon\right]P(\xi) = 0.$$

We note that, for $\xi \to \infty$, the differential equation reduces to

$$\frac{\mathrm{d}^2}{\mathrm{d}\xi^2}P(\xi) - \epsilon P(\xi) = 0.$$

If $\epsilon > 0$ (the case of bound orbits), the asymptotic solution of the equation is therefore

$$P(\xi) = Ce^{\pm\sqrt{\epsilon}\xi},$$

where C is a constant. Of the two solutions, we need to choose the one with the negative exponential because the other diverges. Let then

$$P(\xi) = e^{-\sqrt{\epsilon}\xi}f(\xi),$$

where $f(\xi)$ is a new function. By substituting, we obtain the following differential equation for $f(\xi)$

$$f''(\xi) - 2\sqrt{\epsilon} f'(\xi) + \left[\frac{2}{\xi} - \frac{l(l+1)}{\xi^2} \right] f(\xi) = 0.$$

We seek for a solution $f(\xi)$ in the form of a power series by writing

$$f(\xi) = \xi^p L(\xi) = \xi^p \sum_{k=0}^{\infty} c_k \xi^k \quad (c_0 \neq 0),$$

where p is a real positive number ($p > 0$) since we require that $P(0) = 0$. By substituting we obtain the relation

$$\sum_{k=0}^{\infty} c_k \left[(k+p)(k+p-1) - l(l+1) \right] \xi^{k+p-2} = 2 \sum_{k=0}^{\infty} c_k \left[\sqrt{\epsilon}(k+p) - 1 \right] \xi^{k+p-1}.$$

The term of lowest degree in the first sum (corresponding to $k = 0$) does not have the corresponding one in the second sum. Therefore, we have

$$p(p-1) = l(l+1).$$

This equation of second degree in p has the two solutions $p = l + 1$ and $p = -l$. However, the second one is not acceptable since we must have $p > 0$. It follows that p is an integer given by

$$p = l + 1.$$

By substituting this value for p, we obtain the recurrence relation for the coefficients of the power series

$$c_{k+1} \left[(k+l+2)(k+l+1) \right] = 2c_k \left[\sqrt{\epsilon}(k+l+1) - 1 \right].$$

If the series is not truncated, and since we have

$$\lim_{k \to \infty} \frac{c_{k+1}}{c_k} = 2 \frac{\sqrt{\epsilon}}{k},$$

the function $f(\xi)$ tends to infinity as $e^{2\sqrt{\epsilon}\xi}$, so that the function $P(\xi)$ diverges. To have a finite function, the series must therefore be truncated, i.e. we must have an integer $k_0 \geq 0$ for which

$$\sqrt{\epsilon}(k_0 + l + 1) = 1. \tag{6.15}$$

This expression defines the possible eigenvalues for ϵ:

$$\epsilon = \frac{1}{(k_0 + l + 1)^2} = \frac{1}{n^2},$$

where n is an integer such that

$$n \geq l + 1.$$

Recalling our initial substitutions, we obtain the expression for the energy eigenvalues

$$E = -\frac{e_0^2}{2a_0}\frac{Z^2}{n^2},$$

which coincides with the expression found with Bohr's theory.

Regarding the eigenfunctions, we now determine the differential equation for the power series $L(\xi)$. Recalling its definition and the differential equation for $f(\xi)$, we have

$$\xi\frac{d^2}{d\xi^2}L(\xi) + 2\left(l + 1 - \frac{\xi}{n}\right)\frac{d}{d\xi}L(\xi) + 2\left(1 - \frac{l+1}{n}\right)L(\xi) = 0.$$

This equation can be rewritten as a differential equation for the generalised Laguerre polynomials. This can be done if we introduce a new variable, ρ:

$$\rho = \frac{2}{n}\xi = \frac{2Z}{na_0}r.$$

The differential equation then becomes

$$\rho\frac{d^2}{d\rho^2}L(\rho) + (2l + 2 - \rho)\frac{d}{d\rho}L(\rho) + (n - l - 1)L(\rho) = 0.$$

The generalised Laguerre polynomials[5] are solutions of the differential equation

$$x\frac{d^2}{dx^2}L_p^{(q)}(x) + (q + 1 - x)\frac{d}{dx}L_p^{(q)}(x) + pL_p^{(q)}(x) = 0,$$

hence the function $L(\rho)$ is, except for a proportionality factor, the generalised Laguerre polynomial $L_{n-l-1}^{(2l+1)}$.

Summarizing the previous results, we have found that the reduced radial eigenfunction corresponding to the eigenvalues n and l can be expressed more simply in terms of the variable ρ and is given, apart from a proportionality factor, by

$$P_{nl}(\rho) = e^{-\rho/2}\rho^{l+1}L_{n-l-1}^{(2l+1)}(\rho). \tag{6.16}$$

If we take into account this result, together with what we obtained in the previous section for the angular part of the wave function, we can write the eigenfunctions of the hydrogenic atom in the form

$$\psi_{nlm}(r, \theta, \phi) = N_{nl}e^{-\rho/2}\rho^l L_{n-l-1}^{(2l+1)}(\rho)Y_{lm}(\theta, \phi),$$

where N_{nl} is a factor to be determined by requiring that the eigenfunctions are normalised. Taking into account the relation

$$\int_0^\infty e^{-\rho}\rho^{2l}\left[L_{n-l-1}^{(2l+1)}(\rho)\right]^2\rho^2\,d\rho = \frac{2n[(n+l)!]^3}{(n-l-1)!},$$

---

[5]We follow the conventions of Abramowitz and Stegun (1971).

we have

$$N_{nl} = \left(\frac{Z}{a_0}\right)^{3/2} \frac{2}{n^2} \sqrt{\frac{(n-l-1)!}{[(n+l)!]^3}}. \tag{6.17}$$

The explicit expressions of the eigenfunctions of the hydrogenic atom can be obtained with the formula for the generalised Laguerre polynomials

$$L_p^{(q)}(x) = \sum_{m=0}^{p} (-1)^m \frac{[(p+q)!]^2}{(p-m)!(q+m)!\,m!} x^m. \tag{6.18}$$

The normalised eigenfunctions for the first two levels ($n = 1$ and $n = 2$) are

$$\psi_{100}(r,\theta,\phi) = \left(\frac{Z}{a_0}\right)^{3/2} 2e^{-Zr/a_0} Y_{00}(\theta,\phi),$$

$$\psi_{200}(r,\theta,\phi) = \left(\frac{Z}{a_0}\right)^{3/2} \frac{1}{\sqrt{8}} e^{-Zr/(2a_0)} \left(2 - \frac{Zr}{a_0}\right) Y_{00}(\theta,\phi), \tag{6.19}$$

$$\psi_{21m}(r,\theta,\phi) = \left(\frac{Z}{a_0}\right)^{3/2} \frac{1}{2\sqrt{6}} e^{-Zr/(2a_0)} \frac{Zr}{a_0} Y_{1m}(\theta,\phi).$$

We can now determine the mean values of the powers of r on the radial eigenfunctions using the properties of the generalised Laguerre polynomials. Defining

$$\langle r^k \rangle = \int_0^\infty R_{nl}^2(r) r^k r^2 \, dr = \int_0^\infty P_{nl}^2(r) r^k \, dr,$$

and writing $\langle r^k \rangle$ in units of a_0^k, we have

$$\langle r \rangle = \frac{1}{2Z}[3n^2 - l(l+1)], \qquad \langle r^2 \rangle = \frac{n^2}{2Z^2}[5n^2 + 1 - 3l(l+1)],$$

$$\langle r^3 \rangle = \frac{n^2}{8Z^3}[35n^2(n^2-1) - 30n^2(l+2)(l-1) + 3(l+2)(l+1)l(l-1)],$$

$$\langle r^4 \rangle = \frac{n^4}{8Z^4}[63n^4 - 35n^2(2l^2 + 2l - 3) + 5l(l+1)(3l^2 + 3l - 10) + 12], \tag{6.20}$$

$$\langle r^{-1} \rangle = \frac{Z}{n^2}, \qquad \langle r^{-2} \rangle = \frac{2Z^2}{n^3(2l+1)}, \qquad \langle r^{-3} \rangle = \frac{2Z^3}{n^3 l(l+1)(2l+1)},$$

$$\langle r^{-4} \rangle = \frac{4Z^4[3n^2 - l(l+1)]}{n^5(2l+3)(2l+1)(2l-1)l(l+1)}.$$

It is interesting to see that the results of Bohr's theory can be obtained from the above equations. The case of circular orbits corresponds to assuming, once n is fixed, the maximum possible value for the quantum number l, i.e. $l = n - 1$. For this value we obtain

$$\langle r \rangle = a_0 \frac{2n^2 + n}{2Z},$$

that, for large values of n, coincides with the expression for the radius of Bohr's orbits. It is interesting to calculate the variance $\sigma(r)$ defined as

$$\sigma(r) = \sqrt{\langle r^2 \rangle - \langle r \rangle^2}.$$

Again for $l = n - 1$ we obtain

$$\sigma(r) = a_0 \frac{n\sqrt{2n+1}}{2Z},$$

hence

$$\frac{\sigma(r)}{\langle r \rangle} = \frac{1}{\sqrt{2n+1}}.$$

This shows that for large values of n the eigenfunction of the electron becomes more and more concentrated around the Bohr's orbit.

The eigenfunctions that we have determined depend on three quantum numbers, n, l, and m, which satisfy the relations:

$$n \geq 1, \qquad l \leq n - 1, \qquad |m| \leq l.$$

These three quantum numbers are called, respectively, the principal, azimuthal, and magnetic quantum number. Sometimes the so-called radial quantum number n_r, defined as $n_r = n - l - 1$ is used. As we saw previously, n_r represents the degree of the generalised Laguerre polynomial that appears in the expression for the eigenfunction. n_r also represents the number of values of r where the eigenfunction vanishes (the nodes of the eigenfunction).

For a special circumstance, typical of the Coulomb potential, the eigenvalues of the hydrogenic atom depend only on the principal quantum number n and not on l (the fact that they do not depend on m is a characteristic of the central potential and is related to the spherical symmetry of the Hamiltonian). This means that the eigenvalues of the hydrogenic atom are doubly degenerate (with respect to m and l). To calculate the degeneracy of the level n, it is sufficient to consider that l can have the values $0, 1, \ldots, n - 1$, and that, for each given l, m can have the $(2l + 1)$ values $-l, -l + 1, \ldots, 0, \ldots, l - 1, l$. The degeneracy is therefore

$$g(n) = \sum_{l=0}^{n-1} (2l + 1) = n^2.$$

In relativistic theory, when introducing the spin, the wave functions are characterized by a further quantum number, m_s, the eigenvalue of the operator s_z, projection of the spin along the axis of quantisation z, which can have the two values $\pm\frac{1}{2}$. Taking into account the spin, the degeneracy of the level n is therefore equal to $2n^2$.

6.4 Hydrogen Atom, Relativistic Corrections

Although the Dirac equation for the hydrogenic atom can be solved exactly, we prefer here to apply the perturbation theory to obtain the relativistic corrections to such a system. We start by considering the Dirac equation in the non-relativistic first-order limit (Eq. (5.13)) and we apply the following substitutions: the potential ϕ with Ze_0/r, the charge e with $-e_0$, and the mass m with the reduced mass m_r. Regarding this last substitution, we note that the Dirac equation is valid for a nucleus of infinite mass. The equations for the case of finite mass are quite complex.[6] The substitution $m \rightarrow m_r$ is therefore not exactly justified. It should be considered as an approximation. Taking into account these substitutions, we can rewrite the equation in the following way

$$\mathcal{H}|\psi\rangle = \left(\mathcal{H}_0 + \mathcal{H}'\right)|\psi\rangle = E|\psi\rangle,$$

where $|\psi\rangle$ is, in Dirac's notation, the spinor wave function, and where

$$\mathcal{H}_0 = \frac{p^2}{2m_r} - \frac{Ze_0^2}{r},$$

$$\mathcal{H}' = -\frac{1}{2m_r c^2}\left(E + \frac{Ze_0^2}{r}\right)^2 - \frac{Ze_0^2 \hbar^2}{4m_r^2 c^2}\frac{1}{r^2}\frac{\partial}{\partial r} + \frac{Ze_0^2 \hbar^2}{4m_r^2 c^2}\frac{1}{r^3}\boldsymbol{\sigma} \cdot \boldsymbol{\ell}.$$

We note that we have written the first term in the expression for $\mathcal{H}'$ following the second option in Eq. (5.12).

Now we briefly recall the results of the first-order perturbation theory. Let us start with an Hamiltonian $\mathcal{H}$ which can be written as $\mathcal{H}_0 + \mathcal{H}'$, with $\mathcal{H}' \ll \mathcal{H}_0$. We suppose to have solved the time-independent Schrödinger equation for $\mathcal{H}_0$, finding the eigenvalues E_n and the corresponding eigenvectors $|n\rangle$

$$\mathcal{H}_0|n\rangle = E_n|n\rangle.$$

In order to determine the "perturbation" induced by the Hamiltonian $\mathcal{H}'$ on the eigenvalues E_n, one of the two following approaches has to be followed: (a) if the eigenvalue is not degenerate, the correction ΔE_n to the energy is obtained from the diagonal matrix element

$$\Delta E_n = \langle n|\mathcal{H}'|n\rangle,$$

while the eigenvector stays the same; (b) if instead the eigenvalue is degenerate, one has to calculate the matrix elements

$$\mathcal{H}'_{\nu\nu'} = \langle n, \nu|\mathcal{H}'|n, \nu'\rangle,$$

where ν is another quantum number (or a set of quantum numbers) that is associated with the eigenvectors $|n, \nu\rangle$ of the degenerate space. The eigenvalues and eigenvector of this matrix provide, respectively, the corrections to the energy and the eigenvectors of the total Hamiltonian. Clearly, the calculation is greatly simplified if one finds a basis where the $\mathcal{H}'_{\nu\nu'}$ matrix is diagonal. Otherwise, the calculation can normally be performed only numerically, with the exclusion of matrices of second order, and sometimes of those of third order.

---

[6]See the article: Giachetti and Sorace (2006).

Now we apply the perturbation theory to our particular case. As we saw in the previous section, the Hamiltonian $\mathcal{H}_0$ has, taking also into account the spin, eigenfunctions characterized by the four quantum numbers n, l, m, m_s, whose explicit expression is known. For such eigenfunctions we will use the compact notation $|nlmm_s\rangle$. The energy instead depends only on n, so that for a fixed n, we have $2n^2$ degenerate levels characterized by all possible values of l, m, and m_s. We then need to calculate, in principle, the matrix elements

$$\mathcal{H}'_{lmm_s,l'm'm'_s} = \langle nlmm_s|\mathcal{H}'|nl'm'm'_s\rangle.$$

The Hamiltonian $\mathcal{H}'$ consists of three terms. The first two, acting only on the radial variable r, commute with the operators ℓ^2, ℓ_z and s_z. Therefore, their matrix elements are diagonal with respect to the corresponding quantum numbers. The third term, on the other hand, is not diagonal because it contains the expression $\boldsymbol{\sigma} \cdot \boldsymbol{\ell}$. We can however overcome this drawback by performing a change of base, from the $|nlmm_s\rangle$ to the $|nljm_j\rangle$ base, where j and m_j are the quantum numbers associated to a new operator $\mathbf{j}$ defined by

$$\mathbf{j} = \boldsymbol{\ell} + \mathbf{s}.$$

As shown in the section of this book devoted to the theory of angular momentum (Sect. 7.9), the change of base implies that the new vectors are obtained by appropriate linear combinations of the old vectors involving the Clebsh-Gordan coefficients (or the Wigner's 3-j symbols). For our present purposes it is not necessary to discuss the details of the transformation. We just note that, with the introduction of the operator $\mathbf{j}$, we can find an appropriate expression for the term $\boldsymbol{\ell} \cdot \boldsymbol{\sigma}$. We have in fact

$$j^2 = (\boldsymbol{\ell} + \mathbf{s})^2 = \ell^2 + s^2 + 2\boldsymbol{\ell} \cdot \mathbf{s},$$

from which we obtain, recalling that $\boldsymbol{\sigma} = 2\mathbf{s}$

$$\boldsymbol{\ell} \cdot \boldsymbol{\sigma} = 2\boldsymbol{\ell} \cdot \mathbf{s} = j^2 - \ell^2 - s^2.$$

The operator $\boldsymbol{\ell} \cdot \boldsymbol{\sigma}$ is diagonal in the new base and this implies that the third term in the Hamiltonian $\mathcal{H}'$ is also diagonal, since it contains such an operator multiplied by a function of r.

We now consider a state characterized not only by n, but also by the three quantum numbers l, j, and m_j. The unperturbed energy is, as we know

$$E = -\frac{Z^2 e_0^2}{2a_0} \frac{1}{n^2},$$

where $a_0 = \hbar^2/(m_r e_0^2)$. Let us denote with ΔE_1, ΔE_2, and ΔE_3 the perturbation energy given by the three terms of the Hamiltonian $\mathcal{H}'$. They are, respectively, the correction to the kinetic energy, the Darwin term, and the spin-orbit term. We have for the first term

$$\Delta E_1 = -\frac{1}{2m_r c^2} \frac{Z^2 e_0^4}{a_0^2} \left[\frac{Z^2}{4n^4} - \frac{Za_0\langle r^{-1}\rangle}{n^2} + a_0^2\langle r^{-2}\rangle \right],$$

and recalling the expressions for $\langle r^k \rangle$ given in the previous section (Eq. (6.20)) we obtain

$$\Delta E_1 = -\frac{Z^4 e_0^4}{2m_r c^2 a_0^2}\left[\frac{1}{4n^4} - \frac{1}{n^4} + \frac{2}{n^3(2l+1)}\right].$$

This expression can be rewritten as

$$\Delta E_1 = -\frac{Z^2 e_0^2}{2a_0 n^2}\frac{Z^2 \alpha^2}{n^2}\left(\frac{n}{l+\frac{1}{2}} - \frac{3}{4}\right),$$

where α is the fine-structure constant. For the Darwin term we have

$$\Delta E_2 = -\frac{Z e_0^2 \hbar^2}{4, m_r^2 c^2}\int_0^\infty R_{nl}(r)\frac{1}{r^2}\left(\frac{\mathrm{d}}{\mathrm{d}r}R_{nl}(r)\right)r^2 \,\mathrm{d}r,$$

where we have denoted by $R_{nl}(r)$ the radial function. By solving the integral we have

$$\Delta E_2 = \frac{Z e_0^2 \hbar^2}{8m_r^2 c^2}R_{nl}^2(0).$$

This quantity is typical of a contact term, a result we already obtained in Sect. 5.5 (cf. Eq. (5.15)). It is non-zero only for $l = 0$ states, since they are the only ones for which the radial function is non-zero in the origin. Using Eqs. (6.16), (6.17), and (6.18) we obtain

$$R_{nl}(0) = \left(\frac{P_{nl}(r)}{r}\right)_{r=0} = 2\left(\frac{Z}{na_0}\right)^{3/2}\delta_{l,0},$$

hence

$$\Delta E_2 = \frac{Z^2 e_0^2}{2a_0}\frac{1}{n^2}Z^2\alpha^2\frac{1}{n}\delta_{l,0}.$$

At last, for the spin-orbit term we get

$$\Delta E_3 = \frac{Z e_0^2 \hbar^2}{4m_r^2 c^2}\langle r^{-3}\rangle\left[j(j+1) - l(l+1) - \frac{3}{4}\right],$$

where, for the angular momentum addition rules, the quantum number j can have the two values $(l - \frac{1}{2})$ and $(l + \frac{1}{2})$ if $l \neq 0$, and only the value $j = \frac{1}{2}$ if $l = 0$. Substituting the expression for $\langle r^{-3}\rangle$ (Eq. (6.20)), we have

$$\Delta E_3 = \frac{Z^2 e_0^2}{2a_0}\frac{1}{n^2}Z^2\alpha^2\frac{j(j+1) - l(l+1) - \frac{3}{4}}{nl(l+1)(2l+1)}. \tag{6.21}$$

We note that this expression is undetermined for $l = 0$, because it is of the form $0/0$. A more in-depth analysis shows that in this case $\Delta E_3 = 0$. By summing the contributions of the three terms, and adding it to the unperturbed energy, we obtain for the energy characterized by the quantum numbers n, l, j, and m_j (distinguishing among the three possible cases):

$$E_{nljm_j} = -\frac{Z^2 e_0^2}{2a_0} \frac{1}{n^2} \left[1 + \frac{Z^2 \alpha^2}{n^2} \left(n - \frac{3}{4}\right)\right] \quad (l = 0),$$

$$E_{nljm_j} = -\frac{Z^2 e_0^2}{2a_0} \frac{1}{n^2} \left[1 + \frac{Z^2 \alpha^2}{n^2} \left(\frac{n}{l+1} - \frac{3}{4}\right)\right] \quad \left(l \neq 0, j = l + \frac{1}{2}\right),$$

$$E_{nljm_j} = -\frac{Z^2 e_0^2}{2a_0} \frac{1}{n^2} \left[1 + \frac{Z^2 \alpha^2}{n^2} \left(\frac{n}{l} - \frac{3}{4}\right)\right] \quad \left(l \neq 0, j = l - \frac{1}{2}\right).$$

These three formulae can be summarised in the single one

$$E_{nljm_j} = -\frac{Z^2 e_0^2}{2a_0} \frac{1}{n^2} \left[1 + \frac{Z^2 \alpha^2}{n^2} \left(\frac{n}{j + \frac{1}{2}} - \frac{3}{4}\right)\right].$$

This expression shows that the energy does not depend on the quantum number m_j (as it was obvious to expect, being $[\mathcal{H}, j_z] = 0$) and that, furthermore, it only depends on j (total angular momentum) but not on l (orbital angular momentum). The expression also coincides with the series expansion to second order in α^2 of the exact solution of the Dirac equation for the Coulomb field. This solution, which also contains the rest energy is in fact[7]

$$\mathcal{E}_{nljm_j} = mc^2 \left[1 + \left(\frac{\alpha Z}{n - k + \sqrt{k^2 - \alpha^2 Z^2}}\right)^2\right]^{-1/2},$$

where

$$k = j + \frac{1}{2}.$$

As we have already pointed out, the mass that appears in this formula is the mass of the electron and not its reduced mass. Apart from this difference, the formula we have found and the exact solution differ, for the hydrogen atom, by a quantity of the order of

$$\alpha^6 mc^2 = \alpha^4 \frac{e_0^2}{a_0}.$$

This correction is so small that it is virtually undetectable experimentally. It is not to be confused with the other corrections discussed below. For these reasons, the formula we have found using perturbation theory can, in effect, be regarded as correct. Once all the appropriate corrections are taken into account, the formula is verified experimentally with great precision.

The level structure of the hydrogen atom resulting from the relativistic corrections (the so-called fine structure) is schematically illustrated in Fig. 6.5. The levels have a complex structure and the energy increases, for each value of n, with increasing j. Given that the minimum and maximum of j are $\frac{1}{2}$ and $n - \frac{1}{2}$, the energy difference between the extreme levels is

$$\Delta E = \frac{e_0^2}{2a_0} \frac{\alpha^2}{n^4} (n - 1),$$

---

[7]For the derivation of the equation see, for example, Dirac (1958).

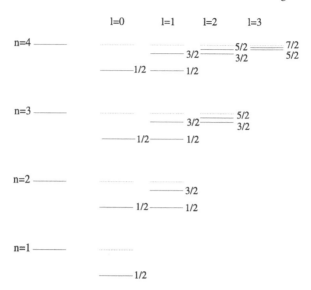

Fig. 6.5 Grotrian diagram of the fine structure of the first 4 levels of the hydrogen atom. The *dotted lines* represent the energies of the unperturbed levels, as they result from the nonrelativistic theory. The value of the quantum number j is provided next to each level. The energies are not to scale

and decreases rapidly with increasing n. In numerical terms, if we consider the Lyα line at 1216 Å, we find that it is split by the fine structure into two components separated by 5.4 mÅ. For the Lyβ line at 1026 Å we also have two components with a separation of 1.1 mÅ.

As we have already pointed out, there are further corrections to the spectrum of the hydrogen atom that are due, on one hand, to the presence of the nuclear spin (hyperfine structure) and, on the other hand, to a purely quantum-mechanical phenomenon, to the so-called self-energy of the electron. Hyperfine structure is discussed in Sects. 9.8 and 9.9. In relation to the other phenomenon, we simply mention the fact that it was highlighted experimentally by W.E. Lamb in 1947. Spectroscopic devices with high resolution show that levels characterized by the same values of the quantum numbers n and j but different values of l have slightly different energies (in contradiction to Dirac's theory which predicts that they should have the same energy). For example, for the $n = 2$ level of the hydrogen atom, a difference in energy equivalent to 1057.8 MHz between the sublevels $l = 0$, $j = \frac{1}{2}$ and $l = 1$, $j = \frac{1}{2}$ is observed. For comparison, the energy difference between the sublevels $l = 1$, $j = \frac{1}{2}$ and $l = 1$, $j = \frac{3}{2}$ is equal to 10968.6 MHz, or about an order of magnitude higher. This effect, which takes the name of Lamb shift, can be explained by assuming that the electron orbiting the nucleus undergoes "virtual" transitions (that do not conserve energy) with the emission of photons which are immediately re-absorbed by the electron. We observe that Heisenberg's uncertainty principle allows, on a small time Δt, that the conservation of energy may be violated by an amount ΔE given by

$$\Delta E \simeq \frac{\hbar}{\Delta t}.$$

The combination of these virtual processes leads to a correction to the energy of the electron which depends on the orbit itself, and is therefore different for states with different l. Detailed calculations, developed by Bethe, are in excellent agreement with the observed values.[8]

6.5 Spectra of Alkaline Metals

After those of the hydrogenic atoms, the simplest spectra are those of the alkaline metals, i.e. of the elements occupying the first column of the periodic system, together with their isoelectronic sequences (Li, Be^+, B^{++}, ..., Na, Mg^+, Al^{++}, ..., K, Ca^+, Sc^{++}, ..., etc.). These atoms (ions) are characterized by the presence of a single "valence electron" (optical electron), i.e. of a single electron that orbits more externally around a charge cloud consisting of the nucleus and of the other electrons. If we assume that the charge cloud has spherical symmetry, the energy levels of the valence electron can be found, as for the hydrogenic atom, by solving the stationary Schrödinger equation in a suitable central potential $V(r)$. The angular part of the eigenfunctions are still given by the spherical harmonics, while the reduced radial function, $P(r)$, obeys Eq. (6.14) which we rewrite here

$$-\frac{\hbar^2}{2m}\frac{d^2}{dr^2}P(r) + \left[V(r) + \frac{\hbar^2 l(l+1)}{2mr^2}\right]P(r) = EP(r).$$

A suitable approximation for $V(r)$ is

$$V(r) = -\frac{a}{r} - \frac{b}{r^2},$$

where a and b are two constants. This expression is the start of an expansion of $V(r)$ in power series of $1/r$. It is particularly appropriate for the alkaline metals because, for large values of r, the potential is nearly Coulombian (in agreement with the fact that the inner electrons shield completely the charge of the nucleus), while, for small values of r, the r^{-2} term (which describes the reduction of the screening effect of the electron cloud) prevails. The constant a is $Z_r e_0^2$, where Z_r (the so-called residual charge number) is given by

$$Z_r = Z - N_e + 1,\tag{6.22}$$

with Z indicating the nucleus charge and N_e the total number of electrons ($Z_r = 1$ for neutral atoms, 2 for singly-ionised atoms, etc.). The constant b can be written as

$$b = \frac{\hbar^2}{2m}\beta,$$

---

[8] For a detailed discussion of the Lamb shift, see Bethe and Salpeter (1957).

β being a dimensionless quantity. With these definitions, the equation for the reduced radial function becomes

$$-\frac{\hbar^2}{2m}\frac{d^2}{dr^2}P(r) + \left[-\frac{Z_r e_0^2}{r} + \frac{\hbar^2 l'(l'+1)}{2mr^2}\right]P(r) = EP(r), \qquad (6.23)$$

where the real number l', defined through the equation

$$l'(l'+1) = l(l+1) - \beta,$$

is traditionally written in the form

$$l' = l - \delta l.$$

The quantity δl is called the Rydberg correction or quantum defect. Equation (6.23) can be solved in full analogy with the hydrogenic case, with a similar change of variables. If we impose that the reduced radial function converges at infinity, we find a relation similar to Eq. (6.15), with l' instead of l, i.e.

$$\sqrt{\epsilon}(k_0 + l' + 1) = \sqrt{\epsilon}(k_0 + l - \delta l + 1) = 1,$$

where k_0 is an integer ≥ 0 giving the order of the polynomial that appears in the reduced radial function. For the energy we obtain

$$E_{nl} = -\frac{Z_r^2 e_0^2}{2a_0}\frac{1}{(n - \delta l)^2},$$

which shows that the energies of the alkaline metals, unlike those of the hydrogenic atoms, also depend on the azimuthal quantum number l, because δl depends on l. The quantum number n is given by

$$n = n_r + l + 1,$$

where n_r, the radial quantum number, coincides with k_0. Given that $n_r \geq 0$, we also have for the alkaline metals that

$$l \leq n - 1.$$

Sometimes it is preferable to write the energies as

$$E_{nl} = -\frac{Z_r^2 e_0^2}{2a_0}\frac{1}{n^{*2}}, \qquad (6.24)$$

where

$$n^* = n - \delta l \qquad (6.25)$$

is a real number that is called the effective quantum number.

The Rydberg correction δl decreases rapidly with increasing l. This means that for large values of l the energies of the levels become more and more similar to the corresponding hydrogenic values. This is not surprising and is interpreted by considering that the orbits with small l are the most elongated, i.e. are the most penetrating within the central electron cloud. For these orbits we therefore expect an energy lower than the corresponding hydrogenic energy. Conversely, for large

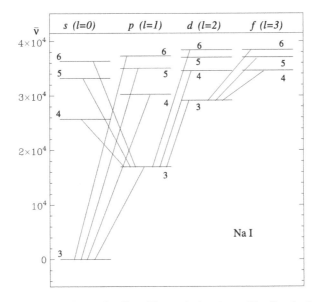

Fig. 6.6 Grotrian diagram of neutral sodium. The *vertical scale* provides directly the energy levels in cm^{-1}. The principal quantum number is shown next to each level. The figure also shows the various series of lines, i.e. the principal series together with the sharp, diffuse, and fundamental ones

values of l, the orbits are nearly circular, and therefore tend to avoid the central area of the atom, where the electric potential differs substantially from the Coulomb potential. For these orbits the energy coincides with the hydrogenic value. Another difference with respect to the hydrogenic case is the fact that the quantum number n corresponding to the ground state (level with lowest energy) is not equal to 1, as for the hydrogenic atoms, but is equal to 2 for lithium, to 3 for sodium, and so on. This is a consequence of the Pauli exclusion principle which will be discussed below (see Sects. 7.1 and 7.7).

The Grotrian diagram of the neutral sodium atom is shown schematically in Fig. 6.6. The spectral lines belonging to the main series of Na I are also shown in the figure. The figure indicates that there are only transitions between terms belonging to adjacent columns. In other words, denoting by Δl the variation of the azimuthal quantum number in the transition, we must have

$$\Delta l = \pm 1. \tag{6.26}$$

A relation of this kind is called a selection rule. We will show in Chap. 12 how this rule can be properly derived by considering the interaction of the atom with the radiation field. For the moment we just introduce the selection rule in a phenomenological way by noting that it is due to the fact that the so-called dipole matrix element between the initial and final states of the transition, i.e. the matrix element

$$\langle \psi_i | \mathbf{r} | \psi_f \rangle,$$

is zero unless the rule (6.26) is verified. Of course, this selection rule is valid not only for the spectrum of the sodium atom, but also for all the spectra of atoms with a single valence electron, including the hydrogenic atoms.

The first classifications proposed to interpret the spectrum of alkali atoms used to indicate the terms having the values of l equal to 0, 1, 2, 3, with the letters (lowercase) "s", "p", "d", "f", respectively. In fact, the so-called "principal series" results from transitions between terms with $l = 1$ and the ground state. This is the reason why the terms with $l = 1$ were given the name "p". Similarly, the series "sharp", due to the combination of terms with $l = 0$ with the lowest $l = 1$ term, justifies the name "s" given to the terms with $l = 0$. The "diffuse" series, connecting the $l = 2$ terms with the lowest $l = 1$ term, provides the name "d" to the $l = 2$ terms. Finally, the "fundamental" series, connecting the $l = 3$ terms with the lowest $l = 2$ term, provides the name "f" to the $l = 3$ terms. For higher values of l, the use is to proceed with the letters in alphabetical order starting from "g", with the exclusion of the letter "j" (reserved, so to say, for angular momenta) and of those already used. In summary:

Values of l	0	1	2	3	4	5	6	7	8	9	10	11	12	...
Denomination	s	p	d	f	g	h	i	k	l	m	n	o	q	...

The jargon of spectroscopy heavily relies on the use of these letters. For example, instead of saying that an atom has an electron whose wave function is characterized by the principal quantum number $n = 3$ and the azimuthal quantum number $l = 1$, one simply refers to an electron $3p$. The fact that $l \leq (n - 1)$ implies that only $1s$, $2s$, $2p$, $3s$, $3p$, $3d$, etc. electrons "exist", whereas electrons such as $1d$ or $3f$ do not exist.

The theoretical results that we have obtained (Eqs. (6.24) and (6.25)) for the spectra of the alkali metals can be compared with laboratory spectroscopic data. For example, considering Na I, it is found that the energies of the ns levels obtained from the equations above are in agreement within 1 % with the experimental data, once an empirical quantum defect $\delta l = 1.36$ is assumed. Similarly, for the np levels a value of $\delta l = 0.87$ is found, and for the nd ones a value of $\delta l = 0.01$.

The above considerations on the spectra of the alkali metals have been made neglecting the presence of the spin. Obviously, also for these atoms one needs to consider relativistic corrections analogous to those seen for the case of the hydrogenic atom. The effect of the third and fourth terms (which depend only on the variable r) in the square bracket of Eq. (5.13) is to provide further corrections to the central potential by e.g. modifying the constant β introduced at the beginning of the section. The last term can now be rewritten in the form

$$\frac{\hbar^2}{4m^2c^2} \frac{1}{r} \frac{d}{dr} V(r) 2\boldsymbol{\ell} \cdot \mathbf{s},$$

and produces a splitting of the levels with $l \neq 0$ (fine structure). If we introduce the quantum number j (that for $l \neq 0$ can have the two values $l + \frac{1}{2}$ or $l - \frac{1}{2}$), we can estimate the energy difference between the two fine-structure levels using

the previous expression obtained for the hydrogenic atoms. A direct application of
Eq. (6.21) gives

$$E_{nlj=l+1/2} - E_{nlj=l-1/2} = \frac{e_0^2}{2a_0}\alpha^2 \frac{Z_{eff}^4}{n^3 l(l+1)},$$

where Z_{eff} is a sort of effective nuclear charge which parameterizes the potential
felt by the valence electron. Various alternative formulae have been proposed to im-
prove the agreement with experimental data. Following a thorough analysis, Landé
proposed to replace in the previous formula Z_{eff}^4 with $Z^2 Z_r^2$ (Z_r being the residual
charge number), and the quantum number n with n^*. The modified expression is
then:

$$E_{nlj=l+1/2} - E_{nlj=l-1/2} = \frac{e_0^2}{2a_0}\alpha^2 \frac{Z^2 Z_r^2}{n^{*3} l(l+1)}.$$

The comparison with experimental data shows a satisfactory agreement, especially
regarding the behaviour with the azimuthal quantum number l.

The fine structure causes the spectra of alkali atoms to be formed by "doublets".
The so-called sodium doublet is particularly well known. It is due to the transition
between the $n = 3$, $l = 0$ ($3s$) and the $n = 3$, $l = 1$ level ($3p$), separated by the
fine structure in two sublevels with $j = \frac{1}{2}$ and $j = \frac{3}{2}$, respectively. The two lines of
the doublet fall respectively at the vacuum wavelengths of 5891.58 and 5897.56 Å,
with a separation of 5.98 Å, or 17.2 cm^{-1}. Note that Landé formula applied to the
$3p$ level provides a separation of 36.6 cm^{-1} (obtained by setting $Z = 11$, $Z_r = 1$,
$n^* = 2.13$, $l = 1$). This value is about twice the experimental one, which clearly
shows the limits of the formula that in many cases can only be used to give an
order of magnitude estimate. The formula, although approximate, shows however
a fundamental characteristic of atomic spectra, namely the fact that the spin-orbit
interaction increases rapidly with increasing Z.

Chapter 7
Atoms with Multiple Valence Electrons

The spectra of atoms having only one valence electron, considered in the previous chapter, are relatively simple and constitute the only examples where the energy levels can be determined by the solution of the one-dimensional Schrödinger equation.

When we consider atoms with more valence electrons, the treatment becomes considerably more complex, and it is necessary to resort to a number of approximations to make the problem mathematically tractable. This chapter is devoted to introducing the physical basis of these approximations as well as the related concepts which form the basis for the complex terminology commonly used in spectroscopy (configurations, terms, multiplets, multiplicity, quantum numbers, etc.).

7.1 The Pauli Exclusion Principle

One of the most important consequences of quantum mechanics is the fact that two particles of the same species (such as two electrons, two protons, two hydrogen atoms, etc.) are in all respects indistinguishable from an observational point of view. Of course, also within classical physics, it is certainly not conceivable that particles of the same nature can have "distinctive signs" making it possible to identify them. However, within classical physics it is always possible, at least in principle, to identify and follow a particle with continuity in time to accurately determine its trajectory, even when it interacts with a particle of the same nature. The situation is completely different in quantum mechanics, where the most adequate representation of a particle is that of a wave packet. If two wave packets that describe identical particles come into interaction with one another (as in a collision between two electrons), when they finally get apart it is impossible to say, both from an observational and from a conceptual point of view, which packet is to be attributed to a particle and which to the other.

On the other hand, when describing a system containing two or more indistinguishable particles, it is necessary to assign to the physical quantities of each particle their own mathematical symbols. For example, for a system composed of N

E. Landi Degl'Innocenti, *Atomic Spectroscopy and Radiative Processes*,
UNITEXT for Physics, DOI 10.1007/978-88-470-2808-1_7, © Springer-Verlag Italia 2014

electrons, we will assign to an electron the coordinates (x_1, y_1, z_1), to another the coordinates (x_2, y_2, z_2), and so on. Obviously, the Hamiltonian, as any other observable of the system, must be symmetrical with respect to the exchange of any two of the indices numbering the electrons (otherwise the electrons would be distinguishable!). If we denote by $\mathcal{S}_{ij}$ the formal operator that, by acting on the dynamic variables of the system, operates the exchange of the particles i and j, we must have, for any observable $\mathcal{O}$,

$$\mathcal{S}_{ij}\mathcal{O} = \mathcal{O}.$$

For the wave function the situation is different in that its phase is not an observable quantity. The invariance condition with respect to the exchange of the two particles is therefore not imposed on the wave function, but rather to its square modulus. If $|\psi\rangle$ is (in Dirac's notation) the wave function of the system of N particles, the condition of invariance for the square modulus is satisfied if

$$\mathcal{S}_{ij}|\psi\rangle = e^{i\alpha}|\psi\rangle,$$

where α is an arbitrary real number. On the other hand, if we apply the exchange operator twice, the wave function must be the same, so we must have

$$\mathcal{S}_{ij}\mathcal{S}_{ij}|\psi\rangle = e^{2i\alpha}|\psi\rangle = |\psi\rangle.$$

Therefore, we obtain

$$e^{i\alpha} = \pm 1,$$

whereby,

$$\mathcal{S}_{ij} = \pm 1.$$

The wave functions for which the sign is positive are called symmetric (with respect to the exchange of particles), while those with the minus sign are called antisymmetric. On the other hand, it follows from the Schrödinger equation that the symmetry of a wave function is constant over time. In fact, the infinitesimal variation $d|\psi\rangle$ in the time dt is given by

$$d|\psi\rangle = \frac{1}{i\hbar}\mathcal{H}|\psi\rangle\,dt,$$

and has the same symmetry as the $|\psi\rangle$ given that the Hamiltonian is symmetric with respect to the exchange of two particles.

Experimental evidence show that the particles having null or integer spin have symmetric wave functions, while those with half-integer spin have antisymmetric wave functions. The first are called Bose-Einstein particles (or more simply bosons), while the latter are called Fermi-Dirac particles, or fermions.

Consider now the special case of N identical, non interacting particles. We shall see how, in this case, we can express the wave function of the entire system through the wave functions of single particles. For such a system, we indicate with x_i the entire set of coordinates (including, where appropriate, the spin coordinates) of the

i-th particle. The Hamiltonian is equal to the sum of N single-particle Hamiltonians, all equal to each other

$$\mathcal{H}(x_1, x_2, \ldots, x_N) = \sum_{i=1}^{N} H(x_i).$$

Denoting by $\psi_a(x)$ the wave functions of the Hamiltonian $H(x)$ and by E_a the corresponding eigenvalues

$$H(x)\psi_a(x) = E_a\psi_a(x),$$

it is simple to verify that the function

$$\Psi(a_1, a_2, \ldots, a_N) = \psi_{a_1}(x_1)\psi_{a_2}(x_2) \cdots \psi_{a_N}(x_N)$$

is a wave function of the total Hamiltonian corresponding to the eigenvalue $(E_{a_1} + E_{a_2} + \cdots + E_{a_N})$, that is

$$\mathcal{H}(x_1, x_2, \ldots, x_N)\Psi(a_1, a_2, \ldots, a_N) = (E_{a_1} + E_{a_2} + \cdots + E_{a_N})\Psi(a_1, a_2, \ldots, a_N).$$

This function does not, however, meet the symmetry requirements. The symmetric solution is obtained with a symmetrisation operation

$$\Psi^S(a_1, a_2, \ldots, a_N) = N_S \sum_{P} P\{\psi_{a_1}(x_1)\psi_{a_2}(x_2) \cdots \psi_{a_N}(x_N)\},$$

where P is the permutation operator which acts on the coordinates of the particles, and where the sum is done over all possible permutations. N_S is a normalisation constant to be determined so that $|\Psi^S|^2 = 1$. The antisymmetric solution is obtained in a similar way by the anti-symmetrisation operation

$$\Psi^A(a_1, a_2, \ldots, a_N) = N_A \sum_{P} (-1)^P P\{\psi_{a_1}(x_1)\psi_{a_2}(x_2) \cdots \psi_{a_N}(x_N)\}, \qquad (7.1)$$

where the sign factor $(-1)^P$ is ± 1 if the permutation is even or odd. Let us now consider an example. From the wave function $\psi_a(x_1)\psi_b(x_2)\psi_c(x_3)$ which represents the state of the whole system where particle 1 occupies the state (of single particle) a, while particles 2 and 3 occupy states b and c, respectively, one gets the anti-symmetric wave function through the equation

$$\Psi^A(a, b, c) = N_A\big[\psi_a(x_1)\psi_b(x_2)\psi_c(x_3) + \psi_a(x_2)\psi_b(x_3)\psi_c(x_1)$$
$$+ \psi_a(x_3)\psi_b(x_1)\psi_c(x_2) - \psi_a(x_2)\psi_b(x_1)\psi_c(x_3)$$
$$- \psi_a(x_1)\psi_b(x_3)\psi_c(x_2) - \psi_a(x_3)\psi_b(x_2)\psi_c(x_1)\big].$$

This eigenfunction now describes a state (of the total system) in which a particle (without specifying which) occupies the single-particle state a, another occupies state b, and the last one state c. Only the anti-symmetrised eigenfunction describes a physical state (of course, if the particles are fermions), while the starting eigenfunction does not describe a physical state as it implies that the particles can be distinguished.

The anti-symmetrising operation can also be obtained by evaluating the so-called Slater determinant of a suitable matrix

$$\Psi^{A}(a_1, a_2, \ldots, a_N) = N_A \, \text{Det} \begin{pmatrix} \psi_{a_1}(x_1) & \psi_{a_2}(x_1) & \cdots & \psi_{a_N}(x_1) \\ \psi_{a_1}(x_2) & \psi_{a_2}(x_2) & \cdots & \psi_{a_N}(x_2) \\ \cdots & \cdots & \cdots & \cdots \\ \psi_{a_1}(x_N) & \psi_{a_2}(x_N) & \cdots & \psi_{a_N}(x_N) \end{pmatrix}.$$

If we recall the rules for the expansion of a determinant, the antisymmetric property (with respect to the exchange of two particles) of the wave function is related to the fact that the determinant of a matrix changes sign if any two rows are exchanged. By the same rules it also follows that if you want to obtain a wave function not identically zero, the single-particle states $a_1, a_2, \ldots, a_N$ must all be distinct. In the opposite case we would in fact obtain a matrix having two or more equal columns, and its determinant would vanish.

What we have shown here is an illustration of the principle discovered empirically by Pauli and named the exclusion principle or Pauli principle: in a system composed of fermions, each quantum state can at most be occupied by a fermion, i.e., each fermion must have a unique set of quantum numbers, different from the set of any other fermion. This principle can be formulated in terms of the so-called occupation number that is, by definition, the number of particles that share the same single-particle quantum state. In the case of fermions, the occupation number can only be 0 or 1. Instead, in the case of bosons, this number is not subject to any limitation.

We finally remark that the normalisation factor introduced in the above formulae is, if the individual $\psi_{a_i}(x_i)$ are normalised,

$$N_S = (N! m_1! m_2! \cdots)^{-1/2} \quad \text{for the symmetric case,}$$
$$N_A = (N!)^{-1/2} \qquad\qquad \text{for the antisymmetric case,}$$

where, if the particles are bosons, $m_1, m_2, \ldots$ denote the occupation numbers of the states.

7.2 The Nonrelativistic Hamiltonian: Good Quantum Numbers

We consider an atom (or ion) with N electrons and a central nucleus having charge number Z. Neglecting relativistic corrections, the total Hamiltonian describing the system can be written in the form

$$\mathcal{H} = \sum_{i=1}^{N} \left(\frac{p_i^2}{2m} - \frac{Ze_0^2}{r_i} \right) + \sum_{i<j} \frac{e_0^2}{r_{ij}}, \tag{7.2}$$

where $\mathbf{r}_i$ is the position vector of the i-th electron (relative to the nucleus), $\mathbf{p}_i$ is its momentum, and r_{ij} is the absolute value of the distance between the i and j electrons:

$$r_{ij} = r_{ji} = |\mathbf{r}_i - \mathbf{r}_j|.$$

The first term in the Hamiltonian is the contribution of the kinetic energy and of the potential energy of the electrons in the field of the nucleus, while the second term is related to the repulsive Coulomb energy among the electrons.

In order to solve the eigenvalue equation for the Hamiltonian $\mathcal{H}$, it is useful to first determine the operators that commute with it, so that we can find a set of quantum numbers which can be assigned in full generality to the quantum states. We first consider the total spin operator $\mathbf{S}$ defined by

$$\mathbf{S} = \sum_{k=1}^{N} \mathbf{s}_k,$$

where $\mathbf{s}_k$ is the spin of the k-th electron. Since we are considering a non-relativistic Hamiltonian which does not contain any spin operator, we clearly have

$$[\mathcal{H}, \mathbf{s}_k] = 0,$$

from which

$$[\mathcal{H}, \mathbf{S}] = 0.$$

The situation is different for the total angular momentum operator $\mathbf{L}$ defined by

$$\mathbf{L} = \sum_{k=1}^{N} \boldsymbol{\ell}_k,$$

given that the single-particle angular momentum $\boldsymbol{\ell}_k$ does not commute with the Hamiltonian, because of the term describing the Coulomb interaction among the electrons. We have in fact

$$[\mathcal{H}, \boldsymbol{\ell}_k] = e_0^2 \sum_{i<j} \left[\frac{1}{r_{ij}}, \boldsymbol{\ell}_k \right].$$

The terms contributing to the sum are those where one of the indices (i or j) is equal to k, since for the others the commutator is null. We therefore have

$$[\mathcal{H}, \boldsymbol{\ell}_k] = e_0^2 \sum_{i \neq k} \left[\frac{1}{r_{ik}}, \boldsymbol{\ell}_k \right].$$

The commutator can be evaluated recalling the definition of the operator $\boldsymbol{\ell}_k$ (see Eq. (6.8)) and considering that

$$\text{grad}^{(k)} \frac{1}{r_{ik}} = \text{grad}^{(k)} \frac{1}{|\mathbf{r}_i - \mathbf{r}_k|} = \frac{\mathbf{r}_i - \mathbf{r}_k}{r_{ik}^3},$$

where $\text{grad}^{(k)}$ denotes the gradient operator with respects to the coordinates of the k-th electron. We thus obtain, with simple algebra

$$[\mathcal{H}, \boldsymbol{\ell}_k] = i e_0^2 \sum_{i \neq k} \frac{\mathbf{r}_k \times \mathbf{r}_i}{r_{ik}^3}.$$

If, on the other hand, we sum over all electrons, i.e. we consider the commutator between the Hamiltonian and the total angular momentum we obtain

$$[\mathcal{H}, \mathbf{L}] = ie_0^2 \sum_k \sum_{i \neq k} \frac{\mathbf{r}_k \times \mathbf{r}_i}{r_{ik}^3} = 0,$$

given that the sum contains pairs of vector products ($\mathbf{r}_k \times \mathbf{r}_i$ and $\mathbf{r}_i \times \mathbf{r}_k$) that cancel out.

Given that the Hamiltonian commutes with both $\mathbf{S}$ and $\mathbf{L}$, if we define the total angular momentum $\mathbf{J}$ as

$$\mathbf{J} = \sum_{k=1}^{N} (\boldsymbol{\ell}_k + \mathbf{s}_k) = \mathbf{L} + \mathbf{S},$$

we obviously have

$$[\mathcal{H}, \mathbf{J}] = 0.$$

Finally, another operator which commutes with the Hamiltonian is the parity operator $\mathcal{P}$, which inverts all the electron coordinates with respect to the origin. The Hamiltonian depends only on the distances between the nucleus and the electrons r_i and the relative distances between the electrons r_{ij}. Therefore it does not change under this inversion, and so we have

$$[\mathcal{H}, \mathcal{P}] = 0.$$

Since the square of the parity operator is the identity, its eigenvalues can only be 1 or -1. The states with eigenvalue 1 are called even, the others odd.[1]

Summarising, we have seen how the Hamiltonian $\mathcal{H}$ commutes with several operators. As a consequence, we have a corresponding number of quantum numbers associated with the eigenvalues of $\mathcal{H}$. The standard convention in spectroscopic work is to define the atomic states having quantum numbers $L = 0, 1, 2$, etc. with the symbols S, P, D, etc. We recall that L is related to the total angular momentum operator $\mathbf{L}$ in the sense that the eigenvalue of the L^2 operator is $L(L + 1)$. The correspondence between letters and numbers is the same as the above-mentioned one for the angular momentum of single particles, i.e.

Values of L	0	1	2	3	4	5	6	7	8	9	10	11	12	...
Denomination	S	P	D	F	G	H	I	K	L	M	N	O	Q	...

Concerning the spin, the convention is to place to the left of the letter corresponding to L the value $(2S + 1)$, equal to the multiplicity, as a superscript. S is the quantum number associated to the total spin, in the sense that the eigenvalue of the operator S^2 is $S(S + 1)$. We therefore have, for $S = 0$, the states ^1S, ^1P, ^1D, etc. For $S = 1/2$

---

[1] The parity of single-particle wave functions is $(-1)^l$, where l is the azimuthal quantum number. This property is contained within Eq. (6.11). The parity operator corresponds, in fact, to the transformation $\theta \to \pi - \theta$, $\phi \to \phi + \pi$.

we have the states ^2S, ^2P, ^2D, and so on. These symbols are read, respectively, "singlet s", "singlet p", "singlet d", "doublet s", "doublet p", "doublet d". The "triplets", "quartets", "quintets" then follow. The value of J is placed as a subscript to the right of the letter which corresponds to L. As usual, we indicate with J the quantum number that corresponds to the total angular momentum in the sense that the eigenvalue of the operator J^2 is $J(J+1)$. Finally, with regards to the parity operator, the odd states (with eigenvalue -1) are identified with a lowercase "o" (for "odd") placed as a superscript to the right of the letter corresponding to L. The full name of an atomic state can therefore be, for example, $^6F^o_{3/2}$ ("sextet f three half odd") or $^3D_3$ ("triplet d three (even)").

This way of naming the energy states is a direct consequence of the property that the Hamiltonian $\mathcal{H}$ commutes with the operators L^2, S^2, J^2, and $\mathcal{P}$. This Hamiltonian is not yet complete because we have totally neglected the relativistic corrections. As we shall see below, the introduction of these corrections implies that L and S cease to be good quantum numbers, so that it becomes necessary to introduce approximate coupling schemes (LS coupling, jj coupling, intermediate coupling). These topics will be discussed in Chap. 9.

7.3 The Central Field Approximation

The analysis of the spectra of complex atoms is based on the so-called "central field (or central potential) approximation" where, as a first approximation, one thinks that each electron moves in a central potential due to the electrostatic interaction with the nucleus and with the remaining electrons. Formally, the approximation means separating the Hamiltonian $\mathcal{H}$ of Eq. (7.2) in two terms, one of zero order ($\mathcal{H}_0$) defining the energies of the basis eigenfunctions, and one of first order ($\mathcal{H}_1$), to be treated as a perturbation term. We write

$$\mathcal{H} = \mathcal{H}_0 + \mathcal{H}_1,$$

where

$$\mathcal{H}_0 = \sum_{i=1}^{N} \left(\frac{p_i^2}{2m} + V_c(r_i) \right), \tag{7.3}$$

$$\mathcal{H}_1 = \sum_{i=1}^{N} \left(-V_c(r_i) - \frac{Ze_0^2}{r_i} \right) + \sum_{i<j} \frac{e_0^2}{r_{ij}}. \tag{7.4}$$

The quantity $V_c(r)$ represents the energy of the electron in the central field. The idea is to choose a function $V_c(r)$ so that

$$\mathcal{H}_1 \ll \mathcal{H}_0,$$

so it is justified to apply the first order perturbation theory. The determination of the function $V_c(r)$ is a complex mathematical problem that can only be solved with approximate numerical methods. Two of these methods are described below. They are

the Thomas-Fermi statistical method and the variational method, further improved
as the Hartree-Fock autoconsistent method.

A priori, we can only establish the boundary conditions for $V_c(r)$. In fact, if we
write

$$V_c(r) = -\frac{\mathscr{Z}(r)e_0^2}{r},$$

we should expect that the function $\mathscr{Z}(r)$ behaves asymptotically as

$$\lim_{r \to 0} \mathscr{Z}(r) = Z, \qquad \lim_{r \to \infty} \mathscr{Z}(r) = Z_r,$$

where Z_r is the residual charge number introduced in Eq. (6.22) ($Z_r = 1$ for neutral
atoms, 2 for ionised atoms, etc.). Most of the discussion that follows does not depend
on the explicit form of the central potential. A detailed knowledge of $V_c(r)$ is only
necessary to establish the quantitative aspects of the atomic spectra.

7.4 The Thomas-Fermi Method

The first method that was developed to determine the central potential of a complex
atom is the Thomas-Fermi statistical method. Being a statistical method, it is more
appropriate for atoms with a large number of electrons, hence for atoms with a large
nuclear charge number Z.

We consider a neutral atom having a nucleus with charge number Z and Z electrons. We assume that the electrons have a spatial distribution with spherical symmetry, and define $n(r)$ as the number of electrons per unit volume. The electrostatic
potential due to the central nucleus and to the electron cloud also has spherical symmetry and obeys the Poisson equation (Eq. (3.4)).

$$\nabla^2 \phi(r) = 4\pi e_0 n(r), \tag{7.5}$$

with the boundary conditions

$$\phi(r) = \frac{Ze_0}{r}, \quad \text{for } r \to 0,$$
$$r\phi(r) = 0, \quad \text{for } r \to \infty. \tag{7.6}$$

We now consider the electrons within a volume element dV at a distance r from the
nucleus. The electron of momentum $\mathbf{p}$ has a corresponding total energy E_t given by

$$E_t = \frac{p^2}{2m} - e_0\phi(r),$$

so, in order for the electron to be bound to the atom, this quantity needs to be negative. This implies that the absolute value of the momentum needs to be lower than
the value p_{max} given by

$$p_{max} = \sqrt{2me_0\phi(r)}.$$

We can assume that the electrons are closely spaced (without violating the Pauli exclusion principle), so that they occupy all the available quantum states with momentum smaller than p_{max}. The number of electrons dN occupying the unit volume dV and having momentum between 0 and p_{max} can be evaluated by calculating the extension of their available phase space and dividing it by h^3, where h is Planck's constant. Considering also the spin (which increases by a factor of 2 the volume of the phase space), we obtain

$$dN = \frac{8\pi}{3h^3} p_{max}^3 \, dV.$$

By substituting the above value for p_{max}, we obtain the following relation between $n(r)$ and $\phi(r)$

$$n(r) = \frac{dN}{dV} = \frac{1}{3\pi^2\hbar^3} \left[2me_0\phi(r) \right]^{3/2}.$$

If we now substitute this expression in Eq. (7.5), we obtain the following differential equation for the potential $\phi(r)$

$$\nabla^2\phi(r) = C\left[\phi(r)\right]^{3/2},$$

where

$$C = \frac{8\sqrt{2}e_0^{5/2}m^{3/2}}{3\pi\,\hbar^3}.$$

This equation can be converted into a dimensionless one by writing

$$\phi(r) = \frac{Ze_0}{r}\chi(r),$$

where $\chi(r)$ is a new function that, for the boundary conditions (7.6), must satisfy the conditions $\chi(0) = 1$ and $\chi(r \to \infty) = 0$. If we recall the expression for the Laplacian operator in spherical coordinates (Eq. (6.7)) we obtain

$$r^{1/2}\frac{d^2}{dr^2}\chi(r) = \mathcal{C}\left[\chi(r)\right]^{3/2},$$

where

$$\mathcal{C} = C(Ze_0)^{1/2} = \frac{8\sqrt{2}Z^{1/2}e_0^3 m^{3/2}}{3\pi\,\hbar^3}.$$

Finally, putting $r = bx$, with x dimensionless, we determine the constant b to simplify as much as possible the differential equation. This is obtained for $b = \mathcal{C}^{-2/3}$, that is

$$b = \frac{1}{2}\left(\frac{3\pi}{4}\right)^{2/3} Z^{-1/3} a_0 = 0.885341 Z^{-1/3} a_0,$$

where a_0 is the radius of the first Bohr orbit. We therefore obtain for $\chi(x)$ the differential equation (known as Thomas-Fermi equation)

$$x^{1/2}\frac{d^2\chi}{dx^2} = \chi^{3/2}.$$

Fig. 7.1 Solution of the
Thomas-Fermi equation

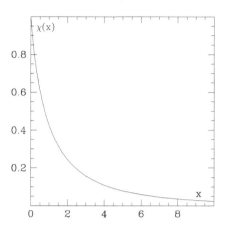

This equation can be solved numerically starting from $x = 0$ (where $\chi(0) = 1$) by
assigning a test value to the derivative $\chi'(0)$ and proceeding by incrementing x. The
value that produces the correct boundary condition to infinity ($\chi(\infty) = 0$) is

$$\chi'(0) = -1.588558,$$

and the corresponding solution is shown in Fig. 7.1.

The solution of the Thomas-Fermi equation can be used to determine the total
ionisation energy for a neutral atom, that is the energy required to remove all the
electrons from the atom, by bringing them to an infinite distance from the nucleus.
The method consists in calculating the electrostatic energy E_{elec} due to the total distribution of charge and to apply the virial theorem to assert that the binding energy
of the atom is equal to half of the electrostatic energy. Indicating with E_{binding} this
binding energy (measured in absolute value) we have

$$E_{\text{binding}} = -\frac{1}{2}E_{\text{elec}},$$

or

$$E_{\text{binding}} = \frac{1}{4}\left[Ze_0^2 \int_0^\infty \frac{1}{r} 4\pi r^2 n(r)\, dr + e_0 \int_0^\infty \phi(r) 4\pi r^2 n(r)\, dr \right],$$

where the first integral in the right-hand side represents the electrostatic energy due
to the interaction of the electron cloud with the nucleus, while the second integral
represents the mutual interaction due to the electron cloud. Substituting the expressions for $\phi(r)$ and $n(r)$, and introducing the variable x, we obtain, by means of
some algebraic manipulations

$$E_{\text{binding}} = \left(\frac{4}{3\pi}\right)^{2/3} \frac{e_0^2}{2a_0} Z^{7/3} \int_0^\infty \frac{(1+\chi)\chi^{3/2}}{x^{1/2}}\, dx.$$

Taking into account the Thomas-Fermi differential equation, the integral can be evaluated by an integration by parts (see Eq. (16.21)). Its value is $-\frac{12}{7}\chi'(0)$, so that, recalling the value of $\chi'(0)$

$$E_{\text{binding}} = -\frac{12}{7}\left(\frac{4}{3\pi}\right)^{2/3}\chi'(0)Z^{7/3}\frac{e_0^2}{2a_0} = 1.53796 Z^{7/3}\frac{e_0^2}{2a_0} = 20.9 Z^{7/3}\ \text{eV}.$$

The dependence with the $7/3$ power of Z is well verified by experimental spectroscopy, which also shows that for the first elements of the periodic system the data are best represented by a numerical coefficient of about 16 instead of 20.9. This discrepancy is not surprising when we recall that the Thomas-Fermi statistical method is especially appropriate to treat atoms with high Z.

Finally, we emphasize that the Thomas-Fermi statistical method provides the value for the potential $\phi(r)$ due to the nucleus and all the electrons, rather than the potential energy $V_c(r)$, to be used in the central field approximation. This latter quantity is in fact the potential energy in which the electron moves in the field created by the nucleus and by all the other electrons (excluding itself, of course). The relation between $\phi(r)$ and $V_c(r)$ can only be given in an approximate way. In their work, Fermi and co-workers often used for $V_c(r)$ the following expression

$$V_c(r) = -\frac{e_0^2}{r} - \frac{(Z-1)e_0^2}{r}\chi(r/b).$$

7.5 The Variational Method and the Hartree-Fock Method

The variational method is mainly used to determine, often approximately, the eigenvalue and eigenfunction of the ground level of an atomic system. However, it can be suitably generalised to extend the calculations also to excited levels. The method is as follows: consider an atomic system described by the Hamiltonian $\mathcal{H}$ and let $|n\rangle$ be its eigenstates and E_n their corresponding eigenvalues. Recalling that the set of eigenvectors $|n\rangle$ is an orthogonal and complete basis, if $|\psi\rangle$ is the normalised eigenfunction that describes an arbitrary physical state of the system, it can be expanded in the form

$$|\psi\rangle = \sum_n c_n |n\rangle.$$

The expectation value of the Hamiltonian on the state $|\psi\rangle$ is therefore given by

$$\langle \mathcal{H} \rangle = \langle \psi | \mathcal{H} | \psi \rangle = \sum_n |c_n|^2 E_n.$$

If we indicate with E_0 the minimum of the eigenvalues E_n, given that $E_n \geq E_0$, we have

$$\langle \mathcal{H} \rangle \geq E_0 \sum_n |c_n|^2 = E_0.$$

The variational method involves calculating the expectation value $\langle \mathcal{H} \rangle$ using some trial wave functions which are dependent on a number of parameters, and to vary these parameters in order to find the minimum value, $\langle \mathcal{H} \rangle_{\min}$. This latter value represents an upper limit for the energy of the ground state of the system E_0. It is as close to the true value as good is the choice of the trial wave functions and the parameters that describe them. Similarly, the wave function that corresponds to the values of the parameters which minimise $\langle \mathcal{H} \rangle$ represents an approximate wave function for the ground state of the system.

We apply the variational method to calculate the energy of the ground state of helium and of the ions belonging to its isoelectronic sequence (H^-, He, Li^+, Be^{++}, etc.).[2] The Hamiltonian of such systems is

$$\mathcal{H} = \frac{p_1^2}{2m} + \frac{p_2^2}{2m} - \frac{Ze_0^2}{r_1} - \frac{Ze_0^2}{r_2} + \frac{e_0^2}{r_{12}}, \tag{7.7}$$

where $Z = 1$ for the negative hydrogen ion H^-, $Z = 2$ for helium, $Z = 3$ for Li^+, and so on.

As test function we take the product of two eigenfunctions (one for each electron), both relative to the ground state of the hydrogenic atom with charge number z, where z is the parameter to be varied. Since each of the two electrons contributes to the screening of the nuclear charge, we should expect that $z < Z$. Recalling Eqs. (6.19) and (6.13), we write

$$\psi(\mathbf{r}_1, \mathbf{r}_2; z) = \frac{z^3}{\pi a_0^3} e^{-z(r_1+r_2)/a_0}. \tag{7.8}$$

We observe in passing that this function is symmetric with respect to the exchange of the two electrons, so it seems to violate the exclusion principle. Indeed, as we shall see later, its correct symmetry is obtained by multiplying it by an antisymmetric spin function. Since the spin is not involved in the problem we are dealing with, we can neglect this complication.

Now that we have introduced the test function, we need to calculate the expectation values of the Hamiltonian (7.7) on this function. Recalling the expression of the Laplacian operator in spherical coordinates (Eq. (6.7)), this leads to the calculation of integrals of the form

$$\mathcal{I}_1 = \int_0^\infty \frac{1}{r^2} \left[\frac{\mathrm{d}}{\mathrm{d}r} \left(r^2 \frac{\mathrm{d}}{\mathrm{d}r} e^{-zr/a_0} \right) \right] e^{-zr/a_0} r^2 \, \mathrm{d}r,$$

$$\mathcal{I}_2 = \int_0^\infty e^{-zr/a_0} \frac{1}{r} e^{-zr/a_0} r^2 \, \mathrm{d}r,$$

$$\mathcal{I}_3 = \int \mathrm{d}^3\mathbf{r}_1 \int \mathrm{d}^3\mathbf{r}_2 \, e^{-z(r_1+r_2)/a_0} \frac{1}{r_{12}} e^{-z(r_1+r_2)/a_0}.$$

---

[2]H^- is the negative hydrogen ion, i.e. an hydrogen atom with an extra electron. Such an ion is stable and provides a significant contribution to the opacity in stellar atmospheres (see Sect. 14.6).

The first two integrals are easily calculated and give

$$\mathcal{I}_1 = -\frac{a_0}{4z}, \qquad \mathcal{I}_2 = \frac{a_0^2}{4z^2}.$$

The calculation of the third integral is more complex and requires the use of an expression that will we will also use later (see Sect. 8.4). Indicating with Θ the angle between the unit vectors $\mathbf{r}_1$ and $\mathbf{r}_2$, and recalling Carnot's theorem we have

$$\frac{1}{r_{12}} = \frac{1}{\sqrt{r_1^2 - 2r_1 r_2 \cos\Theta + r_2^2}}.$$

Denoting by $r_>$ and $r_<$, respectively, the larger and smaller of the two distances r_1 and r_2, one gets

$$\frac{1}{r_{12}} = \frac{1}{r_> \sqrt{1 - 2\frac{r_<}{r_>} \cos\Theta + (\frac{r_<}{r_>})^2}}.$$

If we recall the definition of the generating function of Legendre polynomials

$$\frac{1}{\sqrt{1 - 2\mu x + x^2}} = \sum_{n=0}^{\infty} P_n(\mu) x^n \quad (|x| \le 1),$$

we obtain the expression

$$\frac{1}{r_{12}} = \sum_{n=0}^{\infty} \frac{r_<^n}{r_>^{n+1}} P_n(\cos\Theta). \tag{7.9}$$

Substituting this expression in the integral, and performing the integration on the polar angles of the two electrons, it is easy to see, given the properties of the Legendre functions, that only the term with $n = 0$ contributes to the integral. We then have[3]

$$\mathcal{I}_3 = 16\pi^2 \int_0^\infty dr_1 \int_0^\infty dr_2 \frac{1}{r_>} e^{-2z(r_1+r_2)/a_0} r_1^2 r_2^2,$$

and calculating the integral with elementary methods we get

$$\mathcal{I}_3 = \frac{5\pi^2 a_0^5}{8z^5}.$$

Finally, substituting the values for the integrals, we obtain the expectation value of the Hamiltonian on the test functions

$$\langle \mathcal{H} \rangle = \frac{e_0^2}{a_0} \left(z^2 - 2Zz + \frac{5}{8}z \right).$$

---

[3]In the case we are considering, in which the wave functions of the electrons are spherically symmetric, this expression for the integral $\mathcal{I}_3$ can also be obtained more directly using simple considerations based on Gauss theorem.

Table 7.1 Ionisation potential of the helium atom and a few ions along its isoelectronic sequence

Ion	H$^-$	He	Li$^+$	Be^{++}	B^{3+}	C^{4+}	N^{5+}	O^{6+}
Calculated values (eV)	−0.74	23.1	74.0	152.2	257.6	390.2	550.0	736.9
Experimental values (eV)	0.75	24.6	75.6	153.8	259.3	392.0	551.9	739.1
Error (%)	–	6.1	2.1	1.0	0.7	0.5	0.3	0.3

The first term in parentheses is the contribution of the kinetic energy, the second is due to the Coulomb interaction of the electrons with the nucleus, and the third is due to the Coulomb interaction between the two electrons.

In the spirit of the variational method we need to find the value z that minimises $\langle \mathcal{H} \rangle$. This value is easily found by equating to zero the derivative of $\langle \mathcal{H} \rangle$ with respect to z. The minimum is when

$$z = Z - \frac{5}{16},$$

and we obtain

$$\langle \mathcal{H} \rangle_{\min} = -\frac{e_0^2}{a_0} \left(Z - \frac{5}{16} \right)^2.$$

The corresponding eigenfunction is given by Eq. (7.8) with $(Z - \frac{5}{16})$ instead of z.

The result we have obtained can be used to determine the ionisation potential of the helium atom (and of the ions belonging to its isoelectronic sequence). This potential is in fact given by the equation

$$I + \langle \mathcal{H} \rangle_{\min} = -\frac{e_0^2}{2a_0} Z^2,$$

where the right-hand side represents the energy of the electron in the ground state of the ion formed after ionisation occurs. By substituting the value $\langle \mathcal{H} \rangle_{\min}$ we obtain

$$I = \frac{e_0^2}{2a_0} \left(Z^2 - \frac{5}{4} Z + \frac{25}{128} \right).$$

The comparison with the experimental data is given in Table 7.1.

The table shows that, aside from the ion H$^-$, for which the calculation yields a negative value of the binding energy, the results of the variational method (obtained with a test function containing only one free parameter) are relatively accurate, especially for larger Z. Obviously, better results can be obtained by introducing more sophisticated test functions. With such functions it is found, for instance, that the ion H$^-$ is indeed stable (see Sect. 14.6).

Once the wave function of the ground state is obtained, the calculation can be extended to determine the higher energy levels. For example, to obtain the energy of the first excited level, it is sufficient to calculate the expectation value of the Hamiltonian on the states described by test functions (dependent on one or more parameters) that are orthogonal with respect to the wave function of the ground state. The minimum in the expectation value (obtained by varying the parameters) provides an upper limit for the energy of the first excited level.

The variational method that we have illustrated with a simple model for an atom with two electrons is also used to calculate the energy levels of complex atoms consisting of a large number of electrons. As a first approximation these atoms are described by the nonrelativistic Hamiltonian of Eq. (7.2). For these atoms a test wave function Ψ (containing a number of free parameters) is adopted, and the expectation value of the Hamiltonian on such wave function is calculated

$$\langle \mathcal{H} \rangle = \langle \Psi | \mathcal{H} | \Psi \rangle.$$

The parameters are then varied until a minimum for $\langle \mathcal{H} \rangle$ is obtained.

The first numerical applications of this type were performed in the early 1930s, mainly by Hartree. In these early works, the test function was simply given by the product of N single-particle wave functions, without any antisymmetrisation. The single-particle wave functions were also chosen in such a way as to be orthogonal and normalised. The "best" wave functions were then determined with the variational method, and the charge density calculated by summing the square moduli of the single-particle wave functions. Poisson equation (3.4) was then solved in order to determine the potential $\phi(r)$. Once this potential was calculated, approximate expressions similar to those seen in the previous section were used to estimate the potential energy $V_c(r)$, and the stationary Schrödinger equation was solved to find new single-particle wave functions. These wave functions were again parameterised and the procedure repeated until a self-consistent solution of the problem was found.

Hartree's work was subsequently generalised by various authors (including in particular Fock) by taking into account the indistinguishability of the particles. The result is a complex theory which is now known with the name of Hartree-Fock theory. This theory describes in a quantitative way the structure of the energy levels of the simpler atoms and constitutes a starting point for other more sophisticated calculations, which are in general based on perturbation theory.

7.6 Configurations

The zero-order Hamiltonian $\mathcal{H}_0$, introduced in the central field approximation, is given by the sum of N Hamiltonians formally equal and independent of each other (see Eq. (7.3)). To find its eigenvalues and eigenvectors it is sufficient to solve the Schrödinger equation for the single-particle Hamiltonian. By taking into account also the spin, the solution of this equation is characterized by four quantum numbers n, l, m, and m_s

$$\psi_{nlmm_s} = \frac{1}{r} P_{nl}(r) Y_{lm}(\theta, \phi) \chi_{m_s}, \tag{7.10}$$

where $Y_{lm}(\theta, \phi)$ is the spherical harmonic, and χ_{m_s} is the spin eigenfunction that can either be

$$\begin{pmatrix} 1 \\ 0 \end{pmatrix}, \quad \text{or} \quad \begin{pmatrix} 0 \\ 1 \end{pmatrix},$$

depending on whether the projection of the spin along the quantisation axis is $\frac{1}{2}$ or $-\frac{1}{2}$, respectively. $P_{nl}(r)$ is the solution of the radial Schrödinger equation

$$-\frac{\hbar^2}{2m}\frac{d^2}{dr^2}P_{nl}(r) + \left[V_c(r) + \frac{\hbar^2 l(l+1)}{2mr^2}\right]P_{nl}(r) = W_0(n,l)P_{nl}(r), \qquad (7.11)$$

where $V_c(r)$ is the central potential and $W_0(n,l)$ is the energy eigenvalue that only depends on the two quantum numbers n and l. The four quantum numbers n, l, m, and m_s obey the same restrictions we have seen in the cases of the hydrogenic atoms and the alkali metals, or

$$n \geq l+1, \qquad |m| \leq l, \qquad m_s = \pm\frac{1}{2}.$$

With regard to the first inequality we note that, for an arbitrary potential $V_c(r)$, the meaning of the principal quantum number n is related to the number of nodes of the radial wave function (not counting the origin). This number is in fact given by $(n-l-1)$. With the single particle eigenfunctions we can build the eigenfunctions of the Hamiltonian $\mathcal{H}_0$. They will be characterized by N sets of quantum numbers $a_1, a_2, \ldots, a_N$, each set a_i being formed by the four quantum numbers (n_i, l_i, m_i, m_{si}). This eigenfunction has a corresponding eigenvalue

$$\mathcal{W}_0 = \sum_i W_0(n_i, l_i), \qquad (7.12)$$

which depends only on the quantum numbers n and l, and not on m and m_s. As a consequence, there are in general several distinct eigenfunctions corresponding to the same eigenvalue of the Hamiltonian $\mathcal{H}_0$. The corresponding physical states form, as a whole, a so-called configuration, which can be specified by assigning the number of electrons identified by the pair of quantum numbers n and l, or, more synthetically, by the number of electrons occupying the orbital (nl). In typical spectroscopic notation, a configuration is designated by writing

$$n_1\, l_1^{q_1} n_2\, l_2^{q_2} \cdots n_k\, l_k^{q_k},$$

where q_i is the number of electrons occupying the orbital $(n_i l_i)$, with $\sum_i q_i = N$. For example, a possible configuration of an atom with 5 electrons is the following

$$1s^2 2s^2 2p,$$

where the exponent 1 for the orbital $2p$ is omitted by convention.

The sets of quantum numbers a_i cannot be arbitrary. In fact, due to the Pauli exclusion principle, the set a_i must differ from the set a_j (with $i \neq j$) for at least one of the quantum numbers n, l, m, m_s. This means that there are restrictions on the number of electrons that can occupy a given orbital. From the inequalities written above, we have in fact that an orbital l can have at most Q_l electrons, where

$$Q_l = 2(2l+1).$$

Thus, for example, an orbital s can have at most two electrons, an orbital p six, an orbital d ten, and so on, so a hypothetical configuration such as $1s^3 2s^2$ does not represent any physical state (as it violates the Pauli exclusion principle).

An important characteristic of a configuration is its parity. In the case of single-particle eigenfunctions, we have already seen that the parity is given by $(-1)^l$, where l is the azimuthal quantum number. Such concept is easily generalised to configurations. The parity of a configuration is in fact given by

$$P = (-1)^{\sum_i l_i},$$

and represents the factor by which the wavefunction of any state belonging to the configuration needs to be multiplied if the coordinates of all the electrons are inverted with respect to the origin. Configurations can either be even or odd. For example, the configuration $1s^2 2s^2 2p^2$ is even ($P = 1$), while $1s^2 2s 2p^3$ is an odd configuration ($P = -1$).

Another important concept is the degeneracy. Once a configuration is assigned, we can ask what is the number of distinct quantum states that correspond to it. One way to obtain this number is the following. We start denoting by q_{nl} the number of electrons occupying the orbital (nl) (the so-called subshell of the shell n). To each of these electrons we can assign any pair of quantum numbers (m, m_s) in such a way that each pair differs from the others by at least one of the two quantum numbers. Since the number of distinct pairs is Q_l, the number we are looking for is given by the number of possible combinations of q_{nl} objects from a set of Q_l objects. For the degeneracy we therefore have

$$g = \prod_{nl} \binom{Q_l}{q_{nl}},$$

where the product is extended over all the possible values of the quantum numbers n and l, and where we have introduced the symbol for the binomial coefficient defined by

$$\binom{n}{m} = \frac{n!}{m!(n-m)!} \quad (0 \le m \le n).$$

About this formula it should be noted that, since

$$\binom{n}{n} = 1, \qquad \binom{n}{0} = 1,$$

it is not necessary to explicitly consider the contributions from both the so-called closed subshells (those with $q_{nl} = Q_l$) and the empty subshells (those with $q_{nl} = 0$). For example, for the configuration

$$1s^2 2s^2 2p^4,$$

the degeneracy can be evaluated only considering the open subshell $2p$, obtaining

$$g = \binom{6}{4} = 15.$$

In addition, from one of the properties of the binomial coefficients, namely

$$\binom{n}{m} = \binom{n}{n-m},$$

it follows a rule of symmetry whereby, for example, the degeneracy of the configuration nd^4 is the same as that one of nd^6, etc.

Finally, we note that the degeneracy of a configuration depends critically on the fact that the electrons are "equivalent" or "non-equivalent", that is, if they have equal or different values of n (for equal l). For example, consider the configuration $1s^2 2s^2 2p3p4p$. For this configuration, the degeneracy (due to the three non-equivalent p electrons) is given by

$$g = \binom{6}{1} \cdot \binom{6}{1} \cdot \binom{6}{1} = 216.$$

On the other hand, for the $1s^2 2s^2 2p^2 3p$ configuration, where we have two equivalent p electrons and one non-equivalent, we have

$$g = \binom{6}{2} \cdot \binom{6}{1} = 90,$$

while for the $1s^2 2s^2 2p^3$ configuration, with three equivalent p electrons, the degeneracy is

$$g = \binom{6}{3} = 20.$$

As shown by this example, the degeneracy decreases rapidly with increasing the number of equivalent electrons, an obvious consequence of the Pauli exclusion principle.

7.7 The Principle of Formation of the Periodic Table

The Pauli exclusion principle and the central field approximation are the basis of the so-called principle of formation of the periodic table, with which we can explain, even if not always in a quantitative manner, the fundamental properties of the periodic table of the elements. The principle consists in an imaginary "construction" of atoms (always considered in their ground state) starting from the hydrogen atom and adding, one by one, an electron and a positive charge to the nucleus.

As we know, the hydrogen atom in its ground state has only one electron in the subshell $1s$. By adding an electron and a positive charge to the nucleus, we obtain the helium atom, in which the two electrons can "cohabit" in the subshell $1s$. Obviously, the $1s$ state of the helium atom is different from the $1s$ state of the hydrogen atom, since the (approximate) central potential for the electrons in the helium atom is different from the purely Coulomb potential of the hydrogen atom. The configuration for the ground state of the helium atom is $1s^2$. With helium, both the subshell and the shell (which in this case coincide, being $n = 1$) become closed. The shells are commonly designated with capital letters, starting with the letter K and proceeding in alphabetical order, by following the convention

n	1	2	3	4	5	6	7	...
Symbol	K	L	M	N	O	P	Q	...

So, with the helium atom the K-shell becomes closed. Moving on to the next atom, lithium, the "added" electron has to occupy a state in the L shell. Within this shell, both s and p orbitals are present, but the s ones have lower energy because their orbits are more penetrating (recall the discussion in Sect. 6.5 about the alkali metals). Therefore, the electron occupies the $2s$ subshell, and its radial eigenfunction is much more "expanded" than the radial eigenfunction of the two electrons occupying the K shell. If we recall the results for the hydrogenic atom, the mean value of the radial distance $\langle r \rangle$ for a $2s$ state is 4 times larger than the corresponding value for a $1s$ state and, in this case, the ratio is even greater because the inner electrons experience a higher effective charge. An estimate based on the considerations in Sect. 7.5 produces an additional factor $3 - \frac{5}{16} \simeq 2.7$, so that the ratio assumes a value of the order of 11. As a consequence, the lithium atom has one electron that is more weakly bound than the others, with a ionisation potential of 5.4 eV (to be compared with a value of 13.6 eV for hydrogen and 24.5 for helium). Since the chemical properties of the elements depend only on the peripheral electrons (valence or optical electrons), one can easily understand the reason why lithium is a chemically active element with a tendency to release easily an electron (monovalent element).

The next atom is beryllium, with ground (also called normal) configuration $1s^2 2s^2$, followed by boron, with ground configuration $1s^2 2s^2 2p$. Boron is at the beginning of a "period" of 6 elements with the subshell $2p$ becoming more and more filled. The other five elements of the period are carbon, nitrogen, oxygen, fluorine and neon. With neon, whose ground configuration is $1s^2 2s^2 2p^6$, both the $2p$ subshell and the L shell become closed. The elements with a closed p subshell have, from a chemical point of view, a very low tendency to combine with other elements. Indeed they are called noble gases or inert gases, and, from a physical point of view, are characterized by having a high ionisation potential. The chemical stability is due to the fact that a closed subshell is characterized by a "cloud" of electronic charge having exact spherical symmetry, a consequence of one of the properties of the spherical harmonics

$$\sum_{m=-l}^{l} \left| Y_{lm}(\theta, \phi) \right|^2 = \frac{2l+1}{4\pi} \quad \text{(independent of } \theta \text{ and } \phi\text{)}.$$

In addition, it is difficult to excite electrons belonging to a closed subshell p, since they must be moved to orbitals relatively far away in energy (the $3s$ orbital, for example, in the case of neon). This does not occur for elements having closed s subshells such as beryllium (which still has the property that the charge has a spherically symmetric distribution). In the case of beryllium, it is relatively easy to excite a $2s$ electron to a $2p$ orbital, and this explains why beryllium is more chemically reactive with respect to an element such as neon.

Once a shell is completed, the procedure continues by adding another (weakly bound) electron, then two, three electrons, etc., each of them more strongly bound, until a new shell is filled, reaching maximum chemical stability. This explains the origin of the periodicity of those properties of the elements that depend on the outer electrons, such as the spectroscopic and chemical behaviour.

Table 7.2 shows the experimental ground electronic configurations of all the elements up to uranium ($Z = 92$). To avoid repetitions, the orbitals of the closed shells of the preceding noble gases are omitted. For example, the ground configuration for sodium as shown in the table is $3s$. The full description of the configuration is obtained by including the helium and neon configurations so as to obtain $1s^2 2s^2 2p^6 3s$. Inspection of the tables shows that the "building" of the atoms proceeds regularly until the subshell $3p$ is completely full with argon ($Z = 18$). After argon, one might think that the next subshell to be filled would be the $3d$. Instead, we have a reversal, in the sense that the orbitals $4s$ are filled before the $3d$. This fact is not too surprising when we recall that the Rydberg correction, discussed in the context of the spectra of the alkaline metals, becomes more important for higher nuclear charge. The penetration effect of the orbits is such that the energy of the electron in the $4s$ orbital is lower than the energy in the $3d$ orbital, which is practically circular. The subshell $3d$ therefore begins to fill after the subshell $4s$. We obtain in this way a series of 10 elements ranging from scandium ($Z = 21$) to zinc ($Z = 30$). After zinc, subshell $4p$ becomes filled, and then we proceed following the track shown in Fig. 7.2, where each step is marked with the atomic number of the element with which the next subshell starts to be filled.

It should be noted that, in this filling process of the successive subshells, some irregularities occur. For example, with barium ($Z = 56$) we fill the subshell $6s$ and then we start filling the $5d$ subshell to obtain lanthanum, with ground configuration $5d6s^2$. At this point, instead of continuing with the subshell $5d$, we jump to the subshell $4f$, obtaining a series of 14 elements from cerium ($Z = 58$) to lutetium ($Z = 71$), all having properties similar to those of lanthanum (family of the rare earths or lanthanides). Afterwards, the filling goes on in subshell $5d$. Something very similar occurs with subshells $6d$ and $5f$. The element corresponding to lanthanum is actinium ($Z = 89$), while the family corresponding to the lanthanides is that one of the actinides.

Other minor irregularities are found here and there, especially during the filling of subshells d or f. As it appears from Table 7.2, these irregularities occur for the following elements: chromium, copper, niobium, molybdenum, ruthenium, rhodium, palladium, silver, gadolinium, platinum, gold and thorium. The ground configuration for gold, for example, is $4f^{14}5d^{10}6s$, instead of $4f^{14}5d^96s^2$.

7.8 Configurations of Excited Electrons

With the principle of formation of the periodic table we have found the electron configurations for all the elements in their ground state, i.e. the so-called normal configuration. The excitation of one or more electrons to orbitals of higher energy

Table 7.2 Ground electronic configurations of the elements from hydrogen ($Z = 1$) to uranium ($Z = 92$). The orbitals of the closed shells of the preceding noble gases are omitted for simplicity. For some elements, exceptions to the "rules" described in the text are present. The ground configuration of chromium, for example, is $3d^5 4s$ instead of $3d^4 4s^2$

Z	Symb.	Element	Config.	Z	Symb.	Element	Config.
1	H	Hydrogen	$1s$	41	Nb	Niobium	$4d^4 5s$
2	He	Helium	$1s^2$	42	Mo	Molybdenum	$4d^5 5s$
3	Li	Lithium	$2s$	43	Tc	Technetium	$4d^5 5s^2$
4	Be	Beryllium	$2s^2$	44	Ru	Ruthenium	$4d^7 5s$
5	B	Boron	$2s^2 2p$	45	Rh	Rhodium	$4d^8 5s$
6	C	Carbon	$2s^2 2p^2$	46	Pd	Palladium	$4d^{10}$
7	N	Nitrogen	$2s^2 2p^3$	47	Ag	Silver	$4d^{10} 5s$
8	O	Oxygen	$2s^2 2p^4$	48	Cd	Cadmium	$4d^{10} 5s^2$
9	F	Fluorine	$2s^2 2p^5$	49	In	Indium	$4d^{10} 5s^2 5p$
10	Ne	Neon	$2s^2 2p^6$	50	Sn	Tin	$4d^{10} 5s^2 5p^2$
11	Na	Sodium	$3s$	51	Sb	Antimony	$4d^{10} 5s^2 5p^3$
12	Mg	Magnesium	$3s^2$	52	Te	Tellurium	$4d^{10} 5s^2 5p^4$
13	Al	Aluminium	$3s^2 3p$	53	I	Iodine	$4d^{10} 5s^2 5p^5$
14	Si	Silicon	$3s^2 3p^2$	54	Xe	Xenon	$4d^{10} 5s^2 5p^6$
15	P	Phosphorus	$3s^2 3p^3$	55	Cs	Cesium	$6s$
16	S	Sulfur	$3s^2 3p^4$	56	Ba	Barium	$6s^2$
17	Cl	Chlorine	$3s^2 3p^5$	57	La	Lanthanum	$5d 6s^2$
18	Ar	Argon	$3s^2 3p^6$	58	Ce	Cerium	$4f 5d 6s^2$
19	K	Potassium	$4s$	59	Pr	Praseodymium	$4f^3 6s^2$
20	Ca	Calcium	$4s^2$	60	Nd	Neodymium	$4f^4 6s^2$
21	Sc	Scandium	$3d 4s^2$	61	Pm	Promethium	$4f^5 6s^2$
22	Ti	Titanium	$3d^2 4s^2$	62	Sm	Samarium	$4f^6 6s^2$
23	V	Vanadium	$3d^3 4s^2$	63	Eu	Europium	$4f^7 6s^2$
24	Cr	Chromium	$3d^5 4s$	64	Gd	Gadolinium	$4f^7 5d, 6s$
25	Mn	Manganese	$3d^5 4s^2$	65	Tb	Terbium	$4f^9 6s^2$
26	Fe	Iron	$3d^6 4s^2$	66	Dy	Dysprosium	$4f^{10} 6s^2$
27	Co	Cobalt	$3d^7 4s^2$	67	Ho	Holmium	$4f^{11} 6s^2$
28	Ni	Nickel	$3d^8 4s^2$	68	Er	Erbium	$4f^{12} 6s^2$
29	Cu	Copper	$3d^{10} 4s$	69	Tm	Thulium	$4f^{13} 6s^2$
30	Zn	Zinc	$3d^{10} 4s^2$	70	Yb	Ytterbium	$4f^{14} 6s^2$
31	Ga	Gallium	$3d^{10} 4s^2 4p$	71	Lu	Lutetium	$4f^{14} 5d 6s^2$
32	Ge	Germanium	$3d^{10} 4s^2 4p^2$	72	Hf	Hafnium	$4f^{14} 5d^2 6s^2$
33	As	Arsenic	$3d^{10} 4s^2 4p^3$	73	Ta	Tantalum	$4f^{14} 5d^3 6s^2$
34	Se	Selenium	$3d^{10} 4s^2 4p^4$	74	W	Tungsten	$4f^{14} 5d^4 6s^2$
35	Br	Bromine	$3d^{10} 4s^2 4p^5$	75	Re	Rhenium	$4f^{14} 5d^5 6s^2$
36	Kr	Krypton	$3d^{10} 4s^2 4p^6$	76	Os	Osmium	$4f^{14} 5d^6 6s^2$
37	Rb	Rubidium	$5s$	77	Ir	Iridium	$4f^{14} 5d^7 6s^2$
38	Sr	Strontium	$5s^2$	78	Pt	Platinum	$4f^{14} 5d^9 6s$
39	Y	Yttrium	$4d 5s^2$	79	Au	Gold	$4f^{14} 5d^{10} 6s$
40	Zr	Zirconium	$4d^2 5s^2$	80	Hg	Mercury	$4f^{14} 5d^{10} 6s^2$

Table 7.2 (Continued)

Z	Symb.	Element	Config.	Z	Symb.	Element	Config.
81	Tl	Thallium	$4f^{14}5d^{10}6s^26p$	87	Fr	Francium	$7s$
82	Pb	Lead	$4f^{14}5d^{10}6s^26p^2$	88	Ra	Radium	$7s^2$
83	Bi	Bismuth	$4f^{14}5d^{10}6s^26p^3$	89	Ac	Actinium	$6d7s^2$
84	Po	Polonium	$4f^{14}5d^{10}6s^26p^4$	90	Th	Thorium	$6d^27s^2$
85	At	Astatine	$4f^{14}5d^{10}6s^26p^5$	91	Pa	Protoactinium	$5f^26d7s^2$
86	Rn	Radon	$4f^{14}5d^{10}6s^26p^6$	92	U	Uranium	$5f^36d7s^2$

Fig. 7.2 Schematic diagram illustrating the progressive filling of the subshells according to the principle of formation of the periodic table of the elements. Each step from subshell to subshell is marked with the atomic number of the element with which the next subshell starts to be filled

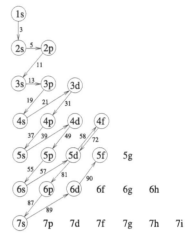

gives rise to additional configurations. Consider for example the carbon atom, having ground configuration $1s^22s^22p^2$. The excitation of one or more electrons can lead to various configurations, such as

$1s2s^22p^3$ excitation of an electron $1s$ to $2p$,

$1s^22s2p^23s$ excitation of an electron $2s$ to $3s$,

$1s^22s^23s3p$ excitation of two electrons $2p$ to $3s$ and $3p$,

$1s^22s^22p3d$ excitation of an electron $2p$ to $3d$.

The configurations resulting from the excitation of inner electrons (belonging to shells or subshells that are closed), such as the first and the second listed above, generally have corresponding energies that are much higher than those of the ground configuration. The spectral lines that originate from these configurations therefore fall in the extreme ultraviolet or in the X-rays. The corresponding spectra are observed in the laboratory mostly in absorption. The configurations that originate from the excitation of two or more outer electrons, as the third in the list, are responsible for the appearance of the so-called anomalous terms. These terms are, as the name itself suggests, an anomaly that is rarely observed and only under particular conditions. The configurations that arise from the excitation of a single electron belonging to an open subshell, as the fourth in the list, are by far the most important ones. Virtually all lines of the visible and near ultraviolet spectrum of a given element originate from transitions between a configuration of this type and the ground configuration or between two configurations of this type.

As we have seen in Sect. 7.6, a configuration corresponds to a number g of different quantum states, described by the wave functions $\Psi(a_1, a_2, \ldots, a_N)$, degenerate eigenfunctions of the Hamiltonian $\mathcal{H}_0$. By considering appropriate linear combinations of these wave functions, it is possible to obtain new functions that are eigenfunctions not only of $\mathcal{H}_0$, but also of a set of operators that commute between themselves as well as with $\mathcal{H}_0$. The most appropriate set is comprised, in most cases, by the operators L^2, L_z, S^2, and S_z. We therefore need to establish, for a given configuration, the possible eigenstates of such operators and, in particular, their corresponding values of the quantum numbers L and S. In other words, we need to find the terms of type L-S (or terms *tout court*) that originate from a given configuration. To solve this problem it is necessary to recall some results of the theory of angular momentum.

7.9 A Summary of Angular Momentum Theory

In quantum mechanics the angular momentum is defined as a vector operator $\mathbf{J}$ with Cartesian components J_x, J_y, and J_z, that satisfy the commutation relations

$$[J_i, J_j] = \sum_k i\epsilon_{ijk} J_k,$$

which may be condensed in the identity

$$\mathbf{J} \times \mathbf{J} = i\mathbf{J}.$$

From these properties we obtain that each component of the angular momentum commutes with its square, i.e.

$$\left[\mathbf{J}, J^2\right] = 0.$$

To find the eigenvalues and eigenvectors of the angular momentum we use the two above equations and take as the maximum set of commuting operators the square of the vector $\mathbf{J}$ and any of its three components, for example the J_z component. The result is that the eigenvectors may be identified by two quantum numbers, j and m, such that j can only be a non-negative integer or a half integer $(0, \frac{1}{2}, 1, \frac{3}{2},$ etc.), and, for a given j, m can only have one of the $(2j + 1)$ values $(-j, -j + 1, \ldots, j - 1, j)$. If we indicate by the symbol $|j, m\rangle$ such (normalised) eigenvectors, the eigenvalues are given by

$$J^2|j, m\rangle = j(j + 1)|j, m\rangle, \qquad J_z|j, m\rangle = m|j, m\rangle.$$

The relative phases of the eigenvectors are then fixed by setting, by convention, that the matrix elements of the so-called shift operators $J_\pm = J_x \pm i J_y$ are real, i.e.

$$\langle j, m \pm 1 | J_\pm | j, m \rangle = \sqrt{j(j+1) - m(m \pm 1)} = \sqrt{(j \pm m + 1)(j \mp m)}.$$

Consider now two angular momentum operators $\mathbf{J}_1$ and $\mathbf{J}_2$ that are commuting. It can be easily proved that the operator sum of the two angular momenta,

$$\mathbf{J} = \mathbf{J}_1 + \mathbf{J}_2,$$

also satisfy the commutation relations characteristic of the angular momenta. Consequently, $\mathbf{J}$ is itself an angular momentum, according to quantum mechanics.

To describe the eigenstates common to both angular momenta we can use two different representations (also called bases). The first basis is the one of the eigenvectors of the four commuting operators J_1^2, J_{1z}, J_2^2, J_{2z}. If $|j_1, m_1\rangle$ are the eigenvectors of $\mathbf{J}_1$ and $|j_2, m_2\rangle$ those of $\mathbf{J}_2$, the eigenvectors of this basis are given by the direct product $|j_1, m_1\rangle |j_2, m_2\rangle$ and can be indicated with the compact symbol $|j_1 j_2 m_1 m_2\rangle$. For them we have

$$J_1^2 |j_1 j_2 m_1 m_2\rangle = j_1(j_1 + 1)|j_1 j_2 m_1 m_2\rangle, \qquad J_{1z}|j_1 j_2 m_1 m_2\rangle = m_1 |j_1 j_2 m_1 m_2\rangle,$$
$$J_2^2 |j_1 j_2 m_1 m_2\rangle = j_2(j_2 + 1)|j_1 j_2 m_1 m_2\rangle, \qquad J_{2z}|j_1 j_2 m_1 m_2\rangle = m_2 |j_1 j_2 m_1 m_2\rangle.$$

Another basis is given by the eigenvectors of the four commuting operators J_1^2, J_2^2, J^2, J_z. Their eigenvectors are indicated with the symbol $|j_1 j_2 J M\rangle$, and are such that

$$J_1^2 |j_1 j_2 J M\rangle = j_1(j_1 + 1)|j_1 j_2 J M\rangle, \qquad J_2^2 |j_1 j_2 J M\rangle = j_2(j_2 + 1)|j_1 j_2 J M\rangle,$$
$$J^2 |j_1 j_2 J M\rangle = J(J + 1)|j_1 j_2 J M\rangle, \qquad J_z |j_1 j_2 J M\rangle = M |j_1 j_2 J M\rangle.$$

Since the two bases describe the same vector space, there is a similarity transformation that relates one basis to the other. We have then

$$|j_1 j_2 J M\rangle = \sum_{m_1 m_2} |j_1 j_2 m_1 m_2\rangle \langle j_1 j_2 m_1 m_2 | j_1 j_2 J M\rangle, \tag{7.13}$$

$$|j_1 j_2 m_1 m_2\rangle = \sum_{J M} |j_1 j_2 J M\rangle \langle j_1 j_2 J M | j_1 j_2 m_1 m_2\rangle. \tag{7.14}$$

The coefficients that appear in these transformations, written here in the form of scalar products, are called Wigner coefficients or more often Clebsh-Gordan coefficients. These coefficients are identically zero unless

$$M = m_1 + m_2,$$

and unless J is not equal to any of the possible values

$$J = |j_1 - j_2|, |j_1 - j_2| + 1, \ldots, j_1 + j_2 - 1, j_1 + j_2.$$

The proof of the first condition is easily obtained by considering the scalar product

$$\langle j_1 j_2 m_1 m_2 | J_{1z} + J_{2z} | j_1 j_2 J M\rangle,$$

and applying the operator $J_{1z} + J_{2z} = J_z$ alternatively on the "bra" or on the "ket". We obtain

$$(m_1 + m_2 - M)\langle j_1 j_2 m_1 m_2 | j_1 j_2 J M \rangle = 0,$$

which shows that, for the Clebsh-Gordan coefficient to be different from zero, we must necessarily have $M = m_1 + m_2$. The proof of the second condition is more complicated and will be given later.

In place of the Clebsh-Gordan coefficients, recent literature often uses the so-called 3-j symbols, introduced by Wigner and defined by the equation

$$\langle j_1 j_2 m_1 m_2 | j_1 j_2 J M \rangle = (-1)^{j_1 - j_2 + M} \sqrt{2J + 1} \begin{pmatrix} j_1 & j_2 & J \\ m_1 & m_2 & -M \end{pmatrix}. \qquad (7.15)$$

The explicit expression of the 3-j symbols (and thus that one of the Clebsh-Gordan coefficients) can be found with a laborious calculation. By defining in an appropriate manner the relative phases of the vectors of the two bases, the 3-j symbols (and the Clebsh-Gordan coefficients) turn out to be real and are given by the expression[4]

$$\begin{pmatrix} a & b & c \\ \alpha & \beta & -\gamma \end{pmatrix} = (-1)^{a-b+\gamma} \frac{1}{\sqrt{2c+1}} \langle ab\alpha\beta | abc\gamma \rangle$$

$$= (-1)^{a-b+\gamma} \Delta(a, b, c)$$

$$\times \sqrt{(a+\alpha)!(a-\alpha)!(b+\beta)!(b-\beta)!(c+\gamma)!(c-\gamma)!}$$

$$\times \sum_\nu (-1)^\nu \big[(a - \alpha - \nu)!(c - b + \alpha + \nu)!(b + \beta - \nu)!$$

$$\times (c - a - \beta + \nu)!\nu!(a + b - c - \nu)!\big]^{-1}, \qquad (7.16)$$

where the index ν takes all the values that give rise to meaningful (i.e. non-negative) factorials, and where the symbol $\Delta(a, b, c)$ is defined by

$$\Delta(a, b, c) = \sqrt{\frac{(a+b-c)!(a+c-b)!(b+c-a)!}{(a+b+c+1)!}}. \qquad (7.17)$$

The 3-j symbols satisfy some important relations that we state here without proof (of course, the Clebsh-Gordan coefficients have similar properties):

(a) the 3-j symbol

$$\begin{pmatrix} a & b & c \\ \alpha & \beta & \gamma \end{pmatrix}$$

is zero unless $\alpha + \beta + \gamma = 0$ and unless a, b and c satisfy the triangular inequality $|a - b| \leq c \leq a + b$.

---

[4]This formula, due to Racah (1942), is particularly symmetric with respect to the exchange of the three angular momenta a, b, and c. For the derivation of an equivalent but less symmetrical formula see, for example, Landi Degl'Innocenti and Landolfi (2004).

Fig. 7.3 Once m_1 and m_2 are defined, the tip of the vector **J**, sum of **J_1** and **J_2**, lies at any point on the circumference drawn at the *top*. Therefore the vector **J** is not uniquely determined

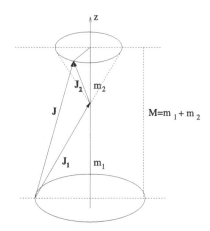

(b) Relations of completeness and orthonormality

$$\sum_{\alpha\beta}(2c+1)\begin{pmatrix} a & b & c \\ \alpha & \beta & \gamma \end{pmatrix}\begin{pmatrix} a & b & c' \\ \alpha & \beta & \gamma' \end{pmatrix} = \delta_{cc'}\delta_{\gamma\gamma'},$$

$$\sum_{c\gamma}(2c+1)\begin{pmatrix} a & b & c \\ \alpha & \beta & \gamma \end{pmatrix}\begin{pmatrix} a & b & c \\ \alpha' & \beta' & \gamma \end{pmatrix} = \delta_{\alpha\alpha'}\delta_{\beta\beta'}. \tag{7.18}$$

(c) By changing the sign of the components of the second row, the 3-j symbol is multiplied by the sign factor $(-1)^{a+b+c}$

$$\begin{pmatrix} a & b & c \\ -\alpha & -\beta & -\gamma \end{pmatrix} = (-1)^{a+b+c}\begin{pmatrix} a & b & c \\ \alpha & \beta & \gamma \end{pmatrix}. \tag{7.19}$$

(d) By interchanging any two columns, the 3-j symbol is multiplied by the same sign factor $(-1)^{a+b+c}$. For example

$$\begin{pmatrix} a & b & c \\ \alpha & \beta & \gamma \end{pmatrix} = (-1)^{a+b+c}\begin{pmatrix} b & a & c \\ \beta & \alpha & \gamma \end{pmatrix}. \tag{7.20}$$

The Clebsh-Gordan coefficients have an immediate physical interpretation. We consider two system for which the squares of their angular momenta were measured, finding the respective values $j_1(j_1+1)$ and $j_2(j_2+1)$, and, separately, the components of the moments along the z axis were also measured, finding the values m_1 and m_2. The square of the absolute value of the Clebsh-Gordan coefficients (the quantity $|\langle j_1 j_2 m_1 m_2 | j_1 j_2 J M\rangle|^2$) is then, according to the principles of quantum mechanics, the probability that the measurement of the square of the angular momentum of the total system gives as a result $J(J+1)$ and the measurement of the z component gives M. As can be seen from Fig. 7.3, the result of such a measurement is not unique even in classical physics. Indeed, while M is equal to $m_1 + m_2$ (a property that is unchanged in quantum mechanics), the value of the magnitude of the resulting angular momentum depends on the reciprocal orientation of the two addenda. To this classical uncertainty, we need to add the typical quantum-mechanical

Fig. 7.4 Schematic diagram showing how the number of states having an assigned value of M varies with M. The diagram relates to the particular case in which $j_1 = 5$ and $j_2 = 4$

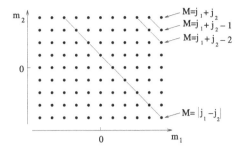

one according to which, when measuring the z component of an angular momentum, the other two components (the x and y components) remain undeterminate.

We can now show that J can have all the possible values (differing by unity) between $|j_1 - j_2|$ and $(j_1 + j_2)$. For this we refer to Fig. 7.4, where the physical states of the system are identified by the points of the grid, each point being characterized by a value of m_1 and a value of m_2. Each of the diagonals sketched in the figure connects states having the same value of $M = m_1 + m_2$. The highest value of M is $(j_1 + j_2)$ and there is only one state having such value of M. If we indicate with $\mathcal{N}(M)$ the number of states with a given value of M, we have that $\mathcal{N}(j_1 + j_2) = 1$. Turning to the lower value of M, we have $\mathcal{N}(j_1 + j_2 - 1) = 2$ and, as M decreases, the number of states increases until the value $M = |j_1 - j_2|$ is reached, for which $\mathcal{N}(|j_1 - j_2|) = 2j_{min} + 1$, where j_{min} is the smallest of j_1 and j_2. When M further decreases, $\mathcal{N}(M)$ remains unchanged up to $M = -(j_1 + j_2)$. This is further illustrated by the histogram of Fig. 7.5, which refers to the particular case $j_1 = 5$, $j_2 = 4$.

The counting of states with a given value of M allows to determine the values of J resulting form the sum of the two angular momenta. The fact that there is a state with the value $M = j_1 + j_2$ implies that there must be a value of J equal to $j_1 + j_2$. This means that we can "delete" from the histogram of Fig. 7.5 a state with $M = j_1 + j_2$, a state with $M = j_1 + j_2 - 1$, and so on, until the cancellation of a state with $M = -j_1 - j_2$. This is equivalent to removing the bottom line from the histogram. At this point we note that there is one, and only one state with $M = j_1 + j_2 - 1$. This implies that a value of J equal to $j_1 + j_2 - 1$ must exist, so we can delete the second row (from the bottom) of the histogram. By repeating this procedure we can easily prove that all the possible values of J are

$$|j_1 - j_2|, |j_1 - j_2| + 1, \ldots, j_1 + j_2 - 1, j_1 + j_2,$$

as we anticipated. To verify it, we can check that the total number of states is effectively what we expected, i.e. that the following relation is verified

$$\sum_{J=|j_1-j_2|}^{j_1+j_2} (2J + 1) = (2j_1 + 1)(2j_2 + 1).$$

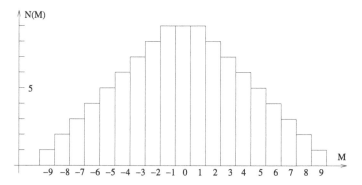

Fig. 7.5 Histogram showing the variation in the number of states with a given M as a function of M. The diagram refers to the case where $j_1 = 5$ and $j_2 = 4$

The sum can be performed by distinguishing the two cases $j_1 \geq j_2$ or $j_1 < j_2$ and by taking into account the identity (valid both for the sum of integers and that of semi-integers)

$$\sum_{i=n_1}^{n_2} i = \frac{n_2(n_2+1) - n_1(n_1-1)}{2}.$$

We obtain for example, for $j_1 \geq j_2$,

$$\sum_{J=j_1-j_2}^{j_1+j_2} (2J+1) = 2\left[\frac{(j_1+j_2+1)(j_1+j_2)}{2} - \frac{(j_1-j_2-1)(j_1-j_2)}{2}\right]$$
$$+ (2j_2+1) = (2j_1+1)(2j_2+1).$$

In the particular case in which the two quantum numbers j_1 and j_2 are equal, it is possible to show that the eigenstates $|j_1 j_2 J M\rangle$ are divided into two groups having opposite symmetry with respect to the interchange of the two angular momenta. In this case, by placing $j_1 = j_2 = j$, we have, recalling Eq. (7.13) and the relation between the Clebsh-Gordan coefficients and the 3-j symbols (Eq. (7.15))

$$|jjJM\rangle = \sum_{m_1 m_2} (-1)^M \sqrt{2J+1} \begin{pmatrix} j & j & J \\ m_1 & m_2 & -M \end{pmatrix} |jjm_1m_2\rangle.$$

By taking into account the property of symmetry of the 3-j symbols with respect to the exchange of two adjacent columns (Eq. (7.20)) we can see that after the exchange of the two angular momenta the eigenstate $|jjJM\rangle$ is multiplied by the factor $(-1)^{2j+J}$. We therefore obtain that the eigenvectors corresponding to eigenvalues $J = 2j, 2j - 2$, etc. are symmetric with respect to the exchange of two angular momenta, while the eigenvectors corresponding to the eigenvalues $J = 2j - 1, 2j - 3$, etc. are antisymmetric. Note that if j is an integer, the group of symmetrical eigenvectors ends with $J = 0$ and that of antisymmetric eigenvectors ends with $J = 1$. On the other hand, if j is a half-integer the opposite occurs.

Fig. 7.6 Schematic diagram
showing the variation with M
of the number of states
having M assigned. The
diagram relates to the case in
which the two particles are
indistinguishable, with both
j_1 and j_2 equal to 3

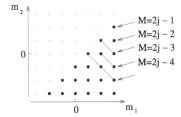

Fig. 7.7 Histogram showing
the variation in the number of
states with a given M as a
function of M. The diagram
relates to the case in which
the two particles are
indistinguishable, with both
j_1 and j_2 equal to 3

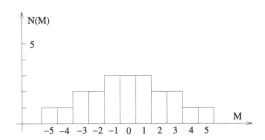

We could have obtained this property by imposing the Pauli exclusion principle to
the grid in Fig. 7.4. Suppose that j_1 and j_2 (with $j_1 = j_2 = j$) are quantum numbers
of an angular momentum (orbital, spin, or total) of two indistinguishable particles,
and that all other possible quantum numbers of the two particles are equal. In this
case, the grid of Fig. 7.4 takes a square form and narrows down, because of the
exclusion principle, into the semi-grid of Fig. 7.6 (which refers to the particular case
$j_1 = j_2 = 3$). If we count the number of states with a given value of M, we obtain
the histogram represented in Fig. 7.7. By repeating the above reasoning, it follows
that the possible values for the total angular momentum are $J = 2j - 1, 2j - 3$, etc.,
until we obtain $J = 1$ if j is an integer, or $J = 0$ if j is a half-integer. These are the
states that satisfy the exclusion principle and are then those having a wave function
that is antisymmetric with respect to the exchange of the two particles.

7.10   Terms Originating from Given Configurations

This section deals with the issue of determining, given a configuration, the possible
values of the quantum numbers L and S. We start by considering the most simple
cases, the configurations with a closed subshell, such as s^2, p^6, d^{10}, f^{14}, etc. Let us
consider e.g. the p^6 case. The possible values of m and m_s for the six electrons are
the following

$$(m, m_s) = \left(-1, -\frac{1}{2}\right), \left(-1, \frac{1}{2}\right), \left(0, -\frac{1}{2}\right), \left(0, \frac{1}{2}\right), \left(1, -\frac{1}{2}\right), \left(1, \frac{1}{2}\right),$$

and the Pauli exclusion principle does not allow any other combination of quantum numbers. We therefore have a single quantum state, characterized by

$$M_L = \sum_i m_i = 0, \qquad M_S = \sum_i m_{si} = 0,$$

that is a state with $L = 0$ and $S = 0$, i.e. a ^1S state.

Similar arguments apply to any other configuration with a closed subshell, so we conclude that a closed subshell always have null angular momentum (both orbital and spin). The corresponding term is a ^1S. To determine the terms of any configuration, we therefore only need to consider the electrons in open subshells, given that all those in closed subshells do not contribute to the angular momentum. We have already seen that such a property holds for the degeneracy of a configuration.

We now consider the configurations with a single electron in an open subshell. Obviously, the orbital (and spin) angular momentum of the configuration is the same as that of the single electron, so that we simply have for an ns configuration (n arbitrary) a term ^2S, for an np configuration a term ^2P, for an nd configuration a term ^2D, and so on.

When we consider configurations with two or more electrons we need to distinguish between equivalent and non-equivalent electrons. For example, the configuration $npn'p$ (with $n \neq n'$), normally indicated with the symbol pp', has terms that differ from those originating from the np^2 configuration, usually written as p^2. Similarly, the configuration d^2d' has terms that differ from those of the d^3 or the $dd'd''$, etc. For the configurations with two non-equivalent electrons, the terms can be obtained by combining (following the angular momentum rules) the orbital and spin angular momenta of the two electrons separately, and then coupling each resulting value of L with each resulting value of S. If we consider for example the configuration dd', L can have the values originating from the addition of two angular momenta each having a value of 2, i.e. the values 0, 1, 2, 3, 4. Similarly, S can assume the values resulting from the sum of two angular momenta each equal to $\frac{1}{2}$, that is either 0 or 1. In the end we obtain the ten possible terms

$$^1S, \, ^1P, \, ^1D, \, ^1F, \, ^1G, \, ^3S, \, ^3P, \, ^3D, \, ^3F, \, ^3G.$$

Similarly, if we consider the pd configuration, L can have the three values 1, 2, 3 and S the two values 0 and 1, so we have the six terms

$$^1P, \, ^1D, \, ^1F, \, ^3P, \, ^3D, \, ^3F.$$

Table 7.3 lists the possible terms for some configurations with two non-equivalent electrons that are often encountered in the spectroscopic analysis.

When considering the terms for two equivalent electrons, we need to recall the properties on the symmetry of the eigenfunctions discussed in the previous section. Given that the total eigenfunction must be antisymmetric with respect to the exchange of two particles, we have two distinct cases: (a) we consider a symmetric eigenfunction for the spin and an antisymmetric eigenfunction for the orbital angular momentum, or (b) an antisymmetric eigenfunction for the spin and a symmetric eigenfunction for the orbital angular momentum. For the spin eigenfunction,

Table 7.3 Terms of configurations with two non-equivalent electrons

Configuration	Terms
ss'	$^1S, {}^3S$
sp	$^1P, {}^3P$
sd	$^1D, {}^3D$
sf	$^1F, {}^3F$
pp'	$^1S, {}^1P, {}^1D, {}^3S, {}^3P, {}^3D$
pd	$^1P, {}^1D, {}^1F, {}^3P, {}^3D, {}^3F$
pf	$^1D, {}^1F, {}^1G, {}^3D, {}^3F, {}^3G$
dd'	$^1S, {}^1P, {}^1D, {}^1F, {}^1G, {}^3S, {}^3P, {}^3D, {}^3F, {}^3G$
df	$^1P, {}^1D, {}^1F, {}^1G, {}^1H, {}^3P, {}^3D, {}^3F, {}^3G, {}^3H$
ff'	$^1S, {}^1P, {}^1D, {}^1F, {}^1G, {}^1H, {}^1I, {}^3S, {}^3P, {}^3D, {}^3F, {}^3G, {}^3H, {}^3I$

Table 7.4 Terms of configurations with two equivalent electrons

Configuration	Terms
s^2	^1S
p^2	$^1S, {}^1D, {}^3P$
d^2	$^1S, {}^1D, {}^1G, {}^3P, {}^3F$
f^2	$^1S, {}^1D, {}^1G, {}^1I, {}^3P, {}^3F, {}^3H$

the $S = 1$ value corresponds to a symmetric eigenfunction, while $S = 0$ to an anti-symmetric eigenfunction. For the orbital angular momentum we have various possibilities. For example, for two equivalent p electrons we can have the three values $L = 2$ (symmetric function), $L = 1$ (antisymmetric), $L = 0$ (symmetric). By taking into account these arguments, we obtain the terms listed in Table 7.4.

The cases with three or more electrons in open subshells are more complicated. The most appropriate way to deal with these types of cases is group theory, although the same results can be obtained with a more simplified approach based on the construction of suitable tables, as described below.

In the case of configurations with three electrons, we still need to distinguish the cases when the electrons are equivalent or not. We can also have intermediate cases, with two equivalent electrons and one not. When at least one of the electrons is non-equivalent, the terms of the configuration can be obtained with the usual sum rules for two angular momenta, by adding the angular momentum of the third electron (the non-equivalent one) to the angular momentum obtained for the other two electrons (either equivalent or not). For example, for the $pp'p''$ configuration we can see from Table 7.3 that the terms arising from the pp' configuration are the following: $^1S, {}^1P, {}^1D, {}^3S, {}^3P, {}^3D$. By adding a non-equivalent p electron, we obtain:

(a) from ^1S: the term ^2P
(b) from ^1P: the terms $^2S, {}^2P, {}^2D$
(c) from ^1D: the terms $^2P, {}^2D, {}^2F$

Table 7.5 Terms of configurations with three electrons, where at least two are non-equivalent

Config.	Terms
$ss's''$	$^2S(2), {}^4S$
$ss'p$	$^2P(2), {}^4P$
$ss'd$	$^2D(2), {}^4D$
spp'	$^2S(2), {}^2P(2), {}^2D(2), {}^4S, {}^4P, {}^4D$
sp^2	$^2S, {}^2P, {}^2D, {}^4P$
spd	$^2P(2), {}^2D(2), {}^2F(2), {}^4P, {}^4D, {}^4F$
$pp'p''$	$^2S(2), {}^2P(6), {}^2D(4), {}^2F(2), {}^4S, {}^4P(3), {}^4D(2), {}^4F$
p^2p'	$^2S, {}^2P(3), {}^2D(2), {}^2F, {}^4S, {}^4P, {}^4D$
$pp'd$	$^2S(2), {}^2P(4), {}^2D(6), {}^2F(4), {}^2G(2), {}^4S, {}^4P(2), {}^4D(3), {}^4F(2), {}^4G$
pdf	$^2S(2), {}^2P(4), {}^2D(6), {}^2F(6), {}^2G(6), {}^2H(4), {}^2I(2), {}^4S, {}^4P(2), {}^4D(3), {}^4F(3), {}^4G(3), {}^4H(2), {}^4I$

(d) from ^3S: the terms $^2P, {}^4P$
(e) from ^3P: the terms $^2S, {}^2P, {}^2D, {}^4S, {}^4P, {}^4D$
(f) from ^3D: the terms $^2P, {}^2D, {}^2F, {}^4P, {}^4D, {}^4F$.

In the end, we obtain the following terms

$$^2S(2), {}^2P(6), {}^2D(4), {}^2F(2), {}^4S, {}^4P(3), {}^4D(2), {}^4F,$$

where the numbers in parentheses indicate how many times a term is present (if more than once). In order to identify each specific term, it is common to write in parenthesis its "progenitor" term. For example, we can distinguish the two ^2S terms as $({}^1P){}^2S$ and $({}^3P){}^2S$, obtained from the ^1P and ^3P terms respectively.

By repeating a similar procedure as we have seen for the $pp'p''$ configuration, we can easily obtain the results shown in Table 7.5 for the most common configurations.

If on the other hand the electrons are equivalent, the procedure to determine the possible L and S values becomes more complex, because we cannot easily apply the angular momentum sum rules to obtain an antisymmetric eigenfunction. In these cases, the terms can be obtained by writing a table with all the possible combinations that satisfy Pauli exclusion principle for the m and m_s values. A practical example for the p^3 configuration is shown in Table 7.6. The table lists, in each cell of the first three columns (one for each possible value of m), the number N_m of the electrons that have that particular value of m. Obviously, such a number can only have the values $N_m = 0, 1, 2$, otherwise the exclusion principle would be violated. Once the N_m values are specified, the value of M_L (shown in the fourth column) is easily obtained by the equation

$$M_L = \sum_m m N_m.$$

We next observe that the cells with $N_m = 0$ or $N_m = 2$ do not contribute to the quantum number M_S. This is obvious for $N_m = 0$, while for $N_m = 2$ this is understood by taking into account the Pauli principle, by which the two electrons must have anti-parallel spin. The contribution to M_S therefore only comes from the cells

Table 7.6 The table shows the procedure to obtain the terms for the p^3 configuration. See the text for a description of the first five columns. The symbols in the last column indicate the terms $(\times = {}^2D, \Delta = {}^2P, + = {}^4S)$

$m = 1$	$m = 0$	$m = -1$	M_L	M_S	Terms
2	1	0	2	$\frac{1}{2}, -\frac{1}{2}$	$\times$
2	0	1	1	$\frac{1}{2}, -\frac{1}{2}$	$\times$
1	2	0	1	$\frac{1}{2}, -\frac{1}{2}$	Δ
1	1	1	0	$\frac{3}{2}, \frac{1}{2}, \frac{1}{2}, \frac{1}{2}, -\frac{1}{2}, -\frac{1}{2}, -\frac{1}{2}, -\frac{3}{2}$	$\times \Delta +$
1	0	2	-1	$\frac{1}{2}, -\frac{1}{2}$	Δ
0	2	1	-1	$\frac{1}{2}, -\frac{1}{2}$	$\times$
0	1	2	-2	$\frac{1}{2}, -\frac{1}{2}$	$\times$

with $N_m = 1$. If we indicate with k the number of such cells, the number of possible values of M_S is then 2^k, because we have two distinct possible orientations for the spin of the electron in each cell. The possible values of M_S are obtained by adding all possible combinations of k addenda equal to $\pm\frac{1}{2}$. So, for example, for $k = 1$ we have two values of M_S equal to $\frac{1}{2}$ and $-\frac{1}{2}$; for $k = 2$ we have the four values of M_S equal to 1, 0, 0, -1; for $k = 3$ we have the 8 values $\frac{3}{2}, \frac{1}{2}, \frac{1}{2}, \frac{1}{2}, -\frac{1}{2}, -\frac{1}{2}, -\frac{3}{2}$; and so on. Once the possible values of M_S are written in the fifth column, the table becomes complete and we can proceed to find the terms. We start by considering the highest value of M_L that is listed in the table ($M_L = 2$). This value is associated with two values of M_S, equal to $\pm\frac{1}{2}$, so we must have a ^2D term, indicated in the last column of the table with the symbol $\times$. Obviously, the other possible values of M_L ($1, 0, -1, -2$) associated with the two $M_S = \pm\frac{1}{2}$ values also correspond to this term. We therefore identify the rows corresponding to the values M_L and M_S with the same symbol in the last column and we mentally remove such states from the table. At this point the largest value of M_L is $M_L = 1$, that is again associated with $M_S = \pm\frac{1}{2}$. We then have a ^2P term, indicated with a new symbol (Δ). Once the states corresponding to this term are pinpointed, we are left with states having $M_L = 0$ and $M_S = \pm\frac{3}{2}, \pm\frac{1}{2}$. We then have a last term, ^4S, indicated with a different symbol ($+$), which completes the list of all possible states. In the end we have found that the p^3 configuration produces three terms, ^2P, ^2D, and ^4S. Another example, for the d^3 configuration, is shown in Table 7.7. From the analysis of this table we obtain the following terms: ^2P, $^2D(2)$, ^2F, ^2G, ^2H, ^4P, and ^4F.

The way in which these tables are constructed and analysed shows quite clearly that the terms of a given configuration with q equivalent electrons are the same as those of the complementary configuration having $(Q - q)$ equivalent electrons, where Q is the maximum number of electrons that can be present in the subshell. For example, the p^2 and p^4 configurations have the same terms. The same holds for the d^3 and d^7 configurations, the f^4 and f^{10} configurations, and so on. To demonstrate this, suppose we have constructed the table for the configuration with q electrons having angular momentum l and exchange each of the numbers contained in the first columns (corresponding to the possible values of m) with its complement to 2 (i.e.

Table 7.7 The table shows the procedure for determining the terms derived from the configuration d^3. The symbols in the last column schematically identify the terms as follows: $\times = {}^2H$, $\Delta = {}^2G$, $+ = {}^4F$, $\ddagger = {}^2F$, $\bigcirc = {}^2D$ (first term), $\odot = {}^2D$ (second term), $\nabla = {}^4P$, $\dagger = {}^2P$.

2	1	0	−1	−2	M_L	M_S	Terms
2	1	0	0	0	5	$\frac{1}{2},-\frac{1}{2}$	$\times$
2	0	1	0	0	4	$\frac{1}{2},-\frac{1}{2}$	$\times$
2	0	0	1	0	3	$\frac{1}{2},-\frac{1}{2}$	$\times$
2	0	0	0	1	2	$\frac{1}{2},-\frac{1}{2}$	$\times$
1	2	0	0	0	4	$\frac{1}{2},-\frac{1}{2}$	Δ
1	1	1	0	0	3	$\frac{3}{2},\frac{1}{2},\frac{1}{2},\frac{1}{2},-\frac{1}{2},-\frac{1}{2},-\frac{1}{2},-\frac{3}{2}$	$\Delta + \ddagger$
1	1	0	1	0	2	$\frac{3}{2},\frac{1}{2},\frac{1}{2},\frac{1}{2},-\frac{1}{2},-\frac{1}{2},-\frac{1}{2},-\frac{3}{2}$	$\Delta + \ddagger$
1	1	0	0	1	1	$\frac{3}{2},\frac{1}{2},\frac{1}{2},\frac{1}{2},-\frac{1}{2},-\frac{1}{2},-\frac{1}{2},-\frac{3}{2}$	$\times \, \Delta +$
1	0	2	0	0	2	$\frac{1}{2},-\frac{1}{2}$	$\bigcirc$
1	0	1	1	0	1	$\frac{3}{2},\frac{1}{2},\frac{1}{2},\frac{1}{2},-\frac{1}{2},-\frac{1}{2},-\frac{1}{2},-\frac{3}{2}$	$\ddagger \bigcirc \nabla$
1	0	1	0	1	0	$\frac{3}{2},\frac{1}{2},\frac{1}{2},\frac{1}{2},-\frac{1}{2},-\frac{1}{2},-\frac{1}{2},-\frac{3}{2}$	$\times \, \Delta +$
1	0	0	2	0	0	$\frac{1}{2},-\frac{1}{2}$	$\ddagger$
1	0	0	1	1	−1	$\frac{3}{2},\frac{1}{2},\frac{1}{2},\frac{1}{2},-\frac{1}{2},-\frac{1}{2},-\frac{1}{2},-\frac{3}{2}$	$\times \, \Delta +$
1	0	0	0	2	−2	$\frac{1}{2},-\frac{1}{2}$	$\times$
0	2	1	0	0	2	$\frac{1}{2},-\frac{1}{2}$	$\odot$
0	2	0	1	0	1	$\frac{1}{2},-\frac{1}{2}$	$\odot$
0	2	0	0	1	0	$\frac{1}{2},-\frac{1}{2}$	$\bigcirc$
0	1	2	0	0	1	$\frac{1}{2},-\frac{1}{2}$	$\dagger$
0	1	1	1	0	0	$\frac{3}{2},\frac{1}{2},\frac{1}{2},\frac{1}{2},-\frac{1}{2},-\frac{1}{2},-\frac{1}{2},-\frac{3}{2}$	$\odot \nabla \dagger$
0	1	1	0	1	−1	$\frac{3}{2},\frac{1}{2},\frac{1}{2},\frac{1}{2},-\frac{1}{2},-\frac{1}{2},-\frac{1}{2},-\frac{3}{2}$	$\ddagger \bigcirc \nabla$
0	1	0	2	0	−1	$\frac{1}{2},-\frac{1}{2}$	$\odot$
0	1	0	1	1	−2	$\frac{3}{2},\frac{1}{2},\frac{1}{2},\frac{1}{2},-\frac{1}{2},-\frac{1}{2},-\frac{1}{2},-\frac{3}{2}$	$\Delta + \ddagger$
0	1	0	0	2	−3	$\frac{1}{2},-\frac{1}{2}$	$\times$
0	0	2	1	0	−1	$\frac{1}{2},-\frac{1}{2}$	$\dagger$
0	0	2	0	1	−2	$\frac{1}{2},-\frac{1}{2}$	$\bigcirc$
0	0	1	2	0	−2	$\frac{1}{2},-\frac{1}{2}$	$\odot$
0	0	1	1	1	−3	$\frac{3}{2},\frac{1}{2},\frac{1}{2},\frac{1}{2},-\frac{1}{2},-\frac{1}{2},-\frac{1}{2},-\frac{3}{2}$	$\Delta + \ddagger$
0	0	1	0	2	−4	$\frac{1}{2},-\frac{1}{2}$	$\times$
0	0	0	2	1	−4	$\frac{1}{2},-\frac{1}{2}$	Δ
0	0	0	1	2	−5	$\frac{1}{2},-\frac{1}{2}$	$\times$

$0 \rightarrow 2$, $1 \rightarrow 1$, $2 \rightarrow 0$). If we indicate with k_m the numbers contained in the original table, the new numbers will be $(2 - k_m)$. The total number of electrons is now given by

$$\sum_{m=-l}^{l} (2 - k_m) = 2(2l + 1) - \sum_{m=-l}^{l} k_m = 2(2l + 1) - q = Q - q.$$

On the other hand, if a row in the original table produced the result M_L, for the new table we have

$$M'_L = \sum_{m=-l}^{l} m(2 - k_m) = -M_L,$$

while the values of M_S remain unchanged. The tables obtained from the two complementary configurations have therefore the same structure and produce exactly the same terms.

Finally, we note that in the case of q equivalent electrons, if we denote by p the number of possible values of m ($p = 2l + 1 = Q/2$), the number of rows N_{rows} that we must consider in the construction of a table like the previous ones is equal to the number of non-negative integer solutions of the equation

$$k_1 + k_2 + \cdots + k_p = q,$$

with $k_i \leq 2$. It is possible to show that such number is given by

$$N_{\text{rows}} = \sum_r \frac{p!}{r!(q - 2r)!(p - q + r)!},$$

with the sum extended to all values of r such that the factorials in the denominator are significant. For example, to build the table for the configuration f^6 we have ($q = 6$, $p = 7$)

$$N_{\text{rows}} = \sum_{r=0}^{3} \frac{7!}{r!(6 - 2r)!(1 + r)!} = 357.$$

With the above procedures, the terms of a given configurations of equivalent electrons can be determined. The results concerning the most common configurations are given in Table 7.8 (which also summarises the results already contained in some of the previous tables).

7.11 The Eigenfunctions of the Non-relativistic Hamiltonian

As we have seen in Sect. 7.6, the configurations are a set or a collection of quantum states described by eigenfunctions of the form $\Psi^A(a_1, a_2, \ldots, a_N)$ that satisfy the Schrödinger equation

$$\mathcal{H}_0 \Psi^A(a_1, a_2, \ldots, a_N) = \mathcal{W}_0 \Psi^A(a_1, a_2, \ldots, a_N),$$

where $\mathcal{H}_0$ is the Hamiltonian of Eq. (7.3) and $\mathcal{W}_0$ is defined in Eq. (7.12). Such quantum states are degenerate, and as we have seen the degeneracy is g. If we consider the action of the Hamiltonian $\mathcal{H}_1$ (Eq. (7.4)), we need, in general, to evaluate all the matrix elements of the form

$$\langle \Psi^A(a_1, a_2, \ldots, a_N) | \mathcal{H}_1 | \Psi^A(a'_1, a'_2, \ldots, a'_N) \rangle,$$

Table 7.8 Terms from configurations with equivalent electrons

Config.	Terms
s	^2S
p, p^5	^2P
p^2, p^4	$^1S, ^1D, ^3P$
p^3	$^2P, ^2D, ^4S$
d, d^9	^2D
d^2, d^8	$^1S, ^1D, ^1G, ^3P, ^3F$
d^3, d^7	$^2P, ^2D(2), ^2F, ^2G, ^2H, ^4P, ^4F$
d^4, d^6	$^1S(2), ^1D(2), ^1F, ^1G(2), ^1I, ^3P(2), ^3D, ^3F(2), ^3G, ^3H, ^5D$
d^5	$^2S, ^2P, ^2D(3), ^2F(2), ^2G(2), ^2H, ^2I, ^4P, ^4D, ^4F, ^4G, ^6S$
f, f^{13}	^2F
f^2, f^{12}	$^1S, ^1D, ^1G, ^1I, ^3P, ^3F, ^3H$
f^3, f^{11}	$^2P, ^2D(2), ^2F(2), ^2G(2), ^2H(2), ^2I, ^2K, ^2L, ^4S, ^4D, ^4F, ^4G, ^4I$
f^4, f^{10}	$^1S(2), ^1D(4), ^1F, ^1G(4), ^1H(2), ^1I(3), ^1K, ^1L(2), ^1N, ^3P(3), ^3D(2), ^3F(4), ^3G(3),$ $^3H(4), ^3I(2), ^3K(2), ^3L, ^3M, ^5S, ^5D, ^5F, ^5G, ^5I$
f^5, f^9	$^2P(4), ^2D(5), ^2F(7), ^2G(6), ^2H(7), ^2I(5), ^2K(5), ^2L(3), ^2M(2), ^2N, ^2O, ^4S, ^4P(2),$ $^4D(3), ^4F(4), ^4G(4), ^4H(3), ^4I(3), ^4K(2), ^4L, ^4M, ^6P, ^6F, ^6H$
f^6, f^8	$^1S(4), ^1P, ^1D(6), ^1F(4), ^1G(8), ^1H(4), ^1I(7), ^1K(3), ^1L(4), ^1M(2), ^1N(2), ^1Q, ^3P(6),$ $^3D(5), ^3F(9), ^3G(7), ^3H(9), ^3I(6), ^3K(6), ^3L(3), ^3M(3), ^3N, ^3O, ^5S, ^5P, ^5D(3),$ $^5F(2), ^5G(3), ^5H(2), ^5I(2), ^5K, ^5L, ^7F$
f^7	$^2S(2), ^2P(5), ^2D(7), ^2F(10), ^2G(10), ^2H(9), ^2I(9), ^2K(7), ^2L(5), ^2M(4), ^2N(2), ^2O,$ $^2Q, ^4S(2), ^4P(2), ^4D(6), ^4F(5), ^4G(7), ^4H(5), ^4I(5), ^4K(3), ^4L(3), ^4M, ^4N, ^6P, ^6D,$ $^6F, ^6G, ^6H, ^6I, ^8S$

that is, all the matrix elements between states belonging to a configuration and the states belonging either to the same configuration or to different configurations. The diagonalisation of the resulting matrix leads to the determination of the energy states of the atomic system (in the non-relativistic approximation).

Obviously, the problem is of great complexity, so it is usual to introduce an approximation that consists in treating the Hamiltonian $\mathcal{H}_1$ as a perturbation of the Hamiltonian $\mathcal{H}_0$ so as to be able to apply the first order perturbation theory. In this approximation, the Hamiltonian $\mathcal{H}_1$ is diagonalised separately for each configuration and the matrix elements between different configurations are omitted. In other words, if our atomic system consists of $\mathcal{N}$ configurations, each having degeneracy $g_i (i = 1, 2, \ldots, \mathcal{N})$, $\mathcal{N}$ matrices having dimension g_i are diagonalised, instead of a single matrix having dimension $\mathcal{D} = \sum_{i=1}^{\mathcal{N}} g_i$.

It should be emphasized that this approach is approximate and that it neglects a phenomenon, called configuration interaction, which can be very important in certain cases. On the other hand, when the interaction between configurations is not negligible, the quantum numbers which characterize the configurations are not good quantum numbers anymore, and a configuration can be assigned to a state only as a zero-order approximation. In this case, one atomic state should be rather described by a linear combination of eigenfunctions of the type $\Psi^{\mathrm{A}}(a_1, a_2, \ldots, a_N)$, with the sum extended to eigenfunctions of differing configurations. However, given that the Hamiltonian $\mathcal{H}_1$ commutes with the parity operator $\mathcal{P}$, the parity of a state remains a good quantum number and the above linear combination is in any case restricted to configurations of the same parity. The interaction between configurations is particularly important for excited states, where the difference between the energies $\mathcal{W}_0$ of neighboring configurations decreases. For low-energy states the interaction between configurations can be neglected in most cases. In the following we will neglect configuration interaction and determine the energies of the states using perturbation theory.

With this approximation, as we have said, the problem of finding the energies of the states that originate from a given configuration implies the diagonalisation of the Hamiltonian $\mathcal{H}_1$ on the degenerate space of the same configuration. This diagonalisation is not however necessary if we are able to determine a basis on which the Hamiltonian $\mathcal{H}_1$ is already diagonal. Indeed this basis exists and is formed by the eigenfunctions that describe the terms, that is by the eigenfunctions of the operators L^2, L_z, S^2, and S_z that we introduced in the previous section. In fact, if $\Psi(\alpha, L, S, M_L, M_S)$ is the eigenfunction of a term (L, S, M_L, M_S) arising from the configuration α, and $\Psi(\alpha, L', S', M'_L, M'_S)$ is the analogous eigenfunction of a term (L', S', M'_L, M'_S), taking into account that the operators L^2, S^2, L_z, S_z commute with $\mathcal{H}_1$, we have

$$0 = \langle \Psi(\alpha, L, S, M_L, M_S) | [L^2, \mathcal{H}_1] | \Psi(\alpha, L', S', M'_L, M'_S) \rangle$$
$$= [L(L+1) - L'(L'+1)] \langle \Psi(\alpha, L, S, M_L, M_S) | \mathcal{H}_1 | \Psi(\alpha, L', S', M'_L, M'_S) \rangle,$$
$$0 = \langle \Psi(\alpha, L, S, M_L, M_S) | [S^2, \mathcal{H}_1] | \Psi(\alpha, L', S', M'_L, M'_S) \rangle$$
$$= [S(S+1) - S'(S'+1)] \langle \Psi(\alpha, L, S, M_L, M_S) | \mathcal{H}_1 | \Psi(\alpha, L', S', M'_L, M'_S) \rangle,$$
$$0 = \langle \Psi(\alpha, L, S, M_L, M_S) | [L_z, \mathcal{H}_1] | \Psi(\alpha, L', S', M'_L, M'_S) \rangle$$
$$= (M_L - M'_L) \langle \Psi(\alpha, L, S, M_L, M_S) | \mathcal{H}_1 | \Psi(\alpha, L', S', M'_L, M'_S) \rangle,$$
$$0 = \langle \Psi(\alpha, L, S, M_L, M_S) | [S_z, \mathcal{H}_1] | \Psi(\alpha, L', S', M'_L, M'_S) \rangle$$
$$= (M_S - M'_S) \langle \Psi(\alpha, L, S, M_L, M_S) | \mathcal{H}_1 | \Psi(\alpha, L', S', M'_L, M'_S) \rangle.$$

These four equations show that the matrix elements of $\mathcal{H}_1$ can be different from zero only if we consider terms having the same values for L, for S, for M_L, and for M_S. With the basis of eigenfunctions $\Psi(\alpha, L, S, M_L, M_S)$, the Hamiltonian is diagonal in blocks, each block being characterized by the four values (L, S, M_L, M_S). In particular, if the configuration does not allow for multiple terms, the Hamiltonian is diagonal. Furthermore, the matrix elements do not depend on M_L and on M_S and can be calculated for any combination of these two quantum numbers. The proof of this fact is simple and is based on the property that the Hamiltonian commutes with the shift operators $L_\pm = L_x \pm \mathrm{i} L_y$, and $S_\pm = S_x \pm \mathrm{i} S_y$.

The search for the energies of the terms can be done by means of a general method, the so-called diagonal sum rule, described in the following chapter. In what follows we present some important results obtained by such a method for two atoms having a particularly simple structure, such as helium and carbon.

7.12 The Helium Atom

As we have seen in Sect. 7.7, the ground configuration for the helium atom is $1s^2$, with a single ^1S term having zero angular momentum. The other most common configurations are obtained by the excitation of one of the two $1s$ electrons to an orbital nl. We therefore have, for example, the excited configurations $1s2s$, $1s2p$, $1s3s$, $1s3p$, $1s3d$, etc. Each of these configurations has two corresponding terms, a singlet and a triplet, of the type ^1L and ^3L.

Consider in particular a configuration $1snl$ (with $n \geq 2$). We denote by $\psi_a(\mathbf{x})$ the wave function of the orbital $1s$ and by $\psi_b(\mathbf{x})$ the wave function of any of the m states of the orbital nl. The wave functions of the singlet and triplet states depend on the spatial coordinates as follows

$$\Psi\left(^1L\right) = \frac{1}{\sqrt{2}}\left[\psi_a(\mathbf{x}_1)\psi_b(\mathbf{x}_2) + \psi_a(\mathbf{x}_2)\psi_b(\mathbf{x}_1)\right],$$

$$\Psi\left(^3L\right) = \frac{1}{\sqrt{2}}\left[\psi_a(\mathbf{x}_1)\psi_b(\mathbf{x}_2) - \psi_a(\mathbf{x}_2)\psi_b(\mathbf{x}_1)\right].$$

The total wave functions are obtained by multiplying the first expression with an antisymmetric spin wave function (corresponding to $S = 0$) and the second expression for a symmetric wave function ($S = 1$), in such a way as to obtain in any case an overall antisymmetric wave function. The explicit expression of the spin eigenfunctions is not relevant as the Hamiltonian $\mathcal{H}_1$ does not depend on the spin. We can obtain the correction $\Delta E(^1L)$ to the energy of the singlet by calculating the matrix element

$$\Delta E\left(^1L\right) = \left\langle \Psi\left(^1L\right)\middle|\mathcal{H}_1\middle|\Psi\left(^1L\right)\right\rangle.$$

The Hamiltonian $\mathcal{H}_1$ can be rewritten (recall Eq. (7.4))

$$\mathcal{H}_1 = f(\mathbf{x}_1) + f(\mathbf{x}_2) + g(\mathbf{x}_1, \mathbf{x}_2),$$

where

$$f(\mathbf{x}_1) = \frac{2e_0^2}{r_1} - V_c(r_1), \qquad f(\mathbf{x}_2) = -\frac{2e_0^2}{r_2} - V_c(r_2), \qquad g(\mathbf{x}_1, \mathbf{x}_2) = \frac{e_0^2}{r_{12}},$$

so we obtain

$$\Delta E\left(^1L\right) = \left\langle \Psi\left(^1L\right)\middle|f(\mathbf{x}_1)\middle|\Psi\left(^1L\right)\right\rangle + \left\langle \Psi\left(^1L\right)\middle|f(\mathbf{x}_2)\middle|\Psi\left(^1L\right)\right\rangle$$
$$+ \left\langle \Psi\left(^1L\right)\middle|g(\mathbf{x}_1, \mathbf{x}_2)\middle|\Psi\left(^1L\right)\right\rangle.$$

Substituting the expression for $\Psi(^1L)$, we see that the first of the three matrix elements consists of four terms which are given by

$$\frac{1}{2}\int d^3\mathbf{x}_1\int d^3\mathbf{x}_2 |\psi_a(\mathbf{x}_1)|^2|\psi_b(\mathbf{x}_2)|^2 f(\mathbf{x}_1),$$

$$\frac{1}{2}\int d^3\mathbf{x}_1\int d^3\mathbf{x}_2 |\psi_a(\mathbf{x}_2)|^2|\psi_b(\mathbf{x}_1)|^2 f(\mathbf{x}_1),$$

$$\frac{1}{2}\int d^3\mathbf{x}_1 \int d^3\mathbf{x}_2 \psi_a^*(\mathbf{x}_1)\psi_b^*(\mathbf{x}_2)\psi_a(\mathbf{x}_2)\psi_b(\mathbf{x}_1)f(\mathbf{x}_1),$$

$$\frac{1}{2}\int d^3\mathbf{x}_1 \int d^3\mathbf{x}_2 \psi_a^*(\mathbf{x}_2)\psi_b^*(\mathbf{x}_1)\psi_a(\mathbf{x}_1)\psi_b(\mathbf{x}_2)f(\mathbf{x}_1).$$

Performing the integration in $d^3\mathbf{x}_2$, the last two integrals cancel out, due to the orthogonality of the wave functions ψ_a and ψ_b, while the first two integrals simplify, thus giving, respectively

$$\frac{1}{2}\mathcal{I}_a = \frac{1}{2}\int d^3\mathbf{x}\left|\psi_a(\mathbf{x})\right|^2 f(\mathbf{x}), \qquad \frac{1}{2}\mathcal{I}_b = \frac{1}{2}\int d^3\mathbf{x}\left|\psi_b(\mathbf{x})\right|^2 f(\mathbf{x}).$$

The second matrix element is calculated in a similar way and leads to the same result. Finally, the third matrix element, which contains the Coulomb interaction, is also made up of four terms. Two of them are given, respectively, by

$$\frac{1}{2}\mathcal{J}_{ab} = \frac{1}{2}\int d^3\mathbf{x}_1 \int d^3\mathbf{x}_2 \left|\psi_a(\mathbf{x}_1)\right|^2 \left|\psi_b(\mathbf{x}_2)\right|^2 \frac{e_0^2}{r_{12}},$$

$$\frac{1}{2}\mathcal{K}_{ab} = \frac{1}{2}\int d^3\mathbf{x}_1 \int d^3\mathbf{x}_2 \psi_a^*(\mathbf{x}_1)\psi_b^*(\mathbf{x}_2)\psi_a(\mathbf{x}_2)\psi_b(\mathbf{x}_1)\frac{e_0^2}{r_{12}},$$

and the other two are the same (as deduced by exchanging the variables of integration $\mathbf{x}_1$ and $\mathbf{x}_2$). Gathering the various contributions we obtain

$$\Delta E\left({}^1L\right) = \mathcal{I}_a + \mathcal{I}_b + \mathcal{J}_{ab} + \mathcal{K}_{ab}.$$

Carrying out the same calculation for the triplet we obtain instead

$$\Delta E\left({}^3L\right) = \mathcal{I}_a + \mathcal{I}_b + \mathcal{J}_{ab} - \mathcal{K}_{ab},$$

which means that the triplet state has lower energy than the singlet (since it can be proven that $\mathcal{K}_{ab}$ is always positive).

The Coulomb interaction integrals that appear in these expressions ($\mathcal{J}_{ab}$, and $\mathcal{K}_{ab}$) are commonly referred to as direct and exchange integrals. The direct integral has an immediate physical interpretation, being equal to the average Coulomb interaction in the form that might have been expected on the basis of elementary considerations. The exchange integral is instead a purely quantum mechanical result and does not have any direct physical interpretation in classical physics.

The result we obtained for the configurations $1s\ nl$ for the helium atom is a particular case (or, more precisely, an extension)[5] of a principle that bears the name of first Hund's rule. This rule states that *among the terms arising from a configuration of equivalent electrons, that one with the lowest energy is the one which has the highest multiplicity (i.e. the maximum value of S)*. The first rule is complemented by the second Hund's rule stating that *for a given multiplicity, the term with the lowest energy is the one which has the largest value of the orbital angular momentum L*. The first rule has a physical qualitative interpretation. In fact, if we consider

---

[5]Hund's rules apply to terms arising from configurations of equivalent electrons. Here they are applied to terms of configurations of the type $1s\ nl$.

Fig. 7.8 Grotrian diagram
for the helium atom. Each
level is identified with the
principal quantum number n
of the optical electron

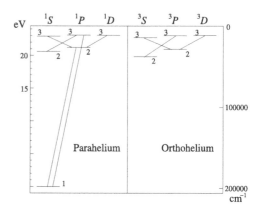

the term with the highest spin, it corresponds to the spin function having maximum symmetry with respect to the exchange of the electrons, i.e. to the most antisymmetric spatial wave function. This means that the electrons are as far away as possible from each other, resulting in a minimum for the energy of the Coulomb repulsion (which is always positive).

The structure of the terms of the helium atom is schematically shown in the Grotrian diagram of Fig. 7.8. In this diagram the system of singlets is drawn separately from that of the triplets. Since the selection rule $\Delta S = 0$ (see below) is particularly well verified for the helium atom, its spectrum was interpreted by early spectroscopists as originating from two distinct elements, which were called respectively parahelium ($S = 0$, system of singlets) and ortohelium ($S = 1$, system of triplets).

A number of selection rules apply to the spectra with more than one optical electron. These rules will be justified theoretically in Chap. 12. For the moment we simply consider them as empirical rules that explain the reason why only certain lines are present in the spectra. There are two different types of selection rules, namely rules regarding the configurations and rules regarding the terms. The first ones are the following

$$(a)\quad \Delta l = \pm 1, \qquad (b)\quad \text{even} \nrightarrow \text{even}, \qquad (b')\quad \text{odd} \nrightarrow \text{odd},$$

while for the second ones we have

$$(c)\quad \Delta S = 0, \qquad (d)\quad \Delta L = \pm 1, 0; \quad 0 \nrightarrow 0.$$

Rule (a) states that allowed transitions are only between configurations that differ from each other for the quantum numbers of a single electron. In addition, the azimuthal quantum number of the electron must differ by unity between the initial and final configurations. Rule (b), known as Laporte's rule, states that allowed transitions are only those between odd and even configurations (and vice versa). Rule (c) expresses the fact that the total spin S should remain unchanged, while rule (d) expresses the fact that the total orbital angular momentum L can vary at most by unity (excluding the transition $0 \to 0$, which is forbidden).

In some cases, the selection rules are not independent. For example, in the case of helium in which an electron always remains in the $1s$ state, the selection rule

Fig. 7.9 Grotrian diagram
for the carbon atom. Each
level is labelled with its
spectroscopic notation

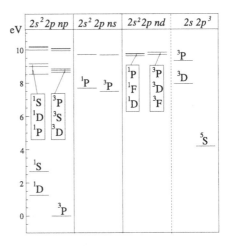

on Δl coincides with the selection rule on ΔL. Additionally, Laporte's rule seems
to be a repetition of the Δl rule. In reality, this rule is more general and can also
be applied when there is interaction between configurations (in fact, even when a
particular configuration cannot be assigned to one state, a definite parity can always
be assigned).

These selection rules help us to understand why the only spectral lines that are
observed in the helium spectrum are those outlined in the diagram of Fig. 7.8. The
helium atom has two so-called metastable levels, or two excited levels that cannot
decay to the ground level, or, more generally, to levels of lower energy. The two
metastable levels for helium are the ^1S and ^3S levels, originating from the $1s2s$
configuration.

7.13 The Carbon Atom

The ground configuration for the carbon atom is $1s^2 2s^2 2p^2$, with the three terms
^1S, ^1D, and ^3P (see Table 7.8). According to the first Hund's rule, the term with the
lowest energy is the ^3P, which therefore is the ground state for the carbon atom.
As we will see in the next chapter, theory predicts that the ^1D term should have an
energy lower than the ^1S term, with a ratio between the intervals given by

$$\frac{^1S - {}^1D}{^1D - {}^3P} = \frac{3}{2}.$$

As shown in the Grotrian diagram of Fig. 7.9, the terms are indeed ordered in energy as predicted by theory, although the ratio between the intervals is 1.13 instead
of 1.50. This difference, caused mainly by configuration interaction, clearly shows
the limits of the approximations that are introduced for the interpretation of spectroscopic data at an elementary level.

The most important configurations with excited electrons are those where one
of the $2p$ electrons moves to an orbital of higher energy. These are of the type

$1s^2 2s^2 2p \; np$ ($n = 3, 4, \ldots$), $1s^2 2s^2 2p \; ns$ ($n = 3, 4, \ldots$), $1s^2 2s^2 2p \; nd$ ($n = 3$, $4, \ldots$), etc. Each of the configurations of the first kind gives rise to six terms, ^1S, ^1P, ^1D, ^3S, ^3P, ^3D. The configurations of the second kind produce two terms, ^1P and ^3P. Finally, the configurations of the third kind produce six terms, ^1P, ^1D, ^1F, ^3P, ^3D, ^3F. The Grotrian diagram for the carbon atom is schematically shown in Fig. 7.9. We see from the figure that the term with lowest energy, among the six originating from the $1s^2 2s^2 2p3p$ configuration, is the ^1P. This fact is not in contradiction with the first Hund's rule (which would predict a lower energy for the term ^3D), as this rule is strictly valid only for configurations with equivalent electrons. Similarly, for the configuration $1s^2 2s^2 2p3d$, the lowest term is the ^1D (and not the ^3F).

The Grotrian diagram also shows some terms belonging to the $1s^2 2s2p^3$ configuration. Such configuration is obtained from the excitation of an innershell electron but in this case, since the difference in energy between the $2s$ and $2p$ orbitals is relatively small, some of the corresponding terms are below the ionisation limit of the atom. The configuration has in principle six terms, ^1P, ^1D, ^3S, ^3P, ^3D, ^5S, three of which are shown in the diagram.

In conclusion of this chapter we note that, despite the huge amount of work dedicated to the observation and analysis of atomic spectra (laboratory and astrophysical), the knowledge of the energy levels of atoms and ions is still far from complete. Sufficiently complete analyses are available only for those atoms (or ions) with few electrons in open shells. The analysis of the metals belonging to the transition groups is still rather incomplete and even more so is that one of the lanthanides and actinides, groups characterized by a considerable complexity in their atomic structure.

Chapter 8
Term Energies

In the previous chapter we introduced the nonrelativistic Hamiltonian of a complex atom and we saw how it can be separated into two parts using the central field approximation: a zero order Hamiltonian whose eigenvectors, in general degenerate, are the states belonging to the different configurations, and a "corrective" Hamiltonian containing various terms including, in particular, the Coulomb repulsion between electrons. By neglecting the interaction between configurations, which is equivalent to consider the corrective Hamiltonian as a perturbation of the zero order Hamiltonian, we have seen, in the particular case of the helium atom, how we can express the energies of the terms by means of integrals which involve single particle eigenfunctions relative to the zero order Hamiltonian. In this chapter we generalise the results obtained for the helium atom to any atom, also using perturbation theory. We will obtain general results that can be directly compared with the spectroscopic data. These results, although approximated, constitute the starting point for the development of more sophisticated treatments that are currently used for the detailed analysis of atomic spectra.

8.1 The Diagonal Sum Rule

Neglecting the interaction between configurations, the energies of the terms of a given configuration are obtained by diagonalising the Hamiltonian $\mathcal{H}_1$ of Eq. (7.4) on the degenerate space of the configuration. This, in principle, implies the diagonalisation of a matrix of order g (the degeneracy of the configuration) having elements of the form

$$\langle \Psi^A(a_1, a_2, \ldots, a_N) | \mathcal{H}_1 | \Psi^A(a'_1, a'_2, \ldots, a'_N) \rangle,$$

where $(a_1, a_2, \ldots, a_N)$ and $(a'_1, a'_2, \ldots, a'_N)$ are two sets of quantum numbers corresponding to the same configuration. As we have seen in Sect. 7.11, the diagonalisation can be avoided by finding suitable linear combinations of the $\Psi^A(a_1, a_2, \ldots, a_N)$ such that they are eigenfunctions of the L^2 and S^2 operators. Such combinations are generally difficult to obtain, so we follow a different method.

E. Landi Degl'Innocenti, *Atomic Spectroscopy and Radiative Processes*,
UNITEXT for Physics, DOI 10.1007/978-88-470-2808-1_8, © Springer-Verlag Italia 2014

To start with, we note that each of the g degenerate wavefunctions $\Psi^A(a_1, a_2, \ldots, a_N)$ is an eigenvector of the operators L_z and S_z, with the corresponding eigenvalues M_L and M_S given by

$$M_L = \sum_{i=1}^{N} m_i, \qquad M_S = \sum_{i=1}^{N} m_{si},$$

where m_i and m_{si} are, respectively, the magnetic quantum number and the quantum number of the projection (along the z axis) of the spin of the i-th electron. Since the Hamiltonian $\mathcal{H}_1$ commutes with the operators L_z and S_z, the Hamiltonian is block-diagonal. Each block (i.e. each single submatrix) is characterized by a suitable pair of values (M_L, M_S).

Suppose now that we have actually performed the diagonalisation of one of these sub-matrices. The energies of the terms compatible with the values of M_L and M_S that characterise the submatrix would then appear on its diagonal. By recalling a fundamental property of matrices, that the trace of a matrix is invariant under a similarity transformation, we obtain the following conclusion: given a pair of values (M_L, M_S), the sum of the energies of all the terms compatible with these values is equal to the sum of the diagonal matrix elements of the form

$$\langle \Psi^A(a_1, a_2, \ldots, a_N) | \mathcal{H}_1 | \Psi^A(a_1, a_2, \ldots, a_N) \rangle,$$

where $\Psi^A(a_1, a_2, \ldots, a_N)$ is an eigenfunction of L_z and S_z that corresponds, respectively, to the eigenvalues M_L and M_S. The energies of the terms are often obtained with these simple considerations by linear combinations of diagonal matrix elements.

We now clarify these considerations with an example. Consider the pp' configuration that, as we have seen, produces the six terms ^1S, ^1P, ^1D, ^3S, ^3P, ^3D. The maximum value for M_L is 2, and the maximum for M_S is 1. These two values of (M_L, M_S) are only compatible with the ^3D term. If we indicate with the same symbol the energy of the term, we can write

$$^3D = [2, 1],$$

where we have introduced the shortened symbol $[M_L, M_S]$ to indicate the sum of the diagonal matrix elements of the Hamiltonian $\mathcal{H}_1$ on all the states Ψ^A that have as eigenvalues of L_z and S_z the values M_L and M_S, respectively. If we proceed, lowering the M_L and M_S values until they both reach zero, we obtain the other equations

$$^3D + {}^3P = [1, 1], \qquad {}^3D + {}^3P + {}^3S = [0, 1], \qquad {}^3D + {}^1D = [2, 0],$$
$$^3D + {}^1D + {}^3P + {}^1P = [1, 0], \qquad {}^3D + {}^1D + {}^3P + {}^1P + {}^3S + {}^1S = [0, 0]. \tag{8.1}$$

Inverting these equations gives, after some simple algebra

$$^3D = [2, 1], \qquad {}^3P = [1, 1] - [2, 1], \qquad {}^3S = [0, 1] - [1, 1],$$
$$^1D = [2, 0] - [2, 1], \qquad {}^1P = [1, 0] - [2, 0] - [1, 1] + [2, 1],$$
$$^1S = [0, 0] - [1, 0] - [0, 1] + [1, 1].$$

Table 8.1 Energies of the terms of the configuration pp' written as sums of diagonal matrix elements of electron states of the form $(m_1^\pm, m_2^\pm)$

Configuration pp'
$^3D = (1^+, 1^+)$
$^3P = (1^+, 0^+) + (0^+, 1^+) - (1^+, 1^+)$
$^3S = (1^+, -1^+) + (0^+, 0^+) + (-1^+, 1^+) - (1^+, 0^+) - (0^+, 1^+)$
$^1D = (1^+, 1^-) + (1^-, 1^+) - (1^+, 1^+)$
$^1P = (1^+, 0^-) + (1^-, 0^+) + (0^+, 1^-) + (0^-, 1^+) - (1^+, 1^-) - (1^-, 1^+) - (1^+, 0^+) - (0^+, 1^+)$
$\quad + (1^+, 1^+)$
$^1S = (1^+, -1^-) + (1^-, -1^+) + (0^+, 0^-) + (0^-, 0^+) + (-1^+, 1^-) + (-1^-, 1^+) - (1^+, 0^-)$
$\quad - (1^-, 0^+) - (0^+, 1^-) - (0^-, 1^+) - (1^+, -1^+) - (0^+, 0^+) - (-1^+, 1^+) + (1^+, 0^+)$
$\quad + (0^+, 1^+)$

The quantities denoted by the symbols $[M_L, M_S]$ are sums of diagonal matrix elements. Each matrix element can be identified using the values of m and m_s of each electron (we recall that the other quantum numbers n and l are already specified by the configuration). Because of this, for two-electron configurations, we use a compact notation of the type[1]

$$\left(m_1^\pm, m_2^\pm\right)$$

to indicate the diagonal matrix element of the Hamiltonian on the state where the electron 1 has magnetic quantum number m_1 and spin quantum number $+1/2$ or $-1/2$, and the electron 2 has magnetic quantum number m_2 and spin quantum number $+1/2$ or $-1/2$. With this notation, it becomes simple to identify all the pairs of the type $(m_1^\pm, m_2^\pm)$ that contribute to the sum $[M_L, M_S]$. In our particular case we have

$$[2, 1] = \left(1^+, 1^+\right), \qquad [1, 1] = \left(1^+, 0^+\right) + \left(0^+, 1^+\right),$$
$$[0, 1] = \left(1^+, -1^+\right) + \left(0^+, 0^+\right) + \left(-1^+, 1^+\right), \qquad [2, 0] = \left(1^+, 1^-\right) + \left(1^-, 1^+\right),$$
$$[1, 0] = \left(1^+, 0^-\right) + \left(1^-, 0^+\right) + \left(0^+, 1^-\right) + \left(0^-, 1^+\right),$$
$$[0, 0] = \left(1^+, -1^-\right) + \left(1^-, -1^+\right) + \left(0^+, 0^-\right) + \left(0^-, 0^+\right) + \left(-1^+, 1^-\right)$$
$$\quad + \left(-1^-, 1^+\right).$$

Substituting these expressions in Eq. (8.1), we obtain the results shown in Table 8.1.

In the end, we have been able to express the energies of the terms as algebraic sums of diagonal matrix elements of the Hamiltonian $\mathcal{H}_1$ on eigenstates of the Hamiltonian $\mathcal{H}_0$. We note that in the final equations in Table 8.1 the matrix element indicated with the symbol $(1^+, 1^-)$ is different from the element indicated with the symbol $(1^-, 1^+)$. This occurs because the two electrons are non-equivalent, and the

---

[1]The symbol is introduced in the classic atomic spectroscopy book of Condon and Shortley (1935).

Table 8.2 Term energies of the configurations p^2, p^3 and p^2p'. The symbol $^2D(2)$ that appears in the box relative to the configuration p^2p' represents the sum of the energies of the two ^2D terms originating from such configuration. The symbol $^2P(3)$ has a similar meaning

---

Configuration p^2

$^3P = (1^+, 0^+)$

$^1D = (1^+, 1^-)$

$^1S = (1^+, -1^-) + (1^-, -1^+) + (0^+, 0^-) - (1^+, 0^+) - (1^+, 1^-)$

Configuration p^3

$^4S = (1^+, 0^+, -1^+)$

$^2D = (1^+, 1^-, 0^+)$

$^2P = (1^+, 1^-, -1^+) + (1^+, 0^+, 0^-) - (1^+, 1^-, 0^+)$

Configuration p^2p'

$^4D = (1^+, 0^+; 1^+)$

$^4P = (1^+, -1^+; 1^+) + (1^+, 0^+; 0^+) - (1^+, 0^+; 1^+)$

$^4S = (1^+, -1^+; 0^+) + (1^+, 0^+; -1^+) + (-1^+, 0^+; 1^+) - (1^+, -1^+; 1^+) - (1^+, 0^+; 0^+)$

$^2F = (1^+, 1^-; 1^+)$

$^2D(2) = (1^+, 0^-; 1^+) + (1^-, 0^+; 1^+) + (1^+, 0^+; 1^-) + (1^+, 1^-; 0^+) - (1^+, 1^-; 1^+)$
$\qquad - (1^+, 0^+; 1^+)$

$^2P(3) = (1^+, 1^-; -1^+) + (1^+, 0^+; 0^-) + (1^+, 0^-; 0^+) + (1^-, 0^+; 0^+) + (1^+, -1^+; 1^-)$
$\qquad + (1^+, -1^-; 1^+) + (1^-, -1^+; 1^+) + (0^+, 0^-; 1^+) - (1^+, 0^-; 1^+) - (1^-, 0^+; 1^+)$
$\qquad - (1^+, 0^+; 1^-) - (1^+, 1^-; 0^+) - (1^+, -1^+; 1^+) - (1^+, 0^+; 0^+) + (1^+, 0^+; 1^+)$

$^2S = (1^+, 0^+; -1^-) + (1^+, 0^-; -1^+) + (1^-, 0^+; -1^+) + (1^+, -1^+; 0^-) + (1^+, -1^-; 0^+)$
$\qquad + (1^-, -1^+; 0^+) + (0^+, 0^-; 0^+) + (0^+, -1^+; 1^-) + (0^+, -1^-; 1^+) + (0^-, -1^+; 1^+)$
$\qquad - (1^+, 1^-; -1^+) - (1^+, 0^+; 0^-) - (1^+, 0^-; 0^+) - (1^-, 0^+; 0^+) - (1^+, -1^+; 1^-)$
$\qquad - (1^+, -1^-; 1^+) - (1^-, -1^+; 1^+) - (0^+, 0^-; 1^+) - (1^+, -1^+; 0^+) - (1^+, 0^+; -1^+)$
$\qquad - (-1^+, 0^+; 1^+) + (1^+, -1^+; 1^+) + (1^+, 0^+; 0^+)$

---

order within parentheses is important. In the case of configurations with two equivalent electrons, however, the order within the parentheses is irrelevant. In such a case matrix elements as $(0^+, 0^+)$ should not appear in the expressions for the energies of the terms because the corresponding wavefunctions refer to states that violate the Pauli exclusion principle.

The method described above for the configuration pp' is based on the general rule that the trace of a matrix is invariant under a similarity transformation. In spectroscopic applications such rule is called the diagonal sum rule. The method can be applied to any configuration and implies some algebra, often laborious. Table 8.2 shows the results for the configurations p^2, p^3, and p^2p'. For the latter case it is necessary to introduce the notation $(m_1^\pm, m_2^\pm; m_3^\pm)$ to distinguish the two equivalent electrons, with indices 1 and 2, from the non-equivalent electron with index 3.

It should be emphasized that in the last case, that of the $p^2 p'$ configuration, it is not possible to use the diagonal sum rule to determine separately the energy of the two ^2D terms and of the three ^2P terms. Only the sum of the energies of repeated terms can be obtained. This is a general limitation of the diagonal sum rule. To determine the individual energies of the repeated terms it is necessary to diagonalise the sub-matrix relative only to those terms.

8.2   Calculation of Diagonal Matrix Elements

We now consider the calculation of the diagonal matrix element

$$\langle \Psi(a_1, a_2, \ldots, a_N) | \mathcal{H}_1 | \Psi(a_1, a_2, \ldots, a_N) \rangle.$$

The Hamiltonian $\mathcal{H}_1$ (see Eq. (7.4)) is composed of the sum of an operator $f(i)$ relative to each electron, and of the sum of an operator $g(i, j)$, relative to all distinct pairs of electrons. $g(i, j)$ is symmetric in the exchange of the i-th electron with the j-th electron. Explicitly

$$\mathcal{H}_1 = \mathcal{F} + \mathcal{G} = \sum_{i=1}^{N} f(i) + \sum_{i<j} g(i, j), \tag{8.2}$$

where

$$f(i) = -V_c(r_i) - \frac{Ze_0^2}{r_i}, \qquad g(i, j) = \frac{e_0^2}{r_{ij}}. \tag{8.3}$$

By recalling that the function $\Psi(a_1, a_2, \ldots, a_N)$ is of the type of a Slater determinant (see Eq. (7.1), with $N_A = 1/\sqrt{N!}$), we have for the first term of $\mathcal{H}_1$

$$\langle \Psi(a_1, a_2, \ldots, a_N) | \mathcal{F} | \Psi(a_1, a_2, \ldots, a_N) \rangle$$

$$= \frac{1}{N!} \sum_{P} \sum_{Q} \sum_{i=1}^{N} (-1)^{P+Q}$$
$$\times \langle P[\psi_{a_1}(x_1)\psi_{a_2}(x_2) \cdots \psi_{a_N}(x_N)] | f(x_i) | Q[\psi_{a_1}(x_1)\psi_{a_2}(x_2) \cdots \psi_{a_N}(x_N)] \rangle,$$

with the sum extended to all possible P and Q permutations of the coordinates of the electrons in the bra and ket, respectively. The sum contains $N \times (N!)^2$ terms. The generic term of the sum is, apart from the sign,

$$\langle \psi_{a_1'}(x_1) | \psi_{a_1''}(x_1) \rangle \langle \psi_{a_2'}(x_2) | \psi_{a_2''}(x_2) \rangle \cdots \langle \psi_{a_i'}(x_i) | f(x_i) | \psi_{a_i''}(x_i) \rangle \cdots$$
$$\times \langle \psi_{a_N'}(x_N) | \psi_{a_N''}(x_N) \rangle,$$

$(a'_1, a'_2, \ldots, a'_N)$ and $(a''_1, a''_2, \ldots, a''_N)$ being two arbitrary (and in general distinct) permutations of $(a_1, a_2, \ldots, a_N)$. Since the eigenfunctions ψ_{a_j} are orthonormal, this term is zero unless $a'_1 = a''_1, a'_2 = a''_2, \ldots, a'_N = a''_N$. We therefore have

$$\langle \Psi(a_1, a_2, \ldots, a_N) | \mathcal{F} | \Psi(a_1, a_2, \ldots, a_N) \rangle$$

$$= \frac{1}{N!} \sum_P \sum_{i=1}^{N} \langle P[\psi_{a_1}(x_1)\psi_{a_2}(x_2)\cdots\psi_{a_N}(x_N)] | f(x_i) |$$

$$P[\psi_{a_1}(x_1)\psi_{a_2}(x_2)\cdots\psi_{a_N}(x_N)] \rangle,$$

so that the sum is made only of $N \times N!$ terms. We now consider the $N!$ terms of the sum that contain the matrix element of $f(x_i)$, with i fixed. $(N-1)!$ of these terms are equal to $\langle \psi_{a_1}(x_i) | f(x_i) | \psi_{a_1}(x_i) \rangle$, $(N-1)!$ are equal to $\langle \psi_{a_2}(x_i) | f(x_i) | \psi_{a_2}(x_i) \rangle$, and so on until the $(N-1)!$ that are equal to $\langle \psi_{a_N}(x_i) | f(x_i) | \psi_{a_N}(x_i) \rangle$. We therefore obtain

$$\langle \Psi(a_1, a_2, \ldots, a_N) | \mathcal{F} | \Psi(a_1, a_2, \ldots, a_N) \rangle = \frac{1}{N} \sum_{i=1}^{N} \sum_{j=1}^{N} \langle \psi_{a_j}(x_i) | f(x_i) | \psi_{a_j}(x_i) \rangle.$$

On the other hand, once the matrix element is calculated, it does not depend anymore on the x_i coordinates of the electron, so that one finally gets

$$\langle \Psi(a_1, a_2, \ldots, a_N) | \mathcal{F} | \Psi(a_1, a_2, \ldots, a_N) \rangle = \sum_{j=1}^{N} \langle \psi_{a_j}(x) | f(x) | \psi_{a_j}(x) \rangle,$$

or, shortening the notations with obvious symbols

$$\langle \Psi(a_1, a_2, \ldots, a_N) | \mathcal{F} | \Psi(a_1, a_2, \ldots, a_N) \rangle = \sum_{j=1}^{N} \langle a_j | f | a_j \rangle. \tag{8.4}$$

Despite the formal complications introduced by the antisymmetrisation of the eigenfunctions, the result that we have obtained for the diagonal matrix element of the operator $\mathcal{F}$ is obvious and intuitive. In practice, we need to sum the diagonal matrix element of the single particle operator f on all the single particle states a_i occupied by the electrons.

We calculate now the diagonal matrix element of the operator $\mathcal{G}$. We have

$$\langle \Psi(a_1, a_2, \ldots, a_N) | \mathcal{G} | \Psi(a_1, a_2, \ldots, a_N) \rangle$$

$$= \frac{1}{N!} \sum_P \sum_Q \sum_{i<j} (-1)^{P+Q}$$

$$\times \langle P[\psi_{a_1}(x_1)\psi_{a_2}(x_2)\cdots\psi_{a_N}(x_N)] | g(x_i, x_j) |$$

$$Q[\psi_{a_1}(x_1)\psi_{a_2}(x_2)\cdots\psi_{a_N}(x_N)] \rangle.$$

As in the previous case for the operator $\mathcal{F}$, if we write in explicit form the generic term of the sum

$$\langle \psi_{a'_1}(x_1) | \psi_{a''_1}(x_1) \rangle \langle \psi_{a'_2}(x_2) | \psi_{a''_2}(x_2) \rangle \cdots$$

$$\times \langle \psi_{a'_i}(x_i)\psi_{a'_j}(x_j) | g(x_i, x_j) | \psi_{a''_i}(x_i)\psi_{a''_j}(x_j) \rangle \cdots \langle \psi_{a'_N}(x_N) | \psi_{a''_N}(x_N) \rangle,$$

we see that the matrix element is non-zero in two cases, i.e. when we have

$$\text{case (a)} \quad a_1' = a_1'', a_2' = a_2'', \ldots, a_i' = a_i'', \ldots, a_j' = a_j'', \ldots, a_N' = a_N'',$$

or when we have

$$\text{case (b)} \quad a_1' = a_1'', a_2' = a_2'', \ldots, a_i' = a_j'', \ldots, a_j' = a_i'', \ldots, a_N' = a_N''.$$

In case (a), the permutation Q is the same as the permutation P, while in case (b) it is the same as P, with in addition the exchange of the quantum numbers of particles i and j. If we indicate such permutation with P', we have that

$$(-1)^{P+P'} = -1,$$

because the two permutations P and P' have different parity. Based on these considerations, we obtain

$$\langle \Psi(a_1, a_2, \ldots, a_N) | \mathcal{G} | \Psi(a_1, a_2, \ldots, a_N) \rangle = \frac{1}{N!} \sum_P \sum_{i<j} \mathcal{A}_{ij} - \frac{1}{N!} \sum_P \sum_{i<j} \mathcal{B}_{ij},$$

where

$$\mathcal{A}_{ij} = \langle P[\psi_{a_1}(x_1)\psi_{a_2}(x_2)\cdots\psi_{a_N}(x_N)] | g(x_i, x_j) | P[\psi_{a_1}(x_1)\psi_{a_2}(x_2)\cdots\psi_{a_N}(x_N)] \rangle,$$

$$\mathcal{B}_{ij} = \langle P[\psi_{a_1}(x_1)\psi_{a_2}(x_2)\cdots\psi_{a_N}(x_N)] | g(x_i, x_j) | P'[\psi_{a_1}(x_1)\psi_{a_2}(x_2)\cdots\psi_{a_N}(x_N)] \rangle.$$

We now draw our attention on a particular pair (i, j). Out of the $N!$ terms of the sum, there are $(N-2)!$ that give as a result

$$\{\langle \psi_{a_1}(x_i)\psi_{a_2}(x_j) | g(x_i, x_j) | \psi_{a_1}(x_i)\psi_{a_2}(x_j) \rangle$$
$$- \langle \psi_{a_1}(x_i)\psi_{a_2}(x_j) | g(x_i, x_j) | \psi_{a_2}(x_i)\psi_{a_1}(x_j) \rangle\},$$

$(N-2)!$ that give a similar result with the pair of states (a_1, a_3) instead of the pair (a_1, a_2), and so on, for all possible ordered pairs of states (a_r, a_s). We then note that, since the operator $g(i, j)$ is symmetric with respect to the exchange of the particles, the ordered pair (a_r, a_s) produces the same result as the ordered pair (a_s, a_r). We therefore obtain

$$\langle \Psi(a_1, a_2, \ldots, a_N) | \mathcal{G} | \Psi(a_1, a_2, \ldots, a_N) \rangle$$
$$= \frac{2}{N(N-1)} \sum_{i<j} \sum_{\text{pairs } a_r, a_s} \{\langle \psi_{a_r}(x_i)\psi_{a_s}(x_j) | g(x_i, x_j) | \psi_{a_r}(x_i)\psi_{a_s}(x_j) \rangle$$
$$- \langle \psi_{a_r}(x_i)\psi_{a_s}(x_j) | g(x_i, x_j) | \psi_{a_s}(x_i)\psi_{a_r}(x_j) \rangle\},$$

where the sum runs over the non-ordered pairs. The result does not depend anymore on the values of the indices i and j, once the matrix elements that appear in this equation are calculated. We note that the sum over i and j implies $N(N-1)/2$ terms. We therefore have, in compact notation

$$\langle \Psi(a_1, a_2, \ldots, a_N) | \mathcal{G} | \Psi(a_1, a_2, \ldots, a_N) \rangle$$
$$= \sum_{\text{pairs } a_r, a_s} \left[\langle a_r, a_s | g | a_r, a_s \rangle - \langle a_r, a_s | g | a_s, a_r \rangle \right]. \qquad (8.5)$$

The matrix elements appearing with the positive sign in the right-hand side are called "direct", while the others are called "exchange". The contribution of the latter to the total energy of an atomic system is called the contribution due to the exchange energy. The direct matrix elements have an immediate physical interpretation, being integrals of the type

$$\int \left|\psi_{a_r}(x_1)\right|^2 \left|\psi_{a_s}(x_2)\right|^2 \frac{e_0^2}{r_{12}} \, dx_1 \, dx_2.$$

The exchange matrix elements, however, are a consequence of the antisymmetric character of the eigenfunctions, i.e. of the exclusion principle, and do not have a classical analogue.

8.3 Single Particle Matrix Elements

In the previous section we have related the diagonal matrix elements of operators of the type $\mathcal{F}$ and $\mathcal{G}$ between states of N particles to matrix elements between states of single particle (for the operator $\mathcal{F}$) or between two particle states (in the case of the operator $\mathcal{G}$). We now calculate explicitly these matrix elements taking into account that the single particle eigenfunctions are of the type (see Eqs. (7.10) and (7.11))

$$\psi_{nlmm_s} = \frac{1}{r} P_{nl}(r) Y_{lm}(\theta, \phi) \chi_{m_s},$$

where $P_{nl}(r)$ is the reduced radial eigenfunction, solution of the Schrödinger equation

$$-\frac{\hbar^2}{2m} \frac{d^2}{dr^2} P_{nl}(r) + \left[V_c(r) + \frac{\hbar^2 l(l+1)}{2mr^2}\right] P_{nl}(r) = W_0(n, l) P_{nl}(r),$$

$V_c(r)$ being the central potential. Recalling the expression for the operator $\mathcal{F}$ given by Eqs. (8.2) and (8.3), the first integral to be evaluated is of the type

$$\int \left|\psi_{nlmm_s}\right|^2 \left[-V_c(r) - \frac{Ze_0^2}{r}\right] r^2 \sin\theta \, dr \, d\theta \, d\phi.$$

Taking into account the property of the spherical harmonics given in Eq. (6.12), we obtain a radial integral, function of n and l, defined by

$$I(n, l) = -\int_0^\infty P_{nl}^2(r) \left[V_c(r) + \frac{Ze_0^2}{r}\right] dr. \tag{8.6}$$

Obviously, the explicit expression for $I(n, l)$ can only be calculated once the potential $V_c(r)$ is known and the reduced radial functions $P_{nl}(r)$ have been determined.

In summary, the results of the previous chapter and those obtained here allow us to write the energy of a given configuration, due to the Hamiltonian $\mathcal{H}_0$ and to the $\mathcal{F}$ part of the Hamiltonian $\mathcal{H}_1$, in the form

$$W = \sum_{\text{subshells}} q_{nl} \left[W_0(n, l) + I(n, l)\right], \tag{8.7}$$

where q_{nl} is the number of electrons in the subshell nl. Such energy is the same for all the g states of the configuration. Therefore, the $\mathcal{F}$ part of the Hamiltonian $\mathcal{H}_1$ does not remove the degeneracy.

8.4   Matrix Elements of the Coulomb Interaction

With regard to the Coulomb interaction, two different types of matrix elements need to be calculated, the direct ones and the exchange ones, traditionally named J and K. These integrals are defined by

$$J(a, b) = \langle a, b | \frac{e_0^2}{r_{12}} | a, b \rangle, \qquad K(a, b) = \langle a, b | \frac{e_0^2}{r_{12}} | b, a \rangle,$$

where a and b are two different sets of quantum numbers, i.e.

$$a = (n_a, l_a, m_a, m_{s_a}), \qquad b = (n_b, l_b, m_b, m_{s_b}).$$

More generally, we calculate a matrix element of the type

$$\langle a, b | \frac{e_0^2}{r_{12}} | c, d \rangle.$$

By writing in explicit form the eigenfunctions, we have

$$\langle a, b | \frac{e_0^2}{r_{12}} | c, d \rangle$$

$$= \delta(m_{s_a}, m_{s_c}) \delta(m_{s_b}, m_{s_d}) \int_0^\infty dr_1 \int_0^\infty dr_2$$

$$\times P_{n_a l_a}(r_1) P_{n_b l_b}(r_2) P_{n_c l_c}(r_1) P_{n_d l_d}(r_2) \int_0^\pi d\theta_1 \int_0^\pi d\theta_2 \int_0^{2\pi} d\phi_1 \int_0^{2\pi} d\phi_2$$

$$\times Y_{l_a m_a}^*(\theta_1, \phi_1) Y_{l_b m_b}^*(\theta_2, \phi_2) Y_{l_c m_c}(\theta_1, \phi_1) Y_{l_d m_d}(\theta_2, \phi_2) \frac{e_0^2}{r_{12}} \sin\theta_1 \sin\theta_2,$$

where the two Kronecker delta are due to the scalar products of the spin eigenfunctions. To perform the integral we recall Eq. (7.9), which we rewrite here

$$\frac{1}{r_{12}} = \sum_{l=0}^\infty \frac{r_<^l}{r_>^{l+1}} P_l(\cos\Theta),$$

where $r_<$ and $r_>$ are, respectively, the smaller and the larger between r_1 and r_2, and where Θ is the angle between the unit vectors $\mathbf{r}_1$ and $\mathbf{r}_2$. Using the cosine rule, we have

$$\cos\Theta = \cos\theta_1 \cos\theta_2 + \sin\theta_1 \sin\theta_2 \cos(\phi_1 - \phi_2),$$

and the Legendre function of argument $\cos\Theta$ can be expressed, using the so-called addition theorem, in the form

$$P_l(\cos\Theta) = \frac{4\pi}{2l+1} \sum_{m=-l}^l Y_{lm}^*(\theta_1, \phi_1) Y_{lm}(\theta_2, \phi_2).$$

We then have

$$\frac{1}{r_{12}} = \sum_{l=0}^{\infty} \frac{r_<^l}{r_>^{l+1}} \frac{4\pi}{2l+1} \sum_{m=-l}^{l} Y_{lm}^*(\theta_1, \phi_1) Y_{lm}(\theta_2, \phi_2).$$

Substituting this expression into the matrix element, we obtain

$$\langle a, b | \frac{e_0^2}{r_{12}} | c, d \rangle$$

$$= e_0^2 \delta(m_{s_a}, m_{s_c}) \delta(m_{s_b}, m_{s_d}) \sum_{l=0}^{\infty} \frac{4\pi}{2l+1}$$

$$\times \left[\int_0^\infty dr_1 \int_0^\infty dr_2 \frac{r_<^l}{r_>^{l+1}} P_{n_a l_a}(r_1) P_{n_c l_c}(r_1) P_{n_b l_b}(r_2) P_{n_d l_d}(r_2) \right]$$

$$\times \sum_{m=-l}^{l} \left[\int_0^\pi \sin\theta_1 \, d\theta_1 \int_0^{2\pi} d\phi_1 Y_{l_a m_a}^*(\theta_1, \phi_1) Y_{l_c m_c}(\theta_1, \phi_1) Y_{lm}^*(\theta_1, \phi_1) \right]$$

$$\times \left[\int_0^\pi \sin\theta_2 \, d\theta_2 \int_0^{2\pi} d\phi_2 Y_{l_b m_b}^*(\theta_2, \phi_2) Y_{l_d m_d}(\theta_2, \phi_2) Y_{lm}(\theta_2, \phi_2) \right].$$

The integrals over the solid angle can be evaluated by recalling the conjugation properties of the spherical harmonics (Eq. (6.11)) and Weyl's theorem, according to which we have, in terms of the 3-j symbols

$$\int_0^\pi \sin\theta \, d\theta \int_0^{2\pi} d\phi Y_{l_1 m_1}(\theta, \phi) Y_{l_2 m_2}(\theta, \phi) Y_{l_3 m_3}(\theta, \phi)$$

$$= \sqrt{\frac{(2l_1+1)(2l_2+1)(2l_3+1)}{4\pi}} \begin{pmatrix} l_1 & l_2 & l_3 \\ 0 & 0 & 0 \end{pmatrix} \begin{pmatrix} l_1 & l_2 & l_3 \\ m_1 & m_2 & m_3 \end{pmatrix}. \tag{8.8}$$

We therefore obtain, for the angular part, the following expression

$$\frac{2l+1}{4\pi} \sqrt{(2l_a+1)(2l_b+1)(2l_c+1)(2l_d+1)} (-1)^{m_a+m+m_b}$$

$$\times \begin{pmatrix} l_a & l_c & l \\ 0 & 0 & 0 \end{pmatrix} \begin{pmatrix} l_b & l_d & l \\ 0 & 0 & 0 \end{pmatrix} \begin{pmatrix} l_a & l_c & l \\ -m_a & m_c & -m \end{pmatrix} \begin{pmatrix} l_b & l_d & l \\ -m_b & m_d & m \end{pmatrix},$$

which can be rewritten in a more symmetric form by changing the summation index m into $-m$, and taking into account the properties of the 3-j symbols (Eqs. (7.19) and (7.20)). Noting also that, being $m_b = m_d + m$, we have

$$(-1)^{m_a+m+m_b} = (-1)^{m_a+2m+m_d} = (-1)^{m_a+m_d},$$

we obtain, still for the angular part,

$$\frac{2l+1}{4\pi} \sqrt{(2l_a+1)(2l_b+1)(2l_c+1)(2l_d+1)} (-1)^{m_a+m_d}$$

$$\times \begin{pmatrix} l_a & l_c & l \\ 0 & 0 & 0 \end{pmatrix} \begin{pmatrix} l_d & l_b & l \\ 0 & 0 & 0 \end{pmatrix} \begin{pmatrix} l_a & l_c & l \\ -m_a & m_c & m \end{pmatrix} \begin{pmatrix} l_d & l_b & l \\ -m_d & m_b & m \end{pmatrix}.$$

We note that this expression is zero unless the two integers $(l_a + l_c + l)$ and $(l_b + l_d + l)$ are both even and unless m is equal both to $(m_a - m_c)$ and to $(m_d - m_b)$. This means, moreover, that the two integers $(m_a - m_c)$ and $(m_d - m_b)$ must be equal. The same expression for the angular part is generally written in a more compact form by introducing the $c^k(a, b)$ symbol, defined by

$$c^k(a, b) = c^k(l_a, m_a; l_b, m_b) = (-1)^{m_a} \sqrt{(2l_a + 1)(2l_b + 1)}$$
$$\times \begin{pmatrix} l_a & l_b & k \\ 0 & 0 & 0 \end{pmatrix} \begin{pmatrix} l_a & l_b & l \\ -m_a & m_b & m \end{pmatrix}.$$

We then obtain

$$\langle a, b| \frac{e_0^2}{r_{12}} |c, d \rangle = e_0^2 \delta(m_{s_a}, m_{s_c}) \delta(m_{s_b}, m_{s_d}) \sum_{k=0}^{\infty} c^k(a, c) c^k(d, b)$$

$$\times \int_0^{\infty} dr_1 \int_0^{\infty} dr_2 \frac{r_<^k}{r_>^{k+1}} P_{n_a l_a}(r_1) P_{n_c l_c}(r_1) P_{n_b l_b}(r_2) P_{n_d l_d}(r_2),$$

where the sum over k, formally extended between 0 and ∞, is in fact limited by the triangular inequality implicitly contained in the $c^k(a, c)$ and $c^k(b, d)$ symbols.

From this general expression, it is easy to obtain, by simple substitution, the corresponding expressions for the direct and exchange matrix elements, traditionally written in the following form

$$J(a, b) = \langle a, b| \frac{e_0^2}{r_{12}} |a, b \rangle = \sum_{k=0}^{\infty} a^k(a, b) F^k(a, b),$$

$$K(a, b) = \langle a, b| \frac{e_0^2}{r_{12}} |b, a \rangle = \delta(m_{s_a}, m_{s_b}) \sum_{k=0}^{\infty} b^k(a, b) G^k(a, b),$$

where

$$a^k(a, b) = c^k(a, a) c^k(b, b)$$
$$= (-1)^{m_a + m_b} (2l_a + 1)(2l_b + 1)$$
$$\times \begin{pmatrix} l_a & l_a & k \\ 0 & 0 & 0 \end{pmatrix} \begin{pmatrix} l_b & l_b & k \\ 0 & 0 & 0 \end{pmatrix} \begin{pmatrix} l_a & l_a & k \\ -m_a & m_a & 0 \end{pmatrix} \begin{pmatrix} l_b & l_b & k \\ -m_b & m_b & 0 \end{pmatrix},$$

$$b^k(a, b) = \left[c^k(a, b) \right]^2 = (2l_a + 1)(2l_b + 1) \begin{pmatrix} l_a & l_b & k \\ 0 & 0 & 0 \end{pmatrix}^2 \begin{pmatrix} l_a & l_b & k \\ -m_a & m_b & m \end{pmatrix}^2,$$

$$F^k(a, b) = e_0^2 \int_0^{\infty} dr_1 \int_0^{\infty} dr_2 \frac{r_<^k}{r_>^{k+1}} P_{n_a l_a}^2(r_1) P_{n_b l_b}^2(r_2), \tag{8.9}$$

$$G^k(a, b) = e_0^2 \int_0^{\infty} dr_1 \int_0^{\infty} dr_2 \frac{r_<^k}{r_>^{k+1}} P_{n_a l_a}(r_1) P_{n_b l_b}(r_1) P_{n_a l_a}(r_2) P_{n_b l_b}(r_2). \tag{8.10}$$

The quantities F^k and G^k can be calculated from the eigenfunctions $P_{nl}(r)$, which depend on the central potential $V_c(r)$. The a^k and b^k coefficients can instead be directly calculated. They have the following symmetry properties with respect to the arguments

$$a^k(l_a, m_a; l_b, m_b) = a^k(l_b, m_b; l_a, m_a) = a^k(l_a, -m_a; l_b, m_b)$$
$$= a^k(l_a, m_a; l_b, -m_b),$$
$$b^k(l_a, m_a; l_b, m_b) = b^k(l_b, m_b; l_a, m_a) = b^k(l_a, -m_a; l_b, -m_b).$$

These results allow us to express the diagonal matrix elements of the Coulomb interaction in the form

$$\langle \Psi(a_1, a_2, \ldots, a_N) | \frac{e_0^2}{r_{12}} | \Psi(a_1, a_2, \ldots, a_N) \rangle$$
$$= \sum_{\text{pairs } a,b} \left[J(a,b) - K(a,b) \right]$$
$$= \sum_{\text{pairs } a,b} \left[\sum_{k=0}^{\infty} a^k(a,b) F^k(a,b) - \delta(m_{s_a}, m_{s_b}) \sum_{k=0}^{\infty} b^k(a,b) G^k(a,b) \right], \quad (8.11)$$

where a and b are any two states of the set $(a_1, a_2, \ldots, a_N)$ and where the sum is over all the $N(N-1)/2$ distinct pairs.

8.5 Sums over Closed Subshells

The sums that express the matrix elements of the Coulomb interaction are considerably simplified when the configuration contains closed subshells. Referring first to the direct integrals, consider an electron belonging to a given subshell (n_b, l_b). We need to evaluate the contribution of the matrix element of the Coulomb interaction originating from all possible pairs formed by such an electron and any of the electrons belonging to a closed subshell (n_a, l_a). To do this we need to compute the following sum

$$S = \sum_{\text{shell } a} J(a,b) = \sum_{m_{s_a} m_a} \sum_k a^k(a,b) F^k(a,b).$$

Substituting the expression for $a^k(a,b)$ of the previous section, we obtain

$$S = (-1)^{m_b} (2l_a + 1)(2l_b + 1) \sum_k \begin{pmatrix} l_a & l_a & k \\ 0 & 0 & 0 \end{pmatrix} \begin{pmatrix} l_b & l_b & k \\ 0 & 0 & 0 \end{pmatrix} \begin{pmatrix} l_b & l_b & k \\ -m_b & m_b & 0 \end{pmatrix}$$
$$\times F^k(a,b) \sum_{m_{s_a} m_a} (-1)^{m_a} \begin{pmatrix} l_a & l_a & k \\ -m_a & m_a & 0 \end{pmatrix}.$$

To evaluate the sum over m_a appearing in this expression, we need to recall the properties of the 3-j symbols. Since we have, from Eq. (7.16)

$$\begin{pmatrix} l_a & l_a & 0 \\ -m_a & m_a & 0 \end{pmatrix} = (-1)^{l_a + m_a} \frac{1}{\sqrt{2l_a + 1}}, \quad (8.12)$$

we can write

$$\sum_{m_a}(-1)^{m_a}\begin{pmatrix}l_a & l_a & k\\ -m_a & m_a & 0\end{pmatrix}$$

$$= (-1)^{l_a}\sqrt{2l_a+1}\times\sum_{m_a}\begin{pmatrix}l_a & l_a & 0\\ -m_a & m_a & 0\end{pmatrix}\begin{pmatrix}l_a & l_a & k\\ -m_a & m_a & 0\end{pmatrix},$$

and considering the property of the 3-j symbols given by Eq. (7.18)

$$\sum_{m_a}(-1)^{m_a}\begin{pmatrix}l_a & l_a & k\\ -m_a & m_a & 0\end{pmatrix} = \delta_{k,0}(-1)^{l_a}\sqrt{2l_a+1}.$$

If we finally take into account that the sum over m_{s_a} produces a factor 2, we get

$$S = (-1)^{l_a+m_b}2(2l_a+1)^{3/2}(2l_b+1)\begin{pmatrix}l_a & l_a & 0\\ 0 & 0 & 0\end{pmatrix}\begin{pmatrix}l_b & l_b & 0\\ 0 & 0 & 0\end{pmatrix}$$

$$\times\begin{pmatrix}l_b & l_b & 0\\ -m_b & m_b & 0\end{pmatrix}F^0(a,b),$$

and recalling again Eq. (8.12) for expressing the 3-j symbols with two zeros in the last column, we obtain

$$\sum_{\text{shell }a}J(a,b) = 2(2l_a+1)F^0(a,b), \tag{8.13}$$

where

$$F^0(a,b) = F^0(n_a,l_a;n_b,l_b) = e_0^2\int_0^\infty dr_1\int_0^\infty dr_2\frac{1}{r_>}P_{n_al_a}^2(r_1)P_{n_bl_b}^2(r_2).$$

As we can see, this expression does not depend on the quantum numbers m_b and m_{s_b} of the electron in the subshell b. We then have that, if the subshell b is also closed

$$\sum_{\text{shells }a,b}J(a,b) = 4(2l_a+1)(2l_b+1)F^0(n_a,l_a;n_b,l_b). \tag{8.14}$$

Similar calculations can be performed for the exchange integrals. Again, consider an electron belonging to the subshell (n_bl_b). We need to evaluate the exchange contribution to the matrix element of the Coulomb interaction originating from all possible pairs formed by such an electron and any of the electrons belonging to a closed subshell (n_a,l_a). To do this we need to calculate the sum

$$\sum_{\text{shell }a}K(a,b)$$

$$= \sum_{m_{s_a}m_a}\delta(m_{s_a},m_{s_b})\sum_k b^k(a,b)G^k(a,b)$$

$$= \sum_{m_a}\sum_k(2l_a+1)(2l_b+1)\begin{pmatrix}l_a & l_b & k\\ 0 & 0 & 0\end{pmatrix}^2\begin{pmatrix}l_a & l_b & k\\ -m_a & m_b & m\end{pmatrix}^2 G^k(a,b).$$

The sum over m_a can be evaluated considering that, being $m = m_a - m_b$ and being m_b fixed, we can formally extend such sum also to m. Recalling the property given in Eq. (7.18), we then have

$$\sum_{\text{shell } a} K(a,b) = (2l_a + 1) \sum_k \begin{pmatrix} l_a & l_b & k \\ 0 & 0 & 0 \end{pmatrix}^2 G^k(a,b), \qquad (8.15)$$

and since this contribution does not depend on m_b and on m_{s_b}, we obtain, if the subshell b is also closed

$$\sum_{\text{shells } a,b} K(a,b) = 2(2l_a + 1)(2l_b + 1) \sum_k \begin{pmatrix} l_a & l_b & k \\ 0 & 0 & 0 \end{pmatrix}^2 G^k(n_a, l_a; n_b, l_b).$$
$$(8.16)$$

We have just discussed the case when the two electrons belong to different subshells a and b. But these results can also be extended to the case of pairs of electrons belonging to the same closed subshell. We obtain

$$\sum_{\text{shell } a} \left[J(a, a') - K(a, a') \right]$$

$$= 2(2l_a + 1)^2 \left[F^0(n_a, l_a; n_a, l_a) - \frac{1}{2} \sum_k \begin{pmatrix} l_a & l_a & k \\ 0 & 0 & 0 \end{pmatrix}^2 F^k(n_a, l_a; n_a, l_a) \right],$$

where we have taken into account the fact that

$$G^k(n_a, l_a; n_a, l_a) = F^k(n_a, l_a; n_a, l_a).$$

We can rewrite the same expression in this equivalent form

$$\sum_{\text{shell } a} \left[J(a, a') - K(a, a') \right] = (2l_a + 1)(4l_a + 1) F^0(n_a, l_a; n_a, l_a)$$

$$- (2l_a + 1)^2 \sum_{k>0} \begin{pmatrix} l_a & l_a & k \\ 0 & 0 & 0 \end{pmatrix}^2 F^k(n_a, l_a; n_a, l_a).$$
$$(8.17)$$

All quantities that we have calculated, relative to the sums over closed subshells, are independent of the quantum numbers of the electrons belonging to the open subshells. We have therefore obtained the important result that the contributions of Coulomb energy from closed subshells do not remove the degeneracy of a given configuration. Such degeneracy can be removed only by a very small fraction of all possible pairs (a, b), i.e. those pairs where both electrons belong to open subshells. A sample calculation for the ground configuration of the silicon atom, $1s^2 2s^2 2p^6 3s^2 3p^2$, is presented in Sect. 16.8.

8.6 Terms Structure

As we have seen, the closed subshells do not remove the degeneracy of a given configuration. They only provide a contribution to the energy that is the same for all

Table 8.3 Values of the coefficients a^k and b^k, for the only possible values of $k = 0, 2$, for the p^2 configuration

l_1	m_1	l_2	m_2	a^0	a^2	b^0	b^2
1	1	1	0	1	$-2/25$	0	$3/25$
1	1	1	1	1	$1/25$	1	$1/25$
1	1	1	-1	1	$1/25$	0	$6/25$
1	0	1	0	1	$4/25$	1	$4/25$

terms. The so-called multiplet structure, i.e. the energy order of the terms and the values of the energy differences between terms, are determined solely by the electrons belonging to the open subshells. They need to be evaluated on a case by case basis, by means of the results contained in Tables 8.1, 8.2, and Eq. (8.11). In the following, we only discuss the calculations for the p^2 configuration, noting that, for this case of two equivalent electrons, there is only a single possible pair. Furthermore, the quantities $F^k(a, b)$ and $G^k(a, b)$ that appear in Eq. (8.11) coincide, so the equation can be written in the form

$$\langle \Psi(a_1, a_2) | \frac{e_0^2}{r_{12}} | \Psi(a_1, a_2) \rangle$$

$$= \sum_{k=0}^{\infty} [a^k(a_1, a_2) - \delta(m_{s_1}, m_{s_2}) b^k(a_1, a_2)] F^k(a_1, a_2).$$

The results of Table 8.2 show that Eq. (8.11) has to be evaluated for various combinations of the quantum numbers of the two electrons. This requires the evaluation of several 3-j symbols, which results in the values of the quantities a^k and b^k that are displayed in Table 8.3. Using such values, we obtain the following results for the energies of the terms

$$^3P = F^0 - \frac{1}{5} F^2, \qquad ^1D = F^0 + \frac{1}{25} F^2, \qquad ^1S = F^0 + \frac{2}{5} F^2.$$

We can see that, since F^2 is positive, the structure of the terms follows the first Hund's rule, already illustrated in Sect. 7.12. According to this rule, the term with the lowest energy is the one having the highest multiplicity (the term ^3P in this case). The order, according to increasing energy, of the other two terms is ^1D, ^1S. Furthermore, we obtain a numerical result for the ratio between the energy intervals. We have in fact

$$\mathcal{R} = \frac{^1S - {}^1D}{^1D - {}^3P} = \frac{3}{2}.$$

This theoretical result can be tested against spectroscopic observations of atoms (or ions) having the configuration p^2 as an open shell (C I, N II, O III, Si I, Ge I, Sn I, etc.). The experimental data show that the $\mathcal{R}$ ratio for such atoms varies between 1.13 and 1.50, in reasonable agreement with the theoretical prediction. The differences can be attributed to configuration interaction and to the relativistic corrections to the Hamiltonian that will be discussed in the following chapter.

Chapter 9
More Details on Atomic Spectra

Spectroscopic measurements normally reach a very high level of accuracy, which means that the non-relativistic approximation introduced in the previous chapter is not sufficient, in the vast majority of cases, to give a quantitatively adequate description of the atomic spectra. In particular, on the basis of this approximation, we cannot explain the presence in the atomic spectra of an important phenomenon such as the fine structure. This phenomenon can be adequately described considering the contribution to the energy of the atom due to the intrinsic angular momentum of the electron. The result is the removal of the degeneracy in the non-relativistic Hamiltonian and the consequent separation of the terms (identified with the quantum numbers L and S) in atomic levels characterized by the quantum number J. In this chapter we describe in detail this phenomenon, together with other phenomena which similarly produce the removal of degeneracy of the atomic states, either due to external agents (such as a magnetic field) or internal ones (nuclear spin). This will lead to the description of other characteristic effects of atomic spectra such as the Zeeman effect, the Paschen-Back effect and the hyperfine structure.

9.1 The Spin-Orbit Interaction

The discussions presented in Chaps. 7 and 8 are based on the approximation of a nonrelativistic Hamiltonian. Now, it necessary to introduce the relativistic corrections and analyse in detail their effect on the atomic levels. Taking into account the results obtained in Sect. 5.5, when we considered the first order non-relativistic limit to the Dirac equation, and in particular Eq. (5.13), we now assume, as a first approximation, that the relativistic corrections are simply described by the Hamiltonian $\mathcal{H}_2$ given by[1]

---

[1]It is the fifth term in the square bracket of Eq. (5.13), that one of the spin-orbit interaction, where we made the substitution $\sigma_i = 2s_i$.

E. Landi Degl'Innocenti, *Atomic Spectroscopy and Radiative Processes*,
UNITEXT for Physics, DOI 10.1007/978-88-470-2808-1_9, © Springer-Verlag Italia 2014

$$\mathcal{H}_2 = \sum_{i=1}^{N} \xi(r_i)\boldsymbol{\ell}_i \cdot \mathbf{s}_i, \tag{9.1}$$

where the sum is extended over al the electrons and where the function $\xi(r)$ is given by

$$\xi(r) = \frac{\hbar^2}{2m^2c^2}\frac{1}{r}\frac{\mathrm{d}}{\mathrm{d}r}V_{\mathrm{c}}(r). \tag{9.2}$$

Actually, Eq. (5.13) has two other nonrelativistic corrections, the p^4 term and the Darwin term. However, these terms do not explicitly contain operators of angular momentum. They only depend on the variable r and on the operator $\partial/\partial r$. Therefore, it is reasonable to think that the effects of such terms can be included into the central potential term $V_{\mathrm{c}}(r)$. We also point out that the Hamiltonian $\mathcal{H}_2$ describes only the spin-orbit interaction, i.e. the interaction between the orbital angular momentum and the intrinsic angular momentum of the electron, and therefore neglects the mutual interaction between the intrinsic angular momenta, such as the spin-spin interaction. Such interaction terms are however only important for the helium atom (and its isoelectronic sequence), and would require a separate discussion. We can say that in practice the Hamiltonian $\mathcal{H}_2$ of Eq. (9.1) provides a good approximation for all atoms with the exception of helium.

We now look at how the various angular momentum operators commute with the Hamiltonian $\mathcal{H}_2$. We start by calculating the commutator of the total orbital angular momentum $\mathbf{L} = \sum_{j=1}^{N}\boldsymbol{\ell}_j$

$$[\mathcal{H}_2, \mathbf{L}] = \sum_{i=1}^{N}\sum_{j=1}^{N}\left[\xi(r_i)\boldsymbol{\ell}_i \cdot \mathbf{s}_i, \boldsymbol{\ell}_j\right].$$

Since the operators relative to a given particle commute with all the operators relative to different particles, the double sum becomes a single sum. We note that the orbital angular momentum operator $\boldsymbol{\ell}_i$ commutes with both the function $\xi(r_i)$ (which depends only on the radial coordinate r_i) and the spin operator $\mathbf{s}_i$. If we recall the commutation properties of the Cartesian components of the $\boldsymbol{\ell}_i$ operator, we therefore obtain, with simple algebra

$$[\mathcal{H}_2, \mathbf{L}] = \mathrm{i}\sum_{i=1}^{N}\xi(r_i)\boldsymbol{\ell}_i \times \mathbf{s}_i.$$

Similarly, the commutator with the total spin $\mathbf{S} = \sum_{j=1}^{N}\mathbf{s}_j$ is

$$[\mathcal{H}_2, \mathbf{S}] = -\mathrm{i}\sum_{i=1}^{N}\xi(r_i)\boldsymbol{\ell}_i \times \mathbf{s}_i,$$

hence the commutator with the total angular momentum $\mathbf{J} = \mathbf{L} + \mathbf{S}$ is zero

$$[\mathcal{H}_2, \mathbf{J}] = 0.$$

In general, when relativistic corrections are introduced, L and S are no longer good quantum numbers (since the corresponding operators L^2 and S^2 do not commute with the Hamiltonian). The quantum number that is still "good" in all generality is J, the total angular momentum. However, if the Hamiltonian $\mathcal{H}_2$ is a perturbation of the nonrelativistic Hamiltonian, i.e. if the following inequalities hold

$$\mathcal{H}_0 \gg \mathcal{H}_1 \gg \mathcal{H}_2,$$

we can then use perturbation theory so that L and S remain good quantum numbers and the effect of the relativistic Hamiltonian can be evaluated by calculating the matrix elements within the subspace identified by a particular pair of (L, S) values and diagonalising then the resulting matrix. This is known as the Russell-Saunders coupling scheme or simply L-S coupling. In this scheme we have a hierarchy of Hamiltonians that are "applied one at a time". We start by solving the Schrödinger equation with the Hamiltonian $\mathcal{H}_0$ of Eq. (7.3) thus finding the energies of the configurations. We then consider the action of the Hamiltonian $\mathcal{H}_1$ of Eq. (7.4) for a given configuration. We treat it as a perturbation, and find the energies of the terms, each characterized by the L and S values. Finally, for a given term, we consider the action of the Hamiltonian $\mathcal{H}_2$ as a perturbation, finding the energy of the levels, each characterized by the value of J.

Within the Russell-Saunders scheme, we therefore need to evaluate the matrix elements of $\mathcal{H}_2$ on the degenerate space of a given term. We have seen in Sect. 7.11 that the eigenfunctions are of the type $\Psi^A(\alpha, L, S, M_L, M_S)$ (recall that the α symbol represents the configuration), and that the degeneracy is $(2L + 1)(2S + 1)$. As we have previously seen, instead of evaluating the matrix elements of $\mathcal{H}_2$ on this basis, it is more convenient to introduce a new basis, where the total angular momentum $\mathbf{J}$ is diagonal. In this way, the Hamiltonian $\mathcal{H}_2$ becomes diagonal since it commutes with $\mathbf{J}$. The eigenfunctions of the new basis will be denoted by the symbols $\Psi^A(\alpha, L, S, J, M)$ and can be obtained with suitable linear combinations of $\Psi^A(\alpha, L, S, M_L, M_S)$ involving the Clebsh-Gordan coefficients or the 3-j symbols (see Sect. 7.9 and in particular Eq. (7.13)). Through the Clebsh-Gordan coefficients we have

$$\Psi^A(\alpha, L, S, J, M) = \sum_{M_L M_S} \langle L S M_L M_S | L S J M \rangle \Psi^A(\alpha, L, S, M_L, M_S),$$

or, in the more compact Dirac notation

$$|\alpha L S J M\rangle = \sum_{M_L M_S} \langle L S M_L M_S | L S J M \rangle |\alpha L S M_L M_S\rangle.$$

To evaluate the effect of the Hamiltonian $\mathcal{H}_2$ it is therefore sufficient to consider the diagonal matrix elements

$$\langle \alpha L S J M | \mathcal{H}_2 | \alpha L S J M \rangle.$$

These computations will be carried out in Sect. 9.3, after a discussion on the matrix elements of some special operators.

9.2 The Wigner-Eckart Theorem and the Projection Theorem

The vector operators that are commonly encountered in theoretical spectroscopy satisfy particular commutation rules with respect to the angular momentum. These rules are intimately connected with the properties of the transformation of the Cartesian components of a vector under rotations in ordinary space and are of the form

$$[J_l, v_k] = i \sum_m \epsilon_{lkm} v_m,$$

where $\mathbf{v}$ is a vector operator and $\mathbf{J}$ is the total angular momentum of the physical system that is being considered, or of the subsystem on which the $\mathbf{v}$ operator acts.

This commutation rule applies to an extensive class of vector operators (actually to all vector operators, as long as the total angular momentum of each subsystem is included in $\mathbf{J}$). Some examples are given below; their proofs are trivial and are not presented here.

(a) The total angular momentum satisfies the commutation rule with respect to itself.
(b) For a system consisting of more subsystems, if $\mathbf{J} = \mathbf{J}_1 + \mathbf{J}_2 + \cdots$, and if each angular momentum commutes with all other angular momenta, then each summand satisfies the commutation rule with respect to itself and with respect to $\mathbf{J}$.
(c) For a system composed of a single particle, both the coordinate $\mathbf{r}$ and the momentum $\mathbf{p}$ satisfy the commutation rule with respect to both the orbital angular momentum $\boldsymbol{\ell}$ and the total angular momentum $\mathbf{j} = \boldsymbol{\ell} + \mathbf{s}$, where $\mathbf{s}$ is the spin.
(d) Still for a system composed of a single particle, $\mathbf{s}$ satisfies the commutation rule with respect to both itself and $\mathbf{j}$.
(e) For a system composed of N particles, if $\mathbf{J} = \boldsymbol{\ell}_1 + \boldsymbol{\ell}_2 + \cdots + \boldsymbol{\ell}_N + \mathbf{s}_1 + \mathbf{s}_2 + \cdots + \mathbf{s}_N$, then all the following operators satisfy the commutation rule with respect to $\mathbf{J}$: the coordinate $\mathbf{r}_i$, the momentum $\mathbf{p}_i$, the orbital angular momentum $\boldsymbol{\ell}_i$, the spin momentum $\mathbf{s}_i$, and the total angular momentum $\mathbf{j}_i$ (all relative to the i-th particle).
(f) The rule is satisfied by any linear combination of vector operators that satisfy the rule itself.
(g) The rule is satisfied by the vector product of two vector operators that satisfy the rule itself.

Finally, we note that, if $\mathbf{v}$ and $\mathbf{w}$ are two vector operators that satisfy the rule, their scalar product $\mathbf{v} \cdot \mathbf{w}$ commutes with $\mathbf{J}$.

An important consequence of the above commutation rule is the so-called Wigner-Eckart theorem. Using this theorem, it is possible to find a particularly simple expression for the matrix elements of the so-called spherical components of the vector operators between eigenstates of the angular momentum. The spherical components of a vector operator, v_q ($q = -1, 0, 1$), are defined, starting from the Cartesian components, as follows

$$v_{-1} = \frac{1}{\sqrt{2}}(v_x - iv_y), \qquad v_0 = v_z, \qquad v_1 = -\frac{1}{\sqrt{2}}(v_x + iv_y). \qquad (9.3)$$

For the matrix elements of such components we have (Wigner-Eckart theorem)

$$\langle \alpha J M | v_q | \alpha' J' M' \rangle = \langle J' 1 M' q | J M \rangle \langle \alpha J \| \mathbf{v} \| \alpha' J' \rangle,$$

where $\langle J' 1 M' q | J M \rangle$ is a Clebsh-Gordan coefficient and where $\langle \alpha J \| \mathbf{v} \| \alpha' J' \rangle$ is the so-called reduced matrix element (independent of M, M', and q) of the vector operator $\mathbf{v}$. If we use the 3-j symbols instead of the Clebsh-Gordan coefficients, the Wigner-Eckart theorem can be rewritten in the form

$$\langle \alpha J M | v_q | \alpha' J' M' \rangle = (-1)^{J'+M+1} \sqrt{2J+1} \begin{pmatrix} J & J' & 1 \\ -M & M' & q \end{pmatrix} \langle \alpha J \| \mathbf{v} \| \alpha' J' \rangle. \quad (9.4)$$

Taking into account the properties of the 3-j symbols, we can see that the matrix element can only be non-zero if the following "selection rules" are satisfied

$$\Delta J = \pm 1, 0, \quad 0 \nrightarrow 0, \quad \Delta M = \pm 1, 0.$$

The Wigner-Eckart theorem, presented here without giving a formal proof, is actually more general. The matrix elements between eigenstates of the angular momentum of irreducible tensor operators of any rank k ($k = 0, 1, 2, \ldots$) can in fact be expressed, using this theorem, in terms of the 3-j symbols and of the reduced matrix elements. The vector operators, once written in spherical components, are irreducible tensor operators of rank 1.

The Wigner-Eckart theorem can be used to demonstrate another important theorem, called the projection theorem, which concerns the diagonal matrix elements with respect to the quantum number J. To prove this theorem, we consider the matrix element of the operator $\mathbf{J} \cdot \mathbf{v}$. It is simple to verify that such scalar product can be written, using spherical components, in the form

$$\mathbf{J} \cdot \mathbf{v} = \sum_q (-1)^q J_{-q} v_q, \quad (9.5)$$

from which we obtain

$$\langle \alpha J M | \mathbf{J} \cdot \mathbf{v} | \alpha' J M' \rangle = \sum_q (-1)^q \langle \alpha J M | J_{-q} v_q | \alpha' J M' \rangle.$$

If we insert between the operators J_{-q} and v_q the identity operator written in the form

$$1 = \sum_{\alpha''} \sum_{J''} \sum_{M''} |\alpha'' J'' M'' \rangle \langle \alpha'' J'' M'' |,$$

taking into account that the operator J_{-q} is diagonal with respect to the quantum numbers α and J, and applying twice the Wigner-Eckart theorem, we obtain

$$\langle \alpha J M | \mathbf{J} \cdot \mathbf{v} | \alpha' J M' \rangle = \sum_{q M''} (-1)^{q+(J+M+1)+(J+M''+1)} (2J+1)$$

$$\times \begin{pmatrix} J & J & 1 \\ -M & M'' & -q \end{pmatrix} \begin{pmatrix} J & J & 1 \\ -M'' & M' & q \end{pmatrix}$$

$$\times \langle \alpha J \| \mathbf{J} \| \alpha J \rangle \langle \alpha J \| \mathbf{v} \| \alpha' J \rangle.$$

We note that the exponent of (-1) can be written as

$$q + (J + M + 1) + (J + M'' + 1) = q + (J + M + 1) - (J + M'' + 1)$$
$$= q + M - M'' = 0.$$

In addition, if we change the sign of the second line of the second 3-j, we reverse the first two columns of the same symbol, and bear in mind the orthonormality property of the 3-j symbols (Eqs. (7.19), (7.20) and (7.18)), we obtain

$$\langle \alpha J M | \mathbf{J} \cdot \mathbf{v} | \alpha' J M' \rangle = \langle \alpha J \| \mathbf{J} \| \alpha J \rangle \langle \alpha J \| \mathbf{v} \| \alpha' J \rangle \delta_{MM'}.$$

This expression, in the particular case where $\mathbf{v} = \mathbf{J}$, can be used to evaluate the reduced matrix element of $\mathbf{J}$. In fact, bearing in mind that the eigenvalue of J^2 is $J(J+1)$ and equalling α to α', we have

$$J(J+1) = \langle \alpha J M | \mathbf{J} \cdot \mathbf{J} | \alpha J M \rangle = \left(\langle \alpha J \| \mathbf{J} \| \alpha J \rangle \right)^2.$$

We now consider the following matrix element

$$\langle \alpha J M | J_q (\mathbf{J} \cdot \mathbf{v}) | \alpha' J M' \rangle.$$

If we insert as done above the identity between the J_q and $(\mathbf{J} \cdot \mathbf{v})$ operators, apply twice the Wigner-Eckart theorem (once in its direct and once in its inverted form), and take into account the previous results, we obtain

$$\langle \alpha J M | J_q (\mathbf{J} \cdot \mathbf{v}) | \alpha' J M' \rangle$$
$$= \langle \alpha J M | J_q | \alpha J M' \rangle \langle \alpha J M' | \mathbf{J} \cdot \mathbf{v} | \alpha' J M' \rangle$$
$$= (-1)^{J+M+1} \sqrt{2J+1} \begin{pmatrix} J & J & 1 \\ -M & M' & q \end{pmatrix} J(J+1) \langle \alpha J \| \mathbf{v} \| \alpha' J \rangle$$
$$= J(J+1) \langle \alpha J M | v_q | \alpha' J M' \rangle.$$

We therefore conclude that, to evaluate the diagonal matrix elements with respect to J, the following operator identity holds

$$J^2 \mathbf{v} = \mathbf{J}(\mathbf{J} \cdot \mathbf{v}),$$

i.e., when $J \neq 0$

$$\mathbf{v} = \frac{\mathbf{J}(\mathbf{J} \cdot \mathbf{v})}{J^2}. \tag{9.6}$$

This equation can be interpreted physically by saying that the vector $\mathbf{v}$ precedes rapidly around the angular momentum $\mathbf{J}$ in such a way that the components of $\mathbf{v}$ perpendicular to $\mathbf{J}$ are on average zero and the only "effective" component that is left is directed along $\mathbf{J}$, as exemplified in Fig. 9.1. From an historical point of view, we note that the projection theorem was commonly used, albeit intuitively, since the first applications of quantum mechanics to atomic physics. The fact that it can be demonstrated in a rigorous manner constitutes a formal proof of the correctness of these applications.

Fig. 9.1 The vector **v** precedes rapidly around **J** so that its components in the direction perpendicular to **J** are on average zero

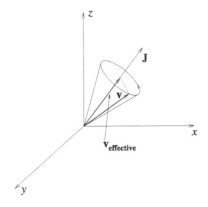

9.3 The Landé Interval Rule

Under the assumption of the Russell-Saunders coupling, the corrections to the term energies ΔE_J due to the Hamiltonian $\mathcal{H}_2$ of Eq. (9.1) are given by diagonal matrix elements of the form

$$\Delta E_J = \sum_{i=1}^{N} \langle \alpha LSJM | \xi(r_i)\boldsymbol{\ell}_i \cdot \mathbf{s}_i | \alpha LSJM \rangle.$$

To calculate these matrix elements, it is convenient to use the basis of the eigenvectors of L^2, S^2, L_z, and S_z, by applying the usual similarity transformation in terms of the Clebsh-Gordan coefficients (Eq. (7.13))

$$\Delta E_J = \sum_{M_L M_S} \sum_{M_L' M_S'} \langle LSJM | LSM_L M_S \rangle$$

$$\times \sum_{i=1}^{N} \langle \alpha LSM_L M_S | \xi(r_i)\boldsymbol{\ell}_i \cdot \mathbf{s}_i | \alpha LSM_L' M_S' \rangle \langle LSM_L' M_S' | LSJM \rangle.$$

In the central matrix element, the operator $\xi(r_i)\boldsymbol{\ell}_i$ acts only on the angular variables, while the operator $\mathbf{s}_i$ acts only on the spin variables. Recalling that the eigenvector $|\alpha LSM_L M_S\rangle$ can also be written as a direct product $|\alpha LM_L\rangle |SM_S\rangle$, we obtain for such matrix element

$$\langle \alpha LSM_L M_S | \xi(r_i)\boldsymbol{\ell}_i \cdot \mathbf{s}_i | \alpha LSM_L' M_S' \rangle$$

$$= \langle \alpha LM_L | \xi(r_i)\boldsymbol{\ell}_i | \alpha LM_L' \rangle \cdot \langle SM_S | \mathbf{s}_i | SM_S' \rangle.$$

The matrix elements in the right-hand side can be evaluated using the projection theorem (Eq. (9.6)). We obtain

$$\langle \alpha LSM_L M_S | \xi(r_i)\boldsymbol{\ell}_i \cdot \mathbf{s}_i | \alpha LSM_L' M_S' \rangle$$

$$= \zeta_i(\alpha, LS) \langle \alpha LM_L | \mathbf{L} | \alpha LM_L' \rangle \cdot \langle SM_S | \mathbf{S} | SM_S' \rangle$$

$$= \zeta_i(\alpha, LS) \langle \alpha LSM_L M_S | \mathbf{L} \cdot \mathbf{S} | \alpha LM_L' M_S' \rangle,$$

where we have introduced the quantity $\zeta_i(\alpha, LS)$ (independent of the quantum numbers M_L and M_S) defined by

$$\zeta_i(\alpha, LS) = \frac{\langle \alpha L M_L | \xi(r_i) \mathbf{L} \cdot \boldsymbol{\ell}_i | \alpha L M_L \rangle \langle S M_S | \mathbf{S} \cdot \mathbf{s}_i | S M_S \rangle}{L(L+1)S(S+1)}. \tag{9.7}$$

We can then write

$$\sum_i \langle \alpha L S M_L M_S | \xi(r_i) \boldsymbol{\ell}_i \cdot \mathbf{s}_i | \alpha L S M_L' M_S' \rangle$$

$$= \zeta(\alpha, LS) \langle \alpha L S M_L M_S | \mathbf{L} \cdot \mathbf{S} | \alpha L S M_L' M_S' \rangle, \tag{9.8}$$

where we have put

$$\zeta(\alpha, LS) = \sum_{i=1}^{N} \zeta_i(\alpha, LS). \tag{9.9}$$

By introducing this expression in the equation for ΔE_J and summing on the Clebsh-Gordan coefficients, we finally get

$$\Delta E_J = \zeta(\alpha, LS)\langle \alpha L S J M | \mathbf{L} \cdot \mathbf{S} | \alpha L S J M \rangle.$$

With these various steps, we have shown that, to calculate the diagonal matrix elements on the basis of the eigenvectors $|\alpha L S J M\rangle$, the Hamiltonian $\mathcal{H}_2$ is equivalent, aside from a proportionality factor, to the operator $\mathbf{L} \cdot \mathbf{S}$. Noting also that

$$J^2 = (\mathbf{L} + \mathbf{S})^2 = L^2 + S^2 + 2\mathbf{L} \cdot \mathbf{S},$$

we obtain

$$\mathbf{L} \cdot \mathbf{S} = \frac{1}{2}(J^2 - L^2 - S^2),$$

so that

$$\Delta E_J = \frac{1}{2}\zeta(\alpha, LS)\big[J(J+1) - L(L+1) - S(S+1)\big].$$

Through this formula we can easily evaluate the term structure, i.e. the energies of the "multiplet levels" in which the term splits due to the relativistic corrections of the Hamiltonian (basically, due to the spin-orbit interaction). In particular, for the energy interval between two successive levels, we have

$$\Delta E_J - \Delta E_{J-1} = \zeta(\alpha, LS)J,$$

a result that is stated as follows: the interval between two successive levels of a multiplet is proportional to the highest J value of the pair. This fact is known as the Landé interval rule. This rule can conveniently be used to verify "a posteriori" how valid the LS coupling is for the description of a particular term of an atom (or ion), by comparing the observed with the predicted energy intervals.

To illustrate these concepts, we consider the special case of the ground term of iron, identified as a^5D in spectroscopic notations (the a symbol represents the ground configuration $1s^2 2s^2 2p^6 3s^2 3p^6 3d^6 4s^2$). Since $L = 2$ and $S = 2$, we can have five different values of J, i.e. $J = 0, 1, 2, 3, 4$, and since the constant $\zeta(\alpha, LS)$

Fig. 9.2 The ground term of the iron atom, a^5D has a negative ζ value, and splits into five fine-structure levels, each characterized by a particular value of J. The energy differences between the levels satisfy the Landé interval rule

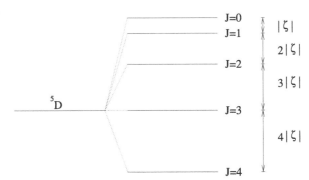

is negative (see below and Sect. 16.9), the multiplet is called inverted and the smaller J values correspond to higher energies. The fine structure of the multiplet is shown in Fig. 9.2, together with the corresponding interval rule. Assuming equal to unity the interval between the $J = 0$ and $J = 1$ levels, theory predicts that the successive intervals should have the values 2, 3, and 4, respectively. The spectroscopic analysis shows that such intervals are equal to 2.05, 3.20, and 4.62, so we can conclude that the LS coupling scheme is reasonably valid to describe this term.

Finally, we note that if one considers the center of gravity (in energy) of a multiplet, defined by

$$\langle \Delta E_J \rangle = \sum_J (2J + 1) \Delta E_J,$$

we obtain

$$\langle \Delta E_J \rangle = \zeta(\alpha, LS) \sum_{J=|L-S|}^{L+S} (2J + 1)\big[J(J + 1) - L(L + 1) - S(S + 1)\big].$$

By means of the elementary formulae that give the sums of powers of the first n integers (or half-integers), one can easily verify that

$$\langle \Delta E_J \rangle = 0,$$

which means that the spin-orbit interaction splits a term in a multiplet without modifying the mean energy.

By applying the projection theorem, we have been able to obtain directly the Landé interval rule, and also an expression for the constant $\zeta(\alpha, LS)$ that quantifies the separation between the levels (Eqs. (9.9) and (9.7)). In some simple cases, this expression can be further developed by elementary methods if we assume (something that is intuitive) that the electrons belonging to closed subshells do not bear any contribution to $\zeta(\alpha, LS)$. This implies that the sum over the electrons appearing in Eq. (9.9) can be restricted to only the optical electrons.

If we consider for example the case of a single electron in the subshell nl, the value of $\zeta(\alpha, LS)$ can easily be determined from Eq. (9.7). Noting that $\mathbf{L} = \boldsymbol{\ell}$ and that $\mathbf{S} = \mathbf{s}$ and recalling the expression of the function $\xi(r)$ (Eq. (9.2)), we obtain

$$\zeta(\alpha, LS) = \zeta_{nl},$$

where

$$\zeta_{nl} = \frac{\hbar^2}{2m^2c^2} \int_0^\infty P_{nl}^2(r) \frac{1}{r} \left[\frac{d}{dr} V_c(r) \right] dr. \tag{9.10}$$

Since $V_c(r)$ is a predominantly increasing function[2] of r, the quantity ζ_{nl} is positive. It can be directly evaluated once the expression for the central potential $V_c(r)$ and that of the corresponding wave functions $P_{nl}(r)$ are known.

In the case of two electrons (equivalent or non-equivalent), we note that the $\mathbf{L} \cdot \boldsymbol{\ell}_1$ operator (as well as similar ones) can be expressed by diagonal operators. We have in fact, since $\boldsymbol{\ell}_2 = \mathbf{L} - \boldsymbol{\ell}_1$,

$$\boldsymbol{\ell}_2^2 = (\mathbf{L} - \boldsymbol{\ell}_1)^2,$$

from which we obtain

$$\mathbf{L} \cdot \boldsymbol{\ell}_1 = \frac{1}{2} \left(L^2 + \ell_1^2 - \ell_2^2 \right).$$

Similarly, we have

$$\mathbf{L} \cdot \boldsymbol{\ell}_2 = \frac{1}{2} \left(L^2 + \ell_2^2 - \ell_1^2 \right), \qquad \mathbf{S} \cdot \mathbf{s}_1 = \frac{1}{2} \left(S^2 + s_1^2 - s_2^2 \right),$$

$$\mathbf{S} \cdot \mathbf{s}_2 = \frac{1}{2} \left(S^2 + s_2^2 - s_1^2 \right).$$

For a term originating from the two-electron configuration $n_1 l_1$ and $n_2 l_2$, recalling that $s_1^2 = s_2^2 = \frac{3}{4}$ and excluding the singlet cases (for $S = 0$ there is no fine structure and the formulae are undetermined), from Eqs. (9.9) and (9.7) we obtain

$$\zeta \left(n_1 l_1 n_2 l_2, {}^3L \right) = \frac{1}{4} \zeta_{n_1 l_1} \frac{L(L+1) + l_1(l_1+1) - l_2(l_2+1)}{L(L+1)}$$

$$+ \frac{1}{4} \zeta_{n_2 l_2} \frac{L(L+1) + l_2(l_2+1) - l_1(l_1+1)}{L(L+1)}. \tag{9.11}$$

In particular, if the two electrons are equivalent (but do not fill the subshell),

$$\zeta \left(nl^2, {}^3L \right) = \frac{1}{2} \zeta_{nl}.$$

When three or more electrons (equivalent or not) are present, the calculation of $\zeta(\alpha, LS)$ cannot be made anymore using elementary methods but must be done using the diagonal sum rule. A particular case is discussed in detail in Sect. 16.9. Using the diagonal sum rule, one can easily prove that, for the configurations having equivalent electrons, the sign of $\zeta(\alpha, LS)$ is positive when the number of electrons occupying the open subshell is lower than half the maximum number of electrons that may occupy the subshell, that is if $q_{nl} < Q_l/2 = (2l + 1)$, with the notation used in Sect. 7.6. Vice versa, if $q_{nl} > Q_l/2$, the value of $\zeta(\alpha, LS)$ is negative.

---

[2]In case of complex atoms, there are indeed restricted intervals in r where the function V_c decreases.

Fig. 9.3 Out of the nine
possible transitions among
the J levels shown in the
figure, only six are allowed by
the selection rule $\Delta J = 0, \pm 1$

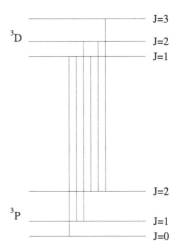

Whenever $\zeta(\alpha, LS)$ is positive, the energy of the levels increases with J, while
the opposite occurs when $\zeta(\alpha, LS)$ is negative. In the first case we say that we are
dealing with regular multiplets, while in the second case we have inverted multiplets, as in Fig. 9.2. All this can be summarized by the empirical rule that takes the
name of third Hund's rule. This rule states that *the multiplets arising from equivalent
electrons are regular when less than half of the subshell is occupied, while they are
inverted when more than half of the subshell is occupied*.[3] For example, if we consider configurations of equivalent p electrons, the configurations p and p^2 give rise
to regular multiplets, the configuration p^3 is undecidable, the configurations p^4 and
p^5 give rise to inverted multiplets, while the configuration p^6, which corresponds
to a closed subshell, produces only one term (^1S) that does not split into a multiplet.

The fine-structure levels of a multiplet can combine with the levels of another
multiplet if they obey the selection rules

$$\Delta J = 0, \pm 1, \quad 0 \nrightarrow 0,$$

which need to be added to those already presented in Sect. 7.13, concerning the
transitions between configurations and terms. This set of spectral lines is called a
multiplet of lines and each multiplet (which corresponds to a pair of terms) has
an appropriate numbering in the spectroscopic tables (the numbering is in practice
arbitrary and is mainly due to historical reasons).

As an example we consider the multiplet number 1 of C I, originating from the
transition between the $2p3s\,^3P$ and the $2p3p\,^3D$ terms. The ^3P term has three fine-
structure levels, $^3P_0$, $^3P_1$, and $^3P_2$, as also does the ^3D term which consists of the
three levels, $^3D_1$, $^3D_2$, and $^3D_3$. Once the selection rules are taken into account, we
obtain a multiplet of six spectral lines, as shown in Fig. 9.3.

---

[3]The case in which exactly half of the subshell is occupied is undecidable, given that theory predicts a null result for ζ. In fact, the experimental values of ζ are very small for these configurations.
The fact that their value is not strictly zero must be attributed to the breakdown of the approximations that we have introduced.

9.4 The j-j Coupling and the Intermediate Coupling

The results of the previous sections were obtained within the LS coupling scheme, based on the "hierarchy" of Hamiltonians synthesized by the inequalities $\mathcal{H}_0 \gg \mathcal{H}_1 \gg \mathcal{H}_2$. Although the LS coupling is appropriate to describe the spectra of numerous elements and ions (and in particular the simpler elements, having a low charge number Z), there are many situations in which it ceases to be valid because the two Hamiltonians $\mathcal{H}_1$ and $\mathcal{H}_2$ are comparable or, in some cases, because the spin-orbit Hamiltonian dominates over the Hamiltonian of Coulomb interaction. To treat these limiting cases, in which

$$\mathcal{H}_0 \gg \mathcal{H}_2 \gg \mathcal{H}_1,$$

we adopt a different scheme, known as j-j coupling.

The Hamiltonian $\mathcal{H}_2$ does not commute with the operators $\mathbf{L}$ and $\mathbf{S}$, but it is simple to demonstrate that it commutes with the total angular momenta $\mathbf{j}_i$ of the single electrons (as well as with the total angular momentum $\mathbf{J}$). The effect of the Hamiltonian $\mathcal{H}_2$ on the degenerate states of the Hamiltonian $\mathcal{H}_0$ (i.e. the configurations), is to split a configuration in a number of terms characterized by the set of quantum numbers j_i, instead of the quantum numbers L and S. As usual, to determine the action of $\mathcal{H}_2$ on the degenerate eigenstates of $\mathcal{H}_0$, it is convenient to introduce a basis on which $\mathcal{H}_2$ is already diagonal. In order to do so, we start with the single-particle functions ψ_{nlmm_s} and consider those linear combinations (obtained with the Clebsh-Gordan coefficients) that are eigenstates of the operators j^2 and j_z. If we indicate by m_j the eigenvalue of the operator j_z, such eigenfunctions are of the form

$$\psi_{nljm_j} = \sum_{mm_s} \langle lsmm_s | lsjm_j \rangle \psi_{nlmm_s}.$$

By means of these single-particle eigenfunctions, it is then possible to obtain eigenfunctions for the total system of N electrons that are antisymmetric (i.e. of the Slater determinant type). Denoting by b_1 the set of quantum numbers (n, l, j, m_j) of the first electron, by b_2 the set of the second electron, and so on, we obtain the eigenfunctions $\Psi^A(b_1, b_2, \ldots, b_N)$, in a very similar way as we have seen for the eigenfunctions $\Psi^A(a_1, a_2, \ldots, a_N)$. While these latter ones are more appropriate for the LS coupling, the present ones are more appropriate for the j-j coupling. The Hamiltonian $\mathcal{H}_2$ is diagonal on this basis, because it can be written in the form

$$\mathcal{H}_2 = \sum_{i=1}^{N} \xi(r_i) \boldsymbol{\ell}_i \cdot \mathbf{s}_i = \sum_{i=1}^{N} \xi(r_i) \frac{1}{2} \left(j_i^2 - \ell_i^2 - s_i^2 \right),$$

hence we obtain

$$\langle \Psi^A(b_1, b_2, \ldots, b_N) | \mathcal{H}_2 | \Psi^A(b_1, b_2, \ldots, b_N) \rangle$$

$$= \frac{1}{2} \sum_{i=1}^{N} \zeta_{n_i l_i} \left[j_i(j_i + 1) - l_i(l_i + 1) - \frac{3}{4} \right],$$

where the quantities ζ_{nl} are the same as those introduced in Eq. (9.10).

Consider now a given configuration. To start with, we note that the contribution of the closed subshells is zero. In fact, for a closed nl subshell, the possible values of j are, if $l \neq 0$, $(l - \frac{1}{2})$ and $(l + \frac{1}{2})$, and we have

$$\sum_{j=l-1/2}^{l+1/2} (2j+1)\left[j(j+1) - l(l+1) - \frac{3}{4} \right] = 0.$$

For example, consider the p^6 configuration, for which the two possible values of j are $\frac{1}{2}$ and $\frac{3}{2}$. We have two electrons with $j = \frac{1}{2}$ (each having one of the two possible projections $m_j = \pm\frac{1}{2}$) and four electrons with $j = \frac{3}{2}$ (each having one of the four possible projections $m_j = \pm\frac{1}{2}, \pm\frac{3}{2}$). This example shows the meaning of the sum that appears in the previous equation.

There remains to consider only the electrons belonging to open subshells. Without expanding the calculations in full generality, we simply consider the case of two non-equivalent p electrons, i.e. the configuration $npn'p$ (with $n \neq n'$). Since both electrons can have $j = \frac{1}{2}$ or $j = \frac{3}{2}$, we have four states with energies $E(j_1, j_2)$ given by

$$E\left(\frac{3}{2}, \frac{3}{2}\right) = \frac{1}{2}(a+b), \qquad E\left(\frac{3}{2}, \frac{1}{2}\right) = \frac{1}{2}a - b,$$

$$E\left(\frac{1}{2}, \frac{3}{2}\right) = -a + \frac{1}{2}b, \qquad E\left(\frac{1}{2}, \frac{1}{2}\right) = -a - b,$$

where

$$a = \zeta_{np}, \qquad b = \zeta_{n'p}.$$

We therefore obtain for the pp' configuration a term structure that is deeply different from the structure obtained with the LS coupling (that, by the way, has six terms instead of four). The same procedure can be applied to the np^2 configuration, of two equivalent electrons. In this case we obtain three terms with energies

$$E\left(\frac{3}{2}, \frac{3}{2}\right) = a, \qquad E\left(\frac{3}{2}, \frac{1}{2}\right) = -\frac{1}{2}a, \qquad E\left(\frac{1}{2}, \frac{1}{2}\right) = -2a,$$

where

$$a = \zeta_{np}.$$

The j-j coupling terms are degenerate. For example, in the case of the configuration p^2, the three terms that we have obtained are respectively degenerate 6 times, 8 times and 1 time. We could then go on and calculate the matrix elements of the Hamiltonian $\mathcal{H}_1$ (i.e. of the Coulomb interaction) on such degenerate space. This is however a rather complex topic and we will not discuss it further here.

As we have already said, in many cases the two Hamiltonians $\mathcal{H}_1$ and $\mathcal{H}_2$ are comparable so that neither the L-S nor the j-j coupling can be applied. These cases are called of intermediate coupling and the atomic structure calculations are much more complex since the two Hamiltonians must be diagonalised simultaneously and not one at a time. In general, the preferred choice is to use the basis of eigenstates of the LS coupling (the base where the operators L^2, L_z, S^2 and S_z are diagonal). The atomic states, characterized in any case by the quantum numbers J and M (recall that the total Hamiltonian commutes with the operator $\mathbf{J}$) can then be expressed, in general, with linear combinations of the form

$$|\alpha J M\rangle = \sum_{LS} \mathcal{C}_{LS} |\alpha L S J M\rangle, \tag{9.12}$$

where the sum is extended to all the L and S values that are compatible with the configuration α and with the value of J. The $\mathcal{C}_{LS}$ are the coefficients of the expansion, obtained with the above diagonalisation. They satisfy the normalisation condition

$$\sum_{LS} |\mathcal{C}_{LS}|^2 = 1.$$

In most cases, there is a particular value of the pair of quantum numbers (L, S) such that the relative coefficient is much greater than all the others. If we specify with (L_0, S_0) such pair of values, it is common practice in spectroscopy to assign to the state the quantum numbers L_0, S_0, even if, in reality, such assignment is only approximate. A very similar situation occurs for the configurations, when configuration interaction is non-negligible. The quantum numbers α, L, S that are assigned to a particular atomic state, either listed in spectroscopic tables or in Grotrian diagrams, must then always be thought of as approximate. The order of the approximation depends on the ratio between the coefficients of the terms of the expansion that are neglected and the coefficient of the dominant term.

9.5 The Zeeman Effect (Classical Approach)

This effect was discovered in 1896 by the Dutch physicist Pieter Zeeman. He observed that, in the presence of a relatively intense magnetic field (of the order of a thousand gauss in the original experiments), the spectral lines of some elements split into various components with special characteristics in terms of intensity and polarisation. The results of the Zeeman experiences can be summarised as follows:

(a) when observing the radiation from a lamp discharge immersed in a magnetic field, a spectral line becomes separated, in the simplest cases, in three components.

Denoting by ν_0 the frequency of the unperturbed line, the three components are located at the frequencies $\nu_0 - \nu_L$, ν_0, and $\nu_0 + \nu_L$, where ν_L is the so-called Larmor frequency[4] given by

$$\nu_L = \frac{e_0 B}{4\pi mc} = 1.3996 \times 10^6 B \text{ s}^{-1}, \tag{9.13}$$

where B is the magnetic field in gauss.

---

[4]In terms of angular frequencies, the Larmor frequency is $\omega_L = 2\pi \nu_L = e_0 B/(2mc)$. It is therefore equal to half the cyclotron frequency which we introduced in Chap. 3 (see Eq. (3.30)).

(b) When observing in a direction parallel to the magnetic field, the central component disappears while the other two components are circularly polarised, one with right polarisation, the other with left polarisation.

(c) When observing in a direction perpendicular to the magnetic field, the three components are linearly polarised. The central component is polarised in a direction parallel to the magnetic field (component π), while the other two are polarised in a direction perpendicular to the magnetic field (components σ).

Zeeman's observations were quickly interpreted by Zeeman himself and by Lorentz within the classical theory of the electron (Quantum Mechanics was not yet born). According to this theory, the atoms are treated as classical oscillators consisting of an electric charge which moves under the action of an elastic restoring force. To interpret the emission of a spectral line at the frequency ν_0, one assumes phenomenologically that the elastic force is such as to give a resonance frequency of the oscillator equal to ν_0. Denoting by $\mathbf{x}$ the coordinate of the charge, it is assumed that, in the absence of external perturbations, the equation of motion of the charge is

$$\frac{d^2\mathbf{x}}{dt^2} = -4\pi^2\nu_0^2\mathbf{x}.$$

When we introduce a magnetic field $\mathbf{B}$, the equation of motion is modified by the presence of the Lorentz force. Assuming that the oscillating charge is an electron with mass m and charge $-e_0$, the equation of motion is therefore

$$\frac{d^2\mathbf{x}}{dt^2} = -4\pi^2\nu_0^2\mathbf{x} - \frac{e_0}{mc}\frac{d\mathbf{x}}{dt} \times \mathbf{B}.$$

If we write the vector $\mathbf{x}$ in terms of its Cartesian components, the differential equations that we have obtained turn out to be coupled. To uncouple them, it is convenient to introduce the components of $\mathbf{x}$ on the three unit vectors $\mathbf{u}_{-1}$, $\mathbf{u}_0$, and $\mathbf{u}_1$ defined by

$$\mathbf{u}_{-1} = \frac{1}{\sqrt{2}}(\mathbf{i} + i\mathbf{j}), \qquad \mathbf{u}_0 = \mathbf{k}, \qquad \mathbf{u}_1 = \frac{1}{\sqrt{2}}(-\mathbf{i} + i\mathbf{j}),$$

where $(\mathbf{i}, \mathbf{j}, \mathbf{k})$ is a Cartesian orthogonal triad with the unit vector $\mathbf{k}$ directed along the magnetic field. Writing then $\mathbf{x} = \sum_\alpha x_\alpha \mathbf{u}_\alpha$, and noting that

$$\mathbf{u}_\alpha \times \mathbf{B} = B\mathbf{u}_\alpha \times \mathbf{u}_0 = -iB\alpha\mathbf{u}_\alpha \quad (\alpha = -1, 0, 1),$$

we obtain for the x_α components the following uncoupled equations

$$\frac{d^2x_\alpha}{dt^2} = -4\pi^2\nu_0^2 x_\alpha + 4\pi i\alpha\nu_L\frac{dx_\alpha}{dt}.$$

Searching for a solution of this equation of the form

$$x_\alpha = A_\alpha e^{-2\pi i\nu_\alpha t},$$

with A_α constant, we obtain for ν_α the second order equation

$$\nu_\alpha^2 + 2\alpha\nu_L\nu_\alpha - \nu_0^2 = 0,$$

Fig. 9.4 Classical oscillator model. The frequency is ν_0 for the linear oscillator, and $\nu_0 \pm \nu_L$ for the circular oscillators

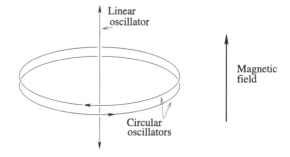

and noting that for typical laboratory magnetic fields $(B < 10^5 \text{ G})$ $\nu_L \ll \nu_0$, we have

$$\nu_\alpha = \nu_0 - \alpha \nu_L.$$

By means of the classical theory of the electron we have obtained a result that perfectly explains Zeeman's observations: under the action of a magnetic field we have obtained three distinct oscillators with frequencies $\nu_0 - \nu_L$, ν_0, and $\nu_0 + \nu_L$. The oscillator at the frequency ν_0 is a linear oscillator directed along the magnetic field, while the other two are circular oscillators that are perpendicular to the magnetic field and of opposite direction, as shown in Fig. 9.4. With these oscillators we can easily explain the characteristics observed by Zeeman in terms of frequency and polarisation properties of the various components.

However, further observations of the Zeeman effect, carried out on an increasing number of spectral lines and with spectroscopic equipment having greater resolving power, showed the existence of a more complex phenomenology (ultimately due to the spin of the electron) that could not be interpreted in the framework of the classical theory developed by Zeeman and Lorentz. For these more complex cases the term "anomalous Zeeman effect" was introduced in order to distinguish them from those of the "normal Zeeman effect", characterized by the experimental facts described above and explained by the classical theory. Today, in light of the modern quantum theory, the distinction between normal and anomalous Zeeman effect is obsolete since the interpretation is done through the quantum theory, which is able to explain consistently the two cases.

9.6 The Zeeman Effect (Quantum Approach)

The effect of a uniform magnetic field on an atomic system can be evaluated by adding to the "unperturbed" Hamiltonian a term that describes the interaction between the system and the magnetic field. Such term was determined, for a single electron, in Sect. 5.4 where we discussed the non-relativistic limit of the Dirac equation. Generalising Eq. (5.11) to the case of N electrons, and recalling the definition of μ_0 (Eq. (5.17)), the magnetic Hamiltonian can then be written as

$$\mathcal{H}_{\mathrm{M}} = \mu_0 (\mathbf{L} + 2\mathbf{S}) \cdot \mathbf{B} + \frac{e_0^2 B^2}{8mc^2} \sum_{i=1}^{N} r_{i\perp}^2,$$

where $r_{i\perp}$ is the component of the position vector of the i-th electron in the plane perpendicular to $\mathbf{B}$. The second term that appears in the r.h.s. of this equation, the so-called diamagnetic term, is negligible, compared to the first. In fact the two terms become comparable for a "critical" value of the magnetic field given by

$$B_c \simeq \frac{\hbar c}{e_0 \langle r_\perp^2 \rangle},$$

and, assuming for $\langle r_\perp \rangle$ a value of the order of the first Bohr orbit, we have, with simple algebra,

$$B_c \simeq \frac{e_0^3 m^2 c}{\hbar^3} = 2.351 \times 10^9 \text{ G}.$$

This value is very high and well above the typical magnetic field strengths achievable in the laboratory. Even in astronomical objects, a magnetic field of the order of B_c is rare. It is thought that such high values of B can only be found in magnetic white dwarfs or in neutron stars (pulsars). On the other hand the magnetic energy $\mu_0 B_c$ associated with B_c is exactly equal to the ionisation energy of the hydrogen atom and, for these magnetic field values, the atomic structure calculations should be carried out *ab initio*. Neglecting then the second term, we assume for the magnetic Hamiltonian the expression

$$\mathcal{H}_M = \mu_0(\mathbf{L} + 2\mathbf{S}) \cdot \mathbf{B} = \mu_0(\mathbf{J} + \mathbf{S}) \cdot \mathbf{B},$$

or, introducing a Cartesian coordinate system such that the z axis is directed along the direction of the magnetic field

$$\mathcal{H}_M = \mu_0 B(J_z + S_z).$$

The magnetic Hamiltonian satisfies the following commutation rules

$$[\mathcal{H}_M, J_z] = 0, \qquad [\mathcal{H}_M, J_x] \neq 0, \qquad [\mathcal{H}_M, J_y] \neq 0,$$

which means that, in general, the quantum number J loses its property of being a good quantum number, while such property is preserved for the magnetic quantum number M. The effect of the magnetic Hamiltonian on the atomic levels can be easily evaluated if this Hamiltonian can be regarded as a perturbation with respect to the unperturbed Hamiltonian $\mathcal{H} = \mathcal{H}_0 + \mathcal{H}_1 + \mathcal{H}_2$. In this case, regardless of the type of coupling valid for the eigenstates of $\mathcal{H}$, it is sufficient to calculate the diagonal matrix elements

$$\langle \alpha J M | \mathcal{H}_M | \alpha J M \rangle = \mu_0 B \langle \alpha J M | J_z + S_z | \alpha J M \rangle,$$

since the non-diagonal matrix elements are zero ($\mathcal{H}_M$ commutes with J_z). Applying the projection theorem (Eq. (9.6)) we have, for $J \neq 0$,

$$\mu_0 B \langle \alpha J M | J_z + S_z | \alpha J M \rangle = \mu_0 B M \left[1 + \frac{\langle \alpha J M | \mathbf{J} \cdot \mathbf{S} | \alpha J M \rangle}{J(J+1)} \right],$$

and for $J = 0$,

$$\mu_0 B \langle \alpha J M | J_z + S_z | \alpha J M \rangle = 0.$$

The (aJ) level therefore becomes split into $(2J + 1)$ sublevels, due to the effect of the magnetic field. Each of such levels (called magnetic sublevels or Zeeman sublevels), is characterized by the magnetic quantum number M and has an extra energy (with respect to the non-magnetic case) given by

$$\Delta E_M = \mu_0 B M g_J,$$

where g_J, defined by the expression (independent of M),

$$g_J = 1 + \frac{\langle \alpha J M | \mathbf{J} \cdot \mathbf{S} | \alpha J M \rangle}{J(J+1)},$$

is a dimensionless quantity called the Landé factor.

Landé factors can easily be calculated only under LS coupling. In fact, in this case, writing the eigenvector $|\alpha J M\rangle$ in the form $|\alpha L S J M\rangle$, and noting that

$$\mathbf{J} \cdot \mathbf{S} = \frac{1}{2} \left[J^2 + S^2 - L^2 \right],$$

we have

$$\langle \alpha L S J M | \mathbf{J} \cdot \mathbf{S} | \alpha L S J M \rangle = \frac{1}{2} \left[J(J+1) + S(S+1) - L(L+1) \right],$$

so that

$$g_J(LS) = 1 + \frac{J(J+1) + S(S+1) - L(L+1)}{2J(J+1)} = \frac{3}{2} + \frac{S(S+1) - L(L+1)}{2J(J+1)}.$$

In intermediate coupling, the calculation of the Landé factors is more complex and involves the coefficients $\mathcal{C}_{LS}$, implicitly defined in Eq. (9.12). With such coefficients, we can generally write

$$g_J = \sum_{LS} |\mathcal{C}_{LS}|^2 g_J(LS),$$

where $g_J(LS)$ has the value given by the previous expression and where the sum is extended to all values of L and S that are compatible with the configuration and with the value of J. Landé factors are easily measured by means of spectroscopic observations, so their values can be used to quantitatively establish how appropriate the LS coupling scheme is for the description of a specific term. This is done by comparing the observed Landé factors for the levels of a term with the values calculated in LS coupling. For example, for the ground term of iron, a^5D, the Landé factors observed spectroscopically are 1.498 $(J = 1)$, 1.494 $(J = 2)$, 1.497 $(J = 3)$, and 1.496 $(J = 4)$, while $g_J(LS)$ is equal to 1.5 for all four levels.

We now consider two levels, one lower with quantum numbers (αJ) and Landé factor g_J and one higher with quantum numbers $(\alpha' J')$ and Landé factor g'_J. The spectral line of frequency ν_0 (in the absence of the field) splits, by the effect of the magnetic field, into various components, each characterized by the pair of quantum numbers M and M', and by the frequency

$$\nu(M, M') = \nu_0 + \nu_L \left(g'_J M' - g_J M \right),$$

where we have introduced the Larmor frequency ν_L, already defined in Eq. (9.13) (also equal to $\mu_0 B/h$). The number of components is considerably reduced (with respect to all the possibilities) because of the selection rule on M

$$\Delta M = \pm 1, 0.$$

Each component is also characterized by a "strength" and by a specific polarisation property. Anticipating a result that will be formally demonstrated later, we observe that the strength of a spectral line is proportional (in the dipole approximation) to the quantity

$$\sum_{q=-1}^{1} |\langle \psi_i | r_q | \psi_f \rangle|^2,$$

where $|\psi_i\rangle$ and $|\psi_f\rangle$ are the wave functions of the initial and final states of the transition, and where r_q is the spherical component of the vector $\mathbf{r}$ defined by

$$\mathbf{r} = \sum_{i=1}^{N} \mathbf{r}_i.$$

If we apply the Wigner-Eckart theorem (from which the selection rules can easily be obtained), we have that the transition between the lower sublevel, characterized by the quantum number M, and the upper sublevel, characterized by the quantum number M', has a strength proportional to

$$|\langle \alpha J M | r_q | \alpha' J' M' \rangle|^2 = (2J+1) \begin{pmatrix} J & J' & 1 \\ -M & M' & q \end{pmatrix}^2 |\langle \alpha J \| \mathbf{r} \| \alpha' J' \rangle|^2,$$

where $q = M - M'$. Eliminating from these expressions all quantities that do not depend on the magnetic quantum numbers, we get the "relative strength" of the components

$$\mathcal{S}_q(M, M') = 3 \begin{pmatrix} J & J' & 1 \\ -M & M' & q \end{pmatrix}^2,$$

where the factor 3 has been introduced so that the strengths are normalised to 1. In fact, for Eq. (7.18) we have

$$\sum_{MM'} \mathcal{S}_q(M, M') = \sum_{MM'} 3 \begin{pmatrix} J & J' & 1 \\ -M & M' & q \end{pmatrix}^2 = 1. \tag{9.14}$$

It is possible to show that the components with $q = 0$ are of type π, while those with $q = \pm 1$ are of type σ, as defined in the previous section. These properties establish the character of polarisation of all the components when observing along a direction that is parallel or perpendicular to the magnetic field.

Once the strengths and shifts of all the components are known, it is possible to construct a diagram, called Zeeman pattern, where a series of vertical segments, one for each component, are drawn at their respective frequency. Each segment has a length proportional to the strength of the single component. The frequencies are

Table 9.1 Frequency shifts (expressed in units of the Larmor frequency ν_L) and relative strengths of the transitions between Zeeman sublevels belonging to the upper $^3P_1$ and lower $^3D_2$ level

q	M'	M	Shift	Strength
1	1	2	$-\frac{5}{6}$	$\frac{6}{10}$
	0	1	$-\frac{7}{6}$	$\frac{3}{10}$
	-1	0	$-\frac{9}{6}$	$\frac{1}{10}$
0	1	1	$\frac{2}{6}$	$\frac{3}{10}$
	0	0	0	$\frac{4}{10}$
	-1	-1	$-\frac{2}{6}$	$\frac{3}{10}$
-1	1	0	$\frac{9}{6}$	$\frac{1}{10}$
	0	-1	$\frac{7}{6}$	$\frac{3}{10}$
	-1	-2	$\frac{5}{6}$	$\frac{6}{10}$

expressed in Lorentz units ν_L, and the segments are drawn upwards for the π components and downwards for the σ components. Consider for example the transition between a lower $^3D_2$ level and a higher $^3P_1$ level. If the two levels can both be considered in LS coupling, the corresponding Landé factors are given by

$$g_J\left(^3D_2\right) = \frac{7}{6}, \qquad g_J\left(^3P_1\right) = \frac{3}{2}.$$

The strengths and shifts of the individual components are shown in Table 9.1, while the possible transitions and the Zeeman diagram are illustrated in Fig. 9.5.

If we consider the particular case of a level in LS coupling originating from a singlet term, the Landé factor is equal to unity, as we get from the expression of $g_J(LS)$ with the substitution $S = 0$ and $L = J$. For a transition between two such levels, the frequencies of the components are given by

$$\nu\left(M, M'\right) = \nu_0 + \nu_L\left(M' - M\right),$$

and we see that all the components with $q = M - M' = 1$ have frequency $\nu_0 - \nu_L$, those with $q = 0$ have ν_0, and those with $q = -1$ have $\nu_0 + \nu_L$.

On the other hand, the sum of the strengths of the components of each type is equal to 1 (see the normalisation equation (9.14)), so we obtain a very simple Zeeman diagram as the one shown in panel (c) of Fig. 9.5. In this case we have the so-called normal Zeeman effect while otherwise we have the anomalous Zeeman effect. The normal Zeeman effect, which corresponds to the absence of spin, is the only case that can also be treated classically.

Finally, we note that the frequency (or wavelength) shifts due to the Zeeman effect are relatively small and that instruments with high resolving power are necessary in order to highlight this effect even for values of the magnetic field of the order of a thousand gauss. If we indicate with $\Delta\nu$ the frequency interval between two Zee-

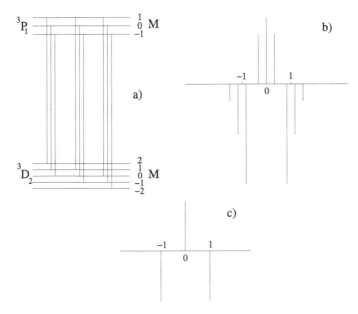

Fig. 9.5 Panel (**a**) shows the transitions between Zeeman sublevels belonging to the upper $^3P_1$ and lower $^3D_2$ level. Panel (**b**) shows the resulting Zeeman pattern, with the π components drawn upwards and the σ components drawn downwards. Panel (**c**) shows the pattern in the case of the normal Zeeman effect

man components, we have that as an order of magnitude $\Delta\nu \simeq \nu_L$. Assuming a field of 10^3 G, we find, for a spectral line at 5000 Å ($\nu_0 = 6 \times 10^{14}$ s^{-1}),

$$\frac{\Delta\nu}{\nu_0} \simeq 2.3 \times 10^{-6}.$$

A spectrometer with a resolving power of the order of 10^6 is therefore necessary to observe the Zeeman effect induced by a magnetic field of 1000 G on a spectral line in the visible.

9.7 The Paschen-Back Effect

The theory of the Zeeman effect developed in the previous section is based on the assumption that the magnetic Hamiltonian $\mathcal{H}_M$ is a perturbation with respect to the total (non-magnetic) Hamiltonian $\mathcal{H}$. For this to be the case, it is obviously necessary that $\mathcal{H}_M$ is a perturbation with respect to $\mathcal{H}_0$, $\mathcal{H}_1$, and $\mathcal{H}_2$. There are, however, some important physical situations in which the magnetic field is sufficiently intense to produce a splitting of the magnetic sublevels comparable to the interval between the different levels originating from a term (fine structure). In the case where the more common L-S coupling scheme is valid, such physical situations give rise to

Fig. 9.6 Splitting of a ^3P level due to the magnetic field in the complete Paschen-Back regime. Each magnetic sublevel is characterized by the pair of values (M_L, M_S). Note that two sublevels are degenerate

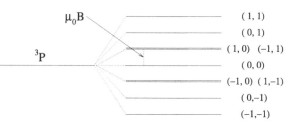

the so-called Paschen-Back effect. It is common to distinguish between the complete Paschen-Back effect, when the following inequality holds

$$\mathcal{H}_2 \ll \mathcal{H}_M,$$

and the incomplete Paschen-Back effect when we have instead

$$\mathcal{H}_2 \simeq \mathcal{H}_M.$$

Note that in the case of the Zeeman effect we have

$$\mathcal{H}_2 \gg \mathcal{H}_M.$$

With increasing magnetic field we have first the Zeeman effect, then the incomplete Paschen-Back, then the complete Paschen-Back effect. The values of the magnetic field at which the transition occurs between the various regimes depend on the particular atomic term that is being considered. If this term is characterized by a certain value of the quantity $\zeta(\alpha, LS)$, we can define a characteristic value of the magnetic field, B_{PB}, such that

$$\mu_0 B_{PB} = \zeta(\alpha, LS).$$

The three different regimes (Zeeman, incomplete Paschen-Back and complete Paschen-Back) are then found for $B \ll B_{PB}$, $B \simeq B_{PB}$, and $B \gg B_{PB}$, respectively.

We start by considering first the complete Paschen-Back effect. To calculate the effect of the magnetic Hamiltonian on the structure of a term, it is convenient to use the degenerate basis $|\alpha LSM_L M_S\rangle$ on which $\mathcal{H}_M$ is diagonal. For the matrix elements of $\mathcal{H}_M$ we find

$$\langle \alpha LSM_L M_S | \mathcal{H}_M | \alpha LSM_L M_S \rangle = \mu_0 B \langle \alpha LSM_L M_S | L_z + 2S_z | \alpha LSM_L M_S \rangle$$
$$= \mu_0 B(M_L + 2M_S).$$

The equations shows that, for example, a ^3P term is split by the magnetic field into seven components, two of which are doubly degenerate (see Fig. 9.6).

We now consider the transition between two different terms and denote by $(\alpha LSM_L M_S)$ the quantum numbers of the lower term and by $(\alpha' L' S' M_L' M_S')$ those of the higher term. The spectral line of frequency ν_0 in the absence of a magnetic field becomes split into several components each characterized by the frequency

$$\nu(M_L M_S, M_L' M_S') = \nu_0 + \nu_L [M_L' - M_L + 2(M_S' - M_S)].$$

The relative strengths of the various components can be derived in a similar way as for the Zeeman effect. In this case, we need to calculate a quantity of the form

$$\sum_{q=-1}^{1} |\langle \alpha L S M_L M_S | r_q | \alpha' L' S' M'_L M'_S \rangle|^2.$$

Noting that r_q is an operator that only acts on the orbital variables and not on the spin, we find

$$\sum_{q=-1}^{1} |\langle \alpha L S M_L M_S | r_q | \alpha' L' S' M'_L M'_S \rangle|^2$$

$$= \sum_{q=-1}^{1} |\langle \alpha L M_L | r_q | \alpha' L' M'_L \rangle|^2 \delta_{SS'} \delta_{M_S M'_S},$$

and applying the Wigner-Eckart theorem we can express the relative strength of the various components with the formula

$$\mathcal{S}_q \left(M_L M_S, M'_L M'_S \right) = 3 \begin{pmatrix} L & L' & 1 \\ -M_L & M'_L & q \end{pmatrix}^2 \delta_{SS'} \delta_{M_S M'_S},$$

which shows that for the complete Paschen-Back effect we have the selection rules

$$\Delta M_S = 0, \qquad \Delta M_L = \pm 1, 0.$$

The expression for the frequencies of the components becomes simplified because of the selection rule on M_S

$$\nu \left(M_L M_S, M'_L M_S \right) = \nu_0 + \nu_{\mathrm{L}} \left(M'_L - M_L \right).$$

The transitions having the same value for ΔM_L have the same frequencies and the sum of their relative strengths is 1. For the complete Paschen-Back effect we therefore obtain the same Zeeman diagram as shown in Fig. 9.5, panel (c), i.e. the diagram for the normal Zeeman effect.

The incomplete Paschen-Back effect is considerably more complex since the spin-orbit ($\mathcal{H}_2$) and the magnetic ($\mathcal{H}_{\mathrm{M}}$) Hamiltonians need to be diagonalised simultaneously. $\mathcal{H}_2$ is diagonal on the vector basis $|\alpha L S J M\rangle$, while $\mathcal{H}_{\mathrm{M}}$ is diagonal on the basis $|\alpha L S M_L M_S\rangle$. Once one of the basis is chosen, all the matrix elements (diagonal and not) need to be calculated, and then the matrix needs to be diagonalised. The problem is partially simplified since both Hamiltonians commute with J_z

$$[\mathcal{H}_2, J_z] = [\mathcal{H}_{\mathrm{M}}, J_z] = 0,$$

so the matrix is block-diagonal, each block being characterized by a particular value of $M = M_L + M_S$. In general, the computations need to be done numerically and show that, as the magnetic field increases, the structure of the levels changes continuously from a typical Zeeman effect to a complete Paschen-Back effect.[5]

---

[5]For a discussion of this topic see, for example, Condon and Shortley (1935) or Landi Degl'Innocenti and Landolfi (2004).

9.8 Hyperfine Structure, Isotope Effect

When spectral lines are examined with spectroscopic devices having very high resolving power, it is found, since the year 1920, that for several elements atomic spectral lines are divided into a number of components, extremely close to each other, having typical intervals in terms of wave number of the order of one hundredth of cm^{-1}, i.e., in terms of wavelength, of the order of some mÅ (for lines in the visible spectrum). Today we know that this so-called hyperfine structure is due to the influence of the atomic nucleus on the energy levels of the atom, an hypothesis originally proposed by Pauli, and later fully confirmed. Such influence has a dual nature being due, on the one hand, to the fact that the isotopes of an element have different mass and nuclear volume (isotope effect) and, secondly, to the presence of the spin of the nucleus (nuclear spin effect). Obviously, the first effect is present only when analyzing spectroscopically an element composed of two or more isotopes, while the second effect is present even for an isotopically pure element (provided that its nuclear spin is not null). In this section we will analyze the first effect. The second effect is discussed in the following section.

The analysis of the isotope effect is simple only in the case of hydrogenic atoms. In fact, we have seen in Sect. 6.1 that the energy of the n-th level of the hydrogenic atom is given by (cf. Eq. (6.4))

$$E_n = -\frac{m_r e_0^4 Z^2}{2\hbar^2} \frac{1}{n^2},$$

where m_r is the reduced mass, i.e.

$$m_r = \frac{m M_n}{m + M_n}, \tag{9.15}$$

with m the electron mass and M_n the mass of the nucleus. If we have two isotopes with masses M_1 and M_2, and denote by ν_1 and ν_2 the frequencies of two corresponding lines of their spectra, we have

$$\frac{\nu_1}{\nu_2} = \frac{(m_r)_1}{(m_r)_2} = \frac{M_1}{M_2} \frac{m + M_2}{m + M_1},$$

or, through a series expansion ($m \ll M_1, M_2$),

$$\frac{\nu_1}{\nu_2} = 1 + m\left(\frac{1}{M_2} - \frac{1}{M_1}\right).$$

Considering for example the case of hydrogen and deuterium, and remembering that the mass of the deuterium nucleus is twice that of hydrogen, we get

$$\frac{\nu_D}{\nu_H} = 1 + \frac{m}{2M_p} \simeq 1 + \frac{1}{3672},$$

where M_p is the proton mass. The frequencies of the deuterium lines are therefore all higher than the corresponding frequencies of the hydrogen lines, their ratio being of the order of 1.0003. Obviously, the opposite occurs for wavelengths. The Hα line of deuterium, for example, is found to be shifted towards shorter wavelengths by 1.79 Å, with respect to the Hα line of hydrogen.

For non-hydrogenic atoms the effect of the mass can be evaluated with the following considerations. The total kinetic energy T of a system of N electrons orbiting around a nucleus of mass M_n is given by

$$T = \sum_{i=1}^{N} \frac{p_i^2}{2m} + \frac{P^2}{2M_n},$$

where $\mathbf{p}_i$ is the momentum of the i-th electron and $\mathbf{P}$ is the momentum of the nucleus. On the other hand, in the rest frame of the atom we must have

$$\mathbf{P} + \sum_{i=1}^{N} \mathbf{p}_i = 0,$$

so, obtaining $\mathbf{P}$ from this equation and substituting it in the expression of the kinetic energy, we find, with simple algebraic transformations

$$T = \sum_{i=1}^{N} \frac{p_i^2}{2m_r} + \sum_{i<j} \frac{\mathbf{p}_i \cdot \mathbf{p}_j}{M_n},$$

where m_r is the reduced mass defined in Eq. (9.15). As shown by this equation, the effect of the finite mass of the nucleus is twofold. On the one hand, the electron mass is replaced by the reduced mass. On the other hand, many additional energy terms appear, as many as the number of possible pairs of electrons. The effect of these terms can be quantitatively evaluated adding a correction term (of nuclear mass) to the atomic Hamiltonian given by

$$\mathcal{H}_{\text{n.m.}} = \sum_{i<j} \frac{\mathbf{p}_i \cdot \mathbf{p}_j}{M_n},$$

and then determining the effect of this atomic Hamiltonian on the levels using perturbation theory. For heavy atoms this mass effect decreases compared to the case of hydrogen, mainly due to the decrease in the relative difference between the masses of the isotopes.

On top of this effect due to the mass, we also have the effect of nuclear volume which can be evaluated, at least as an order of magnitude, in the following way. In the atomic theory developed in the previous chapters we have always assumed that the nucleus is a point-like particle. In fact this is a zero order approximation that leads to a divergent potential at the origin. A better approximation, although crude, is to consider the nucleus as a uniformly charged sphere with a radius r_0 being of the order of the nuclear dimensions ($r_0 \simeq 10^{-13}$ cm). The potential in which the electrons move thus differs, in proximity of the nucleus, from the purely Coulomb potential due to a point-like nucleus. To calculate this potential, we use the Gauss theorem that, for $r \leq r_0$, gives the equation

$$4\pi r^2 E(r) = 4\pi Q(r) = 4\pi Z e_0 \frac{r^3}{r_0^3},$$

Fig. 9.7 The potential
energy of the electron, $V(r)$,
due to the charge of the
nucleus, differs depending on
whether one considers a
nucleus that is point-like or
has a finite dimension with
distributed charge. The graph
illustrates schematically this
difference. The dimensions of
the nucleus, r_0, are greatly
exaggerated in the figure

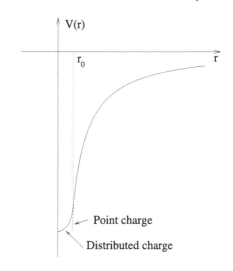

where $E(r)$ is the electric field, $Q(r)$ is the charge contained within the sphere of
radius r, and Z is the nuclear charge number. The electric field is therefore, for
$r \leq r_0$

$$E(r) = Ze_0 \frac{r}{r_0^3},$$

and originates from the potential

$$\phi(r) = -\frac{Ze_0 r^2}{2r_0^3} + C,$$

where C is a constant that can be determined by imposing that for $r = r_0$ we have
$\phi = Ze_0/r_0$. Applying this continuity condition we find

$$\phi(r) = \frac{Ze_0}{2r_0} \left(3 - \frac{r^2}{r_0^2} \right), \quad r \leq r_0.$$

The effect of a finite nucleus is therefore that of introducing a corrective term (of
nuclear volume) to the Hamiltonian, given by

$$\mathcal{H}_{n.v.} = \sum_{i=1}^{N} \left[-\frac{Ze_0^2}{2r_0} \left(3 - \frac{r_i^2}{r_0^2} \right) + \frac{Ze_0^2}{r_i} \right], \quad r_i \leq r_0.$$

Such term is schematically shown in Fig. 9.7.

We can now calculate the variation in the energy of a configuration, ΔE, caused
by this correction to the Hamiltonian. This is done with perturbation theory, noting
that the Hamiltonian $\mathcal{H}_{n.v.}$ is diagonal on the basis of the eigenfunctions of the form
of Eq. (7.1), $\Psi^A(a_1, a_2, \ldots, a_N)$, since it commutes with the operators $\mathbf{l}_i$ and $\mathbf{s}_i$.
We have

$$\Delta E = \frac{Ze_0^2}{2r_0} \sum_{i=1}^{N} \mathcal{I}(a_i),$$

where

$$\mathcal{I}(a_i) = \int_{r \leq r_0} \left| \psi_{n_i l_i m_i m_{si}} (\mathbf{x}) \right|^2 \left(-3 + \frac{r^2}{r_0^2} + \frac{2r_0}{r} \right) d^3\mathbf{x}.$$

To evaluate this integral we note that r_0 is much smaller than the typical scales over which the wave function varies, so we have

$$\mathcal{I}(a_i) = \left| \psi_{n_i l_i m_i m_{si}} (0) \right|^2 \int_{r \leq r_0} \left(-3 + \frac{r^2}{r_0^2} + \frac{2r_0}{r} \right) d^3\mathbf{x},$$

and evaluating the integral with elementary methods we get

$$\mathcal{I}(a_i) = \frac{4\pi}{5} r_0^3 \left| \psi_{n_i l_i m_i m_{si}} (0) \right|^2.$$

Substituting in the expression for ΔE we finally obtain the result

$$\Delta E = \frac{2\pi \, Z e_0^2 r_0^2}{5} \sum_{i=1}^{N} \left| \psi_{n_i l_i m_i m_{si}} (0) \right|^2.$$

As we can see, the correction to the energy is positive, and only the s electrons (i.e. those for which $\psi(0) \neq 0$) contribute to it. Furthermore, the correction depends on r_0, so we obtain different values of ΔE for different isotopes. This is the nuclear volume effect.

9.9 Hyperfine Structure, Effect Due to the Nuclear Spin

A nucleus (more properly a nuclide) with mass number A and charge number Z is composed of A nucleons, namely Z protons and $A - Z$ neutrons. Both protons and neutrons are particles with spin $\frac{1}{2}$ and such spins are added (according to the addition rules of angular momenta) to give the total spin $\mathbf{I}$ of the nucleus. The associated quantum number I is necessarily half-integer if A is odd, and integer if A is even. The value of I that belongs to a given nucleus[6] can be inferred from observations or, in some cases, can be obtained according to models of nuclear structure. Experiments show that all even-even nuclei (i.e. those composed of an even number of protons and an even number of neutrons) have spin zero. The nuclei that do not belong to this class have, in general, $I \neq 0$.

A nuclear magnetic moment is always associated with a nuclear spin. It can be expressed in the form

$$\boldsymbol{\mu}_I = g_N \mu_N \mathbf{I}.$$

---

[6]We refer here to the spin of the nucleus in its ground state. In nuclear reactions the nucleus can be brought to excited levels having, in general, different values of I.

In this formula, μ_N is the so-called nuclear magneton, defined with a formula very similar to that one of the Bohr magneton μ_0 (Eq. (5.17)), with the difference that we now have the proton mass instead of the electron mass

$$\mu_N = \frac{e_0 \hbar}{2 M_p c} = \frac{m}{M_p} \mu_0 \simeq \frac{1}{1836} \mu_0.$$

The quantity g_N is instead a purely numeric factor of the order of unity (different for each nuclide), called gyromagnetic nuclear ratio. For an isolated proton $g_N = 2.79290$, while for an isolated neutron $g_N = -1.91315$. For the deuterium nucleus (composed of a proton and a neutron, having $I = 1$) we have $g_N = 0.85741$, which shows that the value g_N of a nucleus cannot be simply obtained by summing the g_N values of its nucleons.

The nuclear magnetic moment interacts with the electron cloud and the interaction is described, in first approximation, by an Hamiltonian, called of hyperfine structure, of the type[7]

$$\mathcal{H}_{\text{h.s.}} = \mathcal{A}(\alpha, J) \mathbf{J} \cdot \mathbf{I}, \tag{9.16}$$

where $\mathcal{A}(\alpha, J)$ is a quantity which depends on the quantum numbers of the state $|\alpha J M\rangle$. Without going into the details of the theory which allows one to calculate the $\mathcal{A}(\alpha, J)$ value, we can simply provide an order of magnitude estimate for this quantity. We can think of the nuclear magnetic moment as interacting with the magnetic moment of the electron cloud and, since the interaction energy of two dipoles is proportional to the product of the moduli of the dipoles and inversely proportional to the cube of the distance between them, we have, as an order of magnitude (a_0 is the radius of the first Bohr orbit)

$$\mathcal{A}(\alpha, J) \simeq \frac{\mu_N \mu_0}{a_0^3} \simeq \frac{e_0^8 m^2}{c^2 \hbar^4 M_p} = 1.264 \times 10^{-18} \text{ erg},$$

that is an energy value equivalent to 6.37×10^{-3} cm^{-1}, or 191 MHz. Comparing this expression with the analogous one valid for fine structure (see Eq. (5.14)), we reach the conclusion that the typical energies of hyperfine structure are about 2000 times lower than those of fine structure.

The intrinsic angular momentum of the nucleus introduces an additional degree of freedom that must be described by appropriate quantum numbers. An atomic level that in the absence of nuclear spin is represented by the state vector $|\alpha J M\rangle$ must now be described by a new eigenvector of the form $|\alpha J I M M_I\rangle$, where I is the nuclear spin quantum number (in the sense that the eigenvalue of the operator I^2 is $I(I+1)$), and where M_I is the quantum number relative to the operator I_z. Instead of this basis, it is more convenient to consider the basis $|\alpha J I F M_F\rangle$ where

---

[7]Equation (9.16) is the first term of a multipolar expansion and represents the dipole interaction between the spin of the nucleus and that of the electron cloud. It is sometimes necessary to add the next term which describes the quadrupole interaction. This term brings an additional contribution to Eq. (9.17) of the form $\mathcal{B}[K(K+1) - 4I(I+1)J(J+1)/3]$, where $\mathcal{B}$ is a new constant and where $K = F(F+1) - I(I+1) - J(J+1)$.

$\mathbf{F} = \mathbf{J} + \mathbf{I}$ is the total angular momentum (electronic + nuclear) of the atom. The Hamiltonian of hyperfine structure is diagonal on the basis $|\alpha J I F M_F\rangle$. Noting that

$$\mathbf{J} \cdot \mathbf{I} = \frac{1}{2}\left(F^2 - J^2 - I^2\right),$$

we have, for the energy correction due to the nuclear spin,

$$\Delta E(F) = \langle \alpha J I F M_F | \mathcal{H}_{\text{s.i.}} | \alpha J I F M_F \rangle$$
$$= \frac{1}{2}\mathcal{A}(\alpha, J)\left[F(F+1) - J(J+1) - I(I+1)\right]. \tag{9.17}$$

We therefore obtain the result that the hyperfine structure levels have very similar energy intervals as those given by the Landé rule for the fine structure levels.

For example, for the ground state of the hydrogen atom we have $J = \frac{1}{2}$ and $I = \frac{1}{2}$ so that we have the two possibilities, $F = 0$, and $F = 1$, with respective energies

$$\Delta E(F = 0) = -\frac{3}{4}\mathcal{A}, \qquad \Delta E(F = 1) = \frac{1}{4}\mathcal{A}.$$

The interval between the two fine structure levels is $\mathcal{A}$ and has been measured to be equal to 1.420 GHz, i.e. about seven times the order of magnitude value we have estimated. The transition between these hyperfine structure levels gives rise to a spectral line that falls in the radio domain at 21.1 cm. The observation of this line, traditionally known as the "21-cm line", is of fundamental importance in astrophysics to study our own Galaxy, but also other galaxies, and H I regions.

Another particularly important example is that of the ground level $6s\,^2S_{1/2}$ of the isotope 133 of cesium. This isotope has nuclear spin equal to $7/2$, so there are two levels of hyperfine structure, characterized by the quantum numbers $F = 3$ and $F = 4$, respectively. The energy interval between these levels is known with such a precision that, since 1967, the definition of the unit of time in the International System is just based on it. The second is in fact defined as the duration of 9 192 631 770 periods of the radiation emitted by the transition between these two hyperfine levels of the isotope 133 of cesium.

Chapter 10
Laws of Thermodynamic Equilibrium

In the previous chapters we have reviewed the fundamental properties of the radiation field and those of atomic systems. Before we move on to a quantum description of the interaction between these systems, is it necessary to study, in all generality, their physical characteristics in the particular case of thermodynamic equilibrium. On the one hand this study is interesting by itself, since in many cases the physical systems (whether they are the object of laboratory experiments or astronomical observations) can be considered, at least in first approximation, in conditions close to those of thermodynamic equilibrium. On the other hand, the laws that can be derived, being based on the first and second laws of thermodynamics, are extremely general and can therefore provide very useful control methods to test any theory (more or less approximate), that can be used for the description of the systems and their interactions.

10.1 The Principles of Statistical Equilibrium

We start by considering a physical system with $\mathcal{N}$ degrees of freedom described, within classical mechanics, by the dynamic variables q_i and the corresponding conjugate moments p_i $(i = 1, 2, \ldots, \mathcal{N})$. Let $H(q_i, p_i)$ be the Hamiltonian of the system. The statistical properties of the system in thermodynamic equilibrium can be derived using a general principle. Although this principle can be more or less justified by probabilistic arguments,[1] sometimes it is preferred to consider it directly as a postulate. The principle states that the probability dP to find the system in the elementary cell $d\Gamma$ of the phase space is given by the expression

$$dP = A\mathrm{e}^{-\beta H(q_i, p_i)}\, d\Gamma, \tag{10.1}$$

---

[1] A simple formal derivation is presented in Sect. 16.10. For a more in-depth analysis see, for example, Schrödinger (1961).

E. Landi Degl'Innocenti, *Atomic Spectroscopy and Radiative Processes*, 237
UNITEXT for Physics, DOI 10.1007/978-88-470-2808-1_10, © Springer-Verlag Italia 2014

where

$$d\Gamma = dq_1 \, dq_2 \cdots dq_{\mathcal{N}} \, dp_1 \, dp_2 \cdots dp_{\mathcal{N}}, \qquad \beta = \frac{1}{k_B T},$$

with k_B the Boltzmann constant ($k_B = 1.3806 \times 10^{-16} \, \text{erg K}^{-1}$), T the absolute temperature, and A a normalisation constant to be determined by imposing that the integral of dP over the whole phase space is equal to unity

$$\int A e^{-\beta H(q_i, p_i)} \, d\Gamma = 1.$$

We note that the value of the constant A is related to the (arbitrary) zero value of the energy. In fact, if we apply the transformation

$$H(q_i, p_i) \rightarrow H(q_i, p_i) + E_0,$$

with E_0 constant, we obtain

$$A \rightarrow A e^{\beta E_0},$$

and the expression for dP is invariant with respect to E_0.

The same principle can be generalised directly to quantum systems. Let $\mathcal{H}$ be the quantum Hamiltonian of the physical system, and let $E_1, E_2, \ldots, E_n, \ldots$ be the energies of the quantum states (supposed to be nondegenerate)[2] obtained by solving the stationary Schrödinger equation. The probability P_j to find the system in the state j with energy E_j is given by

$$P_j = A e^{-\beta E_j}, \tag{10.2}$$

where A is (again) a normalisation constant to be determined so that

$$\sum_j A e^{-\beta E_j} = 1.$$

The normalisation constant is therefore given by

$$A = \frac{1}{\mathcal{Z}},$$

where the quantity $\mathcal{Z}$, known as the "sum over states", is

$$\mathcal{Z} = \sum_j e^{-\beta E_j}.$$

All the laws of thermodynamics can be derived from these simple considerations if it is assumed, following the fundamental hypothesis of Boltzmann, that the entropy of the system is given by the expression

$$S = -k_B \sum_j P_j \ln P_j.$$

---

[2]If the solution of the stationary Schrödinger equation involves degenerate eigenvalues, the corresponding states must all be individually numbered in the series $E_1, E_2, \ldots, E_n, \ldots$.

We now discuss some of the consequences of the hypotheses we have introduced. Substituting the expression for P_j in this equation, we get

$$S = -k_B \sum_j A e^{-\beta E_j} (\ln A - \beta E_j),$$

that, recalling the previous equations, becomes

$$S = k_B \left[\ln \mathcal{Z} + \beta \sum_j A E_j e^{-\beta E_j} \right].$$

We now notice that the internal energy of the system, traditionally denoted by the symbol U, is obviously given by

$$U = \frac{\sum_j E_j e^{-\beta E_j}}{\sum_j e^{-\beta E_j}} = \frac{\sum_j E_j e^{-\beta E_j}}{\mathcal{Z}} = \sum_j A E_j e^{-\beta E_j},$$

and can also be expressed by the equation

$$U = -\frac{\partial}{\partial \beta} \ln \mathcal{Z}.$$

On the other hand, substituting in the equation defining the entropy we have

$$S = k_B [\ln \mathcal{Z} + \beta U],$$

from which we obtain

$$\ln \mathcal{Z} = \frac{S}{k_B} - \beta U = \beta (ST - U) = -\beta F, \tag{10.3}$$

where $F = U - TS$ is the Helmholtz free energy. Finally, this relation can be used to express the internal energy in the form

$$U = -\frac{\partial}{\partial \beta} \ln \mathcal{Z} = \frac{\partial}{\partial \beta} (\beta F) = F + \beta \frac{\partial F}{\partial \beta}.$$

To establish a more complete connection with the laws of thermodynamics, we must also consider the fact that the thermodynamic transformations in many cases imply changes in the system that are related not only to the temperature but also to other macroscopic variables, such as, for example, the volume V, the pressure P or the number of particles N. Within the above general formalism, these changes result in changes in the energy levels E_j. Therefore, the above formula needs to be more conveniently written in the form

$$U = F + \beta \left(\frac{\partial F}{\partial \beta} \right)_{V,N}.$$

If we consider, for example, an infinitesimal thermodynamic transformation in which a fluid (with a constant number of particles) undergoes a variation of temperature δT and a change in volume δV, the free energy undergoes the change

$$\delta F = \delta (U - TS) = \delta U - S\delta T - T\delta S.$$

On the other hand, according to the first principle of thermodynamics, the variation in internal energy is given by

$$\delta U = \delta Q - P \delta V,$$

where $\delta Q = T \delta S$ is the amount of heat absorbed by the system and where P is the pressure. For the variation of the free energy we then have

$$\delta F = -S \delta T - P \delta V,$$

and we obtain the equations

$$S = -\left(\frac{\partial F}{\partial T}\right)_{V,N}, \qquad P = -\left(\frac{\partial F}{\partial V}\right)_{T,N}.$$

The second equation defines directly the pressure that, using Eq. (10.3), is given by

$$P = \frac{1}{\beta}\left(\frac{\partial \ln \mathcal{Z}}{\partial V}\right)_{\beta,N} = \frac{1}{\beta \mathcal{Z}}\left[\frac{\partial}{\partial V}\left(\sum_j e^{-\beta E_j}\right)\right]_{\beta,N} = -\frac{\sum_j (\frac{\partial E_j}{\partial V})_N e^{-\beta E_j}}{\sum_j e^{-\beta E_j}}.$$

The equations that we have derived are, at the same time, extremely simple and very general. However, their application to practical cases is often complex, especially when one needs to take into account the indistinguishability of the particles or the Pauli exclusion principle. In addition, we point out that the determination of the energies E_j of the quantum states of a system is, except in schematic cases, a problem practically insolvable. However, there are sufficiently simple cases in which the above principle quickly leads to some interesting consequences. These cases are those in which the total Hamiltonian of the system can be expressed as the sum of many independent Hamiltonians, each relative to a different degree of freedom. The simple case of a system of $\mathcal{N}$ free, non-interacting particles is discussed in the following section.

10.2 Maxwell's Velocity Distribution

Consider a physical system consisting of $\mathcal{N}$ non-interacting free particles. To fix ideas we can assume that the particles are electrons, although all the results that we will obtain are independent of the intrinsic nature of the particles (except for the bosonic or fermionic character which is of paramount importance). For $\mathcal{N}$ nonrelativistic electrons moving in a plasma (that is electrically neutral, so the energy of the Coulomb interaction is null), the Hamiltonian of the system is

$$\mathcal{H} = \sum_{i=1}^{\mathcal{N}} \frac{p_i^2}{2m},$$

where $\mathbf{p}_i$ is the momentum operator of the i-th electron. Since the Hamiltonian is the sum of $\mathcal{N}$ independent Hamiltonians, the eigenstates of the overall system can be constructed from the eigenstates of the individual Hamiltonians, i.e. from the single-particle eigenfunctions. As we know, these are characterized by the quantum numbers $\mathbf{p}$, eigenvalue of the momentum, and m_s, projection of the spin along the quantisation axis. If we apply the fundamental principle in the form of Eq. (10.2), we can then express the probability P that the electron 1 is in the state $(\mathbf{p}_1, m_{s1})$, electron 2 in the state $(\mathbf{p}_2, m_{s2}), \ldots$, and electron $\mathcal{N}$ in the state $(\mathbf{p}_\mathcal{N}, m_{s\mathcal{N}})$ through the equation

$$P = Ae^{-\beta(\sum_i \frac{p_i^2}{2m})} = Ae^{-\beta\frac{p_1^2}{2m}}e^{-\beta\frac{p_2^2}{2m}}\cdots e^{-\beta\frac{p_N^2}{2m}}.$$

We can now ask what is the probability, denoted by P', that an electron, say electron 1, occupies the state $(\mathbf{p}, m_s)$ regardless of the states being occupied by the other electrons. If we do not have complications related to the indistinguishability of the particles, such probability is obtained using the usual probabilistic rules, i.e. by fixing in the previous equation the value of $(\mathbf{p}_1, m_{s1})$ to $(\mathbf{p}, m_s)$ and summing over all possible values of $(\mathbf{p}_2, m_{s2}), (\mathbf{p}_3, m_{s3}), \ldots, (\mathbf{p}_\mathcal{N}, m_{s\mathcal{N}})$. Using this method, justified only when the probabilities are independent, we obtain the expression

$$P' = A'e^{-\beta\frac{p^2}{2m}},$$

where A' is a new normalization constant. It should be noted that this result could have been obtained directly by applying the fundamental principle of the previous section (Eq. (10.2)) to a system composed of a single electron (the electron 1) and neglecting all the others. This type of approach is however not justified in general, even if in some cases it can lead to the correct result.

The complexities introduced by the indistinguishability of the particles, which we have ignored to obtain the equation above, are twofold. Firstly, since the electrons are indistinguishable, the state of the system in which the electron 1 occupies the state $(\mathbf{p}_1, m_{s1})$, the electron 2 occupies the state $(\mathbf{p}_2, m_{s2}), \ldots$, the electron $\mathcal{N}$ occupies the state $(\mathbf{p}_\mathcal{N}, m_{s\mathcal{N}})$, is physically indistinguishable from any other state of the system obtained by permuting in an arbitrary way the order of the particles that occupy the different states. To take this into account, either we introduce the occupation number of the states of single particle and require that this number is equal to 0 or 1, or, more simply, we divide the volume of the phase space available to the system by $\mathcal{N}!$, i.e. we perform the substitution

$$d\Gamma \rightarrow \frac{d\Gamma}{\mathcal{N}!}.$$

The $1/\mathcal{N}!$ factor affects the normalisation constant A and its introduction is inessential if we do not consider physical phenomena involving transformations that alter the number of particles making up the system.

Secondly, we need to take into account the Pauli exclusion principle. The complications arising from this principle begin to occur at very high densities, when the number of electrons becomes comparable with the number of quantum states available to the electrons themselves. In such cases the above arguments cannot be followed, as the probability that an electron occupies a given state is not any longer independent of the probability that another electron occupies the same or a different state. In other words, the probabilities are no longer independent of each other and it is essential to introduce the formalism of the occupation numbers. With this formalism the results of the Fermi-Dirac statistics can be obtained from the above principle, as described below in Sect. 10.6. When the system is far from this situation, i.e. is, as we say, far from being degenerate, the formula obtained for P' is correct.

As an application, we now derive the velocity distribution of the electrons, i.e. the probability $d\Pi$ that an electron has a velocity vector within the element of volume $d^3\mathbf{v}$ (of the velocity space) centred around the vector $\mathbf{v}$. To do this, we need to just count the number of quantum states corresponding to $d^3\mathbf{v}$ and use the expression for P' obtained above. It is easier to count the number of states in the momentum space, using the principle according to which the extension in phase-space of a single quantum state is h^3, with h Planck's constant. If $\mathcal{V}$ is the physical volume available to the electrons, the number of states dN with momentum within $d^3\mathbf{p}$ is given by

$$dN = \frac{2\mathcal{V}}{h^3} d^3\mathbf{p},$$

where the factor 2 has been introduced to take into account the two possible orientations of the spin. Noting also that, for non-relativistic electrons, $\mathbf{p} = m\mathbf{v}$, so that $d^3\mathbf{p} = m^3 d^3\mathbf{v}$, we obtain

$$d\Pi = Be^{-\frac{1}{2}\beta mv^2} d^3\mathbf{v},$$

where B is a another constant given by

$$B = \frac{2\mathcal{V}m^3}{h^3} A'.$$

The normalisation constant B is determined requiring that the integral of $d\Pi$ over the entire velocity space is equal to unity. Alternatively, we could have required that such integral is equal to $\mathcal{N}$ (the total number of electrons), or $\mathcal{N}/\mathcal{V}$ (number of electrons per unit of volume). Obviously this would imply changing the physical meaning of $d\Pi$. Recalling that

$$\int_{-\infty}^{\infty} e^{-x^2} dx = \sqrt{\pi},$$

the constant B can easily be calculated. Finally, putting $d\Pi = \mathcal{P}(\mathbf{v}) d^3\mathbf{v}$, we obtain

$$\mathcal{P}(\mathbf{v}) = \left(\frac{\beta m}{2\pi}\right)^{3/2} e^{-\frac{1}{2}\beta mv^2}. \tag{10.4}$$

The probability $P(v) dv$ that the electron has a velocity (in absolute value) between v and $v + dv$ can be obtained from this expression. Noting that $d^3\mathbf{v} = 4\pi v^2 dv$, we have

$$P(v) = \sqrt{\frac{2(\beta m)^3}{\pi}} v^2 e^{-\frac{1}{2}\beta mv^2}.$$

This distribution is known as the Maxwellian distribution of velocities, although the same name is also used to describe the distribution of the velocity vector $\mathcal{P}(\mathbf{v})$ (Eq. (10.4)).

10.3 The Saha-Boltzmann Equation

Consider a physical system consisting of a neutral atom and N electrons contained in a cavity of volume $\mathcal{V}$ that is in thermodynamic equilibrium at the temperature T. The energy levels of the atom can be characterized by two discrete indices r and k. With r we indicate the degree of ionisation that can assume the values $r = 0, 1, 2, \ldots, r_{max}$, where r_{max} is 1 for hydrogen, 2 for helium, and so on. With r fixed, k numbers the energy levels ordered by increasing energy from the ground level of the ion, for which we assume $k = 0$. Each level has an energy $T_{r,k}$ and a degeneracy $g_{r,k}$. We denote by I_1 the first ionisation potential, by I_2 the second ionisation potential, and so on. In addition, we denote by $E_{r,k}$ the energy of the k-th level of the atom r-times ionised and we assume that this energy is measured from the corresponding ground level ($E_{r,0} = 0$ for any r). The energy of the level (r, k), relative to the energy of the ground state of the neutral atom, is then given by

$$T_{r,k} = E_{r,k} + \sum_{i=0}^{r} I_i \quad (r \leq r_{max}),$$

where we have put $I_0 = 0$. Obviously, when the atom is r-times ionised, the cavity contains $(N + r)$ electrons.

If we apply to our physical system the fundamental principle in the form of Eq. (2.10), the probability P that the atom is in the state (r, k) and that the $(N + r)$ electrons have kinetic energies $\epsilon_1, \epsilon_2, \ldots, \epsilon_{N+r}$ is given by

$$P = A g_{r,k} e^{-\beta T_{r,k}} e^{-\beta \epsilon_1} e^{-\beta \epsilon_2} \cdots e^{-\beta \epsilon_{N+r}}.$$

However, if we are simply interested in the probability $P_{r,k}$ that the atom is in the state (r, k), but not in the particular state in which the free electrons are, we need to perform the sum of P over all possible states of such electrons. This sum can be performed easily only if we are far from degeneracy. In this case, in fact, the only effect of the indistinguishability of the electrons is the reduction of the volume of the phase space by a factor $(N + r)!$ and we obtain

$$P_{r,k} = A g_{r,k} e^{-\beta T_{r,k}} \frac{1}{(N+r)!} \int e^{-\beta \epsilon_1} \frac{d\Gamma_1}{h^3} \int e^{-\beta \epsilon_2} \frac{d\Gamma_2}{h^3} \cdots \int e^{-\beta \epsilon_{N+r}} \frac{d\Gamma_{N+r}}{h^3},$$

where $d\Gamma_i$ is the volume element in the phase space available to the i-th electron. The integrals that appear in the right-hand side are all equal. If we consider the nonrelativistic case in which $\epsilon = p^2/(2m)$, these integrals are of the same form as the integral of the previous section used to find the normalisation constant of the Maxwellian distribution of velocities. Repeating the same steps, we find that the single integral, that we denote by $\mathcal{N}_q(T)$, is

$$\mathcal{N}_q(T) = \frac{2\mathcal{V}}{h^3} \left(\frac{2\pi m}{\beta} \right)^{3/2}, \tag{10.5}$$

where the factor 2 originates, as usual, from taking into account the spin. This integral has a precise physical meaning. It represents the number of quantum states available to a free electron in thermodynamic equilibrium at the temperature T. An order of magnitude estimate for this quantity can be obtained as follows: at the temperature T, an electron can have a maximum momentum p_{max} of the order of

$$\frac{p_{max}^2}{2m} = \frac{1}{\beta}.$$

Taking into account the two possible orientations for the spin, the number of quantum states available to the electron is therefore given by

$$\frac{2\frac{4}{3}\pi p_{max}^3 \mathcal{V}}{h^3} = \frac{8\pi\mathcal{V}}{3h^3}\left(\frac{2m}{\beta}\right)^{3/2}.$$

The correct expression is obtained from this estimate by simply multiplying it by a numerical factor equal to $3\sqrt{\pi}/4 \simeq 1.3$.

The probability $P_{r,k}$ can then be expressed using the quantity $\mathcal{N}_q(T)$

$$P_{r,k} = A g_{r,k} e^{-\beta T_{r,k}} \frac{1}{(N+r)!}[\mathcal{N}_q(T)]^{N+r}.$$

We see that the probability is proportional to the number of quantum states available to all the $(N+r)$ electrons. This probability can be rewritten in a more meaningful way by including in the normalisation constant the number of quantum states available to the N electrons (note that N is a constant). Putting

$$A' = A\frac{1}{N!}[\mathcal{N}_q(T)]^N,$$

we obtain

$$P_{r,k} = A' g_{r,k} e^{-\beta T_{r,k}} \frac{N!}{(N+r)!}[\mathcal{N}_q(T)]^r.$$

If the number of particles is very large, the ratio $N!/(N+r)!$ is equal to N^{-r}. Denoting by N_e the electron density ($N_e = N/\mathcal{V}$) and recalling Eq. (10.5), we get the final expression

$$P_{r,k} = A' g_{r,k} e^{-\beta T_{r,k}}\left[\frac{2}{N_e h^3}\left(\frac{2\pi m}{\beta}\right)^{3/2}\right]^r.$$

This equation contains both the Saha and the Boltzmann equations (or laws), and is often called the Saha-Boltzmann equation. The Boltzmann equation is obtained by expressing the ratio between the probabilities $P_{r,k}$ and $P_{r,j}$ relative to two levels k and j of the same ionisation degree r. We have

$$\frac{P_{r,k}}{P_{r,j}} = \frac{g_{r,k}}{g_{r,j}} e^{-\beta(T_{r,k}-T_{r,j})}.$$

Recalling the expression of $T_{r,k}$, and introducing the temperature T instead of β, we get

$$\frac{P_{r,k}}{P_{r,j}} = \frac{g_{r,k}}{g_{r,j}} e^{-(E_{r,k}-E_{r,j})/(k_B T)}. \tag{10.6}$$

The Saha equation is instead obtained by first calculating the probability P_r that the atom is in the ionisation degree r, independently of the level k. Recalling the definition of $T_{r,k}$, we have

$$P_r = \sum_k P_{r,k} = A' u_r(\beta) e^{-\beta T_{r,0}} \left[\frac{2}{N_e h^3} \left(\frac{2\pi m}{\beta} \right)^{3/2} \right]^r,$$

where we have introduced the quantity $u_r(\beta)$, known as the partition function, defined by

$$u_r(\beta) = \sum_k g_{r,k} e^{-\beta E_{r,k}}.$$

If we now write the ratio between the probability that the atom is in two successive ionisation states we obtain

$$\frac{P_{r+1}}{P_r} = \frac{u_{r+1}(\beta)}{u_r(\beta)} e^{-\beta(T_{r+1,0}-T_{r,0})} \frac{2}{N_e h^3} \left(\frac{2\pi m}{\beta} \right)^{3/2},$$

or, recalling that $T_{r+1,0} - T_{r,0} = I_{r+1}$, and introducing the temperature in place of β

$$\frac{P_{r+1}}{P_r} = \frac{u_{r+1}(T)}{u_r(T)} e^{-I_{r+1}/(k_B T)} \frac{2}{N_e h^3} (2\pi m k_B T)^{3/2}. \tag{10.7}$$

This is the Saha equation, or ionisation equation. It was discovered by the Indian physicist M.N. Saha in the early 1920s. It turned out to have a fundamental importance for our understanding of stellar spectra.

About the Saha and Boltzmann equations, we note that the left-hand sides of these equations are often expressed in terms of ratios of populations, rather than in terms of ratios of probabilities. These ratios are equal, an obvious consequence of the probabilistic interpretation of thermodynamics. For example, the Saha equation can also be written, with obvious notations, as

$$\frac{N_{r+1} N_e}{N_r} = \frac{u_{r+1}(T)}{u_r(T)} e^{-I_{r+1}/(k_B T)} \frac{2}{h^3} (2\pi m k_B T)^{3/2},$$

and noting that the right-hand side is a function only of temperature, the equation can also be written as

$$\frac{N_{r+1} N_e}{N_r} = K(T).$$

In this form the Saha equation becomes a special case of a more general equation, known as the law of mass action or equation of Guldberg-Waage, which is used, especially in chemistry, to relate the density (or concentrations) of the reagents of a reaction in thermodynamic equilibrium. For example, for the chemical reaction

$$A + B \rightleftarrows AB,$$

with A and B any elements, we have, in thermodynamic equilibrium

$$\frac{N_A N_B}{N_{AB}} = K(T).$$

Obviously, in the case of the Saha equation, we have a "ionisation reaction", rather than a chemical reaction, i.e. with obvious symbols

$$A_{r+1} + e^- \rightleftarrows A_r.$$

As an application of the Saha equation we now calculate the ionisation degree of a plasma composed of pure hydrogen at the equilibrium temperature T and pressure P_g. Denoting by N_H and N_{H^+} the density of neutral and of ionised hydrogen atoms, respectively, the degree of ionisation x (with $0 \le x \le 1$) is defined by the equation

$$x = \frac{N_{H^+}}{N_H + N_{H^+}}.$$

Form the Saha equation we have

$$\frac{N_{H^+}}{N_H} = \frac{x}{1-x} = e^{-I_H/(k_B T)} \frac{2}{u_H(T)N_e h^3} (2\pi m k_B T)^{3/2},$$

where I_H is the ionisation potential of hydrogen and $u_H(T)$ is the partition function of neutral hydrogen (the partition function of ionised hydrogen is by definition 1). The electron density can be expressed in terms of the total pressure of the gas P_g. For a non-degenerate plasma we have

$$P_g = k_B T (N_H + N_{H^+} + N_e) = k_B T (N_H + 2N_e) = k_B T N_e \frac{1+x}{x},$$

which gives

$$N_e = \frac{x}{1+x} \frac{P_g}{k_B T}.$$

Substituting this value of N_e into the Saha equation, we obtain the following equation of second degree in x

$$\frac{x^2}{1-x^2} = C,$$

where the dimensionless quantity C is given by

$$C = \frac{(\pi m)^{3/2} (2k_B T)^{5/2}}{u_H(T) h^3 P_g} e^{-I_H/(k_B T)}.$$

Solving the equation we find

$$x = \sqrt{\frac{C}{1+C}}.$$

Substituting the numerical values of the constants in C and assuming $u_H = 2$ (a very rough approximation that however does not affect significantly the results) we have, numerically,

$$C = 0.3334 T^{5/2} P_g^{-1} e^{-1.5777 \times 10^5 / T},$$

with T in K and P_g in dyne cm^{-2}. The numerical results obtained from the solution of the equation are shown in Fig. 10.1.

Fig. 10.1 Ionisation degree
of a plasma of pure hydrogen
as a function of the
temperature T (K) and the
gas pressure P_g (dyne cm^{-2}).
The three *curves* are labelled
by the values of the ionisation
degree x. The points A, B,
and C represent the physical
conditions of the solar
photosphere, the centre of the
Sun and the solar corona,
respectively

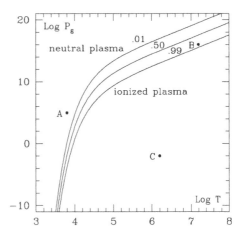

10.4 The Black-Body Radiation

Consider the physical system constituted by the electromagnetic radiation enclosed
in a cavity in thermodynamic equilibrium at the temperature T. Such a physical
system is called a black body and can be realised in the laboratory by drilling a
small hole (so small as not to affect substantially the condition of thermodynamic
equilibrium) in the wall of a furnace internally lined with lampblack and brought
to high temperatures. By analysing spectroscopically the radiation coming from the
hole, one can determine the specific intensity I_ν of the black body radiation. This
quantity is often denoted by the symbol B_ν.

As we shall see, the expression for B_ν can be found by using the general principles of statistical thermodynamics described in Sect. 10.1. This expression confirms
a number of properties which can be inferred classically by applying the first and
second laws of thermodynamics and which are interesting to discuss, especially for
the historical importance of the black-body radiation in the development of modern
physics.

The first statement that can be made classically about blackbody radiation is that
the function B_ν must be a universal function of temperature. We must have, as it can
be proved using the second law of thermodynamics, that $B_\nu = B_\nu(T)$, without any
dependencies on other parameters (such as the nature of the walls of the furnace,
any substance that is located in its interior, etc.). Still using the second principle of
thermodynamics it can also be shown that the black-body radiation is homogeneous,
isotropic and not polarised. If that was not the case it would be possible to design
a cyclic machine that could produce work at the expense of a single heath source,
something that contradicts the second principle. Another consequence of the principles of thermodynamics is the dependence of the total energy density on the fourth
power of the temperature. To show this, consider the system consisting of the electromagnetic radiation contained in a cylinder fitted with a movable piston having
its inner surface perfectly reflecting (see Fig. 10.2). For an infinitesimal thermodynamic transformation due to the expansion of a volume dV we have, according to
the first law of thermodynamics

Fig. 10.2 The radiation field
contained in the cylinder is
subject to an infinitesimal
expansion with exchange of
heat to or from the external
medium

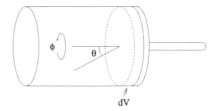

$$dU = \delta Q - \delta L = \delta Q - P\, dV,$$

where U is the internal energy of the system, δQ the amount of heat absorbed by the system in the transformation, δL the work done by the system on the medium and P the pressure. If we denote by u the internal energy density, we have $U = uV$ and therefore, since u depends only on temperature T

$$dU = V\frac{du}{dT}dT + u\, dV.$$

To evaluate the δL we need to calculate the pressure acting on the piston. Taking into account that the radiation is isotropic, i.e. that the energy density of the radiation which propagates within the solid angle $d\Omega$ is given by $u\, d\Omega/(4\pi)$ and that the surface of the piston is perfectly reflecting, we have from Eq. (1.13)

$$P = \int_0^{2\pi} d\phi \int_0^{\pi/2} 2\frac{u}{4\pi}\cos^2\theta \sin\theta\, d\theta,$$

where θ and ϕ are the polar angles defined in Fig. 10.2. Evaluating the integral we obtain, with simple algebra

$$P = \frac{u}{3},$$

hence

$$\delta L = \frac{u}{3}dV.$$

Substituting the expressions for dU and for δL in the equation that formulates the first principle and solving for δQ we obtain

$$\delta Q = V\frac{du}{dT}dT + \frac{4}{3}u\, dV.$$

We now recall that if the transformation is reversible (which implies the absence of friction in the motion of the piston and equal temperatures between the radiation and the heath source furnishing the δQ), the second law of thermodynamics requires that the quantity

$$\frac{\delta Q}{T} = \frac{V}{T}\frac{du}{dT}dT + \frac{4}{3}\frac{u}{T}dV$$

is an exact differential. For this the so-called Schwarz criterion needs to be satisfied, i.e.

$$\frac{\partial}{\partial V}\left(\frac{V}{T}\frac{du}{dT}\right) = \frac{\partial}{\partial T}\left(\frac{4}{3}\frac{u}{T}\right),$$

from which we obtain, with simple algebra

$$\frac{du}{dT} = 4\frac{u}{T}.$$

Solving this differential equation we finally obtain

$$u = aT^4,$$

where a is a constant of integration.

We note that the above considerations also allow us to determine the entropy of the radiation field. Substituting in fact the expression for u we have

$$dS = \frac{\delta Q}{T} = 4aVT^2\,dT + \frac{4}{3}aT^3\,dV,$$

or, by integration

$$S = \frac{4}{3}aT^3V,$$

where, to fix the constant of integration, we have used the third law of thermodynamics, which states that the entropy is zero at $T = 0$. For an adiabatic transformation of the radiation field we then have

$$TV^{1/3} = \text{const.},$$

or, recalling that $P = u/3 = aT^4/3$,

$$pV^{4/3} = \text{const.},$$

that is the usual law for adiabatic transformations with the exponent $\gamma = \frac{4}{3}$.

Beyond the integral law that we have just derived for u, the classical theory also allows to determine the expression of the function $B_\nu(T)$. However, we will see that this expression is physically inconsistent and not in accordance with experiments (except in the limit of low frequencies). As we have seen in Sect. 4.3, a physical system consisting of the radiation field in a cavity of volume $\mathcal{V}$ is described in quantum theory by a Hamiltonian that is equal to the sum of many Hamiltonians of harmonic oscillator, independent of each other, one for each mode of the field. Similar arguments as those developed there can also be made classically. In this case we obtain that, numbering the modes with the integer index k, the classical Hamiltonian can be written in the form

$$H = \sum_k H(p_k, q_k) = \sum_k \gamma_k p_k^2 + \delta_k q_k^2,$$

where q_k is the canonical coordinate relative to the mode k, p_k is the associated kinetic conjugate moment, and γ_k and δ_k are quantities that depend on the index k, but that we do not need to further specify.[3] According to the fundamental principle

---

[3]Recalling the considerations of Sect. 4.2, such quantities can be expressed as a function of the coefficients $c_{k\lambda}$ and $c_{k\lambda}^*$, as well as the frequency (or wavenumber).

of Eq. (10.1), the probability dP that the variable q_1 is between q_1 and $q_1 + dq_1$, the variable p_1 is between p_1 and $p_1 + dp_1$, and so on, is given by

$$dP = Ae^{-\beta H(p_1, q_1)}e^{-\beta H(p_2, q_2)} \cdots e^{-\beta H(p_k, q_k)} \cdots dq_1 \, dp_1 \, dq_2 \, dp_2 \cdots dq_k \, dp_k \cdots,$$

where A is the normalisation constant. Since all the Hamiltonians are independent of each other, the reduced probability $d\Pi$ that the variable q_1 is between q_1 and $q_1 + dq_1$ and the variable p_1 is between p_1 and $p_1 + dp_1$ can be obtained by integrating over all possibilities for the other variables $q_2, p_2, \ldots, q_k, p_k, \ldots$. We have

$$d\Pi = A'e^{-\beta H(p_1, q_1)} \, dq_1 \, dp_1,$$

where A' is a new normalisation constant given by

$$A' = \frac{1}{\int dq_1 \int dp_1 e^{-\beta H(p_1, q_1)}}.$$

The average energy relative to mode 1 is therefore given by

$$\epsilon_1 = \frac{\int dq_1 \int dp_1 H(p_1, q_1)e^{-\beta H(p_1, q_1)}}{\int dq_1 \int dp_1 e^{-\beta H(p_1, q_1)}},$$

and can also be expressed in the form

$$\epsilon_1 = -\frac{d}{d\beta}\left\{\ln\left[\int dq_1 \int dp_1 e^{-\beta H(p_1, q_1)}\right]\right\}.$$

Substituting the expression for the Hamiltonian and performing the integrations we obtain with simple algebra

$$\epsilon_1 = \frac{1}{\beta} = k_B T.$$

This result, obtained for mode 1, is independent of the particular characteristics of the mode itself and therefore applies to all modes. In other words, classically the average energy of an harmonic oscillator is always equal to $k_B T$, whatever the physical properties of the oscillator. By means of this property, and recalling Eqs. (4.16) and (4.13) we obtain for the functions B_ν and u_ν

$$\left[B_\nu(T)\right]_{\text{class}} = \frac{2\nu^2}{c^2}k_B T, \qquad \left[u_\nu(T)\right]_{\text{class}} = \frac{2\nu^2}{c^3}k_B T.$$

This result is in flagrant contradiction with experimental results at high frequencies. Moreover, when integrating the function u_ν over all frequencies we obtain a divergent integral, which means that the energy per unit volume of the radiation field should be infinite. This fact is usually known as the ultraviolet catastrophe. It was to resolve these contradictions that Planck was led to introduce his quantum hypothesis.

We can repeat the same probabilistic arguments, still taking into account the fact that the modes are independent, but introducing the quantum result that the eigenvalues of a generic mode k relative to the frequency ν are given by $nh\nu$, with n an arbitrary integer. The probability $\Pi(n)$ that the mode is in the state with eigenvalue $nh\nu$ (i.e. that we have n photons) is then given by

$$\Pi(n) = A'e^{-\beta nh\nu},$$

where A' is a normalisation constant given by

$$A' = \frac{1}{\sum_{n=0}^{\infty} e^{-\beta nh\nu}}.$$

The average energy of the mode is therefore

$$\epsilon(\nu) = \sum_{n=0}^{\infty} \Pi(n)nh\nu = \frac{\sum_{n=0}^{\infty} nh\nu e^{-\beta nh\nu}}{\sum_{n=0}^{\infty} e^{-\beta nh\nu}},$$

which can be written in the form

$$\epsilon(\nu) = -\frac{d}{d\beta} \ln\left[\sum_{n=0}^{\infty} e^{-\beta nh\nu}\right].$$

The sum is carried out easily by recalling that, for an arbitrary number q of absolute value less than 1 ($|q| < 1$) we have

$$\sum_{n=0}^{\infty} q^n = \frac{1}{1-q}.$$

Thus we obtain, with simple algebra

$$\epsilon(\nu) = \frac{h\nu}{e^{\beta h\nu} - 1}, \tag{10.8}$$

from which we obtain

$$B_\nu(T) = \frac{2h\nu^3}{c^2} \frac{1}{e^{\beta h\nu} - 1}, \qquad u_\nu(T) = \frac{2h\nu^3}{c^3} \frac{1}{e^{\beta h\nu} - 1}.$$

These expressions are in agreement with the experimental facts and the function $B_\nu(T)$ is called the Planck function (although this name is sometimes used for $u_\nu(T)$, or for the associated functions $B_\lambda(T)$ and $u_\lambda(T)$).

The total energy density u can be calculated by simply integrating over frequencies and angles the expression for u_ν. We have

$$u = \oint d\Omega \int_0^\infty \frac{2h\nu^3}{c^3} \frac{1}{e^{\beta h\nu} - 1} d\nu.$$

Putting $x = \beta h\nu$ we obtain, with a change of variable

$$u = \frac{8\pi}{\beta^4 h^3 c^3} \int_0^\infty \frac{x^3}{e^x - 1} dx.$$

The integral is equal to $\pi^4/15$ and we obtain the so-called Stefan-Boltzmann law

$$u = aT^4, \tag{10.9}$$

where the constant a, known as the constant of the radiation density, is given by

$$a = \frac{8\pi^5}{15} \frac{k_B^4}{h^3 c^3} = 7.566 \times 10^{-15} \text{ erg cm}^{-3} \text{ K}^{-4}.$$

As we can see, we have again obtained the expression for the radiation density, previously derived from purely classical considerations. The quantum formulation provides the value for the constant a that was left undetermined in the classical theory.

If we consider the integral over frequencies of the Planck function, i.e. we consider the quantity $B(T)$ defined by

$$B(T) = \int_0^\infty B_\nu(T) \, d\nu,$$

we obtain, with a calculation similar to the previous one,

$$B(T) = \frac{ac}{4\pi} T^4.$$

The quantity $\pi B(T)$ is the flux of the black body radiation, i.e. the amount of energy flowing, per unit time, through the unit surface of a black body (in all directions and at all frequencies). In fact, introducing a coordinate system with the polar axis z directed along the normal to the surface, the flux F is given, with obvious notations, by the expression

$$F = \int_0^\infty d\nu \int_0^{2\pi} d\phi \int_0^{\pi/2} \cos\theta \, B_\nu(T) \sin\theta \, d\theta = \pi B(T),$$

so we have

$$F = \pi B(T) = \frac{ac}{4} T^4.$$

The quantity $ac/4$ is traditionally denoted by the symbol σ and is called the Stefan-Boltzmann constant. We then have

$$F = \sigma T^4, \tag{10.10}$$

where

$$\sigma = \frac{ac}{4} = \frac{2\pi^5}{15} \frac{k_B^4}{h^3 c^2} = 5.670 \times 10^{-5} \text{ erg cm}^{-2} \text{ s}^{-1} \text{K}^{-4}.$$

10.5 Properties of the Black-Body Radiation

The spectrum of the black body radiation, described by the Planck function

$$B_\nu(T) = \frac{2h\nu^3}{c^2} \frac{1}{e^{h\nu/(k_B T)} - 1}, \tag{10.11}$$

has a number of important properties that we are going to discuss. We start by considering the limit of the function at low frequencies, i.e. we suppose that

$$\frac{h\nu}{k_B T} \ll 1.$$

Expanding the exponential in power series we obtain

$$B_\nu(T) = \frac{2\nu^2}{c^2} k_B T.$$

This expression, known as the law of Rayleigh and Jeans, does not contain Planck's constant and.is in fact the black body law which is obtained within the classical theory (as we saw in the previous section). The condition $h\nu \ll k_B T$ implies that the energy of the photons is much less than the thermal energy so that we can neglect the quantum nature of the radiation field. In other words, the energy of the photons behaves as a continuum and the application of the classical theory is fully justified. In astrophysical objects, the law of Rayleigh and Jeans is generally well verified at radio frequencies and at microwaves.

If we consider instead the limit at high frequencies, namely for

$$\frac{h\nu}{k_B T} \gg 1,$$

the Planck equation assumes the form

$$B_\nu(T) = \frac{2h\nu^3}{c^2} e^{-h\nu/(k_B T)}.$$

This expression was derived by Wien on the basis of *ad hoc* arguments that turned out to be not entirely correct to describe the black body radiation at all frequencies. In reality, Wien correctly demonstrated, on the basis of the Doppler effect and of thermodynamic principles, that the black body radiation spectrum should be described by an equation of the type (Wien equation)

$$B_\nu(T) = K\nu^3 f\left(\frac{\nu}{T}\right),$$

with K constant and $f(x)$ an arbitrary function. Wien's arguments were correct, but the *ad-hoc* hypothesis that the $f(x)$ function was of the type $\exp(-x)$ (instead of the type $1/[\exp(x) - 1]$) was later shown to be valid only at high frequencies.

An important property of the black body radiation is the fact that, at a given frequency, the intensity of the radiation monotonically increases with temperature. The proof is simple, as is obtained from the equation

$$\frac{\partial B_\nu(T)}{\partial T} = \frac{2h^2\nu^4}{c^2 k_B T^2} \frac{e^{h\nu/(k_B T)}}{[e^{h\nu/(k_B T)} - 1]^2} > 0.$$

As a consequence, the plot of $B_\nu(T)$ as a function of ν (with T fixed) consists of a curve that, with increasing T, becomes progressively higher enveloping those of lower T as shown in Figs. 10.3 and 10.4.

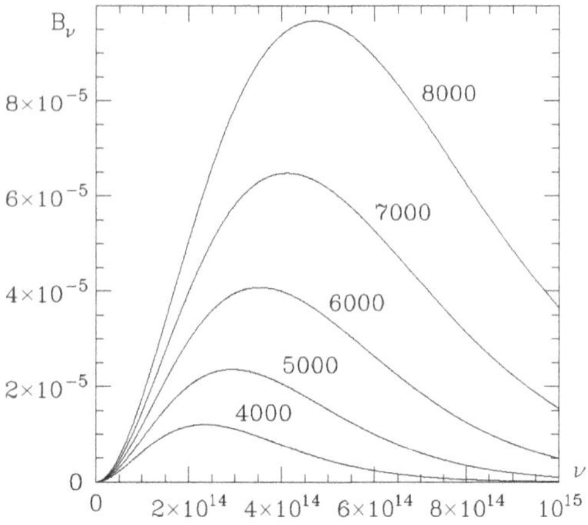

Fig. 10.3 The Planck
function is plotted versus
frequency (in linear scale) for
five different values of the
temperature. B_ν and ν are in
c.g.s. units, and the
temperature is in K

Fig. 10.4 The Planck
function is plotted versus
frequency (in logarithmic
scale) for five different values
of the temperature. Notice
that the differences in the
temperature are much larger
than those of Fig. 10.3. B_ν
and ν are in c.g.s. units, and
the temperature is in K

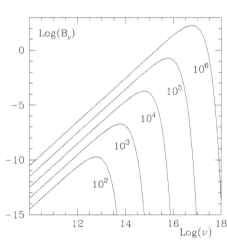

Another important property of the black body radiation is related to the frequency
$\nu_{\max}$ at which the function has a maximum (for a given T). Introducing the reduced
variable $x = h\nu/(k_B T)$, Planck equation becomes

$$B_\nu(T) = \frac{2k_B^3 T^3}{h^2 c^2} b(x),$$

where

$$b(x) = \frac{x^3}{e^x - 1}.$$

The maximum of the function $b(x)$ is obtained by solving the transcendent equation

$$x = 3\left(1 - e^{-x}\right),$$

that has the solution $x_{\max} \simeq 2.82$. Therefore, we have

$$h\nu_{\max} = 2.82 k_B T,$$

or, numerically

$$\nu_{max} = 5.88 \times 10^{10} T \text{ Hz,}$$

where T is in K. This equation shows that the frequency of the maximum increases linearly with increasing temperature, a law which is known as Wien's displacement law.

Finally, we note that, in some cases, the specific intensity of the blackbody radiation is described by the function $B_\lambda(T)$ (instead of $B_\nu(T)$). The two functions are related in this way (see Eq. (4.12))

$$B_\lambda(T) = \frac{c}{\lambda^2} B_\nu(T) = \frac{\nu^2}{c} B_\nu(T).$$

Expressing the function in terms of the wavelength (instead of the frequency), we have

$$B_\lambda(T) = \frac{c_1}{\pi \lambda^5} \frac{1}{e^{c_2/(\lambda T)} - 1},$$

where the constants c_1 and c_2, known as the first and second constants of the radiation, are given by

$$c_1 = 2\pi h c^2 = 3.742 \times 10^{-5} \text{ erg cm}^2 \text{ s}^{-1}, \qquad c_2 = \frac{hc}{k_B} = 1.439 \text{ cm K.}$$

As in the $B_\nu(T)$ case, we can also find the value of the wavelength λ_{max} for which $B_\lambda(T)$ has a maximum for a given T. Putting $y = c_2/T$, we have

$$B_\lambda(T) = \frac{2k_B^5 T^5}{h^4 c^3} \tilde{b}(y),$$

where

$$\tilde{b}(y) = \frac{y^5}{e^y - 1}.$$

The maximum of the function $\tilde{b}(y)$ is obtained by solving the transcendent equation

$$y = 5(1 - e^{-y}),$$

that has the solution $y_{max} \simeq 4.97$. Therefore, we have

$$\lambda_{max} T = \frac{c_2}{4.97},$$

or, numerically

$$\lambda_{max} T = 0.290 \text{ cm K.}$$

It should be noted that the location of the maximum of the function $B_\lambda(T)$ does not correspond to the point where $B_\nu(T)$ has its maximum. This is not surprising, since in one case we consider the energy flowing per unit of frequency interval, and in the other per unit of wavelength interval. The above expressions of ν_{max} and λ_{max} show that the product $\nu_{max}\lambda_{max}$ is about $0.568c$.

10.6 The Fermi-Dirac Statistics

Consider a macroscopic physical system S in thermodynamic equilibrium at the temperature T, composed of a very large number N, of indistinguishable, non-interacting particles obeying the Pauli exclusion principle (gas of fermions). Suppose we have numbered the single-particle states by a discrete index and denote by ϵ_i the energy of the i-th state. Let us fix our attention on a particular single-particle state, marked by the index k, with energy ϵ_k. According to the general principle of Eq. (10.2), the probability $p_0(k)$ that in the k-th state there are no particles, i.e. that the occupation number of such a state is 0, is given by the expression

$$p_0(k) = A \sum_{n_1,n_2,\ldots,n_{k-1},n_{k+1},\ldots} e^{-\beta(n_1\epsilon_1+n_2\epsilon_2+\cdots+n_{k-1}\epsilon_{k-1}+n_{k+1}\epsilon_{k+1}+\cdots)},$$

where A is a constant, $\beta = 1/(k_B T)$, and where the sum is extended to all the solutions of the equation

$$n_1 + n_2 + \cdots + n_{k-1} + n_{k+1} + \cdots = N,$$

with the occupation numbers $n_1, \ldots, n_{k-1}, n_{k+1}, \ldots$ that can only be equal to 0 or 1. The probability $p_1(k)$ that in the k-th state there is one particle is instead given by

$$p_1(k) = Ae^{-\beta\epsilon_k} \sum_{n_1,n_2,\ldots,n_{k-1},n_{k+1},\ldots} e^{-\beta(n_1\epsilon_1+n_2\epsilon_2+\cdots+n_{k-1}\epsilon_{k-1}+n_{k+1}\epsilon_{k+1}+\cdots)},$$

where the sum is extended to all the solutions of the equation

$$n_1 + n_2 + \cdots + n_{k-1} + n_{k+1} + \cdots = N - 1,$$

with $n_1, \ldots, n_{k-1}, n_{k+1}, \ldots$ that can only be equal to 0 or 1.

Consider now the ratio $p_0(k)/p_1(k)$. We have

$$\frac{p_0(k)}{p_1(k)} = e^{\beta\epsilon_k} \frac{\mathcal{Z}(\beta, N)}{\mathcal{Z}(\beta, N-1)},$$

where $\mathcal{Z}(\beta, N)$ is the sum over the states of the (physically unreal) system S′, obtained from the system S by conceptually deleting the k-th elementary state, and in which there are N particles. $\mathcal{Z}(\beta, N-1)$ has a similar meaning, with the difference that $(N-1)$ particles are instead present in S′. Since the physical system contains an extremely large number of elementary states, the distinction between S′ and S is totally inessential, so the quantities $\mathcal{Z}(\beta, N)$ and $\mathcal{Z}(\beta, N-1)$ can be considered in all effects the sums over the states relative to the "initial" physical system S. Denoting by x the ratio $\mathcal{Z}(\beta, N)/\mathcal{Z}(\beta, N-1)$ and taking the natural logarithm, we have

$$\ln x = \ln\big[\mathcal{Z}(\beta, N)\big] - \ln\big[\mathcal{Z}(\beta, N-1)\big].$$

Furthermore, considering that N is a very large number and the values of the energies ϵ_j are fixed (which implies that the volume of the physical system is constant), we can write

$$\ln x = \left(\frac{\partial}{\partial N} \ln\big[\mathcal{Z}(\beta, N)\big]\right)_{\beta,V}.$$

This partial derivative is related to the chemical potential μ, defined in thermodynamics by the equation

$$\mu = \left(\frac{\partial F}{\partial N}\right)_{\beta,V},$$

where F is the Helmholtz free energy. In fact, since (see Eq. (10.3))

$$\ln[\mathcal{Z}(\beta, N)] = -\beta F,$$

we have

$$\ln x = -\beta \left(\frac{\partial F}{\partial N}\right)_{\beta,V} = -\beta \mu.$$

Recalling the definition of x, we then have

$$\frac{p_0(k)}{p_1(k)} = e^{\beta(\epsilon_k - \mu)}.$$

On the other hand, the average occupation number of the elementary state k is given by the expression

$$\bar{n}_k = \frac{p_1(k)}{p_0(k) + p_1(k)} = \frac{1}{\frac{p_0(k)}{p_1(k)} + 1},$$

so we obtain

$$\bar{n}_k = \frac{1}{e^{\beta(\epsilon_k - \mu)} + 1}.$$

This is the fundamental equation describing the statistics of particles that obey the Pauli exclusion principle. It is known as the equation of the Fermi-Dirac statistics. It expresses the average occupation number of the single-particle state having energy ϵ_k as a function of the same energy and of the chemical potential μ. Since the above argument can be repeated for any single-particle state, the index k can be omitted in the previous formula so to obtain

$$\bar{n}(\epsilon) = \frac{1}{e^{\beta(\epsilon - \mu)} + 1}.$$

10.7 The Bose-Einstein Statistics

In a similar way as in the previous section, we now consider a macroscopic physical system in thermodynamic equilibrium at the temperature T, composed of N indistinguishable non-interacting bosons (gas of bosons). Suppose we have numbered the single-particle states with a discrete index and denote by ϵ_i the energy of the i-th state. Let us fix our attention on a particular state, marked by the index k, with energy ϵ_k. According to the general principle of Eq. (10.2), the probability $p_n(k)$ that in the k-th state there are n particles is given by the expression

$$p_n(k) = A e^{-n\beta\epsilon_k} \sum_{n_1, n_2, \ldots, n_{k-1}, n_{k+1}, \ldots} e^{-\beta(n_1\epsilon_1 + n_2\epsilon_2 + \cdots + n_{k-1}\epsilon_{k-1} + n_{k+1}\epsilon_{k+1} + \cdots)},$$

where A is a constant, $\beta = 1/(k_B T)$, and where the sum is extended to all the solutions of the equation

$$n_1 + n_2 + \cdots + n_{k-1} + n_{k+1} + \cdots = N - n,$$

with the numbers $n_1, n_2, \ldots$ that can be equal to an arbitrary integer $0, 1, 2, \ldots$ (i.e. not only equal to 0 and 1 as in the previous case for the fermions).

Consider now the ratio $p_0(k)/p_n(k)$. We have

$$\frac{p_0(k)}{p_n(k)} = e^{n\beta\epsilon_k} \frac{\mathcal{Z}(\beta, N)}{\mathcal{Z}(\beta, N - n)},$$

where the meaning of the quantities $\mathcal{Z}(\beta, N)$ and $\mathcal{Z}(\beta, N - n)$ is the same as in the case discussed in the previous section for the fermions. Denoting by x the ratio $\mathcal{Z}(\beta, N)/\mathcal{Z}(\beta, N - n)$ and taking the natural logarithm, we have

$$\ln x = \ln\left[\mathcal{Z}(\beta, N)\right] - \ln\left[\mathcal{Z}(\beta, N - n)\right],$$

and considering that N is a very large number and the volume is constant, we can write

$$\ln x = n\left(\frac{\partial}{\partial N} \ln\left[\mathcal{Z}(\beta, N)\right]\right)_{\beta, V},$$

or

$$\ln x = -n\beta\mu,$$

where μ is the chemical potential. Recalling the definition of x, we then have

$$\frac{p_0(k)}{p_n(k)} = e^{n\beta(\epsilon_k - \mu)},$$

or

$$\frac{p_n(k)}{p_0(k)} = e^{-n\beta(\epsilon_k - \mu)}.$$

On the other hand, the average occupation number of the state k is given by

$$\bar{n}_k = \frac{\sum_n n p_n(k)}{\sum_n p_n(k)},$$

or, dividing numerator and denominator by $p_0(k)$

$$\bar{n}_k = \frac{\sum_n n e^{-n\beta(\epsilon_k - \mu)}}{\sum_n e^{-n\beta(\epsilon_k - \mu)}}.$$

The ratio of the two sums can easily be calculated following the same procedure used to deduce Eq. (10.8). Putting $z = \beta(\epsilon_k - \mu)$, we have

$$\bar{n}_k = \frac{\sum_n n e^{-nz}}{\sum_n e^{-nz}} = -\frac{d}{dz} \ln\left(\sum_n e^{-nz}\right),$$

and if $z > 0$, which implies $\mu < \epsilon_k$, we obtain

$$\bar{n}_k = \frac{d}{dz} \ln\left(1 - e^{-z}\right) = \frac{1}{e^z - 1}.$$

Repeating the final considerations of the previous section, we finally obtain the fundamental equation describing the Bose-Einstein statistics

$$\bar{n}(\epsilon) = \frac{1}{e^{\beta(\epsilon-\mu)} - 1}.$$

We note that, in the case of the radiation field, the photons behave effectively as bosons, but their total number is not fixed, a priori. The chemical potential μ is therefore null and the previous equation becomes simpler

$$\bar{n}(\epsilon) = \frac{1}{e^{\beta\epsilon} - 1}.$$

This equation coincides with the formula we obtained for the average number of photons per mode of frequency ν, if we replace ϵ with $h\nu$.

Chapter 11
Interaction Between Matter and Radiation

In the previous chapters we have provided a quantum description of the radiation field in vacuum (Chap. 4) and of isolated atomic systems (Chaps. 5–9). We now move on to describe their mutual interaction by introducing some general methods, also based on quantum mechanics, with which we will be able to handle many phenomena associated with the absorption, emission and scattering of radiation, typical of laboratory and astrophysical plasmas. The combination of these methods is now commonly referred to by the name of quantum electrodynamics and represents one of the most successful achievements of theoretical physics, both from the point of view of the precision of the results obtained, and the elegance of the formalism. We will illustrate in this chapter the fundamental concepts and their simplest applications by considering only first order phenomena, i.e. phenomena involving the emission and the absorption of a single photon. We will formally derive the equations for the evolution of the populations of an atomic system in the presence of the radiation field (statistical equilibrium equations) and the equations for the evolution of the radiation field in the presence of an atomic system (equation of radiative transfer). The most relevant second order phenomena (where two photons are involved) will be treated in Chap. 15.

11.1 The Interaction Hamiltonian

Consider a physical system composed of an atom, described by the quantum Hamiltonian $\mathcal{H}_A$, and by the radiation field, described by the quantum Hamiltonian $\mathcal{H}_R$. In the absence of interaction, the Hamiltonian of the system is obviously

$$\mathcal{H}_0 = \mathcal{H}_A + \mathcal{H}_R, \tag{11.1}$$

where, recalling Eq. (4.9),

$$\mathcal{H}_R = \sum_{\mathbf{k}\lambda} \hbar\omega_{\mathbf{k}} \left(a_{\mathbf{k}\lambda}^{\dagger} a_{\mathbf{k}\lambda} + \frac{1}{2} \right),$$

E. Landi Degl'Innocenti, *Atomic Spectroscopy and Radiative Processes*,
UNITEXT for Physics, DOI 10.1007/978-88-470-2808-1_11, © Springer-Verlag Italia 2014

with the sum extended to all possible modes of the radiation.

To obtain the Hamiltonian describing the interaction between the atom and the radiation field we use the principle of minimal coupling described in Sect. 5.2. We operate in the atomic Hamiltonian for each electron the formal substitution (cf. Eq. (5.5))

$$\mathbf{p}_i \to \mathbf{p}_i + \frac{e_0}{c} \mathbf{A}_R(\mathbf{r}_i) \quad (i = 1, 2, \ldots, N),$$

where the index i numbers the N electrons of the atom and where $\mathbf{A}_R(\mathbf{r}_i)$ is the vector potential of the radiation field evaluated at the point where the i-th electron is located. The other substitution, implicit in the principle of minimal coupling, i.e.

$$E_i \to E_i + e_0 \phi_R(\mathbf{r}_i),$$

does not have in this case any effect because the scalar potential ϕ_R of the radiation field is zero. If we assume for $\mathcal{H}_A$ the nonrelativistic expression (Eq. (7.2)), the principle of minimal coupling produces the new Hamiltonian $\mathcal{H}_A'$ given by

$$\mathcal{H}_A' = \sum_{i=1}^{N} \left\{ \frac{1}{2m} \left[\mathbf{p}_i + \frac{e_0}{c} \mathbf{A}_R(\mathbf{r}_i) \right]^2 - \frac{Ze_0^2}{r_i} \right\} + \sum_{i<j} \frac{e_0^2}{r_{ij}}.$$

Putting

$$\mathcal{H}_A' = \mathcal{H}_A + \mathcal{H}^I,$$

with $\mathcal{H}^I$ the interaction Hamiltonian, and noting that $\operatorname{div} \mathbf{A}_R = 0$, we obtain

$$\mathcal{H}^I = \frac{e_0}{mc} \sum_{i=1}^{N} \mathbf{p}_i \cdot \mathbf{A}_R(\mathbf{r}_i) + \frac{e_0^2}{2mc^2} \sum_{i=1}^{N} [\mathbf{A}_R(\mathbf{r}_i)]^2. \tag{11.2}$$

We are interested in obtaining a perturbative expansion of the interaction Hamiltonian, so we neglect the second term (quadratic in the vector potential $\mathbf{A}_R$).[1] Recalling the expansion of the vector potential in terms of the creation and annihilation operators (see Eq. (4.5)), the interaction Hamiltonian is

$$\mathcal{H}^I = \frac{e_0}{m} \sum_{k\lambda} \sqrt{\frac{2\pi\hbar}{\omega_k V}} \left(a_{k\lambda} \sum_i \mathbf{p}_i \cdot \mathbf{e}_{k\lambda} e^{i\mathbf{k}\cdot\mathbf{r}_i} + a_{k\lambda}^\dagger \sum_i \mathbf{p}_i \cdot \mathbf{e}_{k\lambda}^* e^{-i\mathbf{k}\cdot\mathbf{r}_i} \right). \tag{11.3}$$

11.2 The Kinetic Equations

Consider a physical system described by the Hamiltonian

$$\mathcal{H} = \mathcal{H}_0 + \mathcal{H}^I,$$

---

[1] As we will see in Chap. 15, the term in $\mathbf{A}_R^2$ describes second order processes, in particular Thomson scattering.

where $\mathcal{H}_0$ is the unperturbed Hamiltonian and $\mathcal{H}^{\mathrm{I}}$ is the perturbing Hamiltonian (in our case the interaction Hamiltonian). The system evolves in time according to the Schrödinger equation

$$i\hbar\frac{\partial}{\partial t}\big|\psi(t)\big\rangle = \big(\mathcal{H}_0 + \mathcal{H}^{\mathrm{I}}\big)\big|\psi(t)\big\rangle.$$

To solve this equation we use the so-called "method of the variation of the constants". Suppose that we have previously solved the stationary Schrödinger equation for $\mathcal{H}_0$, finding the eigenvectors $|\alpha\rangle$ and the corresponding eigenvalues E_α that satisfy the equation

$$\mathcal{H}_0|\alpha\rangle = E_\alpha|\alpha\rangle. \tag{11.4}$$

We expand the wavefunction $|\psi(t)\rangle$ in series of such eigenvectors (with the relative temporal factor) to obtain

$$\big|\psi(t)\big\rangle = \sum_\beta c_\beta(t)|\beta\rangle e^{-iE_\beta t/\hbar}.$$

The coefficients c_β are function of time and are reduced to some constants in the absence of perturbation (i.e. when $\mathcal{H}^{\mathrm{I}} = 0$).[2] We then substitute this expansion into the Schrödinger equation to obtain

$$\sum_\beta i\hbar\dot{c}_\beta(t)|\beta\rangle e^{-iE_\beta t/\hbar} = \sum_\beta c_\beta(t)\mathcal{H}^{\mathrm{I}}|\beta\rangle e^{-iE_\beta t/\hbar},$$

where we have used the dot (Newton's) notation for the differentiation with respect to time ($\dot{c}_\beta = dc_\beta/(dt)$). We now apply a scalar multiplication of both sides by $\langle\alpha|$. Taking into account the orthonormality property of the eigenvectors $|\beta\rangle$, we have

$$i\hbar\dot{c}_\alpha(t) = \sum_\beta c_\beta(t)\mathcal{H}^{\mathrm{I}}_{\alpha\beta}e^{i\omega_{\alpha\beta}t},$$

where we have put

$$\mathcal{H}^{\mathrm{I}}_{\alpha\beta} = \langle\alpha|\mathcal{H}^{\mathrm{I}}|\beta\rangle,$$

and where we have introduced the so-called Bohr angular frequencies

$$\omega_{\alpha\beta} = \frac{E_\alpha - E_\beta}{\hbar}. \tag{11.5}$$

If we now assume that the interaction Hamiltonian does not depend on time, the previous equation can be integrated between the initial time ($t = 0$) and the generic time t. We have

$$c_\alpha(t) = c_\alpha(0) + \frac{1}{i\hbar}\sum_\beta \mathcal{H}^{\mathrm{I}}_{\alpha\beta}\int_0^t c_\beta(t')e^{i\omega_{\alpha\beta}t'}\,dt'.$$

---

[2] This fact justifies the name of the method (variation of constants) to denote the procedure followed here for the solution of the Schrödinger equation.

Fig. 11.1 The function
behaves, for $t \to \infty$, as
$2\pi t \delta(\omega_{\alpha\gamma})$

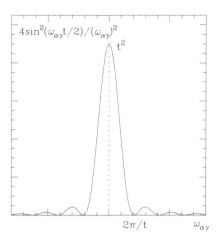

This equation can be used again to express the coefficient $c_\beta(t')$ appearing within the integral. In doing so we obtain the equation

$$c_\alpha(t) = c_\alpha(0) + \frac{1}{i\hbar} \sum_\beta \mathcal{H}_{\alpha\beta}^{\mathrm{I}} \int_0^t c_\beta(0) e^{i\omega_{\alpha\beta}t'} \, dt'$$

$$+ \frac{1}{(i\hbar)^2} \sum_{\beta\gamma} \mathcal{H}_{\alpha\beta}^{\mathrm{I}} \mathcal{H}_{\beta\gamma}^{\mathrm{I}} \int_0^t dt' e^{i\omega_{\alpha\beta}t'} \int_0^{t'} dt'' c_\gamma(t'') e^{i\omega_{\beta\gamma}t''}. \qquad (11.6)$$

Such procedure can be repeated *ad libitum* and we obtain a perturbative expansion with the n-th term containing n matrix elements of the interaction Hamiltonian. If we interrupt the sum at the k-th term, substituting the $c(t)$ coefficient appearing in the last integral with $c(0)$, we obtain a perturbative theory at the order k in the probability amplitude.

We start by considering the particular case of the theory at the lowest order, which reduces Eq. (11.6) to the first row. We also assume that the system is, at time $t = 0$, in the pure state $|\gamma\rangle$. All the $c_\beta(0)$ are null, with the exception of $c_\gamma(0)$ that is equal to unity. Solving the integral we obtain

$$c_\alpha(t) = \frac{1}{\hbar} \mathcal{H}_{\alpha\gamma}^{\mathrm{I}} \frac{1 - e^{i\omega_{\alpha\gamma}t}}{\omega_{\alpha\gamma}},$$

hence

$$|c_\alpha(t)|^2 = \frac{1}{\hbar^2} |\mathcal{H}_{\alpha\gamma}^{\mathrm{I}}|^2 \frac{4\sin^2(\omega_{\alpha\gamma}t/2)}{\omega_{\alpha\gamma}^2}.$$

The function of time that appears in this expression behaves, at the limit for $t \to \infty$, as a Dirac delta (see Eq. (2.9)). This fact can be rigorously proven within the theory of distributions, and is illustrated by the graph in Fig. 11.1. We have

$$\lim_{t \to \infty} \frac{4\sin^2(\omega_{\alpha\gamma}t/2)}{\omega_{\alpha\gamma}^2} = 2\pi t \delta(\omega_{\alpha\gamma}),$$

hence

$$|c_\alpha(t)|^2 = \frac{2\pi}{\hbar^2}|\mathcal{H}^I_{\alpha\gamma}|^2 t\delta(\omega_{\alpha\gamma}). \tag{11.7}$$

This expression shows that the square of the absolute value of the $c_\alpha(t)$ coefficient, which physically represents the probability to find the system in the state $|\alpha\rangle$ at time t, increases linearly with time. Obviously, at the same time the c_γ coefficient must decrease. This feedback effect is however not contained in Eq. (11.7). To determine it, it is convenient to follow a more sophisticated procedure which will lead to the kinetic equations defining the evolution of the system. To do this, we return to Eq. (11.6) and substitute $c_\gamma(t'')$ with $c_\gamma(0)$ (second order perturbative expansion in the probability amplitude). The integrals over time can be evaluated with elementary methods, and we obtain

$$c_\alpha(t) = c_\alpha(0) - \frac{1}{\hbar}\sum_\beta \mathcal{H}^I_{\alpha\beta}c_\beta(0)\frac{e^{i\omega_{\alpha\beta}t}-1}{\omega_{\alpha\beta}}$$
$$+ \frac{1}{\hbar^2}\sum_{\beta\gamma}\mathcal{H}^I_{\alpha\beta}\mathcal{H}^I_{\beta\gamma}c_\gamma(0)\left(\frac{e^{i\omega_{\alpha\gamma}t}-1}{\omega_{\alpha\gamma}\omega_{\beta\gamma}} - \frac{e^{i\omega_{\alpha\beta}t}-1}{\omega_{\alpha\beta}\omega_{\beta\gamma}}\right). \tag{11.8}$$

When calculating the square modulus of $c_\alpha(t)$ we obtain from this equation nine different terms. Of these nine terms we only consider those that are at most quadratic in the matrix elements of $\mathcal{H}^I$, to be consistent with the second order perturbative expansion in the probability amplitude. Taking into account that $\mathcal{H}^I$ is a Hermitian operator, so that $(\mathcal{H}^I_{\alpha\beta})^* = \mathcal{H}^I_{\beta\alpha}$, with some simple algebra we obtain

$$|c_\alpha(t)|^2 = |c_\alpha(0)|^2 - \frac{1}{\hbar}\left[\sum_\beta \mathcal{H}^I_{\alpha\beta}c_\beta(0)c^*_\alpha(0)\frac{e^{i\omega_{\alpha\beta}t}-1}{\omega_{\alpha\beta}} + C.C.\right]$$
$$+ \frac{1}{\hbar^2}\sum_{\beta\delta}\mathcal{H}^I_{\alpha\beta}\mathcal{H}^I_{\delta\alpha}c_\beta(0)c^*_\delta(0)\frac{e^{i\omega_{\alpha\beta}t}-1}{\omega_{\alpha\beta}}\frac{e^{-i\omega_{\alpha\delta}t}-1}{\omega_{\alpha\delta}}$$
$$+ \frac{1}{\hbar^2}\left[\sum_{\beta\gamma}\mathcal{H}^I_{\alpha\beta}\mathcal{H}^I_{\beta\gamma}c_\gamma(0)c^*_\alpha(0)\left(\frac{e^{i\omega_{\alpha\gamma}t}-1}{\omega_{\alpha\gamma}\omega_{\beta\gamma}} - \frac{e^{i\omega_{\alpha\beta}t}-1}{\omega_{\alpha\beta}\omega_{\beta\gamma}}\right) + C.C.\right],$$

where the symbol $[\cdots + C.C.]$ indicates that we need to add to the term in brackets its complex conjugate.

This expression can be greatly simplified if we adopt some approximations with respect to the initial conditions of the system. The approximation that is usually introduced is to suppose that the bilinear quantities in the $c_\alpha(0)$ coefficients are on average zero when the two coefficients refer to different states, or

$$\langle c_\alpha(0)c^*_\beta(0)\rangle = \langle|c_\alpha(0)|^2\rangle\delta_{\alpha\beta}.$$

The average that we perform on the physical system is to be understood as the temporal average to be executed with respect to the initial time, or as a statistical average of many possible different realizations of the system itself.[3] The quantities $\langle |c_\alpha(t)|^2 \rangle$ then represent the diagonal matrix elements of a suitable operator, called the density matrix, and are commonly referred to by the symbol $\rho_\alpha(t)$. This approximation, sometimes called the random phase approximation, is equivalent to saying that the system is not in a pure state but, rather, in a mixture of states such that the non-diagonal elements of the density matrix (also known as quantum interferences, or coherences) are zero. The random phase approximation leads to valid results for a wide class of physical phenomena. But it is not applicable when the phase relations between different states play an important role in the physical phenomenon under study. A particular example is when we want to study the interaction between radiation and an atomic system taking into account also the polarisation phenomena. In this case, the phase relations become essential and the random phase approximation is not applicable anymore but must be replaced by less stringent assumptions. A in-depth study of this topic is discussed in Sect. 16.11 where the evolution equation for the coherences of a physical system is derived.[4]

Introducing the random phase approximation in the equation for $|c_\alpha(t)|^2$ and noting that the term linear in $\mathcal{H}^I$ does not produce any contribution because the diagonal matrix elements $\mathcal{H}^I_{\alpha\alpha}$ are (or can be assumed to be) null.[5] With the new notations we obtain the equation

$$\rho_\alpha(t) = \rho_\alpha(0) + \frac{1}{\hbar^2} \sum_\beta |\mathcal{H}^I_{\alpha\beta}|^2 \frac{4\sin^2(\omega_{\alpha\beta}t/2)}{\omega_{\alpha\beta}^2} \rho_\beta(0)$$

$$- \frac{1}{\hbar^2} \sum_\beta |\mathcal{H}^I_{\alpha\beta}|^2 \frac{4\sin^2(\omega_{\alpha\beta}t/2)}{\omega_{\alpha\beta}^2} \rho_\alpha(0).$$

Going to the limit for $t \to \infty$, recalling the definition of the Bohr angular frequencies (Eq. (11.5)) and noting that all the terms in the right-hand side depend linearly on time, we can write the kinetic equation for the ρ_α quantities in the form

$$\frac{d\rho_\alpha}{dt} = \frac{2\pi}{\hbar} \sum_\beta \rho_\beta |\mathcal{H}^I_{\alpha\beta}|^2 \delta(E_\alpha - E_\beta) - \frac{2\pi}{\hbar} \sum_\beta \rho_\alpha |\mathcal{H}^I_{\alpha\beta}|^2 \delta(E_\alpha - E_\beta). \quad (11.9)$$

11.3 Fermi's Golden Rule

The kinetic equation (11.9) expresses the temporal variation of the ρ_α quantity, i.e. the variation of the probability to find the system in the state $|\alpha\rangle$. In the equation there are two terms: a term with positive sign that describes an increase in ρ_α due to

---

[3]The demonstration that the two types of averages are the same is far from trivial. In statistical mechanics, one usually refers to statements of this type by invoking the so-called ergodic theorem.

[4]For further discussions on polarisation phenomena see Landi Degl'Innocenti and Landolfi (2004).

[5]The diagonal matrix elements $\mathcal{H}^I_{\alpha\alpha}$ are effectively null in the applications that are considered in the following, where the interaction Hamiltonian is the one of Eq. (11.3).

the transitions from $|\beta\rangle \neq |\alpha\rangle$ states to the $|\alpha\rangle$ state, and a negative term describing the decrease of ρ_α due to the transitions out of the $|\alpha\rangle$ state to the $|\beta\rangle \neq |\alpha\rangle$ states. The equation can be simply interpreted by saying that there is a transition probability per unit time between the states $|\alpha\rangle$ and $|\beta\rangle$ given by

$$P_{\alpha\beta} = P_{\beta\alpha} = \frac{2\pi}{\hbar}|\mathcal{H}^I_{\alpha\beta}|^2 \delta(E_\beta - E_\alpha). \tag{11.10}$$

This formula is called the Fermi golden rule. The presence of the Dirac delta is such that the transition probability is different from zero only between isoenergetic states, an obvious manifestation of the principle of conservation of energy. It should be noted that this formula is substantially already contained in Eq. (11.7) and, in principle, it is not therefore necessary to consider the second order Eq. (11.6) to obtain it. As we have already noticed, however, Eq. (11.7) is not sufficient to determine the kinetic equations for the lack of the feedback term.

Writing the kinetic equation in terms of the transition probability per unit time, we have

$$\frac{d\rho_\alpha}{dt} = \sum_\beta \rho_\beta P_{\beta\alpha} - \rho_\alpha \sum_\beta P_{\alpha\beta},$$

from which, summing on α, we get

$$\frac{d}{dt} \sum_\alpha \rho_\alpha = \sum_{\alpha\beta} (\rho_\beta P_{\beta\alpha} - \rho_\alpha P_{\alpha\beta}) = 0.$$

We then obtain the result

$$\sum_\alpha \rho_\alpha = \text{cost.},$$

as we should have obviously expected, according to the probabilistic interpretation of the ρ_α. The constant that represents the normalisation value for the sum of probabilities is in general set to one. The kinetic equations are a system of $\mathcal{N}$ homogeneous differential equations of first degree, where $\mathcal{N}$ is the number of states that occur in the particular physical phenomenon that is considered. In principle, once the transition probabilities are known, the system can be solved starting from appropriate boundary conditions. In stationary situations, i.e. when we have for each state α

$$\frac{d\rho_\alpha}{dt} = 0,$$

the system of differential equations reduces to a homogeneous linear system of $\mathcal{N}$ equations in $\mathcal{N}$ unknowns. The previous condition, i.e.

$$\sum_{\alpha\beta} (\rho_\beta P_{\beta\alpha} - \rho_\alpha P_{\alpha\beta}) = 0,$$

implies that the determinant of the system is zero so the quantities ρ_α are determined apart from a proportionality factor that is fixed by the normalisation condition.

Finally, we note that, for the calculation of the transition probability, the quantity that is essential to be determined is the matrix element of the interaction Hamiltonian between the eigenstates of the unperturbed Hamiltonian. The calculation of this quantity is discussed in the next section for the case we are interested in, namely the interaction between matter and radiation.

11.4 The Matrix Element

We now apply the general considerations carried out in the two previous sections to the interaction between an atomic system and the radiation field. The unperturbed and the interaction Hamiltonian are given, respectively, by Eqs. (11.1) and (11.3). The eigenstates of the unperturbed Hamiltonian, defined by Eq. (11.4), are the direct product of the eigenstates of the atomic system by the eigenstates of the radiation field. We will denote the first with the symbol $|u_n\rangle$ and the corresponding eigenvalues of the energy with the symbol ϵ_n, assuming that we have solved, for the atomic system, the stationary Schrödinger equation

$$\mathcal{H}_A |u_n\rangle = \epsilon_n |u_n\rangle.$$

We now simplify the notations for the radiation field, with respect to those used in Chap. 4. We can think of numbering the modes with an index j which substitutes the pair of indices $(\mathbf{k}, \lambda)$, and, neglecting the vacuum energy, write the Hamiltonian $\mathcal{H}_R$ in the form

$$\mathcal{H}_R = \sum_j h\nu_j a_j^\dagger a_j.$$

We denote the relative eigenvectors by the symbol $|n_1, n_2, \ldots, n_l, \ldots\rangle$ or, more synthetically, by the symbol $|\{n_l\}\rangle$. We then have, with these new notations,

$$\mathcal{H}_R |\{n_l\}\rangle = \left(\sum_j h\nu_j n_j\right) |\{n_l\}\rangle.$$

Furthermore, with the new notations the interaction Hamiltonian (11.3) becomes

$$\mathcal{H}^I = \frac{e_0}{m} \sum_j \sqrt{\frac{h}{2\pi \nu_j \mathcal{V}}} \left(a_j \sum_i \mathbf{p}_i \cdot \mathbf{e}_j e^{i\mathbf{k}_j \cdot \mathbf{r}_i} + a_j^\dagger \sum_i \mathbf{p}_i \cdot \mathbf{e}_j^* e^{-i\mathbf{k}_j \cdot \mathbf{r}_i}\right).$$

The results we have obtained in the previous section can be applied to the case of the matter-radiation interaction with the formal substitutions

$$|\alpha\rangle \rightarrow |u_n, \{n_l\}\rangle, \qquad \rho_\alpha \rightarrow \rho_{n,\{n_l\}}, \qquad E_\alpha \rightarrow E_{n,\{n_l\}} = \epsilon_n + \sum_j h\nu_j n_j,$$

$$\mathcal{H}^I_{\alpha\beta} \rightarrow \langle u_n, \{n_l\} | \mathcal{H}^I | u_m, \{n_l'\}\rangle.$$

An approximation is commonly introduced to calculate the matrix element. It is called the dipole approximation, and consists in substituting the exponentials that appear in the expression for $\mathcal{H}^I$ with unity, i.e.

$$e^{i\mathbf{k}_j \cdot \mathbf{r}_i} \simeq e^{-i\mathbf{k}_j \cdot \mathbf{r}_i} \simeq 1.$$

To understand the degree of approximation introduced, we expand the exponential in series. We have

$$e^{i\mathbf{k}_j \cdot \mathbf{r}_i} = 1 + i\mathbf{k}_j \cdot \mathbf{r}_i - \frac{1}{2}(\mathbf{k}_j \cdot \mathbf{r}_i)^2 + \cdots, \tag{11.11}$$

so the approximation is correct when the following inequality is well satisfied

$$\mathbf{k}_j \cdot \mathbf{r}_i \ll 1.$$

On the other hand, as an order of magnitude we have

$$\mathbf{k}_j \cdot \mathbf{r}_i \simeq \frac{a_0}{\lambda_j},$$

where λ_j is the wavelength of the mode we consider and a_0 is the radius of the first Bohr orbit. We therefore see that the dipole approximation is well justified at the wavelengths of the visible radiation. For example, for $\lambda_j = 5000$ Å, the ratio a_0/λ_j is about 10^{-4}. Within the dipole approximation, putting

$$\mathbf{p} = \sum_i \mathbf{p}_i,$$

the Hamiltonian of the interaction assumes the simplified form

$$\mathcal{H}^{\mathrm{I}} = \frac{e_0}{m} \sum_j \sqrt{\frac{h}{2\pi \nu_j V}} \left(a_j \mathbf{p} \cdot \mathbf{e}_j + a_j^{\dagger} \mathbf{p} \cdot \mathbf{e}_j^* \right).$$

Noting then that the a_j and $a_j^{\dagger}$ operators act only on the eigenstates of the radiation field, while the $\mathbf{p}$ operator acts only on the eigenstates of the atomic system, the matrix element can be written in the form

$$\langle u_n, \{n_l\} | \mathcal{H}^{\mathrm{I}} | u_m, \{n_l'\} \rangle$$
$$= \frac{e_0}{m} \sum_j \sqrt{\frac{h}{2\pi \nu_j V}} \left[\langle \{n_l\} | a_j | \{n_l'\} \rangle \langle u_n | \mathbf{p} | u_m \rangle \cdot \mathbf{e}_j + \langle \{n_l\} | a_j^{\dagger} | \{n_l'\} \rangle \langle u_n | \mathbf{p} | u_m \rangle \cdot \mathbf{e}_j^* \right].$$

We calculate the matrix element of the atomic system first. We transform it from a matrix element of the operator $\mathbf{p}$ in a matrix element of the operator $\mathbf{r}$. To do this, we consider the following commutator

$$[\mathcal{H}_{\mathrm{A}}, \mathbf{r}] = \left[\mathcal{H}_{\mathrm{A}}, \sum_i \mathbf{r}_i \right],$$

and assume, neglecting relativistic corrections, that the atomic Hamiltonian is that one given in Eq. (7.2). Within these approximations we obtain

$$[\mathcal{H}_{\mathrm{A}}, \mathbf{r}] = \left[\sum_k \frac{p_k^2}{2m}, \sum_i \mathbf{r}_i \right] = \sum_i \left[\frac{p_i^2}{2m}, \mathbf{r}_i \right] = -i\frac{\hbar}{m} \sum_i \mathbf{p}_i = -i\frac{\hbar}{m} \mathbf{p},$$

so that

$$\langle u_n | \mathbf{p} | u_m \rangle = i\frac{m}{\hbar} \langle u_n | [\mathcal{H}_{\mathrm{A}}, \mathbf{r}] | u_m \rangle = 2\pi i\frac{m}{h} (\epsilon_n - \epsilon_m) \langle u_n | \mathbf{r} | u_m \rangle.$$

Introducing also the notations

$$v_{nm} = \frac{\epsilon_n - \epsilon_m}{h}, \qquad \mathbf{r}_{nm} = \langle u_n | \mathbf{r} | u_m \rangle,$$

we obtain

$$\langle u_n | \mathbf{p} | u_m \rangle = 2\pi i m v_{nm} \mathbf{r}_{nm}.$$

The quantities v_{nm} are called Bohr (cyclic) frequencies and are simply related to the Bohr angular frequencies introduced in Eq. (11.5) by the relation $v_{nm} = \omega_{nm}/(2\pi)$. The quantity $\mathbf{r}_{nm}$ is called dipole matrix element (between the states $|u_n\rangle$ and $|u_m\rangle$), although rigorously such name should actually be given to the matrix element of the operator $\mathbf{d} = -e_0 \mathbf{r}$, that represents the dipole moment of the electron cloud.

We now consider the matrix elements of the radiation field. We recall the results obtained in Sect. 4.3 for the matrix elements of the operators of creation and annihilation. From Eqs. (4.10) and (4.11) we derive that the matrix element of the annihilation operator $\langle \{n_l\} | a_j | \{n'_l\} \rangle$ is always zero unless the occupation numbers contained in the bra are all equal to the occupation numbers contained in the ket, with the exception of the j mode, for which we must have $n'_j = n_j + 1$. In other words, for the matrix element to be non-zero, we must have

$$n'_1 = n_1, n'_2 = n_2, \ldots, n'_{j-1} = n_{j-1}, n'_j = n_j + 1, n'_{j+1} = n_{j+1}, \ldots$$

Given a set $\{n_l\}$ of occupation numbers, we denote by the symbol $\{n_l + 1_j\}$ the set obtained from the previous one by increasing by one the occupation number of the j mode and keeping all the others the same. We then obtain

$$\langle \{n_l\} | a_j | \{n_l + 1_j\} \rangle = \sqrt{n_j + 1},$$

and with similar considerations

$$\langle \{n_l\} | a_j^\dagger | \{n_l - 1_j\} \rangle = \sqrt{n_j}.$$

By substituting the results we obtained for the matrix elements of the atomic system and for those of the radiation field, we obtain for the non-vanishing matrix elements of the interaction Hamiltonian

$$\langle u_n, \{n_l\} | \mathcal{H}^{\mathrm{I}} | u_m, \{n_l + 1_j\} \rangle = i v_{nm} e_0 \sqrt{\frac{2\pi h (n_j + 1)}{v_j V}} \mathbf{r}_{nm} \cdot \mathbf{e}_j, \qquad (11.12)$$

$$\langle u_n, \{n_l\} | \mathcal{H}^{\mathrm{I}} | u_m, \{n_l - 1_j\} \rangle = i v_{nm} e_0 \sqrt{\frac{2\pi h n_j}{v_j V}} \mathbf{r}_{nm} \cdot \mathbf{e}_j^*. \qquad (11.13)$$

The difference in the energies of the states between which the matrix element is calculated is, respectively,

$$E_{n,\{n_l\}} - E_{m,\{n_l+1_j\}} = \epsilon_n - \epsilon_m - h v_j = h(v_{nm} - v_j),$$

$$E_{n,\{n_l\}} - E_{m,\{n_l-1_j\}} = \epsilon_n - \epsilon_m + h v_j = h(v_{nm} + v_j).$$

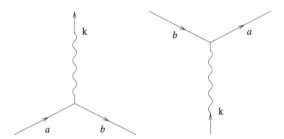

Fig. 11.2 Feynman diagrams for the elementary processes of spontaneous emission (*left*) and absorption (*right*). In the first case, the atomic system goes from the upper level a to the lower level b and a photon is emitted in the mode k. In the second case, the system goes from level b to level a and a photon is absorbed from the mode k

11.5 Elementary Processes

As an application of the formalism developed in the previous section, we now consider the elementary process of spontaneous emission. As initial and final states of the overall system (atom plus radiation) we consider the states described by the state vectors

$$|\Psi_i\rangle = |u_a; 0, 0, \ldots, 0, \ldots\rangle, \qquad |\Psi_f\rangle = |u_b; 0, 0, \ldots, 1_k, \ldots\rangle. \quad (11.14)$$

The initial state $|\Psi_i\rangle$ is the state in which the atom is in the level $|u_a\rangle$ of energy ϵ_a, and the radiation field is in the vacuum state. The final state $|\Psi_f\rangle$ is that one in which the atom is in level $|u_b\rangle$ of energy ϵ_b, while the radiation field is in the state where there is only one photon in the mode denoted by the index k.

To identify in a schematic way the process we are dealing with, it is useful to introduce a diagram such as that one shown on the left panel of Fig. 11.2. Such a diagram is called Feynman diagram, from the name of the American physicist who first introduced such schematic representations of the elementary processes in quantum electrodynamics. In a Feynman diagram, the solid straight lines represent the electrons, while the wavy lines represent the photons. Each interaction between electrons and photons is represented by a junction of two straight lines and a wavy one at a vertex, which represents emission or absorption of a photon by an electron. Thinking that "time" runs from left to right in the diagram, at the vertex the electron "jumps" from the upper level a to the lower level b, while a photon is emitted in the mode k.

The probability per unit time that the process of spontaneous emission occurs is given by Fermi's golden rule

$$P_{fi} = \frac{2\pi}{\hbar} \left| \langle \Psi_f | \mathcal{H}^I | \Psi_i \rangle \right|^2 \delta(E_f - E_i).$$

The matrix element of the interaction Hamiltonian can be calculated using Eq. (11.13). A simple substitution results in

$$\left| \langle \Psi_f | \mathcal{H}^I | \Psi_i \rangle \right|^2 = \frac{2\pi h e_0^2 \nu_{ab}^2}{\nu_k \mathcal{V}} \left| \mathbf{r}_{ba} \cdot \mathbf{e}_k^* \right|^2, \qquad E_f - E_i = h(\nu_k - \nu_{ab}),$$

and we obtain

$$P_{\mathrm{fi}} = \frac{8\pi^3 e_0^2 v_{ab}}{hV} \left| \mathbf{r}_{ba} \cdot \mathbf{e}_k^* \right|^2 \delta(v_k - v_{ab}). \tag{11.15}$$

It is important to clarify that the atomic levels a and b, eigenstates of the atomic Hamiltonian, are in general spatially degenerate and need to be characterized by some internal quantum numbers that we denote by α and β, respectively. For example, the wavefunction of the level a that we have simply denoted by $|u_a\rangle$ (in Dirac notation) is in reality of the type $|\alpha_a J_a M_a\rangle$, where α_a is a collection of quantum numbers that describe the configuration and the term, J_a is the quantum number for the total angular momentum, and M_a the magnetic quantum number. Since M_a can assume any value between $-J_a$ and J_a in steps of one, the level is degenerate g_a times, with $g_a = 2J_a + 1$. The wavefunction of any of the g_a sublevels of level a can therefore be denoted by the symbol $|u_{a\alpha}\rangle$ and, similarly, any of the sublevels of level b by the symbol $|u_{b\beta}\rangle$. Once the sublevel is also specified, both in terms of its initial and final state, the probability of the transition per unit time (Eq. (11.15)) assumes the form

$$P_{\mathrm{fi}} = \frac{8\pi^3 e_0^2 v_{ab}}{hV} \left| \mathbf{r}_{b\beta,a\alpha} \cdot \mathbf{e}_k^* \right|^2 \delta(v_k - v_{ab}), \tag{11.16}$$

where

$$\mathbf{r}_{b\beta,a\alpha} = \langle u_{b\beta} | \mathbf{r} | u_{a\alpha} \rangle.$$

We now perform an average of the squared modulus of the matrix element that appears in Eq. (11.16) over all possible "orientations" of the atom with respect to the unit vector $\mathbf{e}_k^*$, i.e. we consider the average over all values of the degeneracy parameters α and β. It is possible to show that using the Wigner-Eckart theorem the following relation holds[6]

$$\frac{1}{g_a g_b} \sum_{\alpha,\beta} \left| \mathbf{r}_{b\beta,a\alpha} \cdot \mathbf{e}_k^* \right|^2 = \frac{1}{3} |\mathbf{r}_{ba}|^2, \tag{11.17}$$

where

$$|\mathbf{r}_{ba}|^2 = |\mathbf{r}_{ab}|^2 = \frac{1}{g_a g_b} \sum_{\alpha,\beta} |\mathbf{r}_{b\beta,a\alpha}|^2 = \frac{1}{g_a g_b} \sum_{\alpha,\beta} \langle u_{b\beta} | \mathbf{r} | u_{a\alpha} \rangle \cdot \langle u_{a\alpha} | \mathbf{r} | u_{b\beta} \rangle,$$

so that we can conclude that, averaging over the spatial degeneracy, the probability per unit time that the atom decays from any sublevel of level a into any sublevel of level b and that, at the same time, a photon is emitted in the mode k is given by

$$P_{\mathrm{fi}} = \frac{8\pi^3 e_0^2 v_{ab}}{3hV} |\mathbf{r}_{ba}|^2 \delta(v_k - v_{ab}). \tag{11.18}$$

---

[6]This property is formally derived in Sect. 16.12. Intuitively, we can think of the factor $\frac{1}{3}$ as the result of the average over the solid angle of a factor of the type $\cos^2 \theta$, where θ is the angle between the $\mathbf{r}_{ba}$ vector and the polarisation unit vector.

From this expression for P_{fi} we can obtain both the probability π_{ab} per unit time that the atom decays from level a to level b, and the probability $\pi^{(\text{e})}(v, \mathbf{\Omega})$ per unit time that a photon is emitted in any of the two modes (characterized by different polarisations) of frequency v and direction $\mathbf{\Omega}$. To obtain π_{ab} we note that P_{fi} gives the combined probability that the atom decays from any sublevel of level a into any sublevel of level b and, at the same time, a photon is emitted in the mode k. For the known laws of probability we then need to sum P_{fi} over all possible lower sublevels and all possible modes. We then have

$$\pi_{ab} = g_b \sum_k P_{\text{fi}},$$

with the sum extended to all the possible modes of the radiation field. To perform the sum, we recall that the number of modes with frequency between v and $v + dv$ and direction within the solid angle $d\Omega$ is given by Eq. (4.14), or

$$dN = \frac{2\mathcal{V}v^2}{c^3} dv d\Omega.$$

Transforming the sum over the k modes in an integral over frequency and solid angle and recalling Eq. (11.18) gives

$$\pi_{ab} = g_b \oint d\Omega \int_0^\infty \frac{16\pi^3 e_0^2 v_{ab} v^2}{3hc^3} |\mathbf{r}_{ba}|^2 \delta(v - v_{ab}) dv. \qquad (11.19)$$

Finally, evaluating the integral, we have

$$\pi_{ab} = A_{ab},$$

where

$$A_{ab} = \frac{64\pi^4 e_0^2 v_{ab}^3}{3hc^3} g_b |\mathbf{r}_{ba}|^2. \qquad (11.20)$$

The quantity A_{ab} is called the Einstein coefficient for spontaneous emission (or de-excitation) from level a to level b. As a memory aid we note that its expression can be obtained by a semi-classical method. Applying the Lorentz model, we can think of an atom as an electron oscillating at frequency v. If x is the instantaneous amplitude of the oscillation, the magnitude of the electron acceleration is $4\pi^2 v^2 x$ and the electron emits radiation with an average power $\langle W \rangle$ given by the Larmor equation (3.22)

$$\langle W \rangle = \frac{2e_0^2 \langle a^2 \rangle}{3c^3} = \frac{32\pi^4 e_0^2 v^4 \langle x^2 \rangle}{3c^3},$$

where $\langle x^2 \rangle$ is the average of x^2 with respect to time. If we then assume that the harmonic oscillator has an energy $\frac{1}{2}hv$, the characteristic time τ for it to loose its energy is given by

$$\tau = \frac{hv}{2\langle W \rangle},$$

from which we obtain a de-excitation rate per unit time given by

$$\frac{1}{\tau} = \frac{2\langle W \rangle}{h\nu} = \frac{64\pi^4 \nu^3 e_0^2}{3hc^3}\langle x^2 \rangle.$$

As we can see, this expression coincides with that one in Eq. (11.20) if we identify the oscillator frequency ν with the transition frequency ν_{ab}, and the classical quantity $\langle x^2 \rangle$ with $g_b|\mathbf{r}_{ba}|^2$.

We now proceed to determine the quantity $\pi^{(e)}(\nu, \mathbf{\Omega})$, i.e. the probability per unit time that a photon in a mode of frequency ν and direction $\mathbf{\Omega}$ is emitted. For this it is sufficient to multiply Eq. (11.18) by the factor g_b to take into account that the atom can decay in any of the sublevels of the lower level. We obtain

$$\pi^{(e)}(\nu, \mathbf{\Omega}) = \frac{8\pi^3 e_0^2 \nu_{ab}}{3h\mathcal{V}} g_b|\mathbf{r}_{ba}|^2 \delta(\nu - \nu_{ab}).$$

As we can see, this expression contains at the denominator the volume $\mathcal{V}$ of the cavity in which the radiation field has been quantised. This dependence on $\mathcal{V}$ arises from the fact that we are considering the interaction between the radiation field and a single atomic system, which we have supposed being in an arbitrary sublevel of level a. If we consider the more realistic situation in which N_a atoms in level a are actually present in the cavity, and if we assume that each atom emits independently from the others, the right-hand side of the previous equation must be multiplied by N_a and we get

$$\pi^{(e)}(\nu, \mathbf{\Omega}) = \mathcal{N}_a \frac{8\pi^3 e_0^2 \nu_{ab}}{3h} g_b|\mathbf{r}_{ba}|^2 \delta(\nu - \nu_{ab}), \tag{11.21}$$

where we have introduced the quantity $\mathcal{N}_a = N_a/\mathcal{V}$, the number density of atoms in level a.

The results that we found for the Einstein coefficient (Eq. (11.20)) and the probability $\pi^{(e)}(\nu, \mathbf{\Omega})$ (Eq. (11.21)) are related to the choice of the initial and final states $|\Psi_i\rangle$ and $|\Psi_f\rangle$ of Eq. (11.14). We should note that these results do not change if we assume that an arbitrary number of photons (instead of zero) are present in the modes different from mode k, as long as the number of such photons is equal in the initial and final states. This means that an elementary process as that one described in the Feynman diagram of Fig. 11.2 (left panel) is not affected by the possible presence of photons in modes different from k. If we assume instead that n_k photons (instead of zero) are present in the mode k in the initial state $|\Psi_i\rangle$ and $(n_k + 1)$ photons (instead of one) are present in the final state $|\Psi_f\rangle$, the transition probability P_{fi} is still given by the expression in Eq. (11.18) multiplied by the factor $(n_k + 1)$ which originates from the square of the modulus of the matrix element of the interaction Hamiltonian (see Eq. (11.13)). This multiplicative factor is very important as it describes a particular phenomenon known as stimulated emission. The presence of photons in a particular mode "stimulates" the atom to emit photons in the same mode, and such effect increases as the number of photons that are present in the mode increases. The phenomenon of stimulated emission is the principle of operation of the laser[7] and is now the basis of many technological applications. If account is taken of stimulated emission, the discussions that we have carried out previously to find the probabilities per unit time of de-excitation of the atom π_{ab} and of emission of a photon $\pi^{(e)}(\nu, \mathbf{\Omega})$ need to be modified and lead to additional terms. In fact, starting from Eq. (11.19) we obtain, with the addition of the term of stimulated emission

$$\pi_{ab} = g_b \oint d\Omega \int_0^\infty \frac{16\pi^3 e_0^2 \nu_{ab} \nu^2}{3hc^3} |\mathbf{r}_{ba}|^2 [n_\nu(\mathbf{\Omega}) + 1] \delta(\nu - \nu_{ab}) d\nu,$$

where we have introduced the symbol $n_\nu(\mathbf{\Omega})$ to denote the number of photons that are present in the mode of frequency ν and direction $\mathbf{\Omega}$, independently of the polarisation. The contribution originating from the unity term present in the square bracket within the integral has been already evaluated and provides the Einstein coefficient A_{ab}. The other term is evaluated by introducing the average number of photons per mode $\bar{n}_\nu$ defined by

$$\bar{n}_\nu = \frac{1}{4\pi} \oint n_\nu(\mathbf{\Omega}) d\Omega. \tag{11.22}$$

We obtain

$$\pi_{ab} = A_{ab}(1 + \bar{n}_{\nu_{ab}}). \tag{11.23}$$

The effect of stimulated emission is therefore to increase the de-excitation probability of the atom. Similarly, when evaluating the probability per unit time of emission of a photon including stimulated emission, we find that Eq. (11.21) must be substituted with the following one

$$\pi^{(e)}(\nu, \mathbf{\Omega}) = \mathcal{N}_a \frac{8\pi^3 e_0^2 \nu_{ab}}{3h} g_b |\mathbf{r}_{ba}|^2 [1 + n_\nu(\mathbf{\Omega})] \delta(\nu - \nu_{ab}). \tag{11.24}$$

As a further application, we now discuss the elementary process of absorption. As initial and final states of the whole system (atom and radiation) we consider the states described by the vectors

$$|\Psi_i\rangle = |u_b; 0, 0, \ldots, 1_k, \ldots\rangle, \qquad |\Psi_f\rangle = |u_a; 0, 0, \ldots, 0, \ldots\rangle.$$

The initial state $|\Psi_i\rangle$ is the state in which the atom is in the energy level $|u_b\rangle$ while the radiation field is in the state in which there is only one photon in the mode identified by the index k. The final state $|\Psi_f\rangle$ is that one in which the atom is in the energy level $|u_a\rangle$ and the radiation field is in the vacuum state. The process that we are describing is shown in the right panel of Fig. 11.2 and the corresponding probability per unit of time is still given by Fermi's golden rule. The matrix element of the interaction Hamiltonian between the states $|\Psi_i\rangle$ and $|\Psi_f\rangle$ can be calculated with

---

[7]The word *laser* is an acronym that stands for *Light Amplification by Stimulated Emission of the Radiation*.

Eq. (11.12) and we obtain for the transition probability per unit time the analogue of Eq. (11.15) relative to the elementary process of spontaneous emission, i.e.

$$P_{\text{fi}} = \frac{8\pi^3 e_0^2 \nu_{ab}}{h\mathcal{V}} |\mathbf{r}_{ab} \cdot \mathbf{e}_k|^2 \delta(\nu_k - \nu_{ab}).$$

This expression actually coincides with Eq. (11.15). Indeed, since the $\mathbf{r}$ operator is Hermitian, for the matrix elements we have $\mathbf{r}_{ab} = \mathbf{r}_{ba}^*$, hence

$$|\mathbf{r}_{ab} \cdot \mathbf{e}_k|^2 = |\mathbf{r}_{ba} \cdot \mathbf{e}_k^*|^2.$$

We also note that the elementary process that we are considering is exactly the inverse process of that one considered at the beginning of this section. We have therefore verified the important property that an elementary process and its inverse process have the same probability per unit time to occur. This is actually obvious, when one considers that, according to Fermi's golden rule, we must have $P_{\text{fi}} = P_{\text{if}}$.

The result we have obtained for the elementary process can be generalised to the case when we have n_k photons (instead of one) in the initial state and $(n_k - 1)$ photons (instead of zero) in the final state. Still using Eq. (11.12) we can easily find that the expression for the probability P_{fi} given above simply needs to be multiplied by n_k, so we have

$$P_{\text{fi}} = \frac{8\pi^3 e_0^2 \nu_{ab}}{h\mathcal{V}} |\mathbf{r}_{ab} \cdot \mathbf{e}_k|^2 n_k \delta(\nu_k - \nu_{ab}).$$

In a similar way as we did previously, we can now calculate the probability per unit time π_{ba} that the atom goes from level b to level a by absorbing a photon from any mode, as well as the probability per unit time $\pi^{(a)}(\nu, \mathbf{\Omega})$ that a photon is removed from a mode of frequency ν and direction $\mathbf{\Omega}$. Repeating similar steps as done previously, we find

$$\pi_{ba} = \frac{g_a}{g_b} A_{ab} \bar{n}_{\nu_{ab}}, \tag{11.25}$$

$$\pi^{(a)}(\nu, \mathbf{\Omega}) = \mathcal{N}_b \frac{8\pi^3 e_0^2 \nu_{ab}}{3h} g_a |\mathbf{r}_{ba}|^2 n_\nu(\mathbf{\Omega}) \delta(\nu_{ab} - \nu), \tag{11.26}$$

where $\mathcal{N}_b$ is the number density of atoms in level b.

11.6 The Statistical Equilibrium Equations

The discussion of the elementary processes that we have carried out in the previous section allows to obtain the general equations that govern the temporal evolution of the atomic populations in an arbitrary physical system, whether it be a laboratory or an astrophysical plasma. Consider for this a collection of atoms, all of the same species, and suppose, for simplicity, that each of them has only two energy levels, an upper energy level for which we will use the index a and a lower energy level for which we will use the index b. Suppose further that the two levels are degenerate and denote by g_a and g_b their degeneracy. We will also assume the presence of a radiation field that, at the frequency ν_{ab} corresponding to the transition between the two levels ($\nu_{ab} = (\epsilon_a - \epsilon_b)/h$), has an average number of photons per mode given by $\bar{n}_{\nu_{ab}}$. Denoting by N_a and N_b the number of atoms in the levels a and b, respectively, (the so called populations), and taking into account the elementary probabilities π_{ab} and π_{ba} that we found in the previous section, the populations evolve over time according to the equations

$$\frac{dN_a}{dt} = -\pi_{ab} N_a + \pi_{ba} N_b, \qquad \frac{dN_b}{dt} = -\pi_{ba} N_b + \pi_{ab} N_a.$$

Substituting the values of π_{ab} and π_{ba} given by Eqs. (11.23) and (11.25) we obtain the so-called statistical equilibrium equations

$$\frac{dN_a}{dt} = -A_{ab}(1 + \bar{n}_{\nu_{ab}}) N_a + \frac{g_a}{g_b} A_{ab} \bar{n}_{\nu_{ab}} N_b,$$

$$\frac{dN_b}{dt} = -\frac{g_a}{g_b} A_{ab} \bar{n}_{\nu_{ab}} N_b + A_{ab}(1 + \bar{n}_{\nu_{ab}}) N_a.$$

As we can see, the two equations are redundant, because

$$\frac{dN_a}{dt} + \frac{dN_b}{dt} = 0.$$

This means that the total number of atoms is conserved and the equations can only establish the ratio between the two populations.

The statistical equilibrium equations are traditionally written in a different way by introducing, in place of the average number of photons per mode, the average value of the intensity of the radiation field. Recalling Eq. (4.15), which links the number of photons with the specific intensity of the radiation field, and Eq. (11.22), which defines the average number of photons per mode, and defining the average of the intensity of the radiation field over the solid angle as

$$J_\nu = \frac{1}{4\pi} \oint I_\nu(\mathbf{\Omega}) d\Omega,$$

we have

$$J_{\nu_{ab}} = \frac{2h\nu_{ab}^3}{c^2} \bar{n}_{\nu_{ab}}.$$

With this latter quantity, the statistical equilibrium equations can be written in the form

$$\frac{dN_a}{dt} = -A_{ab} N_a - B_{ab} J_{\nu_{ab}} N_a + B_{ba} J_{\nu_{ab}} N_b,$$

$$\frac{dN_b}{dt} = -B_{ba} J_{\nu_{ab}} N_b + A_{ab} N_a + B_{ab} J_{\nu_{ab}} N_a,$$

where we have put

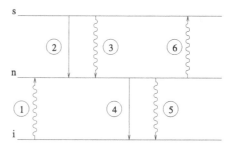

Fig. 11.3 Schematic representation of the various processes that contribute to the equations of statistical equilibrium for a given level n. (1) Absorption from lower levels; (2) spontaneous emission from upper levels; (3) stimulated emission from upper levels; (4) spontaneous emission to lower levels; (5) stimulated emission to lower levels, (6) absorption to upper levels. The *wavy lines* describe processes induced by the radiation field while the *solid lines* indicate spontaneous processes

$$B_{ab} = \frac{c^2}{2h\nu_{ab}^3} A_{ab} = \frac{32\pi^4 e_0^2}{3h^2c} g_b |\mathbf{r}_{ba}|^2, \tag{11.27}$$

$$B_{ba} = \frac{c^2}{2h\nu_{ab}^3} \frac{g_a}{g_b} A_{ab} = \frac{g_a}{g_b} B_{ab} = \frac{32\pi^4 e_0^2}{3h^2c} g_a |\mathbf{r}_{ba}|^2. \tag{11.28}$$

The quantities B_{ab} and B_{ba} are called the Einstein coefficients for stimulated emission and absorption, respectively.

Having introduced these notations, we are now able to write the equations for the statistical equilibrium for an atom with an arbitrary number of energy levels. For the population of a generic level n, denoting by the index i the lower levels (i.e. those of lower energy) and by the index s the higher levels (i.e. those with higher energy), the evolution equation of the population becomes

$$\frac{dN_n}{dt} = \sum_i B_{in} J_{\nu_{ni}} N_i + \sum_s A_{sn} N_s + \sum_s B_{sn} J_{\nu_{sn}} N_s$$
$$- \sum_i A_{ni} N_n - \sum_i B_{ni} J_{\nu_{ni}} N_n - \sum_s B_{ns} J_{\nu_{sn}} N_n. \tag{11.29}$$

The six terms appearing in the above equation are represented schematically in Fig. 11.3, where the solid lines represent spontaneous transitions (proportional to the Einstein coefficients A), while the wavy lines represent transitions induced by the radiation field (proportional to the products of the Einstein coefficients B by the average intensity of the radiation field J). In particular, the wavy lines ending with an arrow directed upwards represent absorption phenomena, while those ending with an arrow directed downwards represent phenomena of stimulated emission. The numbers within the circles in the figure are related to the order of the various terms in Eq. (11.29). In other words, the terms 1 and 6 represent absorption processes produced by the radiation field; the terms 2 and 4 represent processes of spontaneous emission; finally, the terms 3 and 5 represent processes of induced (or stimulated) emission. Note that processes 1, 2, and 3 contribute to populate the level n and are preceded by a plus sign in the equation, while processes 4, 5, and 6 contribute to de-populate the level and are preceded by a minus sign in the equation.

11.7 Einstein Coefficients

The Einstein coefficients, derived in the previous sections for an arbitrary atomic system by means of quantum electrodynamics methods, were introduced by the famous German physicist long before the formalisation of this theory by a physical reasoning of this type. Consider two energy levels of an atomic system and denote by ϵ_a and ϵ_b the energies of the upper and lower levels, respectively, and by g_a and g_b the relative degeneracies. Let ν_{ab} be the frequency of the transition between the two levels:

$$h\nu_{ab} = \epsilon_a - \epsilon_b.$$

If the atomic system is immersed in a radiation field having, at the frequency ν_{ab}, the average intensity $J_{\nu_{ab}}$, then we have a transition probability per unit time from the lower to the upper level given by

$$\pi_{ba} = B_{ba} J_{\nu_{ab}},$$

with B_{ba} independent of the radiation field. This expression is consistent with experiment, since it states, being by hypothesis B_{ba} independent of the radiation field, that the transition probability is proportional to the average intensity of the radiation field, a law that corresponds to the well known phenomenon of absorption. On the other hand, for the transition probability from the upper level to the lower one, the laws of physics—known at the time of Einstein's work—simply resulted in an equation of the type

$$\pi_{ab} = A_{ab},$$

with A_{ab} independent of the radiation field. Indeed, at the time, only the phenomenon of spontaneous emission was known, and not that of induced (or stimulated) emission.

Einstein pointed out that the two expressions for π_{ba} and π_{ab} were incompatible with the laws of thermodynamics. If we assume, in fact, that the atomic system is in a cavity in thermodynamic equilibrium at the temperature T, since the coefficients A_{ab} and B_{ba} are independent of the radiation field, we obtain that at the limit $T \to \infty$ the ratio π_{ba}/π_{ab} also tends to infinity, with the result that all the atoms would be in the upper level, in clear contradiction with the Boltzmann law, which instead states that the ratio of the populations should be equal to the ratio of the statistical weights (see Eq. (10.6) and consider the limit for $T \to \infty$). To solve this contradiction, Einstein postulated that a new term, also proportional to the radiation field, should be added to the equation for the transition probability π_{ab}:

$$\pi_{ab} = A_{ab} + B_{ab} J_{\nu_{ab}}.$$

With a simple thermodynamic reasoning, Einstein was therefore able to predict the existence of stimulated emission, a physical mechanism which was not known at the time.

By means of thermodynamic considerations, it is also possible to establish the relations among the coefficients introduced in the previous equations. According to these, the population of the upper level N_a satisfies the differential equation

$$\frac{\mathrm{d}N_a}{\mathrm{d}t} = -(A_{ab} + B_{ab}J_{\nu_{ab}})N_a + B_{ba}J_{\nu_{ab}}N_b.$$

In stationary conditions we therefore have

$$\frac{N_a}{N_b} = \frac{B_{ba}J_{\nu_{ab}}}{A_{ab} + B_{ab}J_{\nu_{ab}}}.$$

Assuming that we are in thermodynamic equilibrium, and taking the limit $T \to \infty$, we obtain

$$\lim_{T \to \infty} \frac{N_a}{N_b} = \frac{B_{ba}}{B_{ab}}.$$

On the other hand, from the Boltzmann equation we must have

$$\lim_{T \to \infty} \frac{N_a}{N_b} = \frac{g_a}{g_b},$$

so that one gets

$$B_{ba} = \frac{g_a}{g_b} B_{ab}.$$

In general, for an arbitrary T, substituting for N_a/N_b the value given by the Boltzmann equation, writing B_{ba} in terms of B_{ab} and recalling that at thermodynamic equilibrium the average intensity of the radiation field is given by the Planck function ($J_{\nu_{ab}} = B_{\nu_{ab}}(T)$, with $B_\nu(T)$ given by Eq. (10.11)), we obtain, with simple algebra

$$A_{ab} = \frac{2h\nu_{ab}^3}{c^2} B_{ab}.$$

With this thermodynamic reasoning we can then establish the correct relation between the Einstein coefficients. The above equations in fact contain the same results, already expressed in Eqs. (11.27) and (11.28), which we obtained using the principles of quantum electrodynamics. However, we must note that thermodynamics alone is not sufficient to determine the explicit expression of the Einstein coefficients, similarly to what we saw in the previous chapter for the constant a that appears in the Stefan law (Eq. (10.9)).

11.8 The Radiative Transfer Equation

Referring to the same physical system that we introduced at the beginning of Sect. 11.6, we now consider the temporal evolution of the number of photons contained in a mode of the radiation field characterized by the frequency v and direction $\mathbf{\Omega}$. By means of Eqs. (11.24) and (11.26) that express, respectively, the probability that a photon is added to or subtracted from the same mode, we obtain

$$\frac{dn_v(\mathbf{\Omega})}{dt} = -\pi^{(a)}(v, \mathbf{\Omega}) + \pi^{(e)}(v, \mathbf{\Omega}),$$

or

$$\frac{dn_v(\mathbf{\Omega})}{dt} = \frac{8\pi^3 e_0^2 v_{ab}}{3h} |\mathbf{r}_{ba}|^2 \delta(v - v_{ab})\left\{-\mathcal{N}_b g_a n_v(\mathbf{\Omega}) + \mathcal{N}_a g_b [1 + n_v(\mathbf{\Omega})]\right\}.$$

The time derivative that appears in this equation must be interpreted as a total (or Eulerian) derivative, since it expresses the change in the number of photons in time as we follow the photons in their movement. If we denote by s the spatial coordinate measured along the direction of propagation and we consider that the radiation propagates with velocity c, we have

$$\frac{d}{dt} = \frac{\partial}{\partial t} + c\frac{d}{ds}.$$

Performing this substitution and assuming to be in stationary conditions (which is equivalent to neglect the term $\partial/(\partial t)$), we obtain

$$\frac{dn_v(\mathbf{\Omega})}{ds} = \frac{8\pi^3 e_0^2 v_{ab}}{3hc} |\mathbf{r}_{ba}|^2 \delta(v - v_{ab})\left\{-\mathcal{N}_b g_a n_v(\mathbf{\Omega}) + \mathcal{N}_a g_b [1 + n_v(\mathbf{\Omega})]\right\}.$$

The equation that we have found is the so-called radiative transfer equation. Consistently with our hypotheses, it was obtained for an atom with two levels, an upper level a and a lower level b.

The transfer equation is traditionally formulated in a different way by referring to the specific intensity of the radiation field instead of the number of photons per mode. To obtain the standard formulation, it is sufficient to recall Eq. (4.15), which expresses the relation between these two quantities. Multiplying the above equation by the factor $(2hv^3/c^2)$ and rearranging the various terms, we obtain with some simple algebra

$$\frac{d}{ds} I_v(\mathbf{\Omega}) = -k_v^{(a)} I_v(\mathbf{\Omega}) + k_v^{(s)} I_v(\mathbf{\Omega}) + \epsilon_v,$$

where the three quantities $k_v^{(a)}$, $k_v^{(s)}$, and ϵ_v, respectively called the absorption coefficient, the coefficient of stimulated emission (or negative absorption), and the emission coefficient, are given by

$$k_v^{(a)} = \mathcal{N}_b \frac{8\pi^3 e_0^2 v_{ab}}{3hc} g_a |\mathbf{r}_{ba}|^2 \delta(v - v_{ab}),$$

$$k_v^{(s)} = \mathcal{N}_a \frac{8\pi^3 e_0^2 v_{ab}}{3hc} g_b |\mathbf{r}_{ba}|^2 \delta(v - v_{ab}), \qquad (11.30)$$

$$\epsilon_v = \mathcal{N}_a \frac{16\pi^3 e_0^2 v_{ab}^4}{3c^3} g_b |\mathbf{r}_{ba}|^2 \delta(v - v_{ab}).$$

The expressions that we found for the three coefficients are valid for a two-level atom (the " upper" level a and the "lower" level b). Their generalisation to the case of an atom with many levels is trivial and is simply obtained by summing over all possible pairs of levels. For a given level n, denoting as in Fig. 11.3 by the index i any level with lower energy and by the index s any level of higher energy and summing over n we have

$$k_\nu^{(a)} = \sum_n \sum_s \mathcal{N}_n \frac{8\pi^3 e_0^2 \nu_{sn}}{3hc} g_s |\mathbf{r}_{ns}|^2 \delta(\nu - \nu_{sn}),$$

$$k_\nu^{(s)} = \sum_n \sum_i \mathcal{N}_n \frac{8\pi^3 e_0^2 \nu_{ni}}{3hc} g_i |\mathbf{r}_{in}|^2 \delta(\nu - \nu_{ni}), \qquad (11.31)$$

$$\epsilon_\nu = \sum_n \sum_i \mathcal{N}_n \frac{16\pi^3 e_0^2 \nu_{ni}^4}{3c^3} g_i |\mathbf{r}_{in}|^2 \delta(\nu - \nu_{ni}).$$

We finally remark that the transfer equation can be written in the simplified form

$$\frac{d}{ds} I_\nu(\mathbf{\Omega}) = -k_\nu I_\nu(\mathbf{\Omega}) + \epsilon_\nu,$$

where

$$k_\nu = k_\nu^{(a)} - k_\nu^{(s)}.$$

The coefficient k_ν introduced in this way is called absorption coefficient corrected for stimulated emission.[8] Putting

$$S_\nu = \frac{\epsilon_\nu}{k_\nu} = \frac{\epsilon_\nu}{k_\nu^{(a)} - k_\nu^{(s)}},$$

the transfer equation assumes the form

$$\frac{d}{ds} I_\nu(\mathbf{\Omega}) = -k_\nu \left[I_\nu(\mathbf{\Omega}) - S_\nu \right]. \qquad (11.32)$$

The S_ν function is known as the source function. As we shall see in the next Section, an atomic system in thermodynamic equilibrium at the temperature T satisfies the so-called Kirchhoff law

$$S_\nu = B_\nu(T),$$

where $B_\nu(T)$ is the Planck function.

11.9 The Absorption and Emission Coefficients

A simple dimensional analysis of the transfer equation shows that the absorption coefficient and that of stimulated emission have the dimensions of the reciprocal of a length and are thus expressed (in the cgs system of units) in cm^{-1}. The emission

---

[8]It is often improperly called absorption coefficient *tout court*. In reality, this latter name should be reserved for $k_\nu^{(a)}$.

coefficient has instead the dimensions of energy per unit volume and per unit time and is expressed (still in cgs units) in erg cm^{-3} s^{-1}. Integrating in frequency, the contribution of each atomic transition to these coefficients can be related to the Einstein coefficients previously introduced. Referring to the case of a two-level atom and integrating in frequency Eqs. (11.30), we have

$$k_R^{(a)} = \int k_\nu^{(a)} d\nu = \mathcal{N}_b \frac{8\pi^3 e_0^2 \nu_{ab}}{3hc} g_a |\mathbf{r}_{ba}|^2,$$

$$k_R^{(s)} = \int k_\nu^{(s)} d\nu = \mathcal{N}_a \frac{8\pi^3 e_0^2 \nu_{ab}}{3hc} g_b |\mathbf{r}_{ba}|^2, \qquad (11.33)$$

$$\epsilon_R = \int \epsilon_\nu d\nu = \mathcal{N}_a \frac{16\pi^3 e_0^2 \nu_{ab}^4}{3c^3} g_b |\mathbf{r}_{ba}|^2.$$

Comparing these expressions with those of the Einstein coefficients given by Eqs. (11.20), (11.27) and (11.28), we have

$$k_R^{(a)} = \frac{h\nu_{ab}}{4\pi} \mathcal{N}_b B_{ba}, \qquad k_R^{(s)} = \frac{h\nu_{ab}}{4\pi} \mathcal{N}_a B_{ab}, \qquad \epsilon_R = \frac{h\nu_{ab}}{4\pi} \mathcal{N}_a A_{ab}.$$

These equations express the obvious relations that should exist between quantities that appear in the statistical equilibrium equations and the quantities that appear in the transfer equation. Considering for example the emission coefficient, we obviously have that the energy emitted per unit time and per unit volume is given by the number of atoms per unit of volume in the upper level $\mathcal{N}_a$ multiplied by the de-excitation probability of the atom per unit time A_{ab} and by the energy $h\nu_{ab}$ emitted in the transition. The factor 4π in the denominator is due to the fact that the emission coefficient is defined per unit solid angle, while the product of the three previous terms gives the energy emitted in the whole solid angle.

In general, the coefficients of absorption and emission are expressed by a double sum over all atomic levels (see Eqs. (11.31)). Since the levels can be either free or bound, the contributions to the individual coefficients are, as shown in Fig. 11.4: (a) bound-bound transitions; (b) bound-free (or free-bound) transitions; and (c) free-free transitions. The bound-free (and free-free) transitions produce an absorption coefficient that behaves as a continuous function in frequency. Indeed, if we consider the contribution to the absorption coefficient due to the transition between the bound state b and the continuum state c, we can transform the sum over the upper states s that appears in the expression for $k_\nu^{(a)}$ within Eqs. (11.31) in a sum over the free states. Denoting by dn_c the number of free states with energy between ϵ and $\epsilon + d\epsilon$, we have

$$dn_c = D_c(\epsilon) d\epsilon,$$

where $D_c(\epsilon)$ is the energy density of the continuum states. For the absorption coefficient at frequency ν we then have

$$k_\nu^{(a)} = \mathcal{N}_b \frac{8\pi^3 e_0^2 \nu}{3hc} \int |\mathbf{r}_{bc}|^2 \delta(\nu_{cb} - \nu) D_c(\epsilon) d\epsilon.$$

Fig. 11.4 In this schematic
Grotrian diagram, the free
states are represented by the
shaded area. The possible
transitions are:
(**a**) bound-bound transitions;
(**b**) bound-free transitions;
(**c**) free-free transitions

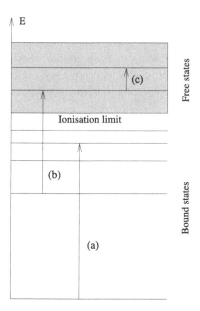

Taking into account the Dirac delta, the integral is performed immediately and leads
to the expression

$$k_\nu^{(a)} = \mathcal{N}_b \frac{8\pi^3 e_0^2 \nu}{3c} |\mathbf{r}_{bc}|^2 D_c(\epsilon_b + h\nu),$$

which can be written in the form

$$k_\nu^{(a)} = \sigma_\nu \mathcal{N}_b, \qquad (11.34)$$

where σ_ν, the cross section for photoionisation from level b, is given by

$$\sigma_\nu = \frac{8\pi^3 e_0^2 \nu}{3c} |\mathbf{r}_{bc}|^2 D_c(\epsilon_b + h\nu).$$

The function σ_ν has a continuous variation with frequency. It has a threshold at the
frequency ν_s such that

$$\epsilon_b + h\nu_s = I,$$

where I is the ionisation potential of the atom. For frequencies $\nu < \nu_s$ the cross
section for photoionisation is zero (photoelectric effect).

For free-free transitions we obtain an expression of the absorption coefficient that
is entirely analogous to the one derived above, with the only difference that now $\mathcal{N}_b$
represents the number of atoms per unit of volume that are present in the free level b.

Finally, regarding the bound-bound transitions, the formalism that we have developed leads to a discontinuous behaviour of the absorption coefficient of the type
of a "comb" of Dirac delta functions. This result originates from our assumption
that the atoms are isolated (i.e. are not interacting with other "perturbing" particles)
and static (i.e. are not moving). Furthermore, in our derivation we stopped at the

second order in the perturbative expansion of the probability amplitude. As we shall see in the next section, when we abandon these approximations we obtain that the Dirac delta functions must be replaced by suitable line profiles (normalised to 1 in frequency) that we denote by the symbol φ. For example, for the transition between the levels a and b, the various coefficients can be written in the form[9]

$$k_{\nu}^{(a)} = k_{R}^{(a)} \varphi(\nu_{ab} - \nu), \qquad k_{\nu}^{(s)} = k_{R}^{(s)} \varphi(\nu_{ab} - \nu), \qquad \epsilon_{\nu} = \epsilon_{R}\varphi(\nu_{ab} - \nu),$$
(11.35)

with

$$\int \varphi(\nu_{ab} - \nu)\mathrm{d}\nu = 1.$$

With these expressions, the source function in the line can be written in the form

$$S_{\nu} = \frac{\epsilon_{\nu}}{k_{\nu}^{(a)} - k_{\nu}^{(s)}} = \frac{\epsilon_{R}}{k_{R}^{(a)} - k_{R}^{(s)}},$$

and recalling the expressions for the coefficients ϵ_R, $k_R^{(a)}$ and $k_R^{(s)}$

$$S_{\nu} = \frac{\mathcal{N}_a A_{ab}}{\mathcal{N}_b B_{ba} - \mathcal{N}_a B_{ab}}.$$

Finally, recalling the relations among the Einstein coefficients, we obtain

$$S_{\nu} = \frac{2h\nu^3}{c^2} \frac{1}{\frac{g_a \mathcal{N}_b}{g_b \mathcal{N}_a} - 1}.$$
(11.36)

At the thermodynamic equilibrium at temperature T the populations are given by the Boltzmann equation (Eq. (10.6)) and the source function becomes

$$S_{\nu} = \frac{2h\nu^3}{c^2} \frac{1}{e^{h\nu_{ab}/(k_B T)} - 1},$$

which practically coincides with the Planck function $B_{\nu}(T)$ being $\nu_{ab} \simeq \nu$. This results shows the validity of the Kirchhoff law in the particular case of a two level atom.

11.10 Profile of the Emission Coefficient

The profile of the emission (or absorption) coefficient due to an atomic transition is not infinitely narrow because there are several causes that contribute to its broadening. Excluding the effects due to the presence of any external electromagnetic

---

[9]In principle, the profiles for emission and stimulated emission can be different from each other and different from that one relative to absorption. The hypothesis to consider them equal, contained in Eqs. (11.35), is known as the approximation of total redistribution in frequency. For an in-depth discussion of this topic see, for example, Mihalas (1978).

fields (Zeeman effect, Stark effect, etc.), and the effect due to hyperfine structure, we have generally three different types of broadening, namely: natural broadening, collisional broadening, and broadening due to thermal motions, also called Doppler broadening.

The natural broadening is ultimately related to the uncertainty principle. Consider for example the transition between the upper level a and the lower level b. If A_{ab} is the Einstein coefficient for spontaneous de-excitation, the average lifetime of the upper level is given by $1/A_{ab}$. For the time-energy uncertainty principle, we then have that the energy of such level presents a broadening ΔE given by

$$\Delta E \frac{1}{A_{ab}} \simeq h,$$

where h is Planck's constant. The transition has therefore a broadening in frequency $\Delta \nu$ given by

$$\Delta \nu = \frac{\Delta E}{h} \simeq A_{ab}.$$

An additional broadening mechanism is caused by the fact that during the emission process the atom is perturbed by collisions with other particles. Although there are various quantum theories that treat these phenomena in a more or less satisfactory way, here we prefer to infer the line profiles resulting from the natural and collisional broadening using a semi-classical, simplified argument. For this we assume that the emitting atom is an harmonic oscillator of frequency ν_0 and that its amplitude of oscillation decays exponentially over time with the decay constant β. If the oscillator is not perturbed by collisions, it emits an electric field of the form

$$E(t) = A e^{-\beta t} \cos(2\pi \nu_0 t - \phi) \quad (t \geq 0),$$

where A and ϕ are the amplitude and the phase of the oscillation. The effect of collisions with electrons (or with other charged particles present in the medium) can be modelled considering that collisions are processes that take place instantaneously (impact approximation) and that each collision introduces in the electric field emitted by the atom a phase shift of random amplitude. As a result of collisions, the electric field can therefore be described by the function

$$E(t) = A e^{-\beta t} \cos\left[2\pi \nu_0 t - \phi(t)\right] \quad (t \geq 0),$$

where $\phi(t)$ is a function having a stochastic character with discontinuities at each instant at which a collision occurs. A possible realisation of the function $E(t)$ is shown in Fig. 11.5.

The spectrum of the radiation emitted by the oscillator is obtained, as we saw in Chap. 2, evaluating the Fourier transform of the function $E(t)$. Introducing the frequency ν in place of the angular frequency ω, we have, from Eq. (2.1)

$$\hat{E}(\nu) = \frac{1}{2\pi} \int_{-\infty}^{\infty} E(t) e^{2\pi i \nu t} dt.$$

Fig. 11.5 Variation with time
of the function $E(t)$ for a
particular realisation of the
collisional processes. In this
particular case, there are
about 10 collisions in 5
oscillation periods of the
electric field. Both the time t
and the function $E(t)$ are
expressed in arbitrary units

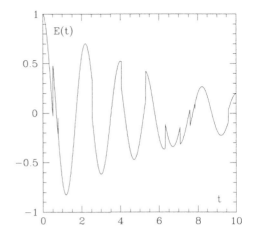

Substituting the expression of $E(t)$ and considering only the Fourier transform at
positive frequencies, we have

$$\hat{E}(\nu) = \frac{A}{4\pi} \int_0^\infty e^{2\pi i(\nu - \nu_0)t - \beta t + i\phi(t)} dt.$$

For the square of the modulus of the Fourier transform we thus have

$$\left|\hat{E}(\nu)\right|^2 = \frac{A^2}{16\pi^2} \int_0^\infty dt_1 \int_0^\infty dt_2 e^{2\pi i(\nu - \nu_0)(t_1 - t_2) - \beta(t_1 + t_2) + i[\phi(t_1) - \phi(t_2)]}.$$

The quantity $\Delta\phi = \phi(t_1) - \phi(t_2)$ has a stochastic character. If during the time interval between t_1 and t_2 (or between t_2 and t_1) no collision occur, $\Delta\phi = 0$ and we
have

$$e^{i\Delta\phi} = 1.$$

If instead many collisions occur within the same interval, taking into account the
fact that the phases introduced by individual collisions are random, we obtain, by
averaging over all possible "collisional histories",

$$e^{i\Delta\phi} = 0.$$

Denoting by f the frequency of collisions (number of collisions per unit time), we
can reasonably assume that the following approximation holds

$$e^{i\Delta\phi} = e^{-f|t_1 - t_2|}.$$

Substituting this expression in the integral we thus have

$$\left|\hat{E}(\nu)\right|^2 = \frac{A^2}{16\pi^2} \int_0^\infty dt_1 \int_0^\infty dt_2 e^{2\pi i(\nu - \nu_0)(t_1 - t_2) - \beta(t_1 + t_2) - f|t_1 - t_2|}.$$

The double integral appearing in this expression can be computed by distinguishing
in the (t_1, t_2) plane two regions, A and B, as illustrated in Fig. 11.6. Noting that the
integral relative to the region B can be obtained from that one relative to region A
by changing sign to the quantity $(\nu - \nu_0)$, we obtain

Fig. 11.6 Region A is that one where $t_1 < t_2$, while in region B we have $t_2 < t_1$

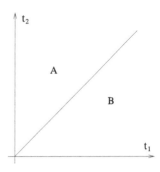

$$\left|\hat{E}(\nu)\right|^2 = \mathcal{A} + \mathcal{B},$$

where

$$\mathcal{A} = \frac{A^2}{16\pi^2} \int_0^\infty dt_1 \int_{t_1}^\infty dt_2 e^{2\pi i(\nu-\nu_0)(t_1-t_2)-\beta(t_1+t_2)-f(t_2-t_1)},$$

and where $\mathcal{B}$ is obtained from $\mathcal{A}$ with the substitution $(\nu - \nu_0) \to (\nu_0 - \nu)$. The double integral is calculated with elementary methods, and adding $\mathcal{B}$ to $\mathcal{A}$, we get

$$\left|\hat{E}(\nu)\right|^2 = \frac{A^2}{64\pi^4} \frac{\beta + f}{\beta} \frac{1}{(\nu - \nu_0)^2 + (\frac{\beta+f}{2\pi})^2}.$$

The quantity β is in general expressed in the form $\beta = \gamma/2$, so that γ represents the decay constant of the square amplitude of the dipole oscillation and can therefore be interpreted as the inverse of the average lifetime of the upper level ($\gamma = A_{ab}$). Introducing γ we then have

$$\left|\hat{E}(\nu)\right|^2 = \frac{A^2}{64\pi^4} \frac{\gamma + 2f}{\gamma} \frac{1}{(\nu - \nu_0)^2 + (\frac{\gamma+2f}{4\pi})^2},$$

and, putting

$$\Gamma = \frac{\gamma + 2f}{4\pi},$$

we obtain the expression for the line profile, normalised in frequency

$$\phi(\nu - \nu_0) = \frac{1}{\pi} \frac{\Gamma}{(\nu - \nu_0)^2 + \Gamma^2}.$$

This function that describes the variation of the profile with frequency is called the Lorentz function (or Lorentzian function). The quantity Γ, called damping constant, contains both a natural and a collisional contribution. It can be written in the form

$$\Gamma = \Gamma_n + \Gamma_c,$$

where

$$\Gamma_n = \frac{\gamma}{4\pi}, \qquad \Gamma_c = \frac{f}{2\pi}.$$

As we already stated, there is a further cause of broadening of the emission (or absorption) coefficient due to the thermal movement of the atoms. We denote by $P(w)dw$ the probability that the component of the velocity of the atom along the direction of the emitted radiation is comprised between w and $w + dw$. By means of Eq. (10.4), this probability can be expressed in the form

$$P(w) = \frac{1}{\sqrt{\pi}\,w_{\mathrm{T}}} e^{-(w/w_{\mathrm{T}})^2},$$

where we have introduced the thermal velocity w_{T} defined by the equation

$$w_{\mathrm{T}} = \sqrt{\frac{2k_{\mathrm{B}}T}{M}},$$

M being the atom mass. Because of the Doppler effect, an atom moving with a velocity component w has (at the lowest relativistic order) an emission profile centred around the frequency ν_0' given by

$$\nu_0' = \nu_0\left(1 + \frac{w}{c}\right).$$

The emission profile due to the collection of atoms is then given by

$$\varphi(\nu - \nu_0) = \int_{-\infty}^{\infty} \frac{1}{\pi} \frac{\Gamma}{(\nu - \nu_0 - \nu_0 w/c)^2 + \Gamma^2} \frac{1}{\sqrt{\pi}\,w_{\mathrm{T}}} e^{-(w/w_{\mathrm{T}})^2} dw.$$

This expression is commonly simplified by introducing the quantities

$$\Delta\nu_{\mathrm{D}} = \nu_0 \frac{w_{\mathrm{T}}}{c}, \qquad a = \frac{\Gamma}{\Delta\nu_{\mathrm{D}}}, \qquad v = \frac{\nu - \nu_0}{\Delta\nu_{\mathrm{D}}},$$

which represent, respectively, the Doppler width of the emission coefficient (in frequency units), the reduced damping constant, and the distance in frequency from the center of the line normalised to the Doppler width. With the change of variable $y = w/w_{\mathrm{T}}$, the above integral can be put in the form

$$\varphi(\nu - \nu_0) = \frac{1}{\sqrt{\pi}\,\Delta\nu_{\mathrm{D}}} H(v, a), \tag{11.37}$$

where the function $H(v, a)$, known as the Voigt function, is defined by

$$H(v, a) = \frac{a}{\pi} \int_{-\infty}^{\infty} \frac{e^{-y^2}}{(v - y)^2 + a^2} dy.$$

The Voigt function has some properties which can be deduced from its general expression:

$$\int_{-\infty}^{\infty} H(v, a)dv = \sqrt{\pi}, \qquad \lim_{a \to 0} H(v, a) = e^{-v^2},$$

$$\lim_{a \to \infty} H(v, a) = \frac{1}{\sqrt{\pi}} \frac{a}{v^2 + a^2}.$$

Fig. 11.7 Plot of the Voigt
function $H(v, a)$ for $a = 0$
(*full line*), $a = 0.2$ (*dotted
line*) and $a = 1$ (*dashed line*)

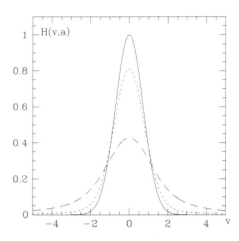

The first property allows one to show with simple algebra that the profile $\varphi(v - v_0)$
is normalised to 1 in frequency

$$\int_{-\infty}^{\infty} \varphi(v - v_0)\mathrm{d}v = 1.$$

The other two properties show that, in the limiting case of negligible damping, the
Voigt function takes the form of a Gaussian. In the opposite limit, in which thermal
broadening is negligible, the Voigt function degenerates into a Lorentzian. In general, the Voigt function is similar to a Gaussian around $v = 0$ and to a Lorentzian in
the wings. Typical examples are shown in Fig. 11.7.

Chapter 12
Selection Rules and Line Strengths

The results presented in the previous chapter show that, due to the effect of the interaction with the radiation field, the transition probability between two energy levels of an atomic system is, in first approximation, proportional to the square of the magnitude of the dipole matrix element evaluated between the atomic eigenfunctions relative to the levels themselves. In this chapter we will see how this property leads in a natural way to the various selection rules that we presented previously, and we will discuss their limits of validity. We will also see how we can evaluate in a quantitative way the relative strengths of the various lines belonging to a fine-structure multiplet that originates from the transitions between two terms. In Sect. 12.5 we present an application regarding the physical principles that justify the presence of forbidden lines in the spectra of some astronomical objects of particular relevance (solar corona, gaseous nebulae, planetary nebulae, etc.).

12.1 Selection Rules for the Quantum Numbers

The transition probability between two quantum states described by the state vectors $|u_m\rangle$ and $|u_n\rangle$ is proportional, within the dipole approximation, to the squared modulus of the matrix element[1] $\langle u_m|\mathbf{r}|u_n\rangle$. Whenever such matrix element is null we have a so-called forbidden transition.

On the contrary, if the matrix element is not null, we have a so-called allowed transition. In almost all cases (with the exception of the Paschen-Back effect) the atomic eigenfunctions can be written in the form $|\alpha J M\rangle$, where α indicates a set of quantum numbers specifying the configuration and other internal physical quantities (such as the L and S values in the case of L-S coupling, or the j values of the single electrons in the case of j-j coupling), J is the eigenvalue of the total angular momentum (of the electrons), and M is the magnetic quantum number. The calculation

---

[1] Since $|\langle u_m|\mathbf{r}|u_n\rangle|^2 = |\langle u_n|\mathbf{r}|u_m\rangle|^2$, the transition probability does not depend on which of the two quantum states appears in the bra or in the ket.

E. Landi Degl'Innocenti, *Atomic Spectroscopy and Radiative Processes*,
UNITEXT for Physics, DOI 10.1007/978-88-470-2808-1_12, © Springer-Verlag Italia 2014

of the matrix element between two such states can easily be done if we consider, instead of the Cartesian components of the vector $\mathbf{r}$, its spherical components, defined in Eq. (9.3). By applying the Wigner-Eckart theorem in the form of Eq. (9.4) we have

$$\langle \alpha J M | r_q | \alpha' J' M' \rangle = (-1)^{J'+M+1} \sqrt{2J+1} \begin{pmatrix} J & J' & 1 \\ -M & M' & q \end{pmatrix} \langle \alpha J \| \mathbf{r} \| \alpha' J' \rangle,$$

where $\langle \alpha J \| \mathbf{r} \| \alpha' J' \rangle$ is the reduced matrix element of the dipole operator. The matrix element is not null (hence the transition is allowed) only if the triangular inequality between the angular momenta of the first row in the 3-j symbol is satisfied, and if the sum of the symbols in the second row is zero. These selection rules must therefore apply

$$\Delta J = \pm 1, 0, \qquad J = 0 \nrightarrow J' = 0, \qquad \Delta M = \pm 1, 0.$$

Whenever the atomic states are described in the L-S coupling scheme, the matrix element that needs to be calculated is

$$\langle \alpha L S J M | r_q | \alpha' L' S' J' M' \rangle.$$

Changing the basis, the eigenvectors $|\alpha L S J M \rangle$ and $|\alpha' L' S' J' M' \rangle$ can be expressed, through suitable Clebsh-Gordan coefficients, in the form

$$|\alpha L S J M \rangle = \sum_{M_L M_S} \langle L S M_L M_S | L S J M \rangle |\alpha L S M_L M_S \rangle,$$

$$|\alpha' L' S' J' M' \rangle = \sum_{M'_L M'_S} \langle L' S' M'_L M'_S | L' S' J' M' \rangle |\alpha' L' S' M'_L M'_S \rangle.$$

Substituting we obtain

$$\langle \alpha L S J M | r_q | \alpha' L' S' J' M' \rangle$$
$$= \sum_{M_L M_S} \sum_{M'_L M'_S} \langle L S M_L M_S | L S J M \rangle$$
$$\times \langle L' S' M'_L M'_S | L' S' J' M' \rangle \langle \alpha L S M_L M_S | r_q | \alpha' L' S' M'_L M'_S \rangle. \qquad (12.1)$$

The operator r_q acts only on the orbital coordinates and not on the spin. The matrix element of r_q in the right-hand side is then given by the expression

$$\langle \alpha L S M_L M_S | r_q | \alpha' L' S' M'_L M'_S \rangle$$
$$= \langle \alpha L M_L | r_q | \alpha' L' M'_L \rangle \langle S M_S | S' M'_S \rangle = \langle \alpha L M_L | r_q | \alpha' L' M'_L \rangle \delta_{SS'} \delta_{M_S M'_S},$$

and, using again the Wigner-Eckart theorem,

$$\langle \alpha L S M_L M_S | r_q | \alpha' L' S' M'_L M'_S \rangle$$
$$= (-1)^{L'+M_L+1} \sqrt{2L+1} \begin{pmatrix} L & L' & 1 \\ -M_L & M'_L & q \end{pmatrix} \langle \alpha L \| \mathbf{r} \| \alpha' L' \rangle \delta_{SS'} \delta_{M_S M'_S}. \qquad (12.2)$$

Substituting this expression in Eq. (12.1), we see that the matrix element is not zero only if these selection rules are satisfied

$$\Delta L = \pm 1, 0, \quad L = 0 \nrightarrow L' = 0, \quad \Delta S = 0, \quad \Delta M_L = \pm 1, 0, \quad \Delta M_S = 0.$$

Finally, for an atom with hyperfine structure, we need to consider the matrix element

$$\langle \alpha J I F M_F | r_q | \alpha' J' I F' M_F' \rangle.$$

With similar steps as described above, we easily obtain the selection rules

$$\Delta F = \pm 1, 0, \qquad F = 0 \nrightarrow F' = 0, \qquad \Delta M_F = \pm 1, 0.$$

12.2 Selection Rules for the Configurations

Consider two distinct configurations of a given atom. As we have seen in Sects. 7.1 and 7.6, one of the g degenerate states of a given configuration can be described by a wavefunction of the type detailed in Eq. (7.1), i.e.

$$\Psi^A(a_1, a_2, \ldots, a_N) = \frac{1}{\sqrt{N!}} \sum_P (-1)^P P \big[\psi_{a_1}(x_1) \psi_{a_2}(x_2) \cdots \psi_{a_N}(x_N) \big],$$

where $(a_1, a_2, \ldots, a_N)$ is a set of quantum numbers that specify the single particle states. The dipole matrix element between two given states (the first belonging to a configuration and the second to the other) is therefore given by

$$\langle \Psi^A(a_1, a_2, \ldots, a_N) | \mathbf{r} | \Psi^A(a_1', a_2', \ldots, a_N') \rangle$$
$$= \frac{1}{N!} \sum_P \sum_Q \sum_{i=1}^N (-1)^{P+Q}$$
$$\times \langle P\big[\psi_{a_1}(x_1) \psi_{a_2}(x_2) \cdots \psi_{a_N}(x_N) \big] | \mathbf{r}_i | Q\big[\psi_{a_1'}(x_1) \psi_{a_2'}(x_2) \cdots \psi_{a_N'}(x_N) \big] \rangle.$$

This sum contains $N \times (N!)^2$ terms. The generic term is proportional to the product

$$\langle \psi_{\bar{a}_1} | \psi_{\bar{a}_1'} \rangle \langle \psi_{\bar{a}_2} | \psi_{\bar{a}_2'} \rangle \cdots \langle \psi_{\bar{a}_i} | \mathbf{r}_i | \psi_{\bar{a}_i'} \rangle \cdots \langle \psi_{\bar{a}_N} | \psi_{\bar{a}_N'} \rangle,$$

being $(\bar{a}_1, \bar{a}_2, \ldots, \bar{a}_N)$ a given permutation of the set $(a_1, a_2, \ldots, a_N)$ and $(\bar{a}_1', \bar{a}_2', \ldots, \bar{a}_N')$ a given permutation of the other set $(a_1', a_2', \ldots, a_N')$. This term is zero unless all the sets $\bar{a}_k$ are equal to the corresponding $\bar{a}_k'$, except for $\bar{a}_i$, which can be different from $\bar{a}_i'$. So we must have

$$\bar{a}_1 = \bar{a}_1', \bar{a}_2 = \bar{a}_2', \ldots, \bar{a}_{i-1} = \bar{a}_{i-1}', \bar{a}_{i+1} = \bar{a}_{i+1}', \ldots, \bar{a}_N = \bar{a}_N'.$$

In order for the matrix element $\langle \Psi^A(a_1, a_2, \ldots, a_n) | \mathbf{r} | \Psi^A(a_1', a_2', \ldots, a_n') \rangle$ to be different from zero, we must then have that $(N-1)$ of the sets belonging to the ensemble $(a_1, a_2, \ldots, a_N)$ coincide with $(N-1)$ of the sets belonging to the other ensemble $(a_1', a_2', \ldots, a_N')$. This implies that if k is the index for which the a_k set does not have the corresponding one among the a_j', the wavefunction of the second state must have the form

$$\Psi^A(a_1', a_2', \ldots, a_N') = \pm \Psi^A(a_1, a_2, \ldots, a_k', \ldots, a_N),$$

where the sign factor $\pm$ is related to the number (even or odd) of exchanges that are needed to order the set $(a_1', a_2', \ldots, a_N')$ in the required form. We therefore have (when the matrix element is not null)

$$
\langle \Psi^A(a_1, a_2, \ldots, a_N) | \mathbf{r} | \Psi^A(a_1', a_2', \ldots, a_n') \rangle
$$
$$
= \pm \langle \Psi^A(a_1, \ldots, a_k, \ldots, a_N) | \mathbf{r} | \Psi(a_1, \ldots, a_k', \ldots, a_N) \rangle
$$
$$
= \pm \frac{1}{N!} \sum_P \sum_Q \sum_{i=1}^{N} (-1)^{P+Q}
$$
$$
\times \langle P[\psi_{a_1}(x_1) \cdots \psi_{a_k}(x_k) \cdots \psi_{a_N}(x_N)] | \mathbf{r}_i |
$$
$$
Q[\psi_{a_1}(x_1) \cdots \psi_{a_k'}(x_k) \cdots \psi_{a_N}(x_N)] \rangle.
$$

In order that the single summand of this sum is not zero, we must have that Q is the same permutation as P and also that, for a given i, the permutation is such as to bring the two unmatched sets of quantum numbers, a_k and a_k', at the i-th position. By so doing we obtain $N(N-1)! = N!$ terms all contributing the same amount, and we have

$$
\langle \Psi^A(a_1, a_2, \ldots, a_N) | \mathbf{r} | \Psi^A(a_1', a_2', \ldots, a_n') \rangle = \pm \langle \psi_{a_k}(x_k) | \mathbf{r}_k | \psi_{a_k'}(x_k) \rangle,
$$

a formula which shows that the matrix element between the two configurations is simply given by the single-particle matrix element evaluated for the single electron that "jumps" from the state a_k to the state a_k'. The dipole transitions are therefore possible only between two configurations that differ in the quantum numbers of one and only one electron. In addition, for the transition to be allowed, it is necessary that, denoting by (n, l, m, m_s) and (n', l', m', m_s') the sets of the quantum numbers of such electron in the initial and final states of the transition, we must have $\mathcal{I} \neq 0$, where

$$
\mathcal{I} = \delta_{m_s m_s'} \int \psi_{nlm}^*(\mathbf{r}) \mathbf{r} \psi_{n'l'm'}(\mathbf{r}) \, d^3 \mathbf{r}.
$$

Recalling the expression of the single-particle wavefunctions (Eq. (7.10)) and Eqs. (9.3), the spherical components $\mathcal{I}_q$ of this integral are

$$
\mathcal{I}_q = \delta_{m_s m_s'} \int_0^\infty P_{nl}(r) P_{n'l'}(r) r \, dr \oint_{4\pi} Y_{lm}^*(\theta, \phi) Y_{l'm'}(\theta, \phi) f_q(\theta, \phi) \, d\Omega,
$$

where $f_q (q = -1, 0, 1)$ are the three spherical components of the unit vector directed along $\mathbf{r}$, i.e.

$$
f_{-1} = \frac{1}{\sqrt{2}} \sin\theta (\cos\phi - i \sin\phi), \qquad f_0 = \cos\theta,
$$
$$
f_1 = -\frac{1}{\sqrt{2}} \sin\theta (\cos\phi + i \sin\phi).
$$

To calculate the angular part of the integral, we recall the explicit expressions of the spherical harmonics $Y_{1q}(\theta, \phi)$ (cf. Eq. (6.13)), for which we have

$$
f_q = \sqrt{\frac{4\pi}{3}} Y_{1q}(\theta, \phi),
$$

and we use the Weyl theorem (Eq. (8.8)) which expresses the integral over the solid angle of the product of three spherical harmonics. Recalling also the conjugation property of the spherical harmonics (Eq. (6.11)), we obtain

$$\oint_{4\pi} Y^*_{lm}(\theta,\phi) Y_{l'm'}(\theta,\phi) f_q(\theta,\phi) \, d\Omega = (-1)^m \sqrt{(2l+1)(2l'+1)}$$
$$\times \begin{pmatrix} l & l' & 1 \\ 0 & 0 & 0 \end{pmatrix} \begin{pmatrix} l & l' & 1 \\ -m & m' & q \end{pmatrix},$$

hence

$$\mathcal{I}_q = \delta_{m_s m'_s} (-1)^m \sqrt{(2l+1)(2l'+1)} \begin{pmatrix} l & l' & 1 \\ 0 & 0 & 0 \end{pmatrix} \begin{pmatrix} l & l' & 1 \\ -m & m' & q \end{pmatrix} \mathcal{R},$$

where $\mathcal{R}$ is the radial integral given by

$$\mathcal{R} = \int_0^\infty P_{nl}(r) P_{n'l'}(r) r \, dr.$$

The equation that we have obtained contains all the selection rules for transitions between configurations. In fact, if we recall the property of the 3-j symbol regarding the inversion of the sign of the quantum numbers appearing in the second row (see Eq. (7.19)), we have

$$\begin{pmatrix} l & l & 1 \\ 0 & 0 & 0 \end{pmatrix} = 0,$$

so that, from the Kronecker delta and the 3-j symbols contained in the expression of $\mathcal{I}_q$, we obtain the selection rules

$$\Delta l = \pm 1, \qquad \Delta m = \pm 1, 0, \qquad \Delta m_s = 0.$$

The dipole transitions are therefore possible only between configurations that differ in their quantum numbers l and l' by one and only one electron. These quantum numbers must in fact obey the relation $\Delta l = l - l' = \pm 1$.

If we consider the parity $\mathcal{P}$ of the two configurations, we see that, being $\mathcal{P} = (-1)^{\sum_i l_i}$, the parity of the two configurations for a dipole transition must be different. We have then obtained Laporte's rule

$$\text{even} \nrightarrow \text{even}, \qquad \text{odd} \nrightarrow \text{odd}.$$

Laporte's rule is very general, because it also applies when configuration interaction is present. In such cases a state of the atomic system is described with an eigenfunction that is a linear combination of eigenfunctions of the type $\psi^A(a_1, a_2, \ldots, a_n)$, all of the same parity. Laporte's rule is therefore valid also when configuration interaction is present.

12.3 Forbidden Transitions

The selection rules that we determined in the previous section apply, all and only, for electric dipole transitions. Any atomic system has, however, other types of electromagnetic multipoles (magnetic dipole, electric quadrupole, etc.) that we have neglected within the formalism developed in Chap. 11. Recall that, in addition to the dipole approximation (that is to perform the substitution $e^{i\mathbf{k}\cdot\mathbf{r}} \to 1$), we introduced two other approximations, by assuming for the atomic Hamiltonian its nonrelativistic expression (a first time to obtain the interaction Hamiltonian via the minimal coupling principle, and a second time to relate the matrix elements of the operator $\mathbf{p}$ to those of the operator $\mathbf{r}$). Since these two approximations neglect the relativistic corrections to the atomic Hamiltonian, it is to be expected that the higher-order multipole contributions play a more important role in complex atoms than in simple atoms.

To estimate these multipole contributions to the transition probability, we start by calculating the order of magnitude of the Einstein coefficient for spontaneous emission for an electric dipole transition. Recalling Eq. (11.20), we introduce two dimensionless quantities, ξ and ζ by putting

$$|\mathbf{r}_{ba}|^2 = \xi^2 a_0^2, \qquad \nu_{ab} = \zeta \frac{e_0^2}{2ha_0}.$$

In this way, the dipole matrix element is expressed in terms of the radius of the first Bohr orbit, while the frequency of the transition is expressed in terms of the frequency that corresponds to the ionization of the hydrogen atom. With simple transformations we obtain

$$A_{\text{e.d.}} = \frac{\alpha^4 c}{6a_0}\zeta^3\xi^2 g_b \simeq 2.677 \times 10^9 \zeta^3\xi^2 g_b \text{ s}^{-1},$$

where α is the fine structure constant. Since the dimensionless factor $\zeta^3\xi^2$ that appears in this equation is generally less than or of the order of 0.1, the Einstein coefficients for the spectral lines that fall in the visible region of the spectrum are typically of the order of 10^7–10^8 s^{-1}.

When the dipole matrix element between two states is null, the matrix element of the magnetic dipole can be non-zero, or the matrix element of the electric quadrupole can be non-zero, and so on. In these cases the transition is forbidden and the Einstein coefficient, which would be zero in the dipole approximation, is much lower, by some orders of magnitude, than the corresponding coefficient for an allowed transition. Without going into a formal derivation, but recalling the classical results that we obtained for the multipolar expansion in Sect. 3.10, we simply mention the fact that the Einstein coefficient for a magnetic dipole transition can be obtained using the same formula valid for the electric dipole, with the substitution of the electric dipole operator $e_0\mathbf{r}$ with the magnetic dipole operator $\boldsymbol{\mu}$. It follows that, as an order of magnitude, we have (with obvious notations)

$$\frac{A_{\text{m.d.}}}{A_{\text{e.d.}}} \simeq \left(\frac{\mu_0}{e_0 a_0}\right)^2 = \frac{\alpha^2}{4} = 1.331 \times 10^{-5},$$

where μ_0 is the so-called Bohr magneton defined in Eq. (5.17). Magnetic dipole transitions therefore have Einstein coefficients that are about 10^5 times lower than those of electric dipole transitions.

Regarding the electric quadrupole transitions, if we recall the series expansion of the exponential $e^{i\mathbf{k}\cdot\mathbf{r}}$ in Sect. 11.4 (in particular Eq. (11.11)), we have

$$\frac{A_{\text{e.q.}}}{A_{\text{e.d.}}} \simeq \left(\frac{a_0}{\lambda}\right)^2,$$

where λ is the wavelength of the transition. In the visible region of the spectrum, electric quadrupole transitions have Einstein coefficients that are about 10^8 lower than those of electric dipole transitions.

The selection rules for magnetic dipole and electric quadrupole transitions are different from those of electric dipole transitions. For the first ones, for example, the $\Delta S = 0$ selection rule is not valid anymore, as is Laporte's rule. For the electric quadrupole transitions, it is the selection rule on J that changes, since in this case we have $\Delta J = 0, \pm1, \pm2$, with the exclusion of the transitions $0 \rightarrow 0$, $0 \rightarrow 1$, $1 \rightarrow 0$, and $\frac{1}{2} \rightarrow \frac{1}{2}$, which are forbidden.

12.4 Semi-forbidden Transitions

The transitions that violate some of the selection rules that we have seen in the previous sections (12.1 and 12.2) due to the fact that the spectroscopic denomination that is assigned to one of the states (or both) is approximate are called semi-forbidden transitions. Consider for example a transition between two atomic states, the first of which is rigorously described by the L-S coupling scheme, while the other by the intermediate coupling scheme. The first state is then described by an eigenvector of the form $|\alpha'L'S'J'M'\rangle$, while the second is described by an eigenvector of the form (see Eq. (9.12) and the related discussion)

$$|\alpha J M\rangle = C_{L_0 S_0}|\alpha L_0 S_0 J M\rangle + \sum_{LS \neq L_0 S_0} C_{LS}|\alpha L S J M\rangle,$$

where L_0 and S_0 are the values of L and S of the (approximate) spectroscopic denomination and where the coefficients of the expansion are such that

$$|C_{LS}| \ll |C_{L_0 S_0}| \simeq 1.$$

If one of the selection rules on L or on S is violated, i.e. we have

$$\Delta L = L_0 - L' \neq \pm1, 0, \quad \text{or} \quad \Delta S = S_0 - S' \neq 0,$$

or if both are violated, the transition could be (naively) considered a forbidden transition, as it would effectively be when also the second state could be rigorously described in L-S coupling. Indeed, if we calculate the matrix element $|\langle\alpha'L'S'J'M'|\mathbf{r}|\alpha J M\rangle|^2$, we find that it can be non-zero since in the series expansion of $|\alpha J M\rangle$ there can be one or more state vectors $|\alpha L S J M\rangle$ for which we have

$$\Delta L = L - L' = \pm1, 0, \qquad \Delta S = S - S' = 0.$$

Fig. 12.1 Schematic
Grotrian diagram for the
ground configuration
$1s^2 2s^2 2p^2$ of doubly ionised
oxygen. The configuration
has three terms, ^1S, ^1D and
^3P (see Table 7.4), the last
being split into three
fine-structure levels. The two
N_1 and N_2 spectral lines are
the so-called forbidden
transitions of "Nebulium"

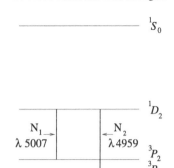

The matrix element is proportional to the corresponding quantity $|C_{LS}|^2$ and there-fore is, in general, much smaller than the matrix element relative to an allowed tran-sition. Typical values of the Einstein coefficients for such semi-forbidden transitions are of the order of 10^3–10^6 s^{-1}.

Similar considerations, developed here for the case of a state for which the in-termediate coupling applies, can be repeated, regarding the selection rules for the configurations, for states in which there is configuration interaction.

12.5 Forbidden Lines in Astronomical Objects

Some forbidden lines belonging to elements cosmically abundant are particularly prominent in the spectra of various astronomical objects, such as the solar corona, gaseous nebulae, planetary nebulae, H I regions, etc. A typical example are the two N_1 and N_2 lines of the [O III] spectrum,[2] which are shown in the schematic Grotrian diagram of Fig. 12.1. The two lines are both forbidden because they violate the selection rule on configurations (which states that the initial and final configurations must differ for the quantum numbers of one electron), the Laporte's rule, and finally also the $\Delta S = 0$ selection rule.

The fact that these lines are particularly prominent in the spectra of nebulae is intimately related with the very low density of these astronomical objects (from 10^{-20} to 10^{-17} g cm^{-3}) and can be adequately explained on the basis of a sim-plified model of the atom, such as that shown in Fig. 12.2, which reproduces the essential points of the physics of atomic excitation in nebulae. The atom can be ex-cited to a higher energy level (level a) from its ground state (level b) by absorption of radiation (typically ultraviolet radiation due to one of the hot young stars that light up the nebula). From level a, the atom can then return to the ground state or can decay to a metastable level m via an allowed transition, both processes taking

---

[2]The symbol [O III] is commonly used to indicate the spectrum of forbidden lines of twice ionised oxygen. This convention of encompassing the symbol of the element by a square bracket applies to the spectrum of forbidden lines of any element or ion.

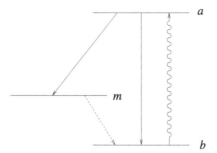

Fig. 12.2 Schematic atomic model showing the possibility to observe forbidden lines in nebular objects. The forbidden line originates from the transition between the metastable level m and the ground level b (*dashed line*). The atom is pumped by the radiation field into level a, from which it can spontaneously decay towards either level b or level m

place by spontaneous de-excitation. Finally, from level m, the atom can return to the ground state still by spontaneous de-excitation, but via a forbidden transition. Note that this simple model neglects the processes of stimulated emission and the absorption processes in the forbidden line and the subordinate line. This is justified as in nebulae the radiation field due to the hot central star is very diluted and, in addition, it is much more intense in the ultraviolet than in the visible.

By applying the statistical equilibrium equations (Eq. (11.29)) to the temporal evolution of the population of level a, we obtain, in stationary conditions, the following equation

$$\frac{dN_a}{dt} = N_b B_{ba} J_{\nu_{ab}} - N_a (A_{ab} + A_{am}) = 0,$$

where we have introduced, with obvious notation, the Einstein coefficients for the single transitions and the average intensity of the radiation field at the frequency of the transition between levels a and b. By solving the equation we get

$$N_a = \frac{B_{ba} J_{\nu_{ab}}}{A_{ab} + A_{am}} N_b.$$

For the evolution of the population of level m we have, still in stationary conditions

$$\frac{dN_m}{dt} = N_a A_{am} - N_m A_{mb} = 0, \tag{12.3}$$

from which we obtain

$$N_m = \frac{A_{am}}{A_{mb}} N_a.$$

Since the transition between levels m and b is forbidden, we have $A_{am} \gg A_{mb}$, hence $N_m \gg N_a$, i.e. the population of the metastable level is much larger than the one of the excited level a. The number of photons emitted per unit time in the transition between levels m and b is given by $N_m A_{mb}$, while the number of those emitted in the transition between levels a and m is given by $N_a A_{am}$. Denoting by $\mathcal{R}$ such ratio, we have

$$\mathcal{R} = \frac{N_m A_{mb}}{N_a A_{am}},$$

and taking into account the previous equation, we see that $\mathcal{R} = 1$. This means that the number of photons emitted in the forbidden transition is equal to the number of photons emitted in the allowed transition.

A more accurate calculation can be performed by taking into account the absorption in the subordinate line. This implies modifying Eq. (12.3) which assumes the form, with obvious notation

$$\frac{dN_m}{dt} = N_a A_{am} - N_m(A_{mb} + B_{ma} J_{v_{am}}) = 0.$$

Solving for the N_m/N_a ratio and substituting the result into the definition of $\mathcal{R}$ we obtain

$$\mathcal{R} = \left(1 + \frac{B_{ma} J_{v_{am}}}{A_{mb}}\right)^{-1},$$

or, taking into account the relations between the Einstein coefficients

$$\mathcal{R} = \left(1 + \frac{A_{am}}{A_{mb}} \frac{g_a}{g_m} \bar{n}_{v_{am}}\right)^{-1},$$

where $\bar{n}_{v_{am}}$ is the averaged number of photons per mode at frequency v_{am}.

Consider a hot star that illuminates with its radiation a nebula. The averaged number of photons per mode at the distance d from the star is given by

$$\bar{n}_{v_{am}} = \bar{n}_* \left(\frac{R_*}{d}\right)^2,$$

where $\bar{n}_*$ is the similar quantity at the surface of the star and where R_* is the stellar radius. Substituting, the expression for $\mathcal{R}$ can be written in the form

$$\mathcal{R} = \left[1 + \left(\frac{d_c}{d}\right)^2\right]^{-1},$$

where the "critical distance", d_c is given by

$$d_c = R_* \sqrt{n_* \frac{A_{am}}{A_{mb}} \frac{g_a}{g_m}}.$$

For example, if the central star has an effective temperature of 2×10^4 K, at typical frequencies of the visible one has $\bar{n}_* \simeq 0.3$, and assuming a ratio of 10^8 between the two Einstein coefficients, we get $d_c \simeq 5 \times 10^3 R_*$. At distances greater than about 5000 stellar radii, the ratio of photons emitted in the two spectral lines is practically equal to unity, as we obtained in the simplified case considered previously.

The above considerations cease to be valid when there are other types of processes, in addition to the radiative ones, that populate or de-populate the atomic levels. As we shall see in the next chapter, collisional processes (here neglected because of the very low density of the medium) cause a drastic decrease in the population of the metastable levels. In laboratory plasmas and stellar atmospheres, where densities are orders of magnitude higher than those of nebular objects, the emission in forbidden lines is actually much lower than the emission in allowed lines, so the forbidden lines are essentially absent in the spectra.

From an historical point of view it is important to note that the interpretation of the two N_1 and N_2 lines was for a long time a real puzzle so that these lines, in the absence of an appropriate explanation, were attributed to a hypothetical element called "Nebulium". Their identification as lines of the [O III] spectrum was due to the American astronomer Bowen in 1928 (Bowen 1928).

In a similar way, the so-called auroral lines, the green line at 5577 Å and the red line at 6300 Å, responsible for the brightness of the sky in polar aurorae, were subsequently identified as forbidden lines of the neutral oxygen spectrum [O I]. These lines are due respectively to the transitions $^1S_0 \rightarrow \, ^1D_2$ and $^1D_2 \rightarrow \, ^3P_2$ occurring within the terms of the ground configuration of neutral oxygen[3] $1s^2 2s^2 2p^4$. A similar situation occurs for the various lines observed in the spectrum of the solar corona (also initially attributed to a hypothetical element, called "coronium"). These lines correspond to forbidden transitions between the lowest terms of various highly-ionised ions such as Fe X, Fe XI, Fe XIII, Fe XIV, Fe XV, Ni XII, Ni XIII, Ni XV, Ni XVI, Ca XII, Ca XIII, Ca XV, Ar X, Ar XIV, etc. The presence of forbidden lines in the auroral and coronal spectra is due to a physical mechanism completely analogous to that seen for the nebular spectra, with the only difference that the pumping mechanism from level b to level a is due, in the first case, to the excitation of the atoms of the upper atmosphere by the high-energy charged particles streaming from the Sun (solar wind), and, in the second case, to the excitation of the atoms by the electrons that, in the corona, have kinetic temperatures of the order of 10^6 K. The interpretation of the auroral lines was given by McLennan (1928) and that of the coronal lines by Edlén (1942).

12.6 Relative Strengths Within Multiplets in L-S Coupling

Consider a fine-structure multiplet of spectral lines due to the transition between two terms, both in L-S coupling. The total spin of the terms is S and the total orbital angular momentum is L for the lower term and L' for the upper term. Starting from the considerations set out in the first section of this chapter, it is possible to relate the dipole matrix elements relative to transitions between any of the lower levels,

---

[3]Remember that the configurations p^2 and p^4, being complementary, give rise to the same structure of terms (see Table 7.8). The Grotrian diagram of the ground configuration of neutral oxygen is therefore structurally equal to that of Fig. 12.1, apart from the obvious differences in the separations between the levels and from the energy reversal in the three $J = 0, 1, 2$ levels of the ^3P term, due to the third Hund's rule.

characterized by the total angular momentum quantum number J, to those relative to any of the higher levels, characterized by the quantum number J'. Returning to Eqs. (12.1) and (2.12) and introducing the 3-j symbols in place of the Clebsh-Gordan coefficients (Eq. (7.15)), we have for the dipole matrix element

$$
\begin{aligned}
&\langle \alpha L S J M | r_q | \alpha' L' S J' M' \rangle \\
&= \sum_{M_L M'_L M_S} (-1)^{L-S+M} \sqrt{2J+1} \\
&\quad \times \begin{pmatrix} L & S & J \\ M_L & M_S & -M \end{pmatrix} (-1)^{L'-S+M'} \sqrt{2J'+1} \begin{pmatrix} L' & S & J' \\ M'_L & M_S & -M' \end{pmatrix} \\
&\quad \times (-1)^{L'+M_L+1} \sqrt{2L+1} \begin{pmatrix} L & L' & 1 \\ -M_L & M'_L & q \end{pmatrix} \langle \alpha L \| \mathbf{r} \| \alpha' L' \rangle.
\end{aligned}
$$

On the other hand, applying directly the Wigner-Eckart theorem to the same matrix element we have, through Eq. (9.4)

$$
\begin{aligned}
&\langle \alpha L S J M | r_q | \alpha' L' S J' M' \rangle \\
&= (-1)^{J'+M+1} \sqrt{2J+1} \begin{pmatrix} J & J' & 1 \\ -M & M' & q \end{pmatrix} \langle \alpha L S J \| \mathbf{r} \| \alpha' L' S J' \rangle.
\end{aligned}
$$

Using these equations we can find a relation between the reduced matrix elements. With simple transformations we obtain

$$
\begin{aligned}
&(-1)^{J'+M+1} \begin{pmatrix} J & J' & 1 \\ -M & M' & q \end{pmatrix} \langle \alpha L S J \| \mathbf{r} \| \alpha' L' S J' \rangle \\
&= \sum_{M_L M'_L M_S} (-1)^{L+M-M'+M_L+1} \sqrt{(2J'+1)(2L+1)} \begin{pmatrix} L & S & J \\ M_L & M_S & -M \end{pmatrix} \\
&\quad \times \begin{pmatrix} L' & S & J' \\ M'_L & M_S & -M' \end{pmatrix} \begin{pmatrix} L & L' & 1 \\ -M_L & M'_L & q \end{pmatrix} \langle \alpha L \| \mathbf{r} \| \alpha' L' \rangle.
\end{aligned}
$$

We now multiply both sides for the following 3-j symbol

$$
\begin{pmatrix} J & J' & 1 \\ -M & M' & q \end{pmatrix},
$$

and sum over M and M'. Recalling the property of the 3-j symbols of Eq. (7.18), for which we have

$$
\sum_{MM'} \begin{pmatrix} J & J' & 1 \\ -M & M' & q \end{pmatrix}^2 = \frac{1}{3},
$$

we obtain

$$
\begin{aligned}
&\frac{1}{3} \langle \alpha L S J \| \mathbf{r} \| \alpha' L' S J' \rangle \\
&= \sum_{MM'M_L M'_L M_S} (-1)^{L-J'-M'+M_L}
\end{aligned}
$$

$$\times \sqrt{(2J'+1)(2L+1)} \begin{pmatrix} J & J' & 1 \\ -M & M' & q \end{pmatrix} \begin{pmatrix} L & S & J \\ M_L & M_S & -M \end{pmatrix}$$

$$\times \begin{pmatrix} L' & S & J' \\ M'_L & M_S & -M' \end{pmatrix} \begin{pmatrix} L & L' & 1 \\ -M_L & M'_L & q \end{pmatrix} \langle \alpha L \|\mathbf{r}\| \alpha' L' \rangle .$$

The sum over the four 3-j symbols that appears in the right-hand side can be expressed using the Racah coefficients, or the equivalent 6-j symbols of Wigner (which are related to the Racah coefficients by a simple sign factor). Such quantities arise naturally within the theory of angular momentum in relation to the addition of three angular momenta.[4] The sign factor that appears in the previous equation, multiplied by the product of the four 3-j symbols, is just the Racah coefficient $W(LS1J'; JL)$ multiplied by the factor $\frac{1}{3}$, so we obtain

$$\langle \alpha L S J \|\mathbf{r}\| \alpha' L' S J' \rangle = \sqrt{(2J'+1)(2L+1)} \, W(LS1J'; JL) \langle \alpha L \|\mathbf{r}\| \alpha' L' \rangle .$$

Alternatively, in terms of the 6-j symbols, taking into account the definition

$$W(abcdef) = (-1)^{a+b+c+d} \begin{Bmatrix} a & b & e \\ d & c & f \end{Bmatrix} ,$$

we have

$$\langle \alpha L S J \|\mathbf{r}\| \alpha' L' S J' \rangle = \sqrt{(2J'+1)(2L+1)} (-1)^{L+S+J'+1} \begin{Bmatrix} L & S & J \\ J' & 1 & L' \end{Bmatrix}$$
$$\times \langle \alpha L \|\mathbf{r}\| \alpha' L' \rangle .$$

Given a transition between two arbitrary levels, the line strength (or strength of the transition) is defined as the quantity[5]

$$S = e_0^2 (2J+1) |\langle \alpha J \|\mathbf{r}\| \alpha' J' \rangle|^2 .$$

Taking into account the Wigner-Eckart theorem and the properties of the 3-j symbols we can easily show that the line strength is symmetric under the exchange of the two levels, being

$$(2J+1) |\langle \alpha J \|\mathbf{r}\| \alpha' J' \rangle|^2 = (2J'+1) |\langle \alpha' J' \|\mathbf{r}\| \alpha J \rangle|^2 .$$

Using the previous results, the strength of a line belonging to a multiplet can therefore be expressed by the following equation

$$S_{JJ'} = e_0^2 (2J+1) |\langle \alpha L S J \|\mathbf{r}\| \alpha' L' S J' \rangle|^2$$
$$= e_0^2 (2J+1)(2J'+1)(2L+1) \begin{Bmatrix} L & S & J \\ J' & 1 & L' \end{Bmatrix}^2 |\langle \alpha L \|\mathbf{r}\| \alpha' L' \rangle|^2 ,$$

or

$$S_{JJ'} = s_{JJ'} S_{\text{mult}} ,$$

---

[4] See for example Brink and Satchler (1968).
[5] The line strength is proportional to the quantity $|\mathbf{r}_{ab}|^2 = |\mathbf{r}_{ba}|^2$ introduced in Chap. 11. As shown in Sect. 16.12, we have $S = e_0^2 (2J_a+1)(2J_b+1)|\mathbf{r}_{ab}|^2$.

where $\mathcal{S}_{\text{mult}}$ is the strength of the multiplet

$$\mathcal{S}_{\text{mult}} = e_0^2 (2L + 1) \left| \langle \alpha L \| \mathbf{r} \| \alpha' L' \rangle \right|^2,$$

and where the relative line strength $s_{JJ'}$ is given by

$$s_{JJ'} = (2J + 1)(2J' + 1) \left\{ \begin{matrix} L & L' & 1 \\ J' & J & S \end{matrix} \right\}^2.$$

Taking into account the following relation valid for the 6-j symbols

$$\sum_k (2k + 1)(2f + 1) \left\{ \begin{matrix} a & b & k \\ c & d & f \end{matrix} \right\} \left\{ \begin{matrix} a & b & k \\ c & d & g \end{matrix} \right\} = \delta_{fg},$$

we obtain

$$\sum_{J'} s_{JJ'} = \frac{2J + 1}{2L + 1}, \qquad \sum_J s_{JJ'} = \frac{2J' + 1}{2L' + 1}.$$

These two formulae express the sum rule, discovered empirically by Ornstein, Burger and Dorgelo, which states that "the sum of the line strengths of a multiplet that originate from a given lower level is proportional to the statistical weight of such level and, similarly, the sum of the strengths of the lines that originate from a given upper level is proportional to the statistical weight of the level itself."

The relative strengths of the different lines of a multiplet can be easily determined by knowing the relevant 6-j symbols or the Wigner coefficients. Often, the definition of the relative strength is slightly modified, noting that

$$\sum_{JJ'} s_{JJ'} = \sum_J \frac{2J + 1}{2L + 1} = \frac{(2L + 1)(2S + 1)}{2L + 1} = 2S + 1.$$

The normalised relative strength $(s_{JJ'})_{\text{norm}}$ can therefore be defined by

$$(s_{JJ'})_{\text{norm}} = \frac{(2J + 1)(2J' + 1)}{2S + 1} \left\{ \begin{matrix} L & L' & 1 \\ J' & J & S \end{matrix} \right\}^2, \quad \text{with} \sum_{JJ'} (s_{JJ'})_{\text{norm}} = 1.$$

The above formulae for the relative strengths of the transitions within a multiplet are known as the Kronig, Sommerfeld, and Hönl formulae. They were originally obtained using the correspondence principle. As an example, consider the multiplet relative to the transition $^3P \rightarrow {}^3D$ of Fig. 9.3. The lower term is composed of three fine-structure levels having $J = 0, 1, 2$, while the upper term is composed of three levels with $J' = 1, 2, 3$. In total, we have six lines corresponding to the transitions $J \rightarrow J'$ of the form $0 \rightarrow 1, 1 \rightarrow 1, 1 \rightarrow 2, 2 \rightarrow 1, 2 \rightarrow 2$ e $2 \rightarrow 3$. The other three transitions are forbidden by the selection rules on J. To obtain the relative strengths

Table 12.1 Normalised relative strengths of the transitions within the $^3P-^3D$ multiplet

Transition	Normalised relative strength
$^3P_0-^3D_1$	$\frac{1}{9}$
$^3P_1-^3D_1$	$\frac{1}{12}$
$^3P_1-^3D_2$	$\frac{1}{4}$
$^3P_2-^3D_1$	$\frac{1}{180}$
$^3P_2-^3D_2$	$\frac{1}{12}$
$^3P_2-^3D_3$	$\frac{7}{15}$

of the transitions within the multiplet we need to calculate the relative 6-j symbols. This can be done using the following formula (Racah 1942)

$$
\begin{Bmatrix} a & b & c \\ d & e & f \end{Bmatrix}
$$

$$
= \Delta(abc)\Delta(aef)\Delta(dbf)\Delta(dec) \sum_z (-1)^z (z+1)!
$$

$$
\times \left[(z-a-b-c)!(z-a-e-f)!(z-d-b-f)!(z-d-e-c)! \right.
$$

$$
\left. \times (a+b+d+e-z)!(b+c+e+f-z)!(a+c+d+f-z)! \right]^{-1},
$$

where the $\Delta(abc)$ symbol is defined in Eq. (7.17). With simple algebra we obtain

$$
\begin{Bmatrix} 1 & 2 & 1 \\ 1 & 0 & 1 \end{Bmatrix} = \frac{1}{3}, \qquad
\begin{Bmatrix} 1 & 2 & 1 \\ 1 & 1 & 1 \end{Bmatrix} = \frac{1}{6}, \qquad
\begin{Bmatrix} 1 & 2 & 1 \\ 2 & 1 & 1 \end{Bmatrix} = -\frac{1}{2\sqrt{5}},
$$

$$
\begin{Bmatrix} 1 & 2 & 1 \\ 1 & 2 & 1 \end{Bmatrix} = \frac{1}{30}, \qquad
\begin{Bmatrix} 1 & 2 & 1 \\ 2 & 2 & 1 \end{Bmatrix} = -\frac{1}{10}, \qquad
\begin{Bmatrix} 1 & 2 & 1 \\ 3 & 2 & 1 \end{Bmatrix} = \frac{1}{5}.
$$

From these values the relative strengths, given in Table 12.1, are obtained and the sum rules can be easily verified.

Very similar formulae exist for the multiplets of hyperfine structure. These can be obtained from the corresponding formulae holding for fine structure multiplets by means of the formal transformations

$$
L \rightarrow J, \qquad S \rightarrow I, \qquad J \rightarrow F.
$$

The relative strengths of the lines of a hyperfine structure multiplet are therefore given by the equation

$$
(s_{FF'})_{\text{norm}} = \frac{(2F+1)(2F'+1)}{2I+1} \begin{Bmatrix} J & J' & 1 \\ F' & F & I \end{Bmatrix}^2, \quad \text{with} \sum_{FF'} (s_{FF'})_{\text{norm}} = 1.
$$

Chapter 13
Non-equilibrium Plasmas

In laboratory and astrophysical plasmas, the conditions of excitation of the atoms are determined not only by their interaction with the electromagnetic field, but also by collisional processes between the atoms and the particles of the plasma. In this chapter we show how it is possible to describe this type of processes and what is their impact on atomic populations.

13.1 The Kinetic Temperature of the Electrons

We have seen in Chap. 10 that at thermodynamic equilibrium the electrons of an electrically neutral plasma have a velocity distribution described by a Gaussian function (the so-called Maxwellian distribution of velocities). The condition of thermodynamic equilibrium is, however, an idealized condition that, in practice, can be realized only with a certain degree of approximation. Both astrophysical and laboratory plasmas that are commonly observed for spectroscopic applications must—just for the fact that they are observable—emit radiation towards the external environment, which necessarily implies a situation of non-equilibrium, at least for their more exterior layers. In such situations, the concept of temperature loses its meaning, as do all the laws of thermodynamic equilibrium. For example, the distribution of the populations of an atomic species between the different states of ionisation and excitation cannot be determined anymore by the Saha-Boltzmann law but must be determined by solving the statistical equilibrium equations. In principle it is therefore to be expected that under non-equilibrium conditions the distribution of the velocities of the electrons differs from the Maxwellian distribution.

However, there is a wide range of physical conditions in which, despite an overall non-equilibrium, the velocity distribution of the electrons is effectively Maxwellian. This is due to the fact that the collisional processes, which cause the redistribution of kinetic energy between the various electrons and therefore tend to establish a condition of equilibrium, are much more effective than the processes that are opposed to the establishment of the condition of equilibrium.

E. Landi Degl'Innocenti, *Atomic Spectroscopy and Radiative Processes*,
UNITEXT for Physics, DOI 10.1007/978-88-470-2808-1_13, © Springer-Verlag Italia 2014

The processes of the first type are the elastic electron-electron collisions (and also the elastic electron-atom collisions that are, however, less effective). The processes of the second type are the inelastic electron-atom collisions, in which an electron transfers part of its kinetic energy which is converted into internal energy (excitation or ionisation) of the atomic system, or the superelastic electron-atom collisions in which the inverse processes occur (the electron gains kinetic energy due to de-excitation or recombination of the atom). Without pretending to give a rigorous proof of this fact, we simply develop some order of magnitude considerations to show that, in general, the mean free times between two successive processes of the first type are much shorter than the mean free times of the processes of the second type. This justifies, albeit not quite rigorously, that the velocity distribution of the electrons can be considered as Maxwellian to within a good approximation.

Denoting by N_e the electron density and by σ_E the cross section for elastic electron-electron collisions, the mean free time between two elastic collisions τ_E is given by

$$\tau_E \simeq \frac{1}{N_e \sigma_E v},$$

where v is the typical velocity of the electrons. Similarly, denoting by N_a the density of the atoms and by σ_A the cross section for inelastic (or superelastic) electron-atom collisions, the mean free time between two collisions of this type is given by

$$\tau_A \simeq \frac{1}{N_a \sigma_A v}.$$

From the two previous equations we obtain

$$\frac{\tau_E}{\tau_A} \simeq \frac{N_a \sigma_A}{N_e \sigma_E}.$$

The cross section σ_E can be estimated in the following way. Suppose that an electron having kinetic energy ϵ approaches another electron at rest. We can assume that the two electrons collide only if the incident electron will have a distance from the other electron less than a critical value b_c given by the equation

$$\frac{e_0^2}{b_c} = \epsilon.$$

In this case, in fact, the energy due to the Coulomb repulsion becomes comparable to the kinetic energy and we have an appreciable exchange of energy between the two particles. Solving for b_c and averaging over the energy of the particles we get

$$\sigma_E \simeq b_c^2 \simeq \frac{e_0^4}{\langle \epsilon^2 \rangle} \simeq \frac{e_0^4}{\langle \epsilon \rangle^2}.$$

Now we introduce the parameter T_e, the kinetic temperature of the electrons (or electron temperature), with the relation

$$\langle \epsilon \rangle \simeq k_B T_e.$$

By substitution we obtain, as an order of magnitude,

$$\sigma_E \simeq \frac{e_0^4}{k_B^2 T_e^2} \simeq 2.8 \times 10^{-6} T_e^{-2} \text{ cm}^2,$$

with T_e in K. If we consider for example a temperature range typical of stellar atmospheres (4×10^3 K $< T_e < 2 \times 10^4$ K), the cross section σ_E varies between 10^{-13} and 10^{-14} cm^2. The actual calculation of the cross section σ_A is more complex and has to be performed by means of a quantum-mechanical approach. The result is that, for the same temperature range, the value of σ_A is approximately of the order of 10^{-21} or 10^{-22} cm^2. We therefore obtain, as an order of magnitude

$$\frac{\tau_E}{\tau_A} \simeq 10^{-8} \frac{N_a}{N_e},$$

and even in the presence of a weakly ionised plasma with $N_e/N_a = 10^{-4}$, we still obtain a value of the order of 10^{-4} for this ratio.

We can conclude that an electron undergoes a large number of elastic collisions before suffering an inelastic (or superelastic) one, so that such collision will not be able to alter appreciably the Maxwellian distribution of velocity. The above considerations lead us to the conclusion that at a given point of a typical stellar atmosphere we can uniquely define a parameter T_e (kinetic temperature of the electrons) that characterizes the velocity distribution of the electrons. This parameter maintains a well defined operational definition, unlike the thermodynamic temperature T that completely loses its significance in non-equilibrium conditions.

13.2 Electron-Atom Collisions

Consider the collision between an electron having kinetic energy ϵ and an atom of a given atomic species. If the atom is, before the collision, in the energy level $|u_b\rangle$, it could be excited by the collision with the electron to the level $|u_a\rangle$ of higher energy.[1] For this process to occur, it is necessary that the relation $\epsilon \geq (\epsilon_a - \epsilon_b)$ is satisfied. After the collision, the electron is found to have a kinetic energy ϵ' given by

$$\epsilon' = \epsilon - (\epsilon_a - \epsilon_b).$$

Obviously, the inverse process can also occur, i.e. the collision is followed by the de-excitation of the atomic level $|u_a\rangle$ to the level $|u_b\rangle$. In this case, the energy of the colliding electron is given, after the collision, by

$$\epsilon' = \epsilon + (\epsilon_a - \epsilon_b).$$

The processes of the first type are called inelastic electron-atom collisions, while those of the second type are called superelastic electron-atom collisions (although some authors prefer to speak of collisions of the first and of the second kind, respectively).

---

[1] As in Chap. 11, we use here the index a to denote the upper level and the index b to denote the lower level.

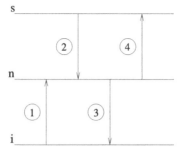

Fig. 13.1 Schematic representation of the collisional processes that contribute to the statistical equilibrium equations of a given level (the intermediate level in the figure). (1) Inelastic collisions from lower levels; (2) superelastic collisions from higher levels; (3) superelastic collisions to lower levels; (4) inelastic collisions to higher levels

The effect of the collisions on the atomic populations can be conveniently described by means of statistical equilibrium equations similar to those which we introduced in Chap. 11 for the interaction of an atom with the radiation field. For the population of a given level n, denoting by the index i the lower levels (i.e. those with lower energy) and by the index s the upper levels (i.e. those having higher energy), the equation describing the evolution of the system, if there are only collisions, is written in the form

$$\frac{dN_n}{dt} = \sum_i N_i C_{in}^{(A)} + \sum_s N_s C_{sn}^{(S)} - \sum_i N_n C_{ni}^{(S)} - \sum_s N_n C_{ns}^{(A)}. \tag{13.1}$$

The four terms appearing in this equation are shown in the diagram of Fig. 13.1. The quantities $C_{ba}^{(A)}$ and $C_{ab}^{(S)}$ appearing in this equation are called collisional rates. They are due, respectively, to inelastic and superelastic collisions. These quantities are obviously proportional to the density of the colliding particles. They also depend on the velocity distribution of the particles and on atomic properties related to the wavefunctions of the two levels between which the transition occurs. For electronic collisions, the rate for inelastic collisions from level b to level a can be expressed by means of the cross section

$$C_{ba}^{(A)} = N_e \int_{v_0}^{\infty} \sigma_{ba}(v) f(v) v \, dv,$$

where N_e is the electron density, $f(v)$ is the velocity distribution of the electrons, and $\sigma_{ba}(v)$ is the cross section for collisional excitation relative to the velocity v. The limit of integration v_0 is the threshold velocity, i.e. the minimum electron velocity for the electron to be able to excite the atom from level b to level a. It is given by

$$\frac{1}{2}mv_0^2 = \epsilon_a - \epsilon_b.$$

Similarly, the rate for superelastic collisions is given by

$$C_{ab}^{(S)} = N_e \int_0^\infty \sigma_{ab}(v) f(v) v \, dv,$$

where $\sigma_{ab}(v)$ is the cross-section for collisional de-excitation.

13.3 The Einstein-Milne Relations

When the velocity distribution of the colliding electrons is Maxwellian, it can be proved, by means of thermodynamic considerations, that the two collisional rates introduced in the previous section are simply related. These thermodynamic considerations, due to Milne, are very similar to those previously developed by Einstein to determine the relations between the coefficients involved in the statistical equilibrium equations for the interaction between atoms and radiation (Einstein coefficients, see Sect. 11.7). For this reason, these relations are called Milne or Einstein-Milne relations.

Consider an atom consisting of only two levels, a and b, subject to collisions by a plasma of electrons with density N_e. If the system is in thermodynamic equilibrium at the temperature T, we can invoke the so-called principle of detailed balance to assert that the number of collisional transitions (due to the electrons) that occur between level a and level b are exactly balanced by the number of collisional transitions (also due to the electrons) that occur between level b and level a. In other words, at thermodynamic equilibrium conditions, a perfect balance must hold for any process that contributes to populate or de-populate the atomic levels regardless of the number and of the characteristics of the physical processes that are simultaneously in operation (radiative processes, collisional processes still with electrons, but among other pairs of levels, collisional processes with other atomic species, etc.). Otherwise, in fact, it would be possible to construct an ideal machine, working in cycle, which could produce work at the expense of a single source, which would contradict the second law of thermodynamics. If we denote then by $\tilde{N}_a$ and $\tilde{N}_b$ the populations of the levels a and b in thermodynamic equilibrium, we must have, writing the evolution equation for the population of level a,

$$0 = \frac{dN_a}{dt} = \tilde{N}_b C_{ba}^{(A)} - \tilde{N}_a C_{ab}^{(S)}.$$

Solving this equation and using the Boltzmann equation to express the ratio $\tilde{N}_b / \tilde{N}_a$ (Eq. (10.6)), we obtain, in thermodynamic equilibrium at the temperature T,

$$\frac{C_{ab}^{(S)}}{C_{ba}^{(A)}} = \frac{\tilde{N}_b}{\tilde{N}_a} = \frac{g_b}{g_a} e^{(\epsilon_a - \epsilon_b)/(k_B T)}.$$

On the other hand, the two collisional rates depend only on atomic factors and on the velocity distribution of the electrons. The result that we have obtained thus continues to be valid even outside thermodynamic equilibrium, as long as the velocity distribution of the electrons is Maxwellian. If we are under these conditions, far less restrictive than the thermodynamic equilibrium, and if we denote by T_e the kinetic temperature of the electrons, we obtain the Einstein-Milne relation

$$\frac{C_{ab}^{(S)}}{C_{ba}^{(A)}} = \frac{g_b}{g_a} e^{(\epsilon_a - \epsilon_b)/(k_B T_e)}. \tag{13.2}$$

13.4 The Two-Level Atom in Non-equilibrium Conditions

Consider a two-level atom that interacts with a radiation field having, at the frequency v corresponding to the transition, the mean intensity J_v. Let the atom be subject to collisions with a population of electrons having kinetic temperature T_e. Taking into account both collisional processes (Eq. (13.1)) and radiative processes (Eq. (11.29)), the statistical equilibrium equation for the population of the upper level is

$$\frac{dN_a}{dt} = -N_a \left(A_{ab} + B_{ab} J_v + C_{ab}^{(S)} \right) + N_b \left(B_{ba} J_v + C_{ba}^{(A)} \right).$$

In stationary conditions, solving the equation we obtain

$$\frac{N_b}{N_a} = \frac{A_{ab} + B_{ab} J_v + C_{ab}^{(S)}}{B_{ba} J_v + C_{ba}^{(A)}}.$$

We now substitute this result into the expression for the source function given by Eq. (11.36). Taking into account the relations between the Einstein coefficients (Eqs. (11.27) and (11.28)) and the Einstein-Milne relations between the collisional rates (Eq. (13.2)), with some algebra we obtain

$$S_v = \frac{J_v + \varepsilon B_v(T_e)}{1 + \varepsilon}, \tag{13.3}$$

where B_v is the Planck function and where we have introduced the quantity ε defined by

$$\varepsilon = \frac{C_{ab}^{(S)} (1 - e^{-hv/(k_B T_e)})}{A_{ab}}.$$

Apart from a correction factor of the order of unity, ε represents the ratio between the number of de-excitations of the upper level due to superelastic collisions and the number of de-excitations due to spontaneous emission. The general expression that we have found for S_v allows us to write, inverting Eq. (11.36), the ratio of the populations N_b/N_a in the form

$$\frac{N_b}{N_a} = \frac{g_b}{g_a} \left(\frac{2hv^3}{c^2 S_v} + 1 \right).$$

If we introduce $\bar{n}_v = J_v c^2/(2hv^3)$ as the average number of photons per mode at frequency v, and $\bar{n}_v(T_e) = B_v(T_e)/(2hv^3)$ as the average number of photons per mode relative, at the same frequency, to the blackbody radiation of temperature T_e, we obtain

$$\frac{N_b}{N_a} = \frac{g_b}{g_a}\left(\frac{1+\varepsilon}{\bar{n}_\nu + \varepsilon\bar{n}_\nu(T_e)} + 1\right). \tag{13.4}$$

The above expressions (Eqs. (13.3) and (13.4)) assume a special form in three limiting cases of particular importance.

(a) The first case is when $\varepsilon \gg 1$. Substituting in the expressions for the source function and for the ratio between the populations we get

$$S_\nu = B_\nu(T_e), \qquad \frac{N_b}{N_a} = \frac{g_b}{g_a}e^{h\nu/(k_B T_e)}.$$

In this case the collisions are extremely effective and are able to thermalise the atomic populations at the kinetic temperature. For the ratio of populations we obtain the Boltzmann equation (Eq. (10.6)), while for the source function we obtain the Planck function, both relative to the temperature T_e. This limiting case is known as local thermodynamic equilibrium (LTE).

(b) The second case is when $\varepsilon \ll 1$ and, at the same time, $\varepsilon B_\nu(T_e) \ll J_\nu$. Substituting in the same equations we get

$$S_\nu = J_\nu, \qquad \frac{N_b}{N_a} = \frac{g_b}{g_a}\left(\frac{1}{\bar{n}_\nu} + 1\right).$$

This time the collisions have a completely negligible role and the source function is just the average over the solid angle of the incoming radiation. The atom simply behaves as a scattering centre of the radiation. For the atomic populations, defining a suitable "radiation temperature" T_r through the equation

$$\bar{n}_\nu = \frac{1}{e^{h\nu/(k_B T_r)} - 1},$$

we obtain

$$\frac{N_b}{N_a} = \frac{g_b}{g_a}e^{h\nu/(k_B T_r)},$$

which shows that the atomic populations are in equilibrium with the radiation temperature. The parameter T_r that we have so defined is however a completely *ad hoc* parameter. Indeed, for an arbitrary radiation field, there is a different T_r value for each frequency.

(c) Finally, the third case is when the inequalities $\varepsilon \ll 1$ and $\varepsilon B_\nu(T_e) \gg J_\nu$ hold. Again substituting we obtain

$$S_\nu = \varepsilon B_\nu(T_e), \qquad \frac{N_b}{N_a} = \frac{g_b}{g_a}\left(\frac{1}{\varepsilon\bar{n}_\nu(T_e)} + 1\right).$$

This is an intermediate case in which, although the collisions are not very effective in de-populating the upper level, the kinetic temperature is so high and the radiation field is so diluted that actually the collisions (and not the radiative processes) are populating the upper level.

The three cases that we have schematically described here are suitable to describe, in a qualitative way, the conditions of excitation of an atom which is located, respectively, in the photosphere, in the chromosphere and in the solar corona. For laboratory plasmas, the most common physical situations are those described by case (a) (plasma with high densities, as discharge lamps) or by case (b) (plasmas of low densities, for experiments of optical pumping with lasers).

Chapter 14
Radiative Transfer

The radiation that propagates in an extended medium is subject to continuous emission and absorption processes that modify its intensity and spectral distribution. These phenomena are governed by the equation of radiative transfer that we have formally obtained in Chap. 11 from the principles of quantum electrodynamics. By solving this equation, it is possible to relate the observed properties of the radiation emerging from an extended medium with the intrinsic properties of the plasma responsible for its emission. In this chapter, we explore this issue by analysing in detail the prototype case for which the theory of radiative transfer was developed, namely the case of stellar atmospheres. The considerations presented here can be easily extended to deal with specific problems encountered in the analysis of the propagation of radiation in non-astrophysical environments, such as laboratory plasmas or the Earth's atmosphere.

14.1 Formal Solution of the Radiative Transfer Equation

In Sect. 11.8 we have seen that the equation of radiative transfer for the specific intensity of the radiation field that propagates, at frequency ν along the direction $\boldsymbol{\Omega}$ inside a plasma takes the form (Eq. (11.32))

$$\frac{\mathrm{d}}{\mathrm{d}s}I_\nu(\boldsymbol{\Omega}) = -k_\nu\big[I_\nu(\boldsymbol{\Omega}) - S_\nu\big],$$

where s is the spatial coordinate measured along the direction $\boldsymbol{\Omega}$, k_ν is the absorption coefficient (corrected for stimulated emission) and S_ν is the source function. In general, k_ν and S_ν are function of the coordinate s and, if we assume that these quantities are known, the radiative transfer equation can easily be solved.

We now introduce, in place of the geometrical coordinate s, the so-called specific optical depth (function of ν) by the equation

$$\mathrm{d}\tau_\nu = -k_\nu\,\mathrm{d}s.$$

As shown by the equation, the optical depth is defined in the direction opposite to that of the propagation of the radiation, which reflects the point of view of an

E. Landi Degl'Innocenti, *Atomic Spectroscopy and Radiative Processes*,
UNITEXT for Physics, DOI 10.1007/978-88-470-2808-1_14, © Springer-Verlag Italia 2014

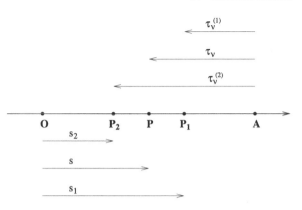

Fig. 14.1 The radiation propagates along the direction from point P$_2$ to point P$_1$. The spatial coordinate is measured starting from point O while the optical depth is measured starting from point A in the direction opposite to that of the propagation

observer receiving the radiation. Considering the plasma contained within a fixed geometrical thickness, for example between the points P$_1$ and P$_2$ of Fig. 14.1, characterized by the coordinates s_1 and s_2 (with $s_1 > s_2$), the corresponding optical thickness is, by simple integration of the above equation

$$\tau_\nu(\text{P}_1, \text{P}_2) = \int_{s_2}^{s_1} k_\nu(s)\,\mathrm{d}s.$$

The optical thickness depends on frequency. A fixed geometrical optical thickness is defined as being optically thin when $\tau_\nu \ll 1$ or optically thick when $\tau_\nu \gg 1$. From a physical point of view, a medium is optically thin at the frequency ν when a photon of that frequency has a negligible probability to be absorbed as it travels across it. Conversely, if the medium is optically thick, the photon has a probability practically equal to unity to be absorbed within the medium.

For a fixed direction $\mathbf{\Omega}$, we can simplify the notations by omitting such argument in the notation for the intensity. Dividing the transfer equation for k_ν and changing the sign, we have

$$\frac{\mathrm{d}I_\nu}{\mathrm{d}\tau_\nu} = I_\nu - S_\nu.$$

To solve this equation we multiply both sides for the $\mathrm{e}^{-\tau_\nu}$ factor. We obtain

$$\mathrm{e}^{-\tau_\nu}\frac{\mathrm{d}I_\nu}{\mathrm{d}\tau_\nu} = \mathrm{e}^{-\tau_\nu}I_\nu - \mathrm{e}^{-\tau_\nu}S_\nu,$$

or

$$\frac{\mathrm{d}}{\mathrm{d}\tau_\nu}\left(\mathrm{e}^{-\tau_\nu}I_\nu\right) = -\mathrm{e}^{-\tau_\nu}S_\nu.$$

With reference to Fig. 14.1, we integrate this equation between the points P$_2$ and P$_1$ that have corresponding optical depths $\tau_\nu^{(2)}$ and $\tau_\nu^{(1)}$, with $\tau_\nu^{(1)} < \tau_\nu^{(2)}$. We obtain

$$\mathrm{e}^{-\tau_\nu^{(1)}}I_\nu\left(\tau_\nu^{(1)}\right) - \mathrm{e}^{-\tau_\nu^{(2)}}I_\nu\left(\tau_\nu^{(2)}\right) = -\int_{\tau_\nu^{(2)}}^{\tau_\nu^{(1)}} S_\nu(\tau_\nu)\mathrm{e}^{-\tau_\nu}\,\mathrm{d}\tau_\nu,$$

or

$$I_\nu\left(\tau_\nu^{(1)}\right) = I_\nu\left(\tau_\nu^{(2)}\right)e^{-\left(\tau_\nu^{(2)}-\tau_\nu^{(1)}\right)} + \int_{\tau_\nu^{(1)}}^{\tau_\nu^{(2)}} S_\nu(\tau_\nu)e^{-\left(\tau_\nu-\tau_\nu^{(1)}\right)}\,d\tau_\nu. \qquad (14.1)$$

This result is easily interpreted in the following way. The intensity at P_1 is given by the intensity at P_2 (the boundary condition) multiplied by the attenuation factor due to the absorption between points P_2 and P_1, with in addition the contribution due to the emission within the interval between the two points. The contribution due to the infinitesimal optical depth element $d\tau_\nu$ centred at the generic point P is multiplied by the relative attenuation factor due to the absorption between points P and P_1. In particular, if we consider the radiation emerging from a plasma and we put $\tau_\nu^{(1)} = 0$, the previous equation can be written in the form

$$I_\nu(0) = I_\nu(\tau_\nu)e^{-\tau_\nu} + \int_0^{\tau_\nu} S_\nu\left(\tau_\nu'\right)e^{-\tau_\nu'}\,d\tau_\nu'.$$

In many cases, especially in astrophysics, we have to deal with plasmas that are practically infinite in one direction (consider, for example, a stellar atmosphere where we are interested in expressing the emerging intensity as a function of the local properties of the atmosphere itself). In such cases, one must consider the limit of the above equation for $\tau_\nu \to \infty$, and, assuming that we have

$$\lim_{\tau_\nu \to \infty} I_\nu(\tau_\nu)e^{-\tau_\nu} = 0,$$

we obtain

$$I_\nu(0) = \int_0^\infty S_\nu(\tau_\nu)e^{-\tau_\nu}\,d\tau_\nu. \qquad (14.2)$$

The above mathematical limit is in practice always met because otherwise one would obtain the absurd result that the intensity emerging from the medium has an infinite value. Equation (14.2) expresses in all generality the intensity emerging from a semi-infinite medium, i.e. from a medium that is undefined in the direction opposite to that of the emerging radiation. This equation is the basis of the quantitative interpretation of stellar spectra.

14.2 Radiative Transfer in Stellar Atmospheres

To determine the specific intensity of the radiation emitted from a stellar atmosphere, a number of approximations are generally introduced. They are used to simplify the problem from a mathematical point of view in order to obtain certain analytical results that are valid as a zero order approximation. The first of these approximations consists in neglecting the curvature of the surface layers of the star due to its spherical shape. This approximation is in general well justified because the thickness of the atmosphere (defined as the surface layer from which the observed radiation originates) is much smaller than the radius of the star. For the Sun,

Fig. 14.2 Schematic
representation of a
plane-parallel atmosphere in
which the physical properties
depend only on the height z
and the radiation field only on
z and on the angle θ (the
heliocentric angle)

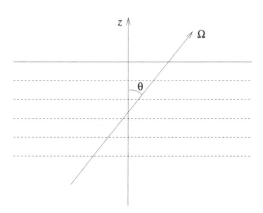

for example, the thickness H is less than or of the order of a thousand km, so we
have ($R_\odot$ is the radius of the Sun)

$$\frac{H}{R_\odot} \leq \frac{10^3 \text{ km}}{7 \times 10^5 \text{ km}} \simeq 1.4 \times 10^{-3}.$$

We next assume that the physical properties of the atmosphere depend only on the
height z (measured from an origin that we do not need to specify at the moment) and
not on the other two coordinates x and y. The radiation field, which in stationary
conditions depends on the frequency, on the point P and the direction $\mathbf{\Omega}$, in this
case depends only on the height z and on the angle θ (called the heliocentric angle
in the case of the Sun), defined as in Fig. 14.2. When we introduce this further
approximation we say that we are dealing with a plane-parallel atmosphere.

Denoting by $I_\nu(z, \mu)$ the specific intensity of the radiation that propagates along
the direction identified by the angle θ (with $\mu = \cos\theta$), the transfer equation becomes

$$\mu \frac{\mathrm{d}}{\mathrm{d}z} I_\nu(z, \mu) = -k_\nu \left[I_\nu(z, \mu) - S_\nu \right],$$

and if we assume that we are in Local Thermodynamic Equilibrium (LTE)

$$\mu \frac{\mathrm{d}}{\mathrm{d}z} I_\nu(z, \mu) = -k_\nu \left[I_\nu(z, \mu) - B_\nu(T) \right],$$

where $B_\nu(T)$ is the Planck function, which only depends on the local temperature.

The transfer equation can be formally solved by introducing the specific optical
depth t_ν, measured along the vertical in the sense of increasing depth (note that this
quantity differs from that defined by the symbol τ_ν in the previous section which
refers to the optical depth measured along the beam). Putting

$$\mathrm{d}t_\nu = -k_\nu \, \mathrm{d}z,$$

and using the results of the previous section we have that the emergent intensity is
given by

$$I_\nu(0, \mu) = \int_0^\infty B_\nu(T) \mathrm{e}^{-t_\nu/\mu} \frac{\mathrm{d}t_\nu}{\mu}. \tag{14.3}$$

This expression can be conveniently approximated in order to obtain some qualitative results. If we assume, for example, that the Planck function is linear in t_ν, i.e. that

$$B_\nu(t_\nu) = a_\nu + b_\nu t_\nu,$$

with a_ν and b_ν constants, we obtain with simple integrations

$$I_\nu(0, \mu) = a_\nu + b_\nu \mu = B_\nu(t_\nu = \mu).$$

The so-called Eddington-Barbier approximation consists in assuming that this identity is valid in general (although, rigorously, it is only valid when the Planck function is linear in t_ν). With this approximation we thus have

$$I_\nu(0, \mu) \simeq B_\nu(t_\nu = \mu).$$

If we assume that the temperature in the stellar atmosphere is a given function of the geometrical height z, i.e. $T = T(z)$, to determine the emerging intensity using the Eddington-Barbier approximation it is sufficient to calculate the height $\tilde{z}$ for which $t_\nu = \mu$ and we obtain

$$I_\nu(0, \mu) \simeq B_\nu\big[T(\tilde{z})\big].$$

Since in general the temperature decreases with z in a stellar atmosphere, we should expect two different phenomena:

(a) for a fixed frequency, the intensity emitted by the star is greater at the center ($\mu = 1$) than at the limb ($\mu \to 0$). This phenomenon, known as limb darkening, is observable only on the Sun (as it is impossible with current technologies to spatially resolve the radiation from other stars). This phenomenon is due to the fact that, when observing at the center of the Sun, we can penetrate into deeper layers of the solar atmosphere. When observing at the limb, instead, we observe the outer layers, which are cooler (and therefore less bright).

(b) For a fixed μ, since the absorption coefficient is a function of frequency, we have a lower intensity at those frequencies for which the absorption coefficient is higher and a higher intensity at the frequencies for which the absorption coefficient is lower. Obviously, "lower" and "higher" are here to be understood in a relative sense, i.e. with respect to an "average" Planck function not specified. In other words, one can think that a stellar spectrum is constituted by a "modulated" Planck function with an increase at the frequencies where the absorption coefficient is small and a decrease at those where the absorption coefficient is large. In this way, we can readily explain the discontinuities in the continuum spectrum which are observed at the limits of the series (typical is the sudden decrease in the continuum at wavelengths shorter than 3647 Å, the so-called Balmer discontinuity). Similarly, the absorption coefficient has a rapid variation within a frequency range centered around a spectral line, with a very high value at the center of the line and a much lower value in the wings of the line. This explains, qualitatively, the presence of absorption lines in stellar spectra and suggests that, when emission lines are observed instead, the temperature should increase with height (stellar chromospheres).

Fig. 14.3 The radiation emitted by the stellar surface element identified by the angle θ is inclined by the same angle with respect to the local vertical

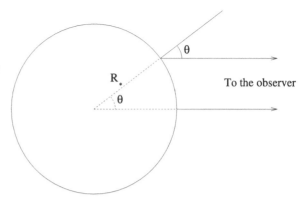

Regarding stars other than the Sun, as we said, it is not possible to observe the variation of the intensity with μ. What is observed instead is the average intensity of the radiation on the stellar disk $\bar{I}_\nu$ defined by (see Fig. 14.3), with R_* the stellar radius

$$\pi R_*^2 \bar{I}_\nu = \int_0^{\pi/2} d\theta \int_0^{2\pi} d\phi I_\nu(0, \theta) R_*^2 \cos\theta \sin\theta,$$

or

$$\bar{I}_\nu(0) = 2 \int_0^1 I_\nu(0, \mu)\mu\, d\mu.$$

Substituting in this equation the above formal solution and inverting the order of the integrations, we obtain

$$\bar{I}_\nu(0) = 2 \int_0^\infty dt_\nu B_\nu(T) \int_0^1 e^{-t_\nu/\mu}\, d\mu.$$

The integral in $d\mu$ can be expressed in terms of known functions. With the substitution $w = 1/\mu$ we have

$$\int_0^1 e^{-t_\nu/\mu}\, d\mu = \int_1^\infty e^{-wt_\nu} \frac{1}{w^2}\, dw.$$

Recalling the definition of the exponential integral functions

$$E_n(x) = \int_1^\infty \frac{e^{-xt}}{t^n}\, dt \quad (n \ge 1), \tag{14.4}$$

we obtain

$$\int_0^1 e^{-t_\nu/\mu}\, d\mu = E_2(t_\nu),$$

so that

$$\bar{I}_\nu = 2 \int_0^\infty B_\nu(T) E_2(t_\nu)\, dt_\nu. \tag{14.5}$$

We obtain the analogue of the Eddington-Barbier approximation by assuming that $B_\nu(T)$ is a linear function of t_ν. In this case, taking into account that

$$\int_0^\infty E_n(x)\,dx = \frac{1}{n}, \qquad \int_0^\infty x E_n(x)\,dx = \frac{1}{n+1},$$

we obtain

$$\bar{I}_\nu = B_\nu\left(t_\nu = \frac{2}{3}\right).$$

The Eddington-Barbier approximation for the mean intensity emitted by a stellar atmosphere is then

$$\bar{I}_\nu \simeq B_\nu\left(t_\nu = \frac{2}{3}\right).$$

14.3 The Grey Atmosphere

Consider a plane-parallel atmosphere in local thermodynamic equilibrium. As shown by Eqs. (14.3) and (14.5), the emerging intensity can be expressed with an integral that involves the knowledge of the Planck function (i.e. of the temperature) at the different optical depths. When the variation of the temeparature with depth is known (together with the variation of other physical quantities such as the pressure), we say that we have a "model of a stellar atmosphere".

More or less sophisticated models of a stellar atmosphere can be constructed, depending on the amount of physical information that is introduced in the description of the atmosphere itself. The simplest of these models (and also the first from the historical point of view) is the so-called model of the grey atmosphere. In this model we consider, as a starting point, a plane-parallel atmosphere in local thermodynamic equilibrium and radiative equilibrium. With regard to the latter concept, we note that energy can flow through a stellar atmosphere by means of three distinct physical mechanisms: radiation, convection, and conduction. The third is in many cases negligible. A stellar atmosphere is said to be in radiative equilibrium when the energy flows only via radiation.

Consider a given height z in the stellar atmosphere. The net energy flowing per unit time through the unit surface at frequency ν is given by

$$F_\nu(z) = 2\pi \int_{-1}^{1} \mu I_\nu(z, \mu)\,d\mu.$$

The quantity $F_\nu(z)$ is called monochromatic flux. The radiation coming from the interior contributes to it with a positive sign ($\mu > 0$), while the one coming from the exterior contributes to it with a negative sign ($\mu < 0$). The monochromatic flux at the surface is related to the quantity $\bar{I}_\nu$, previously introduced, by the relation

$$F_\nu(0) = \pi \bar{I}_\nu.$$

The condition of radiative equilibrium implies that the integral of the monochromatic flux over all frequencies is constant, i.e. is independent of z. In formulae, defining the total flux F by the equation

$$F = \int_0^\infty F_\nu \, d\nu = 2\pi \int_0^\infty d\nu \int_{-1}^1 \mu I_\nu(z, \mu) \, d\mu, \tag{14.6}$$

the radiative equilibrium hypothesis implies

$$\frac{dF}{dz} = 0.$$

The value of F is in general parametrised through the effective temperature T_{eff}, defined by the relation (see Eq. (10.10))

$$F = \sigma T_{\text{eff}}^4,$$

where σ is the Stefan-Boltzmann constant. It represents the temperature that a blackbody should have to irradiate the same flux as the star. The flux is also related to the luminosity L_* and to the stellar radius R_* via the relation

$$F = \frac{L_*}{4\pi R_*^2}.$$

For the Sun, for which

$$L_\odot = 3.845 \times 10^{33} \text{ erg s}^{-1}, \qquad R_\odot = 6.9626 \times 10^{10} \text{ cm},$$

we have

$$F = 6.312 \times 10^{10} \text{ erg cm}^{-2} \text{s}^{-1}, \qquad T_{\text{eff}} = 5776 \text{ K}.$$

The further hypothesis which is introduced in the model of the grey atmosphere (which justifies its name) consists in assuming that the absorption coefficient k_ν is independent of frequency. This assumption greatly simplifies the problem from a mathematical point of view but is not at all realistic from a physical point of view. Obviously, the model that is obtained in this way must be considered as a sort of zero order model for a real stellar atmosphere.

In the grey atmosphere we can define, instead of the specific optical depth t_ν, a "universal" optical depth t, and the transfer equation becomes

$$\mu \frac{d}{dt} I_\nu(t, \mu) = I_\nu(t, \mu) - B_\nu(t).$$

Integrating the transfer equation in $d\nu$ and defining

$$I(t, \mu) = \int_0^\infty I_\nu(t, \mu) \, d\nu, \qquad B(t) = \int_0^\infty B_\nu(t) \, d\nu,$$

we obtain

$$\mu \frac{d}{dt} I(t, \mu) = I(t, \mu) - B(t).$$

As we can see, the hypothesis that the absorption coefficient k_ν is independent of frequency allows one to write a single transfer equation for the quantities integrated in frequency. This is the fundamental simplification of the grey atmosphere.

From $I(t, \mu)$, the intensity integrated in frequency, its various moments can be defined by integrating over the solid angle. The moment of order n, $M_n(t)$, is defined by

$$M_n(t) = \frac{1}{4\pi} \oint \mu^n I(t, \mu) \, d\Omega = \frac{1}{2} \int_{-1}^{1} \mu^n I(t, \mu) \, d\mu.$$

The zero-order moment is the mean intensity of the radiation field and is denoted by the symbol $J(t)$

$$J(t) = M_0(t) = \frac{1}{2} \int_{-1}^{1} I(t, \mu) \, d\mu. \tag{14.7}$$

The first-order moment is proportional to the flux of radiating energy. In fact, we have, recalling Eq. (14.6)

$$F(t) = 4\pi M_1(\tau) = 2\pi \int_{-1}^{1} \mu I(t, \mu) \, d\mu.$$

Finally, the second-order moment is proportional to the radiation pressure and is denoted by the symbol $K(t)$

$$K(t) = M_2(t) = \frac{1}{2} \int_{-1}^{1} \mu^2 I(t, \mu) \, d\mu.$$

Integrating in $d\mu$ the transfer equation divided by 2, we obtain

$$\frac{1}{4\pi} \frac{dF(t)}{dt} = J(t) - B(t),$$

and using the hypothesis of radiative equilibrium ($F = \text{const.}$) we have

$$J(t) = B(t). \tag{14.8}$$

Multiplying the transfer equation by $\mu/2$ and integrating in $d\mu$ we have

$$\frac{dK(t)}{dt} = \frac{F}{4\pi},$$

which, once solved, gives

$$K(t) = \frac{Ft}{4\pi} + C,$$

where C is a constant to be determined using the boundary conditions.

We note that for $t \to \infty$, i.e. at the base of the atmosphere, we would expect the radiation field to become isotropic. Under this hypothesis, and even under the less restrictive hypothesis that the intensity varies linearly with μ according to the expression

$$I(t, \mu) = a(t) + b(t)\mu,$$

with $a(t)$ and $b(t)$ independent of μ, the quantities J and K can be related to each other, being

$$K(t) = \frac{1}{3} J(t).$$

If we assume that this relation is valid for any value of t (and not only for $t \to \infty$) we adopt the so-called Eddington approximation, so that the problem of the grey atmosphere can be solved analytically. In fact, using the previous relations, we have

$$B(t) = J(t) = 3K(t) = \frac{3}{4\pi} Ft + C',$$

with $C' = 3C$. To determine the constant C' we use the boundary conditions relative to the surface of the star ($t = 0$). If the star is isolated (i.e. does not belong to a double or multiple system), the flux at the surface can be calculated using the equation (obtained from the formal solution of the transfer equation)

$$F = 2\pi \int_0^1 \mu I(0, \mu) \, d\mu = 2\pi \int_0^1 d\mu \mu \int_0^\infty B(t) e^{-t/\mu} \frac{dt}{\mu}.$$

Substituting the expression for $B(t)$ and performing the calculation we obtain

$$C' = \frac{F}{2\pi},$$

so that we have for $B(t)$

$$B(t) = \frac{3F}{4\pi} \left(t + \frac{2}{3} \right).$$

Finally, recalling that

$$B(t) = \frac{\sigma}{\pi} T^4(t), \qquad F = \sigma T_{\text{eff}}^4,$$

we obtain the law expressing the dependence of the temperature on the optical depth for the grey atmosphere (in the Eddington approximation)

$$T(t) = T_{\text{eff}} \sqrt[4]{\frac{3}{4} \left(t + \frac{2}{3} \right)}. \tag{14.9}$$

In particular, we see that at the surface of the atmosphere we have

$$T(0) = 0.841 T_{\text{eff}},$$

and that, for $t = \frac{2}{3}$, we obtain $T = T_{\text{eff}}$.

The centre-to-limb variation of the emerging intensity can also be determined from the expression of $B(t)$. We have, in fact,

$$I(0, \mu) = \int_0^\infty \frac{3F}{4\pi} \left(t + \frac{2}{3} \right) e^{-t/\mu} \frac{dt}{\mu} = \frac{3F}{4\pi} \left(\mu + \frac{2}{3} \right).$$

Defining the limb darkening ratio $r(\mu)$ by the equation

$$r(\mu) = \frac{I(0, \mu)}{I(0, 1)},$$

we obtain

$$r(\mu) = \frac{3\mu + 2}{5}.$$

This limb darkening variation can be compared to the results made available by solar observations. The difference between the theoretical and observed value of $r(\mu)$ is always within 5 % (see Fig. 14.4).

Fig. 14.4 Comparison
between the limb darkening
ratio resulting from the grey
atmosphere model (*full line*)
and the observed solar values
(*points*)

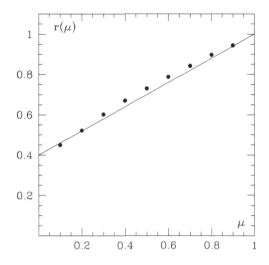

14.4 The Hopf Equation

The model of the grey atmosphere can be solved exactly from the mathematical
point of view, although this implies the numerical solution of an integral equation.
The exact solution is obtained by writing the function $B(t)$ in the form

$$B(t) = \frac{3F}{4\pi}\big[t + q(t)\big], \tag{14.10}$$

where $q(t)$ is a suitable function, known as Hopf function. We should expect that
this function has a small variation centred around the $\frac{2}{3}$ value, which is obtained
from the Eddington approximation.

We fix a specific value of t in the atmosphere and express separately the intensity of the radiation for beams propagating upwards and downwards. Recalling the
formal solution of the transfer equation (Eq. (14.1)), we have

$$I(t,\mu) = \int_t^\infty B(t')e^{-(t'-t)/\mu}\frac{dt'}{\mu} \quad (\mu > 0),$$

$$I(t,\mu) = \int_0^t B(t')e^{-(t-t')/(-\mu)}\frac{dt'}{-\mu} \quad (\mu < 0).$$

Using these expression we can find the mean value of the intensity of the radiation
field over the solid angle $J(t)$, as defined in Eq. (14.7). Recalling that in the grey
atmosphere $B(t) = J(t)$ (see Eq. (14.8)), we obtain

$$B(t) = \frac{1}{2}\left[\int_0^1 d\mu \int_t^\infty B(t')e^{-(t'-t)/\mu}\frac{dt'}{\mu} + \int_{-1}^0 d\mu \int_0^t B(t')e^{-(t-t')/(-\mu)}\frac{dt'}{-\mu}\right].$$

t	$q(t)$	t	$q(t)$	t	$q(t)$
0.	0.577	0.3	0.663	1.5	0.705
0.01	0.588	0.4	0.673	2.0	0.708
0.03	0.601	0.5	0.680	2.5	0.709
0.05	0.611	0.6	0.686	3.0	0.710
0.1	0.628	0.8	0.694	5.0	0.710
0.2	0.650	1.0	0.699	∞	0.710

Table 14.1 Values of the Hopf function $q(t)$ for a selection of values of t

This equation can be transformed by exchanging the order of integration and applying the substitution $w = 1/\mu$ in the first integral, and the substitution $w = -1/\mu$ in the second integral. By doing so, we obtain

$$B(t) = \frac{1}{2}\left[\int_t^\infty dt'\, B(t') \int_1^\infty e^{-w(t'-t)}\frac{dw}{w} + \int_0^t dt'\, B(t') \int_1^\infty e^{-w(t-t')}\frac{dw}{w}\right],$$

from which, recalling the definition of the integro-exponential functions given by Eq. (14.4), we obtain

$$B(t) = \frac{1}{2}\int_0^\infty B(t')\mathrm{E}_1\left(|t - t'|\right)dt'.$$

We now introduce the Hopf function (Eq. (14.10)) and we find the following integral equation, known as the Hopf equation

$$t + q(t) = \frac{1}{2}\int_0^\infty [t' + q(t')]\mathrm{E}_1\left(|t - t'|\right)dt'.$$

This equation can be solved numerically to obtain the results shown in Table 14.1. As we can see, the Hopf function is monotonic and increases from the 0.577 value (for $t = 0$) to the 0.710 value (for $t \to \infty$). It effectively differs very little from the $\frac{2}{3}$ value, obtained by means of the Eddington approximation.

The exact solution for the function $T(t)$ of a grey atmosphere can be expressed using the Hopf function. Repeating the steps that led to Eq. (14.9), we easily find that such equation must be substituted by the following one

$$T(t) = T_{\text{eff}}\sqrt[4]{\frac{3}{4}[t + q(t)]}.$$

This function is plotted in Fig. 14.5 (full line), together with the function of Eq. (14.9) (dashed line), obtained through the Eddington approximation. The two curves differ very little, with a maximum percentage variation of 3.5 % at $t = 0$.

14.5 Realistic Models of Stellar Atmospheres

The model of the grey atmosphere that we have discussed in the two previous sections is a rough approximation for a stellar atmosphere since the absorption coefficient is actually a variable function of frequency and can be assumed constant only

Fig. 14.5 Variation of the temperature as a function of t in a grey atmosphere. The *full line* is the exact solution while the *dashed line* is the solution obtained with the Eddington approximation

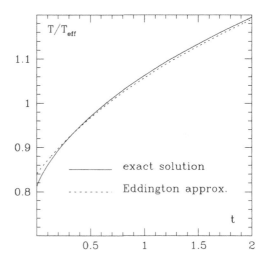

in reasonably narrow frequency intervals. The advantage of the model is only to provide an analytical (or semi-analytical) approximation of the physical structure of the atmosphere. From this point of view it can be compared to the polytropic models of stellar interiors.

With the advent of modern computers it has been possible, since the end of the 1950s, to build more realistic models of stellar atmospheres. These models involve the self-consistent solution of a set of differential equations and are based, in general, on the usual approximation of the plane-parallel atmosphere in local thermodynamic equilibrium. Together with the equation of radiative transfer for the specific intensity

$$\mu\frac{d}{dz}I_\nu(z,\mu) = -k_\nu\big[I_\nu(z,\mu) - B_\nu(z)\big],$$

the equation of radiative equilibrium, that of hydrostatic equilibrium, and the perfect gas law are considered, namely

$$\frac{dF}{dz} = \frac{d}{dz}\int_0^\infty d\nu 2\pi \int_{-1}^1 \mu I_\nu(z,\mu)\,d\mu = 0,$$

$$\frac{dP}{dz} = -\rho g, \qquad P = \frac{\rho}{\bar{\mu}m_H}k_B T.$$

In these equations, P is the pressure of the gas in the stellar atmosphere, ρ is its density, g is the gravity at the stellar surface, $\bar{\mu}$ is the mean molecular weight and m_H is the unit of atomic weight. Considering P and T as independent variables, and assuming that we know how k_ν and $\bar{\mu}$ are related to P and T (see the next section for the expression of k_ν as a function of these two variables), we can numerically solve the equations by taking into account the suitable boundary conditions. Such conditions are the following ones

$$F = \sigma T_{eff}^4,$$

which sets the value of the radiation flux;

$$I_\nu(0, \mu < 0) = 0,$$

which means that the star is isolated, hence not externally illuminated.

From the solution of the equations one obtains the model of the stellar atmosphere, i.e. a table of $P(z)$ and $T(z)$ values. The model depends explicitly only on three parameters: the effective temperature T_{eff}, the gravity at the stellar surface g, and a set of numbers $\{A_i\}$ which characterise the chemical abundances of the elements. The dependence on this latter parameter is contained in the $k_\nu(P, T)$ and $\bar{\mu}(P, T)$ functions.

14.6 The Continuum Spectrum

For the analysis of stellar spectra it is often necessary to calculate, as a function of frequency, the specific intensity that emerges from a stellar atmosphere described by an appropriate theoretical model.[1] Once the functions $P(z)$ and $T(z)$ are given, the integrals appearing in the r.h.s. of Eq. (14.3) or (14.5) need to be evaluated. The greatest difficulty in the calculation is in expressing the relation between the optical depth and the height, i.e. in finding the function

$$t_\nu = t_\nu(z).$$

Since this relation is derived from the integration of the differential equation

$$dt_\nu = -k_\nu(z)\,dz,$$

the problem is ultimately reduced in finding the expression of the absorption coefficient k_ν for a given height z.

In the spectral regions where no line is present, the absorption coefficient (also called opacity in astrophysics) is solely due to the bound-free, free-free and scattering processes. The processes that contribute most to the opacity in stellar atmospheres are the following ones:

(a) photo-ionisation of the hydrogen atom. An hydrogen atom in a bound state absorbs a photon and becomes ionised. Such process can schematically be written as a "reaction" of the type $H + h\nu \rightarrow H^+ + e^-$;
(b) free-free transitions between states of positive energy of the hydrogen atom, i.e. processes of inverse *Bremsstrahlung*: $H^+ + e^- + h\nu \rightarrow H^+ + e^-$;
(c) photo-ionisation of the negative hydrogen ion: $H^- + h\nu \rightarrow H + e^-$;
(d) processes of the type (a) and (b) for the helium atom and any other element relatively abundant in stellar atmospheres (O, C, N, Si, Mg, Ne, Fe, etc.);

---

[1]In the case of the Sun, given its close proximity and the possibility to perform very detailed observations. appropriate empirical models are also available. These models explicitly take into account the fact that the atmosphere is not strictly in radiative equilibrium. For an in-depth discussion see Landi Degl'Innocenti (2008).

(e) Thomson scattering on free electrons and Rayleigh scattering on atoms or ions.

The contribution to the absorption coefficient for each process can be computed using the general methods described in Sect. 11.9, with the exception of the processes related to step (e) for which the classical results of Chap. 3 can be used. Without analysing in detail all the processes, we explicitly consider here only the contribution to the absorption coefficient due to the H^- ion. This can be regarded as the prototype of the various processes and in any case it is the most important for the atmosphere of the Sun and the stars of solar type.

As we have already noted in Sect. 7.5, a hydrogen atom and a free electron can "combine" to produce a stable negative ion having a binding energy of the order of 0.75 eV. The resulting state, in analogy to the ground state of helium, is a state of the type $1s^2\,^1S_0$. The stability of this ion was predicted theoretically by Bethe (1929) by means of a variational calculation similar to that described in Sect. 7.5 but containing a greater number of free parameters. The value currently accepted for the binding energy of the H^- ion is equal to 0.75416 eV. A photon having a wavelength less than the threshold value of 1.6438 μm is able to ionize the H^- ion and the relative absorption coefficient is given by Eq. (11.34), or

$$k_\nu^{(a)} = \sigma_\nu \mathcal{N}_{H^-},$$

where $\mathcal{N}_{H^-}$ is the density of the H^- ions (number of ions per unit volume) and σ_ν is the cross section of the process. The theoretical calculation of the cross section implies, as we have seen, the evaluation of the dipole matrix element between the initial and final states of the transition and this, in turn, implies the knowledge of the eigenfunctions of the bound state and of the free states of the H^- ion. Detailed calculations performed by Chandrasekhar and collaborators in the 1950s give for σ_ν the values plotted in Fig. 14.6. The cross section shows a maximum of about 1.4 a_0^2 (a_0 being the radius of the first Bohr orbit) around 8500 Å and is slightly lower than this value throughout the whole visible spectrum (the region where the intensity of the radiation field is maximum for solar-type stars).

Once σ_ν is known, the problem of finding the value of the absorption coefficient becomes the problem of determining the density of the H^- ions in an atmosphere characterized by given values of P and T. This is in turn a very general problem that, as we shall see, involves the ionisation balance of all the elements present in the atmosphere. The details of the calculation, developed in the following, can be considered as a characteristic example of similar calculations that need to be carried out to determine the density of all the chemical species that are present in a stellar atmosphere.

Suppose we know the abundances of all the elements relative to hydrogen, and denote these abundances with the symbol A_{He} for helium, and A_{M_i} for the generic element characterised by the index i. If we also suppose, for simplicity, that in the atmosphere the maximum degree of ionisation of helium and of all other elements is 1, and we neglect the contribution of the "minor" species (such as molecules and

Fig. 14.6 Photo-ionisation
cross section of the negative
ion of the hydrogen atom.
The cross section is in units
of a_0^2, while the wavelength is
in μm

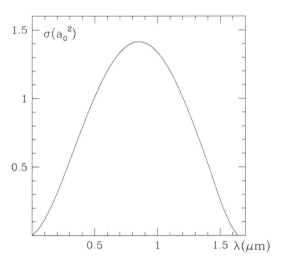

the H^- ion itself, which is generally a good approximation),[2] we have by Dalton's
law of partial pressures

$$P = P_e + (P_H + P_{H^+}) + (P_{He} + P_{He^+}) + \sum_i (P_{M_i} + P_{M_i^+}),$$

where P_e is the electron pressure, P_H the pressure due to the neutral hydrogen atoms,
P_{H^+} the pressure due to the ionised hydrogen atoms, and so on. On the other hand,
since the plasma is neutral, we must have

$$P_e = P_{H^+} + P_{He^+} + \sum_i P_{M_i^+}.$$

We now introduce the ionisation ratios for the single species

$$x = \frac{P_{H^+}}{P_H + P_{H^+}}, \qquad y = \frac{P_{He^+}}{P_{He} + P_{He^+}}, \qquad z_i = \frac{P_{M_i^+}}{P_{M_i} + P_{M_i^+}},$$

that, according to the Saha equation (Eq. (10.7)), are known functions of T and P_e.
Substituting in the previous equations and introducing the abundances relative to
hydrogen we have

$$P = P_e + (P_H + P_{H^+})\left(1 + A_{He} + \sum_i A_{M_i}\right),$$

$$P_e = (P_H + P_{H^+})\left(x + y A_{He} + \sum_i z_i A_{M_i}\right),$$

---

[2]An exception are the stars of late spectral type (cool stars) for which the molecules play an important role.

from which, eliminating $(P_H + P_{H^+})$, we get

$$P = P_e \left(1 + \frac{1 + A_{He} + \sum_i A_{M_i}}{x + y A_{He} + \sum_i z_i A_{M_i}} \right).$$

Regarding this equation, we note that since all the abundances are negligible with respect to those of hydrogen and helium, the sum over i at the numerator can be omitted. The analogous sum at the denominator cannot be omitted, however, since, due to the low ionisation potential of the metals, some of the z_i may be close to unity while x and y are nearly null. For this reason P is a function of P_e that is very sensitive to the A_{M_i} abundances, something that is intuitive from a physical point of view. Returning now to the above equation and recalling that x, y and z_i are functions of T and P_e, we have

$$P = P(T, P_e).$$

This equation can be inverted numerically to obtain

$$P_e = P_e(T, P).$$

Once P_e is known, the partial pressures of the single species can easily be obtained. For example, the partial pressure of neutral hydrogen is

$$P_H = (1 - x)(P_H + P_{H^+}) = (1 - x)\frac{P - P_e}{1 + A_{He} + \sum_i A_{M_i}}.$$

Finally, we can determine, using the Saha equation, the ratio r between the partial pressure of the H^- ion and the partial pressure of neutral hydrogen. We have

$$N_{H^-} = \frac{P_{H^-}}{k_B T} = \frac{r P_H}{k_B T} = \frac{r(1 - x)}{k_B T} \frac{P - P_e}{1 + A_{He} + \sum_i A_{M_i}}.$$

It should be noted that this result was obtained by assuming that the density of the negative hydrogen atoms was much less than the densities of the neutral or ionised hydrogen atoms (otherwise the explicit contribution of the partial pressure of the H^- ions should have been included in the starting equations for P and P_e). This approximation is fully justified for stellar atmospheres where the fraction of H^- ions is negligible compared to the total hydrogen (in the solar atmosphere, for example, the ratio N_{H^-}/N_H varies between 10^{-9} and 10^{-7}).

We have therefore solved the problem of determining the contribution of the H^- ion to the absorption coefficient k_ν. In reality we also need to take into account the fact that the absorption coefficient we have calculated is $k_\nu^{(a)}$ and not $k_\nu = k_\nu^{(a)} - k_\nu^{(s)}$. However, when in LTE, we simply have

$$k_\nu = k_\nu^{(a)} \left[1 - e^{-h\nu/(k_B T)} \right],$$

so that the correction due to the stimulated emission is easily introduced. All the other contributions to k_ν can be obtained with similar calculations, and the absorption coefficient then becomes determined as a function of T and P, i.e. as a function of height z. This allows us to solve the transfer equation and determine the continuum spectrum of the radiation emitted by the star.

14.7 Spectral Lines in Local Thermodynamic Equilibrium

Around a spectral line in local thermodynamic equilibrium, the transfer equation
assumes the form

$$\mu \frac{dI(\nu, \mu)}{dz} = -\big[k_\nu + k_R \varphi(\nu - \nu_0)\big]\big[I(\nu, \mu) - B_\nu(T)\big],$$

where k_R is the line absorption coefficient corrected for stimulated emission and integrated in frequency,[3] and where $\varphi(\nu - \nu_0)$ is the profile (normalised in frequency)
given by Eq. (11.37). The continuum absorption coefficient k_ν is practically constant around a spectral line (the width of a line is typically a fraction of an Å while
k_ν varies on a scale of the order of hundreds of Å). We can therefore put $k_\nu = k_c$,
with k_c independent of frequency. We then define the optical depth in the continuum
t_c by the equation

$$dt_c = -k_c \, dz.$$

With this definition the transfer equation assumes the form

$$\mu \frac{dI(\nu, \mu)}{dt_c} = \big[1 + \eta_0 H(\nu, a)\big]\big[I(\nu, \mu) - B_\nu(T)\big],$$

where we have put

$$\eta_0 = \frac{k_R}{k_c \sqrt{\pi} \Delta \nu_D}.$$

This equation can be solved numerically for a given atmospheric model. The
calculation of the quantity η_0 can be done in a similar way to what was done in
the previous section for the calculation of the continuum absorption coefficient. It
is just necessary to calculate the density of the absorbing atoms (ionized atoms of
iron, for example, in the case of a line belonging to the spectrum of Fe II) and deduce, using the Boltzmann equation, the fraction of these atoms in the lower level of
the transition. The knowledge of such a quantity, together with that of the Einstein
coefficients for the transition, and to that of k_c and $\Delta \nu_D$, allows one to obtain η_0
at all heights. Regarding the Voigt profile, i.e. the function $H(\nu, a)$, it is then necessary to know the value of the reduced damping constant a and the value of the
Doppler width $\Delta \nu_D$. In general, these quantities also vary with z because both the
pressure (and therefore the frequency of collisions) and the temperature vary. The
Voigt profile is therefore a function of z and this should be taken into account in the
numerical calculation.

Although the quantitative analysis of spectral line profiles requires, in many
cases, a numerical solution of the transfer equation, an analytical solution can be
found by introducing a number of simplifying assumptions. These assumptions, although not strictly verified in stellar atmospheres, can give a qualitative idea about
the mechanisms that contribute to characterize the shapes of the line profiles observed in stellar spectra. Suppose then that:

---

[3] At local thermodynamic equilibrium we have $k_R = k_R^{(a)}\{1 - \exp[-h\nu_0/(k_B T)]\}$, with $k_R^{(a)}$ defined
in Eq. (11.33).

(a) the ratio η_0 between the absorption coefficient in the line and the absorption coefficient in the continuum is constant with the optical depth t_c;
(b) the Voigt profile $H(v, a)$ is constant with t_c;
(c) the Planck function $B_\nu(T)$ is a linear function of t_c

$$B_\nu(T) = B_0(1 + \beta t_c),$$

with β constant. When we introduce these assumptions we are dealing with a so-called Milne-Eddington atmosphere.

Substituting the expression of $B_\nu(T)$ in the formal solution of the transfer equation we obtain for the emerging intensity

$$I_\nu(0, \mu) = B_0\left[1 + \frac{\beta\mu}{1 + \eta_0 H(v, a)}\right].$$

We note that, far from the centre of the line, the function $H(v, a)$ tends to zero and the intensity to a constant value (which represents the intensity of the continuum next to the line) given by

$$I_c = B_0(1 + \beta\mu).$$

Conversely, if we consider the limit of a very strong line ($\eta_0 \to \infty$), the intensity around line center tends to a saturation value (which is a lower limit) given by

$$I_s = B_0.$$

We can therefore define a sort of "universal profile" r_ν by subtracting the saturation value from the intensity and then normalising to the continuum intensity. We have

$$r_\nu = \frac{I_\nu(0, \mu) - I_s}{I_c} = \frac{\beta\mu}{1 + \beta\mu}\frac{1}{1 + \eta_0 H(v, a)}.$$

The $\beta\mu/(1 + \beta\mu)$ factor is related to the thermodynamic characteristics of the atmosphere (through β) and the value of the heliocentric angle (through μ). The remaining factor gives the form of the line profile. It depends on frequency only through the parameter v (recall that $v = (\nu - \nu_0)/\Delta\nu_D$). Denoting such factor by $p(v)$, i.e. putting

$$p(v) = \frac{1}{1 + \eta_0 H(v, a)},$$

we note that for weak lines ($\eta_0 \ll 1$) we have, by a series expansion

$$p(v) \simeq 1 - \eta_0 H(v, a).$$

in this case we are far from saturation and the line profile is a sort of "negative" of the profile of the absorption coefficient. For strong lines ($\eta_0 \gg 1$), instead, we obtain that $p(v)$ tends to zero for a whole range of values of v such that $\eta_0 H(v, a) \gg 1$. In this case we have extreme saturation. The line profile is very different from that of the absorption coefficient, having a nearly flat central profile (in the so-called core of the line) and very extended wings (that depend strongly on the value of the damping constant a). Figure 14.7 shows the shape of the profile $p(v)$ for different values of the line strength η_0 and for $a = 0.1$.

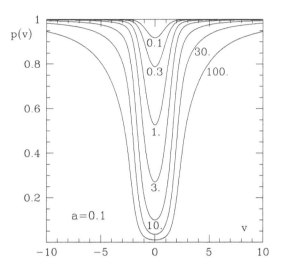

Fig. 14.7 Graph of the profile of $p(v)$ as a function of the reduced wavelength v, for different values of the line strength η_0. The plot is obtained for a damping constant $a = 0.1$

14.8 Spectral Lines in Non-equilibrium Conditions

The local thermodynamic equilibrium hypothesis allows one to determine with a certain simplicity the line profiles once a model stellar atmosphere is known. This hypothesis is however not always verified because the strongest lines are formed in the outer layers of a stellar atmosphere, where the density is lower and the collisions with the electrons in the plasma are not sufficient to thermalise the atomic populations (chromosphere). In the case of the solar atmosphere, for example, the hypothesis of local thermodynamic equilibrium cannot be applied for the calculation of the profiles of the Ca II H and K lines, of the spectral lines of hydrogen (such as e.g. H α), of the Na I D lines, and several other spectral lines in the visible and in the ultraviolet. In these cases the transfer equation is the same as that written in the previous section with the difference that, in place of the Planck function $B_\nu(T)$, one must consider the source function S_ν. However, as shown by Eq. (13.3) for the two-level atom case, the source function depends on the radiation field itself and the formal solution of the transfer equation is not sufficient anymore to solve the problem.

In these cases we resort to more complex calculations which involve the so-called non-LTE theory.[4] Simplifying as much as possible the method, the problem can be addressed as follows. Suppose that we know, as a function of the height z, a zero order approximation for the populations of the levels of a given atomic species (for example the populations of the levels of the ionised calcium atom, if we are interested in calculating the profiles of the Ca II H and K lines). One could start, for example, from the populations obtained from the Saha-Boltzmann equations using the values of P and T given by the model atmosphere. Once the populations are known, we then locally compute the coefficients appearing in the radiative transfer

---

[4]For an in-depth discussion see, for example, Mihalas (1978).

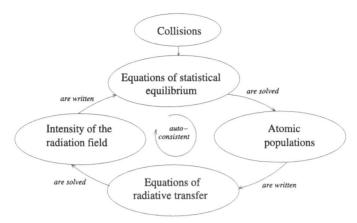

Fig. 14.8 To solve the coupled equations of the statistical equilibrium and radiative transfer we follow the auto-consistent loop method shown here

equation (written for each direction in the stellar atmosphere). Solving with numerical methods the transfer equation, we can then determine the radiation field present at any point of the atmosphere (as a function of frequency and direction). At this point we consider the statistical equilibrium equations for the atomic populations taking into account both radiative and collisional processes. These equations form, for each height z, a system with known coefficients (since the radiation field is known) and the system may be solved with appropriate numerical methods. This gives a first-order approximation for the populations and the process is repeated until we reach convergence of the solution as exemplified in Fig. 14.8. The solution for the radiation field at the surface of the stellar atmosphere provides the profiles of the spectral lines of interest.

Although this method is, in principle, simple and direct, it requires the knowledge of a number of numerical techniques on which we cannot delve here. Such techniques have been developed since the early 1950s and are still evolving.

Chapter 15
Second Order Processes

We have shown in Chap. 11 how one can describe the quantum interaction between the radiation field and material systems. We used an appropriate theoretical framework, known as quantum electrodynamics, to obtain, with an expansion at the lowest perturbative order, a number of results about the more elementary processes that occur in the interaction between radiation and atomic systems, namely the processes of absorption and emission (spontaneous and stimulated) of photons. Quantum electrodynamics has however a much wider scope. The purpose of this chapter is to deduce further consequences, obtained by extending the perturbation theory to the second order. These consequences affect the interaction of radiation with electrons, free or bound, either non-relativistic or relativistic. In particular, we will analyse the scattering processes of free non-relativistic electrons (Thomson scattering), of bound non-relativistic electrons (Rayleigh and Raman scattering) and of relativistic electrons (Compton effect). The study of this latter effect will lead us to obtain the properties of the cross section (equation of Klein-Nishina) and of the polarisation of the emitted radiation. We will also develop some considerations with respect to the exchange of energy between photons and relativistic electrons, which are the basis of the inverse Compton effect. The topics covered in this chapter are a valuable complement to the ones we already developed in Chap. 3 using exclusively classical (non-quantum) electrodynamics.

15.1 Introduction

Within the framework of quantum electrodynamics we discussed in Sect. 11.5 the elementary processes of interaction between radiation and material systems. In that section we saw how the interaction Hamiltonian between an atomic system and the radiation field leads to the appearance of particular transitions described by the Feynman diagrams of Fig. 11.2. These diagrams refer to the processes of absorption or emission in which a single photon is involved and, for this reason, are rightly called processes of the first order. The probability of transition per unit time is given

E. Landi Degl'Innocenti, *Atomic Spectroscopy and Radiative Processes*,
UNITEXT for Physics, DOI 10.1007/978-88-470-2808-1_15, © Springer-Verlag Italia 2014

by Fermi's golden rule (Eq. (11.10)) and is null if the conservation of energy between the initial the final state of the transition is not verified.

The conservation of energy, and the fact that a single photon can be involved in the process, considerably limits the number of physical phenomena that can be described by first-order processes. If we consider, for example, the interaction between photons and free electrons, we easily see that first order processes are not possible. Consider in fact the hypothetical absorption of a photon by a free electron and, in the first instance, let us assume that the energies involved are non-relativistic. When we evaluate the transition probability per unit time, the matrix element is not null only if the conservation of the momentum is verified (see Sect. 16.13). Therefore we must have

$$\mathbf{p}_f = \mathbf{p}_i + \hbar \mathbf{k},$$

where $\mathbf{p}_i$ and $\mathbf{p}_f$ are, respectively, the initial and final momenta of the free electron, and where $\mathbf{k}$ is the wave vector of the absorbed photon. On the other hand, for the conservation of energy, implicit in Fermi's golden rule, we must also have

$$\frac{p_f^2}{2m} = \frac{p_i^2}{2m} + \hbar \omega,$$

where $\omega = ck$. Substituting the conservation of the momentum in this equation and introducing the angle θ, formed by the directions of the vectors $\mathbf{p}_i$ and $\mathbf{k}$, we obtain the following kinematic relation

$$\hbar \omega = 2mc^2 \left(1 - \frac{p_i \cos \theta}{mc} \right).$$

The result we have obtained with our non-relativistic hypothesis is however not acceptable, as the energy of the photon results in being of the order of the rest energy of the electron, which is in contradiction with our assumption. We can then try to drop this hypothesis by using the relativistic formula for energy conservation, namely

$$\sqrt{c^2 p_f^2 + m^2 c^4} = \sqrt{c^2 p_i^2 + m^2 c^4} + \hbar \omega.$$

Substituting again, after a few algebraic steps we obtain the equation

$$p_i^2 \sin^2 \theta + m^2 c^2 = 0,$$

that cannot in any way be satisfied. The example we have considered clearly illustrates the implicit limitations in quantum electrodynamics, when only first order processes are accounted for.

Another example is the interaction of a photon of a given frequency with an atomic system whose energy levels have energies ϵ_n ($n = 0, 1, 2, \ldots$). If we assume for example that the atom is initially in the ground energy level, having energy ϵ_0, and if the frequency ω is such that none of the following relations is satisfied

$$\hbar \omega = \epsilon_n - \epsilon_0 \quad (n = 1, 2, \ldots),$$

then the atom is completely "transparent" to the radiation itself, since no absorption process that preserves energy is possible.

In the previous examples, it is possible to assess the consequences of the interaction between the atomic system and the radiation field by considering, in the spirit of perturbation theory, processes of order higher than the first. We can for example consider processes of the second order, which imply the presence in the amplitude of the transition probability of two operators of the type[1] $\mathbf{A_R}$, or we could consider processes of the third order, and so on. Obviously, the transition probability decreases as higher orders are considered. Also, the corresponding processes involve an increasing number of photons (as many as the order of the process that we are considering). In this chapter we will focus on the processes of the second order. We note that within the non-covariant theory we developed in Chap. 11, there are two different possible ways to take into account these processes.

The first possibility consists in recalling that the interaction Hamiltonian contains two terms, one linear and one quadratic in the operator $\mathbf{A_R}$. The quadratic term was systematically neglected in Chap. 11, because in that case we were only interested in first order processes. This quadratic term is given by the expression (cf. Eq. (11.2))

$$\mathcal{H}^I_{\text{quad}} = \frac{e_0^2}{2mc^2} \sum_{i=1}^{N} \left[\mathbf{A_R}(\mathbf{r}_i) \right]^2, \tag{15.1}$$

where $\mathbf{r}_i$ $(i = 1, \ldots, N)$ is the coordinate of the i-th electron of the atomic system. The transition probability per unit of time for the physical processes induced by this term is given by the usual Fermi golden rule (Eq. (11.10))

$$P_{\alpha\beta} = \frac{2\pi}{\hbar} \left| \left(\mathcal{H}^I_{\text{quad}} \right)_{\alpha\beta} \right|^2 \delta(E_\alpha - E_\beta).$$

The second possibility is to consider in the interaction Hamiltonian only the term linear in the operator $\mathbf{A_R}$, but expand to the second order the probability amplitude. In this case we go back to Eq. (11.6) and assume that the first-order processes do not contribute and that, at the initial time $t = 0$, the physical system is in the state $|\gamma\rangle$ $(c_\gamma(0) = 1)$. By simple integration we get

$$c_\alpha(t) = \frac{1}{\hbar^2} \sum_\beta \mathcal{H}^I_{\alpha\beta} \mathcal{H}^I_{\beta\gamma} \left(\frac{e^{i\omega_{\alpha\gamma}t} - 1}{\omega_{\alpha\gamma}\omega_{\beta\gamma}} - \frac{e^{i\omega_{\alpha\beta}t} - 1}{\omega_{\alpha\beta}\omega_{\beta\gamma}} \right).$$

The first of the two terms in brackets is by far the dominant one when we consider final states α such that $\omega_{\alpha\gamma} \to 0$, i.e. states iso-energetic with the initial one. We can therefore write

$$c_\alpha(t) = \frac{1}{\hbar} \mathcal{M}_{\alpha\gamma} \frac{e^{i\omega_{\alpha\gamma}t} - 1}{\omega_{\alpha\gamma}},$$

where

$$\mathcal{M}_{\alpha\gamma} = \sum_\beta \frac{\mathcal{H}^I_{\alpha\beta} \mathcal{H}^I_{\beta\gamma}}{\hbar\omega_{\beta\gamma}}. \tag{15.2}$$

---

[1] Recall that $\mathbf{A_R}$ is the quantum operator associated with the vector potential of the radiation field. Its explicit expression in terms of operators of creation and annihilation is given by Eq. (4.5).

If we consider now the squared modulus of $c_\alpha(t)$, i.e. the probability that at time t the system is in the state $|\alpha\rangle$, we obtain

$$|c_\alpha(t)|^2 = \frac{1}{\hbar^2}|\mathcal{M}_{\alpha\gamma}|^2 \frac{4\sin^2(\omega_{\alpha\gamma}t/2)}{\omega_{\alpha\gamma}^2}.$$

On the other hand, as we have often seen in this volume, we have

$$\lim_{t\to\infty} \frac{4\sin^2(\omega_{\alpha\gamma}t/2)}{\omega_{\alpha\gamma}^2} = 2\pi t\delta(\omega_{\alpha\gamma}),$$

so we obtain a probability per unit time for the transition from the state $|\gamma\rangle$ to the state $|\alpha\rangle$ given by

$$P_{\alpha\gamma} = \frac{2\pi}{\hbar}|\mathcal{M}_{\alpha\gamma}|^2\delta(E_\alpha - E_\gamma), \tag{15.3}$$

where the matrix element $\mathcal{M}_{\alpha\gamma}$ is given by Eq. (15.2) or, through a more significant expression, by

$$\mathcal{M}_{\alpha\gamma} = \sum_\beta \frac{\mathcal{H}_{\alpha\beta}^I \mathcal{H}_{\beta\gamma}^I}{E_\alpha - E_\beta} = \sum_\beta \frac{\mathcal{H}_{\alpha\beta}^I \mathcal{H}_{\beta\gamma}^I}{E_\gamma - E_\beta}. \tag{15.4}$$

Equation (15.3) can be regarded as a direct generalisation of Fermi's golden rule to second-order processes. Again, the presence of the Dirac delta function means that the transition probability is different from zero only between states having the same energy. Considering the structure of the equation, and in particular that one of the matrix element $\mathcal{M}_{\alpha\gamma}$, we can interpret the transition process as a succession of two subsequent events. We have first a transition in which the system goes from the initial state $|\alpha\rangle$ to the intermediate state $|\beta\rangle$. In such state the energy is not conserved, and the state must be considered as a kind of "virtual state" in which the system can exist only for a very short time interval Δt (according to the uncertainty principle we have that $\Delta t \Delta E \simeq \hbar$, with ΔE equal to the energy difference between the initial or the final state and the intermediate state). Subsequently, the physical system goes into the final state $|\gamma\rangle$. We can therefore think that the transition occurs "via" the intermediate state. In the following, we will refer to Eq. (15.3) with the name of the generalised Fermi golden rule.

Before we can move on to applications, we have to see how the formalism introduced in Chap. 11 needs to be modified in order to study the processes of interaction between radiation and relativistic electrons. The only difference consists in the fact that, to obtain the interaction Hamiltonian, we must apply the principle of minimal coupling (Eq. (5.5)) to the Dirac Hamiltonian for the free particle instead of the Schrödinger Hamiltonian. Recalling that the Dirac Hamiltonian $\mathcal{H}_D$ is given by (see Eq. (5.1))

$$\mathcal{H}_D = c\boldsymbol{\alpha} \cdot \mathbf{p} + \beta mc^2,$$

denoting the new interaction Hamiltonian with the symbol $\mathcal{H}^{\mathrm{I}}_{\mathrm{rel}}$ and considering the interaction of the radiation field with a system of N electrons, we have

$$\mathcal{H}^{\mathrm{I}}_{\mathrm{rel}} = e_0 \sum_{i=1}^{N} \boldsymbol{\alpha} \cdot \mathbf{A}_{\mathrm{R}}(\mathbf{r}_i), \tag{15.5}$$

where $\boldsymbol{\alpha}$ are the Dirac matrices and $\mathbf{r}_i$ is the coordinate of the i-th electron.

15.2 Thomson Scattering (Quantum Approach)

Consider a photon of frequency ω interacting with a free electron at rest and suppose that the photon energy $\hbar\omega$ is much lower than mc^2, the rest energy of the electron. As we will see in more detail later on, when we discuss Compton scattering, under this limit the process of scattering takes place in a coherent way, i.e. the scattered photon has the same energy as the incident photon and the electron undergoes a negligible recoil. We are interested in calculating the cross section for the process in which a photon having wavenumber $\mathbf{k}$ (with $k = \omega/c$) and polarisation characterised by the unit vector $\mathbf{e}$ (in general complex) is scattered by an angle Θ, producing a photon with wavenumber $\mathbf{k}'$, frequency ω', and polarisation characterised by the unit vector $\mathbf{e}'$ (also generally complex). To do this, we calculate the transition probability per unit time using Fermi's golden rule in the form

$$P_{\mathrm{fi}} = \frac{2\pi}{\hbar} \left| \langle \Psi_{\mathrm{f}} | \mathcal{H}^{\mathrm{I}}_{\mathrm{quad}} | \Psi_{\mathrm{i}} \rangle \right|^2 \delta(E_{\mathrm{i}} - E_{\mathrm{f}}),$$

where $|\Psi_{\mathrm{i}}\rangle$ and $|\Psi_{\mathrm{f}}\rangle$ are, respectively, the initial and final eigenstates of the whole system formed by the electron and the radiation field and $\mathcal{H}^{\mathrm{I}}_{\mathrm{quad}}$ is the "quadratic" interaction Hamiltonian given by Eq. (15.1), for the case of a single electron ($N = 1$). Taking into account the expansion of the vector potential $A_{\mathrm{R}}(\mathbf{r})$ in terms of creation and annihilation operators (Eq. (4.5)), the Hamiltonian is

$$\mathcal{H}^{\mathrm{I}}_{\mathrm{quad}} = \frac{\pi \hbar e_0^2}{mV} \sum_{\mathbf{k}\lambda} \sum_{\mathbf{k}'\lambda'} \frac{1}{\sqrt{\omega_{\mathbf{k}}\omega_{\mathbf{k}'}}} \left[a_{\mathbf{k}\lambda} \mathbf{e}_{\mathbf{k}\lambda} e^{i\mathbf{k}\cdot\mathbf{r}} + a^{\dagger}_{\mathbf{k}\lambda} \mathbf{e}^{*}_{\mathbf{k}\lambda} e^{-i\mathbf{k}\cdot\mathbf{r}} \right]$$
$$\times \left[a_{\mathbf{k}'\lambda'} \mathbf{e}_{\mathbf{k}'\lambda'} e^{i\mathbf{k}'\cdot\mathbf{r}} + a^{\dagger}_{\mathbf{k}'\lambda'} \mathbf{e}^{*}_{\mathbf{k}'\lambda'} e^{-i\mathbf{k}'\cdot\mathbf{r}} \right].$$

The initial and final states of the transition are the following: from the point of view of the electromagnetic field, in the initial state we have a photon of frequency ω, wavenumber $\mathbf{k}$ (with $\omega = ck$), and polarisation characterised by the unit vector $\mathbf{e}$, while in the final state we have a photon of frequency ω', wavenumber $\mathbf{k}'$ (with $\omega' = ck'$), and polarisation characterised by the unit vector $\mathbf{e}'$. From the point of view of the atomic system (a single electron) we suppose that in the initial and final states the electron is described by the wavefunctions $\psi_{\mathrm{i}}(\mathbf{r})$ and $\psi_{\mathrm{f}}(\mathbf{r})$, respectively. Using the formalism introduced in Chap. 11, we have

$$|\Psi_{\mathrm{i}}\rangle = \left| \psi_{\mathrm{i}}(\mathbf{r}); 0, 0, \ldots, 1_{\mathbf{k}\mathbf{e}}, \ldots, 0, \ldots \right\rangle,$$

$$|\Psi_{\mathrm{f}}\rangle = \left| \psi_{\mathrm{f}}(\mathbf{r}); 0, 0, \ldots, 0, \ldots, 1_{\mathbf{k}'\mathbf{e}'}, \ldots \right\rangle.$$

Considering the form of the state vectors and the expression of the interaction Hamiltonian, we see that, among the infinite number of terms of the double sum that appears in the expression of $\mathcal{H}^I_{\text{quad}}$, only two provide a non-zero contribution to the matrix element $\langle\Psi_f|\mathcal{H}^I_{\text{quad}}|\Psi_i\rangle$. The first term is the one containing within the first square bracket the operator of annihilation relative to the mode $(\mathbf{ke})$ and in the second square bracket the operator of creation relative to the mode $(\mathbf{k'e'})$. The second term is the one where the order of the same operators is inverted. In other words, to calculate the matrix element we can apply to the Hamiltonian $\mathcal{H}^I_{\text{quad}}$ the following formal substitution

$$\mathcal{H}^I_{\text{quad}} \to \frac{\pi\hbar e_0^2}{mV\sqrt{\omega\omega'}}\left[a_{\mathbf{ke}}\mathbf{e}\,e^{i\mathbf{k}\cdot\mathbf{r}}\cdot a^\dagger_{\mathbf{k'e'}}\mathbf{e'}^*e^{-i\mathbf{k'}\cdot\mathbf{r}} + a^\dagger_{\mathbf{k'e'}}\mathbf{e'}^*e^{-i\mathbf{k'}\cdot\mathbf{r}}\cdot a_{\mathbf{ke}}\mathbf{e}\,e^{i\mathbf{k}\cdot\mathbf{r}}\right].$$

The matrix element can now be easily calculated. Taking into account Eqs. (4.10) and (4.11) to evaluate the matrix elements of the operators of creation and annihilation, we obtain

$$\langle\Psi_f|\mathcal{H}^I_{\text{quad}}|\Psi_i\rangle = \frac{2\pi\hbar e_0^2}{mV\sqrt{\omega\omega'}}(\mathbf{e}\cdot\mathbf{e'}^*)\langle\psi_f(\mathbf{r})\big|e^{i(\mathbf{k}-\mathbf{k'})\cdot\mathbf{r}}\big|\psi_i(\mathbf{r})\rangle.$$

We now need to evaluate the matrix element on the wavefunctions of the electron. In the case of a free electron, such wavefunctions have the form of plane waves, i.e.

$$\psi_i(\mathbf{r}) = \frac{1}{\sqrt{V}}e^{i\mathbf{p}_i\cdot\mathbf{r}/\hbar}, \qquad \psi_f(\mathbf{r}) = \frac{1}{\sqrt{V}}e^{i\mathbf{p}_f\cdot\mathbf{r}/\hbar},$$

where $\mathbf{p}_i$ and $\mathbf{p}_f$ are the momenta of the electron in the initial and final states, respectively. We therefore obtain

$$\langle\psi_f(\mathbf{r})\big|e^{i(\mathbf{k}-\mathbf{k'})\cdot\mathbf{r}}\big|\psi_i(\mathbf{r})\rangle = \frac{1}{V}\int e^{i(-\mathbf{p}_f/\hbar+\mathbf{k}-\mathbf{k'}+\mathbf{p}_i/\hbar)\cdot\mathbf{r}}\,d^3\mathbf{r}.$$

This integral is equal to V if the final momentum of the electron $\mathbf{p}_f$ is such that the total momentum of the system is conserved, i.e. if

$$\mathbf{p}_f = \mathbf{p}_i + \hbar\mathbf{k} - \hbar\mathbf{k'},$$

and is equal to zero otherwise. We have therefore obtained (as one would expect) that in the scattering process the total momentum is conserved. On the other hand, the electron is initially at rest ($\mathbf{p}_i = 0$) and the energy of the photon is much less than the rest energy of the electron. Therefore, the effect of the variation in the momentum of the electron $\Delta\mathbf{p} = \mathbf{p}_f$ on the conservation of energy (hence on the final frequency of the photon) can certainly be neglected. More formally, we can think that the electron does not change its status during the scattering, so that we have $\psi_f(\mathbf{r}) = \psi_i(\mathbf{r})$, and using the dipole approximation for which

$$e^{i(\mathbf{k}-\mathbf{k'})\cdot\mathbf{r}} \simeq 1,$$

we obtain

$$\langle\psi_f(\mathbf{r})\big|e^{i(\mathbf{k}-\mathbf{k'})\cdot\mathbf{r}}\big|\psi_i(\mathbf{r})\rangle \simeq \langle\psi_i(\mathbf{r})\big|e^{i(\mathbf{k}-\mathbf{k'})\cdot\mathbf{r}}\big|\psi_i(\mathbf{r})\rangle \simeq \langle\psi_i(\mathbf{r})|\psi_i(\mathbf{r})\rangle = 1.$$

These last arguments could be repeated for the case of a bound electron in an atom or molecule. In this case we need to impose that the final state of the electron is the same as the initial one. We therefore obtain, for either a free or a bound electron

$$\langle \Psi_f | \mathcal{H}_{\text{quad}}^I | \Psi_i \rangle = \frac{2\pi \hbar e_0^2}{mV\sqrt{\omega\omega'}} (\mathbf{e} \cdot \mathbf{e}'^*),$$

and, for the transition probability per unit time

$$P_{fi} = \frac{8\pi^3 \hbar e_0^4}{m^2 \omega\omega' V^2} |\mathbf{e} \cdot \mathbf{e}'^*|^2 \delta(\hbar\omega - \hbar\omega').$$

We now need to obtain from this expression the cross section for scattering of the initial photon. We start with the usual procedure of dividing the transition probability per unit time by the flux of the incoming particles. Since we have only one photon in the normalisation volume V, the flux is given by c/V. We then need to perform a sum over the final states. This sum depends on the cross section that we intend to define. We are interested in the cross section with fixed polarisation $\sigma(\Theta, \mathbf{e}, \mathbf{e}')$ for the scattering of a photon in the solid angle $d\Omega$ centred on the direction $\mathbf{\Omega}$ which forms an angle Θ with the direction of the initial photon. The number of final photon states having direction within $d\Omega$ and angular frequency within ω' and $\omega' + d\omega'$ is given by the right-hand side of Eq. (4.3) divided by 2, i.e.

$$dN = \frac{V}{8\pi^3 c^3} \omega'^2 \, d\omega' \, d\Omega,$$

so we obtain

$$\sigma(\Theta, \mathbf{e}, \mathbf{e}') \, d\Omega = \frac{\hbar e_0^4}{m^2 c^4 \omega} |\mathbf{e} \cdot \mathbf{e}'^*|^2 \, d\Omega \int \omega' \delta(\hbar\omega - \hbar\omega') \, d\omega'.$$

The presence of the Dirac delta has the obvious consequence that $\omega' = \omega$ so we obtain the final expression for the differential cross section

$$\sigma(\Theta, \mathbf{e}, \mathbf{e}') = \frac{e_0^4}{m^2 c^4} |\mathbf{e} \cdot \mathbf{e}'^*|^2,$$

or, recalling the definition of the classical radius of the electron (Eq. (3.26)),

$$\sigma(\Theta, \mathbf{e}, \mathbf{e}') = r_c^2 |\mathbf{e} \cdot \mathbf{e}'^*|^2. \tag{15.6}$$

As we can see, the differential cross section depends explicitly only on the direction of the polarisation unit vectors and not on the direction of the final photon, the dependence on the latter parameter being implicitly contained in the same unit vectors.

The formula can be used to obtain the polarisation characteristics of the scattered radiation (the radiation diagram) and the total scattering cross section. All the results are consistent with those obtained from classical electrodynamics in Sect. 3.5. For example, we now show how the previous equation can be used to obtain the cross section $\sigma(\Theta)$ for the scattering of the photon in the angle Θ independently of the polarisation properties of the initial and final photons. To obtain this quantity we must perform a sum over the final polarisations and an average over the initial polarisations. We introduce then two real unit vectors of polarisation $\mathbf{e}_1$ and $\mathbf{e}_2$, perpendicular to the direction of the initial photon and perpendicular to each other, and, similarly, two real unit vectors $\mathbf{e}_1'$ and $\mathbf{e}_2'$ perpendicular to the direction of the final photon and perpendicular to each other. We have

$$\sigma(\Theta) = \frac{1}{2} \sum_{i=1,2} \sum_{j=1,2} \sigma\left(\Theta, \mathbf{e}_i, \mathbf{e}_j'\right) = \frac{r_c^2}{2} \mathcal{S},$$

where

$$\mathcal{S} = \sum_{i=1,2} \sum_{j=1,2} \left(\mathbf{e}_i \cdot \mathbf{e}_j'\right)^2.$$

To evaluate this sum, we first note that, given two arbitrary vectors $\mathbf{v}$ and $\mathbf{w}$, and their dyadic product $\mathbf{T} = \mathbf{vw}$, we have

$$\mathrm{Tr}(\mathbf{T}) = \sum_i (\mathbf{vw})_{ii} = \sum_i v_i w_i = \mathbf{v} \cdot \mathbf{w}.$$

Given this property, the sum $\mathcal{S}$ can be written in the form

$$\mathcal{S} = \mathrm{Tr}\left[(\mathbf{e}_1\mathbf{e}_1 + \mathbf{e}_2\mathbf{e}_2) \cdot \left(\mathbf{e}_1'\mathbf{e}_1' + \mathbf{e}_2'\mathbf{e}_2'\right)\right].$$

Introducing the unit vectors $\mathbf{u}$ and $\mathbf{u}'$ parallel to the direction of the initial and final photons, respectively, it follows that $(\mathbf{e}_1, \mathbf{e}_2, \mathbf{u})$ and $(\mathbf{e}_1', \mathbf{e}_2', \mathbf{u}')$ are two Cartesian orthogonal triads. We have, therefore,

$$\mathbf{e}_1\mathbf{e}_1 + \mathbf{e}_2\mathbf{e}_2 + \mathbf{u}\mathbf{u} = \mathbf{U}, \qquad \mathbf{e}_1'\mathbf{e}_1' + \mathbf{e}_2'\mathbf{e}_2' + \mathbf{u}'\mathbf{u}' = \mathbf{U}, \qquad (15.7)$$

where $\mathbf{U}$ is the identity tensor ($\mathbf{U}_{ij} = \delta_{ij}$). Substituting in the expression for $\mathcal{S}$ and taking into account that $\mathrm{Tr}(\mathbf{U}) = 3$, we have

$$\mathcal{S} = \mathrm{Tr}\left[(\mathbf{U} - \mathbf{u}\mathbf{u}) \cdot \left(\mathbf{U} - \mathbf{u}'\mathbf{u}'\right)\right]$$
$$= \mathrm{Tr}(\mathbf{U}) - \mathrm{Tr}(\mathbf{u}\mathbf{u}) - \mathrm{Tr}\left(\mathbf{u}'\mathbf{u}'\right) + \left(\mathbf{u} \cdot \mathbf{u}'\right)\mathrm{Tr}(\mathbf{u}\mathbf{u}') = 1 + \cos^2\Theta,$$

and substituting this result into the expression for $\sigma(\Theta)$

$$\sigma(\Theta) = \frac{r_c^2}{2}\left(1 + \cos^2\Theta\right). \qquad (15.8)$$

Finally, by integrating over the solid angle, we obtain the same classical result of Sect. 3.6

$$\int \sigma(\Theta)\, d\Omega = \frac{8\pi}{3} r_c^2 = \sigma_T.$$

15.3 The Rayleigh and Raman Scattering (Quantum Approach)

We now discuss the Rayleigh and Raman scattering. We consider an atomic system that in its initial state is described by the wavefunction[2] $\psi_i(\mathbf{r})$. A photon of frequency ω, wavenumber $\mathbf{k}$, and polarisation characterised by the unit vector $\mathbf{e}$ is incoming on the atomic system. The photon is scattered, producing a new photon with

---

[2]To simplify notations, we assume that the atomic system is composed of a single optical electron.

frequency ω', wavenumber $\mathbf{k}'$, and polarisation characterised by the unit vector $\mathbf{e}'$. The state of the atomic system is described, after the scattering, by the wavefunction $\psi_f(\mathbf{r})$. If the final state is the same as the initial one, or, more generally, if the energies of the final and initial states are the same, we have Rayleigh scattering. On the contrary, if the energy of the final state is different we have Raman scattering. In any case, denoting by ϵ_i and ϵ_f the energies of the atomic system in their initial and final state, the frequency of the scattered photon is related to that one of the incoming photon by

$$\omega' = \omega + \frac{\epsilon_i - \epsilon_f}{\hbar}.$$

Obviously, $\omega' = \omega$ for Rayleigh scattering, while for Raman scattering ω' can be either larger or smaller than ω. If $\omega' < \omega$ we have the so-called Stokes transitions, while if $\omega' > \omega$ we have the so-called anti-Stokes transitions.

The transition probability of the process can be calculated using the generalised Fermi golden rule (Eqs. (15.3) and (15.4)). Denoting as usual by $|\Psi_i\rangle$ and $|\Psi_f\rangle$ the initial and final states of the whole physical system (the atomic system and the radiation field), and by $|\Psi_n\rangle$ an arbitrary state of the same system (intermediate or virtual state), the transition probability per unit time is given by

$$P_{fi} = \frac{2\pi}{\hbar}|\mathcal{M}_{fi}|^2 \delta(E_f - E_i),$$

where

$$\mathcal{M}_{fi} = \sum_n \frac{\langle \Psi_f | \mathcal{H}^I | \Psi_n \rangle \langle \Psi_n | \mathcal{H}^I | \Psi_i \rangle}{E_i - E_n}.$$

The initial and final states of the whole system are described by the state vectors

$$|\Psi_i\rangle = |\psi_i(\mathbf{r}); 0, 0, \ldots, 1_{\mathbf{ke}}, \ldots, 0, \ldots\rangle,$$

$$|\Psi_f\rangle = |\psi_f(\mathbf{r}); 0, 0, \ldots, 0, \ldots, 1_{\mathbf{k'e'}}, \ldots\rangle.$$

Recalling the expression of the interaction Hamiltonian, we obtain that the only intermediate states that contribute to the sum are of the type

$$|\Psi_n\rangle = |\psi_n(\mathbf{r}); 0, 0, \ldots, 0_{\mathbf{ke}}, \ldots, 0_{\mathbf{k'e'}}, \ldots\rangle,$$

or

$$|\Psi_n\rangle = |\psi_n(\mathbf{r}); 0, 0, \ldots, 1_{\mathbf{ke}}, \ldots, 1_{\mathbf{k'e'}}, \ldots\rangle,$$

where $|\psi_n(\mathbf{r})\rangle$ is any state of the atomic system, whose energy is denoted by ϵ_n. As we have seen for the Thomson scattering, also in this case we have that the radiation field can have two intermediate states, corresponding, respectively, to the annihilation of the photon $(\mathbf{ke})$ from the second Hamiltonian and the creation of the photon $(\mathbf{k'e'})$ from the first Hamiltonian, as well as the opposite case (when the actions of the two Hamiltonians are inverted). The differences in the energies that appear in the denominator of the expression for $\mathcal{M}_{fi}$ are given, in the two cases, by

$$E_i - E_n = \epsilon_i - \epsilon_n + \hbar\omega, \quad \text{and} \quad E_i - E_n = \epsilon_i - \epsilon_n - \hbar\omega'.$$

The matrix element can now be calculated directly from the expressions of the interaction Hamiltonian and of the operator $\mathbf{A}_R(\mathbf{r})$. Using the dipole approximation for the matrix elements relative to the atomic system, we have

$$\mathcal{M}_{\mathrm{fi}} = \frac{2\pi \hbar e_0^2}{m^2 \mathcal{V}\sqrt{\omega\omega'}} \sum_n \left[\frac{\langle \psi_{\mathrm{f}}|\mathbf{p}\cdot\mathbf{e}'^*|\psi_n\rangle \langle \psi_n|\mathbf{p}\cdot\mathbf{e}|\psi_i\rangle}{\epsilon_i - \epsilon_n + \hbar\omega} \right.$$
$$\left. + \frac{\langle \psi_{\mathrm{f}}|\mathbf{p}\cdot\mathbf{e}|\psi_n\rangle \langle \psi_n|\mathbf{p}\cdot\mathbf{e}'^*|\psi_i\rangle}{\epsilon_i - \epsilon_n - \hbar\omega'} \right].$$

The calculation of the cross section then proceeds according to the scheme previously followed for Thomson scattering. The transition probability per unit time must be divided by the flux of the incoming photons and then the sum over the final photon states must be carried out. The differential cross section depends now also on the initial and final electronic states (as well as on the directions and the polarisation of the photons), and therefore requires in its definition the additional indexes f and i. We have, after simple transformations

$$\sigma_{\mathrm{fi}}(\Theta, \mathbf{e}, \mathbf{e}') = r_{\mathrm{c}}^2 \frac{\omega'}{\omega} \left| \frac{1}{m} \sum_n \left[\frac{\langle \psi_{\mathrm{f}}|\mathbf{p}\cdot\mathbf{e}'^*|\psi_n\rangle \langle \psi_n|\mathbf{p}\cdot\mathbf{e}|\psi_i\rangle}{\epsilon_i - \epsilon_n + \hbar\omega} \right. \right.$$
$$\left. \left. + \frac{\langle \psi_{\mathrm{f}}|\mathbf{p}\cdot\mathbf{e}|\psi_n\rangle \langle \psi_n|\mathbf{p}\cdot\mathbf{e}'^*|\psi_i\rangle}{\epsilon_i - \epsilon_n - \hbar\omega'} \right] \right|^2. \tag{15.9}$$

This equation is referred to as the Kramers-Heisenberg equation and describes both Rayleigh and Raman scattering. In the case of Rayleigh scattering, where the final and initial electronic states coincide, (and therefore the frequency ω coincides with the frequency ω'), a second contribution that originates from the quadratic interaction Hamiltonian $\mathcal{H}_{\mathrm{quad}}^I$ needs to be added. This contribution is exactly the same as the one we have calculated for Thomson scattering (Eq. (15.6)), so the Kramers-Heisenberg equation becomes

$$\left[\sigma_{\mathrm{fi}}(\Theta, \mathbf{e}, \mathbf{e}')\right]_{\mathrm{Ray}} = r_{\mathrm{c}}^2 \left| (\mathbf{e}\cdot\mathbf{e}'^*)\delta_{\mathrm{fi}} + \frac{1}{m} \sum_n \left[\frac{\langle \psi_{\mathrm{f}}|\mathbf{p}\cdot\mathbf{e}'^*|\psi_n\rangle \langle \psi_n|\mathbf{p}\cdot\mathbf{e}|\psi_i\rangle}{\epsilon_i - \epsilon_n + \hbar\omega} \right. \right.$$
$$\left. \left. + \frac{\langle \psi_{\mathrm{f}}|\mathbf{p}\cdot\mathbf{e}|\psi_n\rangle \langle \psi_n|\mathbf{p}\cdot\mathbf{e}'^*|\psi_i\rangle}{\epsilon_i - \epsilon_n - \hbar\omega'} \right] \right|^2. \tag{15.10}$$

The Kramers-Heisenberg equation is very general, but considering the approximations we introduced to obtain it, it can be used only far away from resonances. These occur for particular values of ω (and ω') such that one of the denominators of the fractions contained within the square brackets of Eqs. (15.9) and (15.10) is null. For such values, the equation diverges. This is due to the fact that we have neglected damping phenomena. In classical terms, we have neglected the damping force due to the reaction on the atom of the emitted radiation. A coherent introduction of these phenomena within quantum electrodynamics is a complex topic, beyond the scope of this volume. We also note that near resonances one of the two summands in the square brackets clearly prevails over the other (and also over the term in $(\mathbf{e}\cdot\mathbf{e}'^*)$ of Eq. (15.10)). In the most common case where the initial atomic

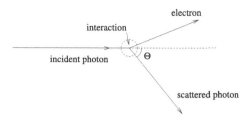

Fig. 15.1 Geometry of the Compton effect. A photon of frequency ω incident on an electron at rest is scattered, at frequency ω', along the direction defined by the angle Θ. The electron suffers a recoil and gains kinetic energy. The plane containing the three directions is called the scattering plane

state is the ground state of the atom, the first term in the square brackets prevails and the Kramers-Heisenberg equation can be written in the following simplified form

$$\sigma_{fi}\left(\Theta, \mathbf{e}, \mathbf{e}'\right) = \frac{r_c^2}{m^2} \frac{\omega'}{\omega} \left| \sum_n \frac{\langle \psi_f | \mathbf{p} \cdot \mathbf{e}'^* | \psi_n \rangle \langle \psi_n | \mathbf{p} \cdot \mathbf{e} | \psi_i \rangle}{\epsilon_i - \epsilon_n + \hbar\omega} \right|^2 .$$

15.4 Compton Scattering, Kinematic Aspects

Having analysed the scattering processes in the non-relativistic case, we now turn to the relativistic case. Consider a photon interacting with a free electron at rest. If the photon has sufficient energy, part of this energy is transferred to the electron and its frequency decreases. At the same time the photon is deflected by an angle Θ with respect to the initial direction, as shown in Fig. 15.1. The kinematic relations among the various quantities involved can be easily determined from the momentum and energy conservation. We denote respectively by $\mathbf{k}$ and ω the wavenumber and frequency of the photon before impact and with $\mathbf{k}'$ and ω' the same quantities after the collision. We also denote by $\mathbf{p}'$ and ϵ' the momentum and energy of the electron after the collision. Obviously, we have

$$\omega = ck, \qquad \omega' = ck', \qquad \epsilon' = \sqrt{c^2 p'^2 + m^2 c^4}.$$

According to the conservation laws we must have, since the electron is initially at rest,

$$\hbar\mathbf{k} = \hbar\mathbf{k}' + \mathbf{p}', \qquad \hbar\omega + mc^2 = \hbar\omega' + \epsilon'.$$

Eliminating ϵ' and squaring we obtain

$$c^2 p'^2 = \hbar^2\left(\omega - \omega'\right)^2 + 2mc^2\hbar\left(\omega - \omega'\right),$$

while getting $\mathbf{p}'$ from the equation for the momentum conservation and squaring we have

$$c^2 p'^2 = \hbar^2\left(\omega^2 + \omega'^2 - 2\omega\omega' \cos\Theta\right).$$

From the last two equation we obtain, with simple algebra

$$\omega\omega'(1 - \cos\Theta) = \frac{mc^2}{\hbar}(\omega - \omega'), \qquad (15.11)$$

so we can express ω' as a function of ω by the equation

$$\frac{\omega'}{\omega} = \frac{mc^2}{mc^2 + \hbar\omega(1 - \cos\Theta)}.$$

The kinematic relation between ω' and ω is usually written in a more meaningful way. Dividing both sides of Eq. (15.11) by the product $\omega\omega'$ we have

$$\frac{1}{\omega'} - \frac{1}{\omega} = \frac{\hbar}{mc^2}(1 - \cos\Theta), \qquad (15.12)$$

and introducing instead of the angular frequencies ω and ω' the respective wavelengths λ and λ', defined by

$$\lambda = \frac{2\pi c}{\omega}, \qquad \lambda' = \frac{2\pi c}{\omega'},$$

we obtain

$$\lambda' - \lambda = \lambda_C(1 - \cos\Theta),$$

where the quantity λ_C, referred to as the Compton wavelength of the electron, is given by

$$\lambda_C = \frac{h}{mc} \simeq 2.426 \times 10^{-10} \text{ cm}.$$

As we can see, the increase of the photon wavelength $\Delta\lambda = \lambda' - \lambda$ is always limited between 0 and $2\lambda_C$, independently of the wavelength of the incoming photon. The relative variation $\Delta\lambda/\lambda$ is of the order of 5×10^{-6} for visible radiation and is therefore difficult to observe. In this limit of long wavelengths, the scattering is practically coherent, as we anticipated in Sect. 15.2 when we discussed Thomson scattering. The relative variations become significant only at short wavelengths, where the photon energy becomes comparable to the rest energy of the electron. Indeed the Compton effect was discovered by irradiating X-rays onto electrons.

Finally, we note that the electron, initially at rest, during the collision gains form the photon an energy ΔE given by

$$\Delta E = \epsilon' - mc^2 = \hbar(\omega - \omega'),$$

which can also be expressed with simple algebra in the form

$$\Delta E = \hbar\omega \frac{\hbar\omega(1 - \cos\Theta)}{mc^2 + \hbar\omega(1 - \cos\Theta)}.$$

15.5 Compton Scattering, Dynamic Aspects

In the previous section we have discussed the kinematics of Compton scattering. In this section we examine the same phenomenon in the framework of quantum electrodynamics. We will obtain the laws of conservation of momentum and energy, and determine the general expression for the cross section of the process. Recalling the considerations developed in the introductory paragraph of this chapter, we use the generalised Fermi rule (Eq. (15.3)) to express the transition probability per unit time from the initial state $|\Psi_i\rangle$ having energy E_i, to the final state $|\Psi_f\rangle$ with energy E_f. Furthermore, since in this case relativistic phenomena play a crucial role, we use for the interaction Hamiltonian the relativistic expression of Eq. (15.5). The probability per unit time is then given by

$$P_{fi} = \frac{2\pi}{\hbar} |\mathcal{M}_{fi}|^2 \delta(E_i - E_f), \tag{15.13}$$

where

$$\mathcal{M}_{fi} = \sum_n \frac{(\mathcal{H}^I_{rel})_{fn}(\mathcal{H}^I_{rel})_{ni}}{E_i - E_n}. \tag{15.14}$$

The matrix elements of this equations are given by

$$\left(\mathcal{H}^I_{rel}\right)_{fn} = \langle \Psi_f | \mathcal{H}^I_{rel} | \Psi_n \rangle, \qquad \left(\mathcal{H}^I_{rel}\right)_{ni} = \langle \Psi_n | \mathcal{H}^I_{rel} | \Psi_i \rangle,$$

where $|\Psi_n\rangle$ is any intermediate (or virtual) state of the whole system and the sum is extended to all possible intermediate states. We will use this equation to evaluate the cross section for the Compton effect in the general relativistic case, in which the energy of the photon is comparable with that of the electron at rest. We have seen that in the non-relativistic case, previously discussed, the total cross section for the process is σ_T, the Thomson cross section.

We begin by calculating the probability per unit time that a transition occurs between the initial state which contains, from the point of view of the radiation field, a photon of frequency ω, wavenumber $\mathbf{k}$ (with $\omega = ck$), and polarisation characterised by the unit vector $\mathbf{e}$, and the final state, which contains a photon of frequency ω', wavenumber $\mathbf{k}'$ (with $\omega' = ck'$) and polarisation characterised by the unit vector $\mathbf{e}'$. From the point of view of the atomic system (composed in this case by a free electron) we assume that in its initial state the electron has momentum $\mathbf{p}$, energy ϵ, and spin characterised by the index r, while in its final state has momentum $\mathbf{p}'$, energy ϵ', and spin characterised by the index s. Obviously, we have

$$\epsilon = \sqrt{c^2 p^2 + m^2 c^4}, \qquad \epsilon' = \sqrt{c^2 p'^2 + m^2 c^4}.$$

The initial and final states are therefore described by the state vectors

$$|\Psi_i\rangle = \left| \psi_{\mathbf{p}}^{(r)}(\mathbf{x}); 0, 0, \ldots, 1_{k\mathbf{e}}, \ldots, 0, \ldots \right\rangle,$$

$$|\Psi_f\rangle = \left| \psi_{\mathbf{p}'}^{(s)}(\mathbf{x}); 0, 0, \ldots, 1_{k'\mathbf{e}'}, \ldots, 0, \ldots \right\rangle,$$

where $\psi_{\mathbf{p}}^{(r)}(\mathbf{x})$ and $\psi_{\mathbf{p}'}^{(s)}(\mathbf{x})$ are eigenfunctions of the Dirac Hamiltonian relative, respectively, to the electron in the initial and final state, of the type of the spinors introduced in Sect. 5.1. The interaction Hamiltonian is given by Eq. (15.5), that, considering the case of a single electron and recalling the expression of the vector potential operator given by Eq. (4.5), becomes

$$\mathcal{H}_{rel}^I = \sum_{k\lambda} e_0 c \sqrt{\frac{2\pi\hbar}{\omega_k \mathcal{V}}} \boldsymbol{\alpha} \cdot \left(a_{k\lambda} \mathbf{e}_{k\lambda} e^{i\mathbf{k}\cdot\mathbf{x}} + a_{k\lambda}^\dagger \mathbf{e}_{k\lambda}^* e^{-i\mathbf{k}\cdot\mathbf{x}} \right),$$

where $\boldsymbol{\alpha}$ are the Dirac matrices. The energies of the initial and final states are given by

$$E_i = \epsilon + \hbar\omega, \qquad E_f = \epsilon' + \hbar\omega',$$

and the presence of the Dirac delta in the expression for the transition probability ensures that energy is conserved.

We are now going to consider what are the possible intermediate states that contribute to the sum, namely those states for which the matrix elements $(\mathcal{H}_{rel}^I)_{fn}$ and $(\mathcal{H}_{rel}^I)_{ni}$ are non-null. Regarding the radiation field, since the interaction Hamiltonian is linear in the operators of creation and annihilation, we see immediately that there are only two photon states that are contributing to the sum, namely

$$|0, 0, \ldots, 0_{ke}, \ldots, 0_{k'e'}, \ldots\rangle, \qquad |0, 0, \ldots, 1_{ke}, \ldots, 1_{k'e'}, \ldots\rangle.$$

In the first case the second Hamiltonian (that one that first operates on $|\Psi_i\rangle$) annihilates the photon $(\mathbf{k}, \mathbf{e})$, while the first one creates the photon $(\mathbf{k}', \mathbf{e}')$. In the second case, on the contrary, the second Hamiltonian creates the photon $(\mathbf{k}', \mathbf{e}')$, while the first one annihilates the photon $(\mathbf{k}, \mathbf{e})$. Taking this into account and recalling the formulae of the matrix elements of the operators of creation and annihilation, we obtain for the matrix element $\mathcal{M}_{fi}$ of Eq. (15.14)

$$\mathcal{M}_{fi} = \frac{2\pi e_0^2 c^2 \hbar}{\mathcal{V}\sqrt{\omega\omega'}}(A + B),$$

where

$$A = \sum_{qt} \langle \psi_{p'}^{(s)}(\mathbf{x}) | (\boldsymbol{\alpha} \cdot \mathbf{e}'^*) e^{-i\mathbf{k}'\cdot\mathbf{x}} | \psi_q^{(t)}(\mathbf{x}) \rangle \frac{1}{\epsilon + \hbar\omega - \epsilon_q}$$
$$\times \langle \psi_q^{(t)}(\mathbf{x}) | (\boldsymbol{\alpha} \cdot \mathbf{e}) e^{i\mathbf{k}\cdot\mathbf{x}} | \psi_p^{(r)}(\mathbf{x}) \rangle,$$

$$B = \sum_{qt} \langle \psi_{p'}^{(s)}(\mathbf{x}) | (\boldsymbol{\alpha} \cdot \mathbf{e}) e^{i\mathbf{k}\cdot\mathbf{x}} | \psi_q^{(t)}(\mathbf{x}) \rangle \frac{1}{\epsilon - \hbar\omega' - \epsilon_q}$$
$$\times \langle \psi_q^{(t)}(\mathbf{x}) | (\boldsymbol{\alpha} \cdot \mathbf{e}'^*) e^{-i\mathbf{k}'\cdot\mathbf{x}} | \psi_p^{(r)}(\mathbf{x}) \rangle.$$

We now take into account the fact that the normalised wavefunctions of the free electrons are of the form

$$\psi_p^{(r)} = \frac{1}{\sqrt{\mathcal{V}}} e^{i\mathbf{p}\cdot\mathbf{x}/\hbar} W_r(\mathbf{p}), \qquad \psi_{p'}^{(s)} = \frac{1}{\sqrt{\mathcal{V}}} e^{i\mathbf{p}'\cdot\mathbf{x}/\hbar} W_s(\mathbf{p}'),$$

$$\psi_q^{(t)} = \frac{1}{\sqrt{\mathcal{V}}} e^{i\mathbf{q}\cdot\mathbf{x}/\hbar} W_t(\mathbf{q}),$$

where the quantities W are the spinors introduced in Sect. 5.1, and that we rewrite here as

$$W_1(\mathbf{q}) = C \begin{pmatrix} \epsilon_\mathbf{q} + mc^2 \\ 0 \\ cq_z \\ cq_+ \end{pmatrix}, \qquad W_2(\mathbf{q}) = C \begin{pmatrix} 0 \\ \epsilon_\mathbf{q} + mc^2 \\ cq_- \\ -cq_z \end{pmatrix},$$

$$W_3(\mathbf{q}) = C \begin{pmatrix} -cq_z \\ -cq_+ \\ \epsilon_\mathbf{q} + mc^2 \\ 0 \end{pmatrix}, \qquad W_4(\mathbf{q}) = C \begin{pmatrix} -cq_- \\ cq_z \\ 0 \\ \epsilon_\mathbf{q} + mc^2 \end{pmatrix},$$

with

$$C = \frac{1}{\sqrt{2\epsilon_\mathbf{q}(\epsilon_\mathbf{q} + mc^2)}}, \qquad \epsilon_\mathbf{q} = \sqrt{c^2 q^2 + m^2 c^4}.$$

We recall also that $W_1(\mathbf{q})$ and $W_2(\mathbf{q})$ are positive-energy spinors, while $W_3(\mathbf{q})$ and $W_4(\mathbf{q})$ are negative-energy spinors.

The integrals in $d^3\mathbf{x}$ over the wavefunctions, implicitly contained in the matrix elements, are trivial and lead to the conservation of momentum. The sum over $\mathbf{q}$ that appears in the term A provides a non-zero contribution only if $\mathbf{q} = (\mathbf{p} + \hbar\mathbf{k})$ and also $\mathbf{q} = (\mathbf{p}' + \hbar\mathbf{k}')$, while the similar sum that appears in the term B gives a non-zero contribution only if $\mathbf{q} = (\mathbf{p} - \hbar\mathbf{k}')$ and also $\mathbf{q} = (\mathbf{p}' - \hbar\mathbf{k})$. In any case, one must have

$$\mathbf{p} + \hbar\mathbf{k} = \mathbf{p}' + \hbar\mathbf{k}'.$$

In what follows we will use the symbols $\mathbf{g}$ and $\mathbf{h}$ to indicate the momentum of the electron in the intermediate state corresponding, respectively, to the terms A and B. We then put

$$\mathbf{g} = \mathbf{p} + \hbar\mathbf{k}, \qquad \mathbf{h} = \mathbf{p} - \hbar\mathbf{k}'. \tag{15.15}$$

The corresponding energies of the electron in the intermediate states are given by $\pm\epsilon_\mathbf{g}$ e $\pm\epsilon_\mathbf{h}$, where

$$\epsilon_\mathbf{g} = \sqrt{c^2(\mathbf{p} + \hbar\mathbf{k})^2 + m^2 c^4}, \qquad \epsilon_\mathbf{h} = \sqrt{c^2(\mathbf{p} - \hbar\mathbf{k}')^2 + m^2 c^4}, \tag{15.16}$$

and where the sign $\pm$ must be chosen depending on whether states with positive ($t = 1, 2$) or negative ($t = 3, 4$) energy are considered. It is easy to verify that, in any case, in the intermediate state energy is not conserved. As we have seen previously, the intermediate state must therefore be considered as a virtual state in which the electron exists only during an extremely short time Δt such that $\Delta t \Delta E \simeq \hbar$, where $\Delta E = E_i - E_n$ is the energy difference between the initial (or final) state and the intermediate state. Sometimes, and especially in quantum electrodynamics, we refer to virtual states as those states that do not conserve energy and which are therefore outside the so-called energy-shell. The physical processes that we are analysing here are conveniently illustrated by the two typical Feynman diagrams of Fig. 15.2.

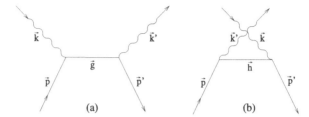

Fig. 15.2 Feynman diagrams for the Compton effect. Diagram (**a**) describes the process where the photon of momentum **k** is absorbed (destroyed) and that one of momentum **k**′ is emitted (created). Diagram (**b**) describes the process where the same phenomena occur in the reverse order. In the intermediate virtual state, the electron has momentum **g** in case (**a**) and **h** in case (**b**)

Returning to our problem, and noting that the integrals in d^3x of the exponentials are equal to $\mathcal{V}$ when momentum is conserved, the expression for $\mathcal{M}_{\mathrm{fi}}$ is

$$
\mathcal{M}_{\mathrm{fi}} = \frac{2\pi e_0^2 c^2 \hbar}{\mathcal{V}\sqrt{\omega\omega'}} \left(\sum_t W_s^\dagger(\mathbf{p}')(\boldsymbol{\alpha}\cdot\mathbf{e}'^*) W_t(\mathbf{g}) \frac{1}{\epsilon + \hbar\omega \mp \epsilon_{\mathbf{g}}} W_t^\dagger(\mathbf{g})(\boldsymbol{\alpha}\cdot\mathbf{e}) W_r(\mathbf{p}) \right.
$$
$$
\left. + \sum_t W_s^\dagger(\mathbf{p}')(\boldsymbol{\alpha}\cdot\mathbf{e}) W_t(\mathbf{h}) \frac{1}{\epsilon - \hbar\omega' \mp \epsilon_{\mathbf{h}}} W_t^\dagger(\mathbf{h})(\boldsymbol{\alpha}\cdot\mathbf{e}'^*) W_r(\mathbf{p}) \right),
$$

(15.17)

where, in front of the factors $\epsilon_{\mathbf{g}}$ and $\epsilon_{\mathbf{h}}$, we must take the minus sign for $t = 1, 2$ and the plus sign for $t = 3, 4$. At this point it is important to note that the sum over the virtual states must be extended to the negative-energy states, although these states are fully occupied, according to the hypothesis of the Fermi sea. Referring, for example, to the first Feynman diagram and the case where $t = 3, 4$, the intermediate state is in effect one in which the photon (**k**, **e**) is absorbed by an electron in the Fermi sea, which creates a virtual electron-positron pair. In the intermediate state we therefore have two electrons (the original and the "new" one) and a positron. In the transition to the final state, the original electron decays into the hole of the Fermi sea, annihilating the positron and emitting the photon (**k**′, **e**′). The end results are the same either if the sum over the intermediate states is extended to the negative-energy electronic states, or if we introduce as intermediate states also the positronic states (which, however, would require a formally more in-depth discussion, well beyond the scope of this volume).

We now perform the sum over the spin states of the intermediate electron. Referring to the first of the two terms, one must evaluate the matrix $\mathcal{S}$ given by the sum

$$
\mathcal{S} = \sum_{t=1,2} W_t(\mathbf{g}) \frac{1}{\epsilon + \hbar\omega - \epsilon_{\mathbf{g}}} W_t^\dagger(\mathbf{g}) + \sum_{t=3,4} W_t(\mathbf{g}) \frac{1}{\epsilon + \hbar\omega + \epsilon_{\mathbf{g}}} W_t^\dagger(\mathbf{g}).
$$

To write this matrix in a more convenient form, we define the matrices P_+ and P_-, called projection operators, by the equations

$$
P_+ = \sum_{t=1,2} W_t(\mathbf{g}) W_t^\dagger(\mathbf{g}), \qquad P_- = \sum_{t=3,4} W_t(\mathbf{g}) W_t^\dagger(\mathbf{g}). \tag{15.18}
$$

Using these matrices we can write

$$S = \frac{P_+}{\epsilon + \hbar\omega - \epsilon_{\mathbf{g}}} + \frac{P_-}{\epsilon + \hbar\omega + \epsilon_{\mathbf{g}}}. \tag{15.19}$$

We also note that, since the spinors are the normalised eigenvectors of an Hermitian matrix, we have

$$P_+ + P_- = 1.$$

On the other hand we also have

$$H_{\mathbf{g}} P_+ = \epsilon_{\mathbf{g}} P_+, \qquad H_{\mathbf{g}} P_- = -\epsilon_{\mathbf{g}} P_-,$$

where $H_{\mathbf{g}}$ is the Dirac Hamiltonian corresponding to the momentum $\mathbf{g}$, i.e.

$$H_{\mathbf{g}} = c\boldsymbol{\alpha} \cdot \mathbf{g} + \beta mc^2.$$

By summing the previous equations we get

$$H_{\mathbf{g}}(P_+ + P_-) = \epsilon_{\mathbf{g}}(P_+ - P_-),$$

from which we obtain, being $P_+ + P_- = 1$,

$$P_+ - P_- = \frac{H_{\mathbf{g}}}{\epsilon_{\mathbf{g}}}.$$

This equation, together with the one for $(P_+ + P_-)$, can be used to write the two matrices P_+ and P_- in the form

$$P_+ = \frac{\epsilon_{\mathbf{g}} + H_{\mathbf{g}}}{2\epsilon_{\mathbf{g}}}, \qquad P_- = \frac{\epsilon_{\mathbf{g}} - H_{\mathbf{g}}}{2\epsilon_{\mathbf{g}}}. \tag{15.20}$$

By substituting this result into Eq. (15.19), after some simple algebra we can express the sum S, usually called in quantum field theory the "propagator", by the equation

$$S = \frac{H_{\mathbf{g}} + \epsilon + \hbar\omega}{(\epsilon + \hbar\omega)^2 - \epsilon_{\mathbf{g}}^2}.$$

Following similar steps for the sum over the spin states in the second term of Eq. (15.17), we can write the matrix element $\mathcal{M}_{\mathrm{fi}}$ as

$$\mathcal{M}_{\mathrm{fi}} = \frac{2\pi e_0^2 c^2 \hbar}{V\sqrt{\omega\omega'}} \mathcal{R}_{\mathrm{fi}}, \tag{15.21}$$

where

$$\mathcal{R}_{\mathrm{fi}} = W_s^\dagger(\mathbf{p}')(\boldsymbol{\alpha} \cdot \mathbf{e}'^*)\frac{H_{\mathbf{g}} + \epsilon + \hbar\omega}{(\epsilon + \hbar\omega)^2 - \epsilon_{\mathbf{g}}^2}(\boldsymbol{\alpha} \cdot \mathbf{e})W_r(\mathbf{p})$$

$$+ W_s^\dagger(\mathbf{p}')(\boldsymbol{\alpha} \cdot \mathbf{e})\frac{H_{\mathbf{h}} + \epsilon - \hbar\omega'}{(\epsilon - \hbar\omega')^2 - \epsilon_{\mathbf{h}}^2}(\boldsymbol{\alpha} \cdot \mathbf{e}'^*)W_r(\mathbf{p}). \tag{15.22}$$

The expression that we have found for $\mathcal{M}_{\mathrm{fi}}$ is very general and, evaluating its square modulus, we can determine the transition probability for any combination of kinematic parameters, spin states, and polarisation states of the initial and final electrons and photons. Before proceeding any further, it is however important to point out a fundamental property of the above quantities. This property is related to the invariance of physical phenomena under gauge transformations. If we perform the transformation

$$\boldsymbol{\alpha} \cdot \mathbf{e} \to \boldsymbol{\alpha} \cdot \mathbf{e} + C(\boldsymbol{\alpha} \cdot \mathbf{u} - 1),$$

where C is an arbitrary constant and $\mathbf{u}$ is the unit vector pointing along the direction of the initial photon ($\mathbf{u} = \mathbf{k}/k$), the quantity $\mathcal{R}_{\mathrm{fi}}$ does not change. Similarly, if we perform the transformation

$$\boldsymbol{\alpha} \cdot \mathbf{e}'^{*} \to \boldsymbol{\alpha} \cdot \mathbf{e}'^{*} + C'\big(\boldsymbol{\alpha} \cdot \mathbf{u}' - 1\big),$$

where C' is another arbitrary constant and $\mathbf{u}'$ is the unit vector of the direction of the final photon ($\mathbf{u}' = \mathbf{k}'/k'$), the quantity $\mathcal{R}_{\mathrm{fi}}$ does not change. This property is demonstrated in Sect. 16.14.

Returning to the evaluation of the quantity $\mathcal{M}_{\mathrm{fi}}$ of Eq. (15.21), we note that the spin states of the electrons are quantities that, in general, are of minor physical interest. It is therefore appropriate to calculate, in the square modulus of the matrix element, the sum over the spin states of the final electron and the average over the spin states of the initial electron. To calculate this quantity, later denoted by $\langle|\mathcal{M}_{\mathrm{fi}}|^2\rangle$, it is convenient to introduce two matrices, $\mathcal{A}$ and $\mathcal{B}$, defined by

$$\mathcal{A} = \big(\boldsymbol{\alpha} \cdot \mathbf{e}'^{*}\big) \frac{H_{\mathbf{g}} + \epsilon + \hbar\omega}{(\epsilon + \hbar\omega)^2 - \epsilon_{\mathbf{g}}^2} (\boldsymbol{\alpha} \cdot \mathbf{e}),$$

$$\mathcal{B} = (\boldsymbol{\alpha} \cdot \mathbf{e}) \frac{H_{\mathbf{h}} + \epsilon - \hbar\omega'}{(\epsilon - \hbar\omega')^2 - \epsilon_{\mathbf{h}}^2} (\boldsymbol{\alpha} \cdot \mathbf{e}'^{*}). \tag{15.23}$$

With these definitions, the matrix element $\mathcal{M}_{\mathrm{fi}}$ becomes

$$\mathcal{M}_{\mathrm{fi}} = \frac{2\pi e_0^2 c^2 \hbar}{V\sqrt{\omega\omega'}} W_s^{\dagger}(\mathbf{p}')(\mathcal{A} + \mathcal{B}) W_r(\mathbf{p}),$$

so we obtain

$$\langle|\mathcal{M}_{\mathrm{fi}}|^2\rangle = \frac{4\pi^2 e_0^4 c^4 \hbar^2}{V^2 \omega\omega'} \frac{1}{2} \sum_{r,s=1,2} W_s^{\dagger}(\mathbf{p}')(\mathcal{A} + \mathcal{B}) W_r(\mathbf{p}) W_r^{\dagger}(\mathbf{p})\big(\mathcal{A}^{\dagger} + \mathcal{B}^{\dagger}\big) W_s(\mathbf{p}').$$

To calculate the sums over r and s we can use the results about the projection operators that we obtained previously (see Eqs. (15.18) and (15.20)). Taking into account that

$$\sum_{r=1,2} W_r(\mathbf{p}) W_r^{\dagger}(\mathbf{p}) = \frac{\epsilon + H_{\mathbf{p}}}{2\epsilon}, \qquad \sum_{s=1,2} W_s(\mathbf{p}') W_s^{\dagger}(\mathbf{p}') = \frac{\epsilon' + H_{\mathbf{p}'}}{2\epsilon'},$$

and using the definition of the trace of a matrix (by which a scalar product of the form $W^{\dagger}\mathcal{X}W$ can be written in the form $\mathrm{Tr}(WW^{\dagger}\mathcal{X})$, with $\mathcal{X}$ an arbitrary matrix), we get

$$\langle|\mathcal{M}_{\mathrm{fi}}|^2\rangle = \frac{4\pi^2 e_0^4 c^4 \hbar^2}{V^2 \omega\omega'} \frac{1}{8\epsilon\epsilon'} \mathcal{T}, \tag{15.24}$$

where

$$T = \mathrm{Tr}\big[(\epsilon' + H_{\mathbf{p}'})(\mathcal{A} + \mathcal{B})(\epsilon + H_{\mathbf{p}})(\mathcal{A}^\dagger + \mathcal{B}^\dagger)\big]. \tag{15.25}$$

The calculation of this expression is, in general, rather complex, even though it does not present any particular conceptual difficulty. Some formal simplifications can be obtained using the formalism of the quadrivectors and of the so-called γ matrices, as shown in Sect. 16.15. Without introducing this formalism, we will see in the next section how the trace T can be computed in the simplified case where the electron is assumed initially at rest. We will then obtain the well-known Klein-Nishina equation.

15.6 The Klein-Nishina Equation

Consider the case where the initial electron is at rest. To further simplify the calculation, we also assume that the polarisation unit vectors $\mathbf{e}$ and $\mathbf{e}'$ are real. Recalling Eq. (15.15), for the various terms appearing in Eq. (15.25) we have

$$\epsilon = mc^2, \qquad \epsilon' = mc^2 + \hbar(\omega - \omega'), \qquad \mathbf{g} = \hbar\mathbf{k}, \qquad \mathbf{h} = -\hbar\mathbf{k}',$$

from which, considering Eq. (15.16), we obtain the expressions appearing in the denominator of the matrices $\mathcal{A}$ and $\mathcal{B}$ of Eq. (15.23), that is

$$(\epsilon + \hbar\omega)^2 - \epsilon_{\mathbf{g}}^2 = 2mc^2\hbar\omega, \qquad (\epsilon - \hbar\omega')^2 - \epsilon_{\mathbf{h}}^2 = -2mc^2\hbar\omega'.$$

Considering the explicit forms of $H_{\mathbf{g}}$ and $H_{\mathbf{h}}$ we obtain

$$\mathcal{A} = (\boldsymbol{\alpha}\cdot\mathbf{e}')\frac{(\boldsymbol{\alpha}\cdot\mathbf{k})\hbar c + (1+\beta)mc^2 + \hbar\omega}{2mc^2\hbar\omega}(\boldsymbol{\alpha}\cdot\mathbf{e}),$$

$$\mathcal{B} = (\boldsymbol{\alpha}\cdot\mathbf{e})\frac{(\boldsymbol{\alpha}\cdot\mathbf{k}')\hbar c - (1+\beta)mc^2 + \hbar\omega'}{2mc^2\hbar\omega'}(\boldsymbol{\alpha}\cdot\mathbf{e}').$$

Noting that $H_{\mathbf{p}} = \beta mc^2$ (since the electron is initially at rest, hence $\mathbf{p} = 0$), the trace of Eq. (15.25) is

$$T = \mathrm{Tr}\big[(\epsilon' + H_{\mathbf{p}'})(\mathcal{A} + \mathcal{B})(1+\beta)mc^2(\mathcal{A}^\dagger + \mathcal{B}^\dagger)\big]. \tag{15.26}$$

We can now simplify the previous relations by noting that the quantity $(1+\beta)mc^2$ in the numerator of the expressions for $\mathcal{A}$ and $\mathcal{B}$ (and of their adjoints) can simply be omitted. This is because the α matrices anticommute with the matrix β, so we have, for an arbitrary vector $\mathbf{v}$

$$(1+\beta)(\boldsymbol{\alpha}\cdot\mathbf{v})(1+\beta) = (\boldsymbol{\alpha}\cdot\mathbf{v})(1-\beta)(1+\beta) = (\boldsymbol{\alpha}\cdot\mathbf{v})(1-\beta^2) = 0,$$

where the last step is due to the fact that $\beta^2 = 1$. Considering then the sum $\mathcal{A} + \mathcal{B}$ and taking into account this simplification we have, with simple algebra

$$\mathcal{A} + \mathcal{B} = \frac{1}{2mc^2}\big[(\boldsymbol{\alpha}\cdot\mathbf{e}')(\boldsymbol{\alpha}\cdot\mathbf{u})(\boldsymbol{\alpha}\cdot\mathbf{e}) + (\boldsymbol{\alpha}\cdot\mathbf{e})(\boldsymbol{\alpha}\cdot\mathbf{u}')(\boldsymbol{\alpha}\cdot\mathbf{e}')\big]$$

$$+ \frac{1}{2mc^2}\big[(\boldsymbol{\alpha}\cdot\mathbf{e}')(\boldsymbol{\alpha}\cdot\mathbf{e}) + (\boldsymbol{\alpha}\cdot\mathbf{e})(\boldsymbol{\alpha}\cdot\mathbf{e}')\big],$$

where $\mathbf{u}$ and $\mathbf{u}'$ are the unit vectors directed, respectively, along $\mathbf{k}$ and $\mathbf{k}'$. On the other hand, for the anticommutation properties of the $\boldsymbol{\alpha}$ matrices, we have

$$(\boldsymbol{\alpha} \cdot \mathbf{e}')(\boldsymbol{\alpha} \cdot \mathbf{e}) + (\boldsymbol{\alpha} \cdot \mathbf{e})(\boldsymbol{\alpha} \cdot \mathbf{e}') = 2(\mathbf{e}' \cdot \mathbf{e}),$$

so we obtain

$$\mathcal{A} + \mathcal{B} = \frac{1}{2mc^2}(2a + \mathcal{E} + \mathcal{F}), \tag{15.27}$$

where we have introduced the quantities

$$a = \mathbf{e}' \cdot \mathbf{e}, \qquad \mathcal{E} = (\boldsymbol{\alpha} \cdot \mathbf{e}')(\boldsymbol{\alpha} \cdot \mathbf{u})(\boldsymbol{\alpha} \cdot \mathbf{e}), \qquad \mathcal{F} = (\boldsymbol{\alpha} \cdot \mathbf{e})(\boldsymbol{\alpha} \cdot \mathbf{u}')(\boldsymbol{\alpha} \cdot \mathbf{e}'). \tag{15.28}$$

Taking into account these results and writing for convenience the matrix $(\epsilon' + H_{\mathbf{p}'})$ in the form

$$\epsilon' + H_{\mathbf{p}'} = mc^2(b + \beta + \boldsymbol{\alpha} \cdot \mathbf{d}), \tag{15.29}$$

where the dimensionless quantities b and $\mathbf{d}$ are defined by

$$b = 1 + \frac{\hbar\omega - \hbar\omega'}{mc^2}, \qquad \mathbf{d} = \frac{\hbar}{mc}(\mathbf{k} - \mathbf{k}') = \frac{\hbar\omega}{mc^2}\mathbf{u} - \frac{\hbar\omega'}{mc^2}\mathbf{u}', \tag{15.30}$$

we obtain, by substitution of Eqs. (15.27) and (15.29) in Eq. (15.26)

$$\mathcal{T} = \frac{1}{4}\,\mathrm{Tr}\{(b + \beta + \boldsymbol{\alpha} \cdot \mathbf{d})(2a + \mathcal{E} + \mathcal{F})(1 + \beta)(2a + \mathcal{E}^\dagger + \mathcal{F}^\dagger)\}.$$

We now consider the product $\mathcal{P}$ defined by

$$\mathcal{P} = (2a + \mathcal{E} + \mathcal{F})(1 + \beta)(2a + \mathcal{E}^\dagger + \mathcal{F}^\dagger).$$

Taking into account the anticommutation properties of the Dirac matrices, it can be transformed in the following way

$$\begin{aligned}
\mathcal{P} &= (2a + \mathcal{E} + \mathcal{F})(2a + \mathcal{E}^\dagger + \mathcal{F}^\dagger) + \beta(2a - \mathcal{E} - \mathcal{F})(2a + \mathcal{E}^\dagger + \mathcal{F}^\dagger) \\
&= 4a^2 + 2a(\mathcal{E}^\dagger + \mathcal{F}^\dagger + \mathcal{E} + \mathcal{F}) + \mathcal{E}\mathcal{E}^\dagger + \mathcal{F}\mathcal{F}^\dagger + \mathcal{E}\mathcal{F}^\dagger + \mathcal{F}\mathcal{E}^\dagger \\
&\quad + \beta\left[4a^2 + 2a(\mathcal{E}^\dagger + \mathcal{F}^\dagger - \mathcal{E} - \mathcal{F}) - \mathcal{E}\mathcal{E}^\dagger - \mathcal{F}\mathcal{F}^\dagger - \mathcal{E}\mathcal{F}^\dagger - \mathcal{F}\mathcal{E}^\dagger\right].
\end{aligned}$$

We now note that the two products $\mathcal{E}\mathcal{E}^\dagger$ and $\mathcal{F}\mathcal{F}^\dagger$ are both equal to the unit matrix. This result is obtained from the anticommutation properties of the $\boldsymbol{\alpha}$ matrices by which, if $\mathbf{v}$ and $\mathbf{w}$ are two arbitrary vectors, we have

$$(\boldsymbol{\alpha} \cdot \mathbf{v})(\boldsymbol{\alpha} \cdot \mathbf{v}) = v^2, \tag{15.31}$$

$$(\boldsymbol{\alpha} \cdot \mathbf{v})(\boldsymbol{\alpha} \cdot \mathbf{w}) = -(\boldsymbol{\alpha} \cdot \mathbf{w})(\boldsymbol{\alpha} \cdot \mathbf{v}) + 2\mathbf{v} \cdot \mathbf{w}, \tag{15.32}$$

so that, for example, applying three times Eq. (15.31)

$$\begin{aligned}
\mathcal{E}\mathcal{E}^\dagger &= (\boldsymbol{\alpha} \cdot \mathbf{e}')(\boldsymbol{\alpha} \cdot \mathbf{u})(\boldsymbol{\alpha} \cdot \mathbf{e})(\boldsymbol{\alpha} \cdot \mathbf{e})(\boldsymbol{\alpha} \cdot \mathbf{u})(\boldsymbol{\alpha} \cdot \mathbf{e}') \\
&= (\boldsymbol{\alpha} \cdot \mathbf{e}')(\boldsymbol{\alpha} \cdot \mathbf{u})(\boldsymbol{\alpha} \cdot \mathbf{u})(\boldsymbol{\alpha} \cdot \mathbf{e}') = (\boldsymbol{\alpha} \cdot \mathbf{e}')(\boldsymbol{\alpha} \cdot \mathbf{e}') = 1.
\end{aligned}$$

Taking into account this result, we have that the trace $\mathcal{T}$ is given by

$$\begin{aligned}
\mathcal{T} = \frac{1}{4}\,\mathrm{Tr}\{&(b + \beta + \boldsymbol{\alpha} \cdot \mathbf{d})\left[4a^2 + 2 + 2a(\mathcal{E}^\dagger + \mathcal{F}^\dagger + \mathcal{E} + \mathcal{F}) + \mathcal{E}\mathcal{F}^\dagger + \mathcal{F}\mathcal{E}^\dagger\right] \\
&+ (b + \beta + \boldsymbol{\alpha} \cdot \mathbf{d})\beta\left[4a^2 - 2 + 2a(\mathcal{E}^\dagger + \mathcal{F}^\dagger - \mathcal{E} - \mathcal{F}) - \mathcal{E}\mathcal{F}^\dagger - \mathcal{F}\mathcal{E}^\dagger\right]\}.
\end{aligned}$$

We now note that the traces of matrices obtained from the product of α and β matrices are all zero, unless such product does not contain an even (or null) number of α matrices and an even (or null) number of β matrices. Taking into account this property, as well as the fact that the $\mathcal{E}$ and $\mathcal{F}$ matrices are obtained from the product of three α matrices, and that $\beta^2 = 1$, we have

$$\mathcal{T} = \frac{1}{4}\mathrm{Tr}\{4(b+1)a^2 + (b-1)(2 + \mathcal{E}\mathcal{F}^\dagger + \mathcal{F}\mathcal{E}^\dagger)$$
$$+ 2a(\boldsymbol{\alpha}\cdot\mathbf{d})(\mathcal{E}^\dagger + \mathcal{F}^\dagger + \mathcal{E} + \mathcal{F})\}.$$

We now need to recall a few properties of the trace of a matrix. Given an arbitrary matrix M we have $\mathrm{Tr}(M) = [\mathrm{Tr}(M^\dagger)]^*$, and for any M and N matrices $\mathrm{Tr}(MN) = \mathrm{Tr}(NM)$. If we now also recall that the α matrices are Hermitian, that the trace of the unit matrix is 4, and that the quantities b and $\mathbf{d}$ are real, we obtain

$$\mathcal{T} = 4(b+1)a^2 + \frac{1}{2}(b-1)\{4 + \mathrm{Re}[\mathrm{Tr}(\mathcal{E}\mathcal{F}^\dagger)]\} + a\,\mathrm{Re}\{\mathrm{Tr}[(\boldsymbol{\alpha}\cdot\mathbf{d})(\mathcal{E}^\dagger + \mathcal{F}^\dagger)]\}.$$
$$(15.33)$$

We now calculate the traces of the matrices that are left in this expression. Recalling Eqs. (15.28), we have

$$\mathrm{Tr}(\mathcal{E}\mathcal{F}^\dagger) = \mathrm{Tr}[(\boldsymbol{\alpha}\cdot\mathbf{e}')(\boldsymbol{\alpha}\cdot\mathbf{u})(\boldsymbol{\alpha}\cdot\mathbf{e})(\boldsymbol{\alpha}\cdot\mathbf{e}')(\boldsymbol{\alpha}\cdot\mathbf{u}')(\boldsymbol{\alpha}\cdot\mathbf{e})].$$

Taking into account that the unit vector $\mathbf{u}$ is orthogonal to the unit vector $\mathbf{e}$, that the unit vector $\mathbf{u}'$ is orthogonal to the unit vector $\mathbf{e}'$, and recalling Eq. (15.32), we have

$$\mathrm{Tr}(\mathcal{E}\mathcal{F}^\dagger) = \mathrm{Tr}[(\boldsymbol{\alpha}\cdot\mathbf{e}')(\boldsymbol{\alpha}\cdot\mathbf{e})(\boldsymbol{\alpha}\cdot\mathbf{u})(\boldsymbol{\alpha}\cdot\mathbf{u}')(\boldsymbol{\alpha}\cdot\mathbf{e}')(\boldsymbol{\alpha}\cdot\mathbf{e})]$$
$$= \mathrm{Tr}[(\boldsymbol{\alpha}\cdot\mathbf{e}')(\boldsymbol{\alpha}\cdot\mathbf{e})(\boldsymbol{\alpha}\cdot\mathbf{e}')(\boldsymbol{\alpha}\cdot\mathbf{e})(\boldsymbol{\alpha}\cdot\mathbf{u})(\boldsymbol{\alpha}\cdot\mathbf{u}')],$$

where we have used successively the cyclic property of the trace. We now apply to the first pair of scalar products the substitution

$$(\boldsymbol{\alpha}\cdot\mathbf{e}')(\boldsymbol{\alpha}\cdot\mathbf{e}) = -(\boldsymbol{\alpha}\cdot\mathbf{e})(\boldsymbol{\alpha}\cdot\mathbf{e}') + 2a,$$

where we have used Eq. (15.32) (again) and the definition of a given by Eq. (15.28). Using also Eq. (15.31) we obtain

$$\mathrm{Tr}(\mathcal{E}\mathcal{F}^\dagger) = 2a\,\mathrm{Tr}[(\boldsymbol{\alpha}\cdot\mathbf{e}')(\boldsymbol{\alpha}\cdot\mathbf{e})(\boldsymbol{\alpha}\cdot\mathbf{u})(\boldsymbol{\alpha}\cdot\mathbf{u}')] - \mathrm{Tr}[(\boldsymbol{\alpha}\cdot\mathbf{u})(\boldsymbol{\alpha}\cdot\mathbf{u}')]. \quad (15.34)$$

We now calculate the other trace. We have

$$\mathrm{Tr}[(\boldsymbol{\alpha}\cdot\mathbf{d})(\mathcal{E}^\dagger + \mathcal{F}^\dagger)] = \mathrm{Tr}[(\boldsymbol{\alpha}\cdot\mathbf{d})(\boldsymbol{\alpha}\cdot\mathbf{e})(\boldsymbol{\alpha}\cdot\mathbf{u})(\boldsymbol{\alpha}\cdot\mathbf{e}')]$$
$$+ \mathrm{Tr}[(\boldsymbol{\alpha}\cdot\mathbf{d})(\boldsymbol{\alpha}\cdot\mathbf{e}')(\boldsymbol{\alpha}\cdot\mathbf{u}')(\boldsymbol{\alpha}\cdot\mathbf{e})].$$

Recalling the definition of the vector $\mathbf{d}$ (Eq. (15.30)), we obtain four terms

$$\mathrm{Tr}[(\boldsymbol{\alpha}\cdot\mathbf{d})(\mathcal{E}^\dagger + \mathcal{F}^\dagger)] = \frac{\hbar\omega}{mc^2}\{\mathrm{Tr}[(\boldsymbol{\alpha}\cdot\mathbf{u})(\boldsymbol{\alpha}\cdot\mathbf{e})(\boldsymbol{\alpha}\cdot\mathbf{u})(\boldsymbol{\alpha}\cdot\mathbf{e}')]$$
$$+ \mathrm{Tr}[(\boldsymbol{\alpha}\cdot\mathbf{u})(\boldsymbol{\alpha}\cdot\mathbf{e}')(\boldsymbol{\alpha}\cdot\mathbf{u}')(\boldsymbol{\alpha}\cdot\mathbf{e})]\}$$
$$- \frac{\hbar\omega'}{mc^2}\{\mathrm{Tr}[(\boldsymbol{\alpha}\cdot\mathbf{u}')(\boldsymbol{\alpha}\cdot\mathbf{e})(\boldsymbol{\alpha}\cdot\mathbf{u})(\boldsymbol{\alpha}\cdot\mathbf{e}')]$$
$$+ \mathrm{Tr}[(\boldsymbol{\alpha}\cdot\mathbf{u}')(\boldsymbol{\alpha}\cdot\mathbf{e}')(\boldsymbol{\alpha}\cdot\mathbf{u}')(\boldsymbol{\alpha}\cdot\mathbf{e})]\}.$$

Using Eq. (15.32) and taking into account the cyclic property of the trace and the orthogonality of the polarisation unit vectors with respect to the direction of propagation, we obtain

$$\mathrm{Tr}\big[(\boldsymbol{\alpha}\cdot\mathbf{d})(\mathcal{E}^{\dagger}+\mathcal{F}^{\dagger})\big]=\frac{\hbar\omega'-\hbar\omega}{mc^{2}}\big\{\mathrm{Tr}\big[(\boldsymbol{\alpha}\cdot\mathbf{e})(\boldsymbol{\alpha}\cdot\mathbf{e}')\big]$$
$$+\,\mathrm{Tr}\big[(\boldsymbol{\alpha}\cdot\mathbf{e}')(\boldsymbol{\alpha}\cdot\mathbf{e})(\boldsymbol{\alpha}\cdot\mathbf{u})(\boldsymbol{\alpha}\cdot\mathbf{u}')\big]\big\}. \qquad (15.35)$$

We can now finally evaluate the expression of $\mathcal{T}$. Recalling the definition of b (Eq. (15.30)) and substituting Eqs. (15.34) and (15.35) into Eq. (15.33), we obtain, with some algebra, that

$$\mathcal{T}=8a^{2}+\frac{\hbar\omega-\hbar\omega'}{mc^{2}}\bigg\{4a^{2}+2-\frac{1}{2}\,\mathrm{Re}\big\{\mathrm{Tr}\big[(\boldsymbol{\alpha}\cdot\mathbf{u})(\boldsymbol{\alpha}\cdot\mathbf{u}')\big]\big\}$$
$$-\,a\,\mathrm{Re}\big\{\mathrm{Tr}\big[(\boldsymbol{\alpha}\cdot\mathbf{e})(\boldsymbol{\alpha}\cdot\mathbf{e}')\big]\big\}\bigg\}.$$

Considering that, for two arbitrary vectors $\mathbf{v}$ and $\mathbf{w}$, we have

$$\mathrm{Tr}\big[(\boldsymbol{\alpha}\cdot\mathbf{v})(\boldsymbol{\alpha}\cdot\mathbf{w})\big]=4\mathbf{v}\cdot\mathbf{w},$$

we obtain

$$\mathrm{Tr}\big[(\boldsymbol{\alpha}\cdot\mathbf{u})(\boldsymbol{\alpha}\cdot\mathbf{u}')\big]=4\cos\Theta,\qquad\mathrm{Tr}\big[(\boldsymbol{\alpha}\cdot\mathbf{e})(\boldsymbol{\alpha}\cdot\mathbf{e}')\big]=4\mathbf{e}\cdot\mathbf{e}'=4a,$$

where Θ is the angle defined in Fig. 15.1 and where we have used the definition of a (Eq. (15.28)). After simple substitutions of these results, we get

$$\mathcal{T}=8a^{2}+2\frac{\hbar\omega-\hbar\omega'}{mc^{2}}(1-\cos\Theta).$$

Finally, noting that, according to the kinematic relations, the quantity $(1-\cos\Theta)$ can be written in the form (see Eq. (15.12))

$$1-\cos\Theta=\frac{mc^{2}}{\hbar}\bigg(\frac{1}{\omega'}-\frac{1}{\omega}\bigg),$$

and recalling that $a=\mathbf{e}\cdot\mathbf{e}'$, the expression for $\mathcal{T}$ can be written in the more meaningful form

$$\mathcal{T}=2\bigg[\frac{\omega}{\omega'}+\frac{\omega'}{\omega}-2+4(\mathbf{e}\cdot\mathbf{e}')^{2}\bigg].$$

To finish the calculation of the transition probability, the expression for $\mathcal{T}$ needs to be substituted into Eq. (15.24). Recalling that $\epsilon=mc^{2}$, we have

$$\langle|\mathcal{M}_{\mathrm{fi}}|^{2}\rangle=\frac{\pi^{2}e_{0}^{4}c^{2}\hbar^{2}}{\mathcal{V}^{2}m\omega\omega'\epsilon}\bigg[\frac{\omega}{\omega'}+\frac{\omega'}{\omega}-2+4(\mathbf{e}\cdot\mathbf{e}')^{2}\bigg].$$

Recalling Eq. (15.13), the transition probability per unit time is

$$P_{\mathrm{fi}} = \frac{2\pi^3 e_0^4 c^2 \hbar}{V^2 m \omega \omega' \epsilon'} \left[\frac{\omega}{\omega'} + \frac{\omega'}{\omega} - 2 + 4(\mathbf{e} \cdot \mathbf{e}')^2 \right] \delta(E_{\mathrm{i}} - E_{\mathrm{f}}).$$

The calculation of the cross section proceeds according to the method that we have already used in Sect. 15.2 to find the Thomson cross section and in Sect. 15.3 to obtain the Rayleigh and Raman cross sections. The transition probability per unit time must be divided by the flux of the incident photons, given by the usual expression c/V. We then need to sum over the final states. The interesting quantity is the cross section at fixed polarisation (i.e. at $\mathbf{e}$ and $\mathbf{e}'$ fixed) $\sigma(\Theta, \mathbf{e}, \mathbf{e}')$ for the scattering of a photon in the solid angle $d\Omega$ centered around a direction $\mathbf{\Omega}$ which forms an angle Θ with the direction of the initial photon. The number of final photon states with direction contained in the solid angle $d\Omega$ and angular frequency between ω' and $\omega' + d\omega'$ is given by the right hand side of Eq. (4.3) divided by 2, or

$$dN = \frac{V}{8\pi^3 c^3} \omega'^2 \, d\omega' \, d\Omega.$$

We therefore obtain

$$\sigma(\Theta, \mathbf{e}, \mathbf{e}') \, d\Omega = \frac{e_0^4 \hbar}{4 m c^2 \omega} \, d\Omega \int \frac{\omega'}{\epsilon'} \left[\frac{\omega}{\omega'} + \frac{\omega'}{\omega} - 2 + 4(\mathbf{e} \cdot \mathbf{e}')^2 \right] \delta(E_{\mathrm{i}} - E_{\mathrm{f}}) \, d\omega'.$$

The presence of the Dirac delta has the obvious consequence that all the quantities within the integral need to be evaluated in such a way that energy is conserved. Moreover, the presence of the delta introduces a multiplicative factor, given by the change of variable from ω' to E_{f}, which results in

$$\left| \frac{\partial \omega'}{\partial E_{\mathrm{f}}} \right| = \frac{1}{|\partial E_{\mathrm{f}}/\partial \omega'|}.$$

On the other hand, being

$$\epsilon' = \sqrt{m^2 c^4 + c^2 \mathbf{p}'^2} = \sqrt{m^2 c^4 + \hbar^2 \omega^2 + \hbar^2 \omega'^2 - 2\hbar^2 \omega \omega' \cos \Theta},$$

we have

$$E_{\mathrm{f}} = \hbar \omega' + \epsilon' = \hbar \omega' + \sqrt{m^2 c^4 + \hbar^2 \omega^2 + \hbar^2 \omega'^2 - 2\hbar^2 \omega \omega' \cos \Theta},$$

so that

$$\frac{\partial E_{\mathrm{f}}}{\partial \omega'} = \hbar + \frac{\hbar^2 (\omega' - \omega \cos \Theta)}{\epsilon'} = \hbar \frac{\epsilon' + \hbar \omega' - \hbar \omega \cos \Theta}{\epsilon'}.$$

Recalling the kinematic relations, according to which

$$\epsilon' + \hbar \omega' = mc^2 + \hbar \omega, \qquad mc^2 + \hbar \omega (1 - \cos \Theta) = \frac{mc^2 \omega}{\omega'},$$

we obtain

$$\frac{\partial E_{\mathrm{f}}}{\partial \omega'} = \frac{\hbar m c^2 \omega}{\epsilon' \omega'}.$$

Substituting this result and recalling the definition of the classical radius of the electron ($r_c = e_0^2/(mc^2)$), we finally obtain the Klein-Nishina equation

$$\sigma\left(\Theta, \mathbf{e}, \mathbf{e}'\right) = \frac{r_c^2 \, \omega'^2}{4 \, \omega^2} \left[\frac{\omega}{\omega'} + \frac{\omega'}{\omega} - 2 + 4\left(\mathbf{e} \cdot \mathbf{e}'\right)^2\right]. \tag{15.36}$$

In many cases, we are interested in the cross section $\sigma(\Theta)$ for the scattering of the photon in the angle Θ, independently of the polarisation property of both the initial and the final photon. To obtain this quantity we must sum over the final polarisations and average over the initial polarisations, following the same procedure that we have already used in Sect. 15.2 to obtain the Thomson cross section. We introduce two real unit vectors of polarisation $\mathbf{e}_1$ and $\mathbf{e}_2$. They are perpendicular to the direction of the initial photon and perpendicular to each other. Similarly, we introduce two real unit vectors $\mathbf{e}'_1$ and $\mathbf{e}'_2$, perpendicular to the direction of the final photon and perpendicular to each other. We need to calculate the quantity

$$\sigma(\Theta) = \frac{1}{2} \sum_{i=1,2} \sum_{j=1,2} \sigma\left(\Theta, \mathbf{e}_i, \mathbf{e}'_j\right).$$

The result of this sum (see Sect. 15.2 for the calculation of the sum over i and j of the quantity $(\mathbf{e}_i \cdot \mathbf{e}'_j)^2$) leads to a new expression for the Klein-Nishina cross section which is independent of the polarisation properties of the photons

$$\sigma(\Theta) = \frac{r_c^2 \, \omega'^2}{2 \, \omega^2} \left(\frac{\omega}{\omega'} + \frac{\omega'}{\omega} - \sin^2 \Theta\right). \tag{15.37}$$

In this expression (as in the previous one), the dependence on Θ is effectively contained also in the ω'/ω and ω/ω' ratios, which, recalling the kinematic relations, are given as a function of Θ by the expressions

$$\frac{\omega'}{\omega} = \frac{1}{1 + \varepsilon(1 - \cos\Theta)}, \qquad \frac{\omega}{\omega'} = 1 + \varepsilon(1 - \cos\Theta), \tag{15.38}$$

where we have introduced the quantity ε, the energy of the initial photon, written in terms of the rest energy of the electron

$$\varepsilon = \frac{\hbar\omega}{mc^2}. \tag{15.39}$$

Figure 15.3 illustrates, by means of a radiation diagram, how the cross section varies as a function of Θ for various values of the parameter ε. As shown by the figure, the cross section for forward scattering remains constant with varying photon energy, while the cross section in the other directions decreases significantly (and monotonically) with increasing energy. At the limit for very high energies, the cross section is more and more "focused" around $\Theta = 0$.

Obviously, the two ratios ω/ω' and ω'/ω tend to 1 for $\epsilon \to 0$, and Eq. (15.37) reproduces exactly the result we have obtained for the Thomson cross section, either classically (Sect. 3.5) or within quantum mechanics, but in the non-relativistic approximation (Sect. 15.2, Eq. (15.8)).

Fig. 15.3 The diagram shows how the cross section $\sigma(\Theta)$ varies as a function of Θ, for various values of the parameter ε. Each curve corresponds to the ε value indicated next to it. The cross section in the direction indicated by the angle Θ is proportional to the segment that connects the origin of the diagram with the curve

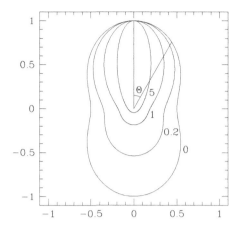

15.7 The Total Cross-Section for Compton Scattering

Using Eqs. (15.38), we can rewrite the Klein-Nishina equation for the differential cross section (Eq. (15.37)) by explicitly introducing the dependence on the angle Θ

$$\sigma(\Theta) = \frac{1}{2}r_c^2 \frac{1}{[1+\varepsilon(1-\cos\Theta)]^2}$$
$$\times \left[\frac{1}{1+\varepsilon(1-\cos\Theta)} + 1 + \varepsilon(1-\cos\Theta) - \sin^2\Theta\right],$$

where ε is the energy of the incoming photon divided by the rest energy of the electron. We are now interested in calculating the total cross section, i.e. the integral

$$\sigma_C = \oint_{4\pi} \sigma(\Theta)\,d\Omega = 2\pi \int_0^\pi \sigma(\Theta)\sin\Theta\,d\Theta.$$

Substituting the expression of $\sigma(\Theta)$ and introducing the variable

$$x = 1 + \varepsilon(1-\cos\Theta),$$

we obtain

$$\sigma_C = \pi r_c^2 \frac{1}{\varepsilon} \int_1^{1+2\varepsilon} \frac{1}{x^2}\left[\frac{1}{x} + x + \left(\frac{x-1}{\varepsilon}\right)^2 - 2\frac{x-1}{\varepsilon}\right] dx,$$

or, with some simple algebra

$$\sigma_C = \pi r_c^2 \frac{1}{\varepsilon^3} \int_1^{1+2\varepsilon}\left[1 + (\varepsilon^2 - 2\varepsilon - 2)\frac{1}{x} + (2\varepsilon + 1)\frac{1}{x^2} + \frac{\varepsilon^2}{x^3}\right] dx.$$

The integrals that appear in this expression are simple and can easily be calculated. We obtain

$$\sigma_C = \pi r_c^2 \left[\frac{4}{\varepsilon^2} + \frac{2(1+\varepsilon)}{(1+2\varepsilon^2)} + \frac{\varepsilon^2 - 2\varepsilon - 2}{\varepsilon^3}\ln(1+2\varepsilon)\right],$$

Fig. 15.4 Total cross section of the Compton effect, normalised to its non-relativistic value (Thomson cross section), as a function of the energy of the incident photon, normalised to the rest mass of the electron

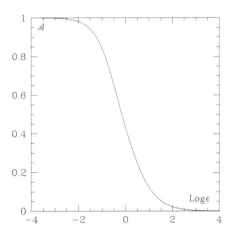

and, normalising the cross section to its non-relativistic value (Thompson cross section) given by $\sigma_T = 8\pi r_c^2/3$, we have

$$\sigma_C = \sigma_T \mathcal{A}(\varepsilon),$$

where

$$\mathcal{A}(\varepsilon) = 3\left[\frac{1}{2\varepsilon^2} + \frac{1+\varepsilon}{4(1+2\varepsilon)^2} + \frac{\varepsilon^2 - 2\varepsilon - 2}{8\varepsilon^3}\ln(1+2\varepsilon)\right].$$

$\mathcal{A}(\epsilon)$ is a monotonically decreasing function of its argument. It is shown in Fig. 15.4. Its series expansion for low energies ($\varepsilon \ll 1$) is

$$\mathcal{A}(\varepsilon) = 1 - 2\varepsilon + \frac{26}{5}\varepsilon^2 + \cdots.$$

On the other hand, for high energies the function tends asymptotically to zero as

$$\mathcal{A}(\varepsilon) \simeq \frac{3}{8\varepsilon}\left(\ln\varepsilon + \ln 2 + \frac{3}{16}\right).$$

A numerical computation shows that the critical value ε_c for which the cross section is equal to half of its non-relativistic value is given by $\varepsilon_c = 0.69016$.

15.8 Polarisation Properties of Compton Scattering

Consider the Klein-Nishina equation for the cross section at fixed polarisation $\sigma(\Theta, \mathbf{e}, \mathbf{e}')$ given by Eq. (15.36)

$$\sigma(\Theta, \mathbf{e}, \mathbf{e}') = \frac{r_c^2}{4}\frac{\omega'^2}{\omega^2}\left[\frac{\omega}{\omega'} + \frac{\omega'}{\omega} - 2 + 4(\mathbf{e}\cdot\mathbf{e}')^2\right].$$

We want to analyse the polarisation properties of the scattered photons assuming that the incoming photons are a "natural" mix by 50 % of photons polarised along the real unit vector $\mathbf{e}_1$ and by 50 % of photons polarised along the real unit vector $\mathbf{e}_2$, orthogonal to the previous one. The scattering cross section now only depends on the polarisation of the scattered photon. Denoting such cross section by the symbol $\sigma_n(\theta, \mathbf{e}')$, we have

$$\sigma_n(\Theta, \mathbf{e}') = \frac{r_c^2}{4} \frac{\omega'^2}{\omega^2} \left(\frac{\omega}{\omega'} + \frac{\omega'}{\omega} - 2 + 4S \right),$$

where

$$S = \frac{1}{2} \sum_{i=1,2} (\mathbf{e}_i \cdot \mathbf{e}')^2.$$

The quantity S can be calculated with a procedure similar to the one adopted in Sect. 15.2. We have

$$S = \frac{1}{2} [1 - (\mathbf{e}' \cdot \mathbf{u})^2].$$

Now, we write the polarisation unit vector $\mathbf{e}'$ as a linear combination of the two real polarisation unit vectors $\mathbf{e}'_1$ and $\mathbf{e}'_2$, defined in a way that $\mathbf{e}'_1$ is perpendicular to the scattering plane while $\mathbf{e}'_2$ is in the scattering plane (see Fig. 3.7).[3] We put then

$$\mathbf{e}' = c_1 \mathbf{e}'_1 + c_2 \mathbf{e}'_2$$

with

$$c_1^2 + c_2^2 = 1.$$

Noting that $\mathbf{e}'_1 \cdot \mathbf{u} = 0$ and that $\mathbf{e}'_2 \cdot \mathbf{u} = -\sin\Theta$, we obtain

$$S = \frac{1}{2} \left(1 - c_2^2 \sin^2\Theta \right).$$

The scattering cross section for a photon that is polarised in a direction perpendicular to the scattering plane ($c_2^2 = 0$) is therefore given by

$$\sigma_n(\Theta, \mathbf{e}'_1) = \frac{r_c^2}{4} \frac{\omega'^2}{\omega^2} \left(\frac{\omega}{\omega'} + \frac{\omega'}{\omega} \right),$$

while that one for a photon polarised along the scattering plane ($c_2^2 = 1$) is given by

$$\sigma_n(\Theta, \mathbf{e}'_2) = \frac{r_c^2}{4} \frac{\omega'^2}{\omega^2} \left(\frac{\omega}{\omega'} + \frac{\omega'}{\omega} - 2\sin^2\Theta \right).$$

The scattered radiation is therefore linearly polarised along the direction perpendicular to the scattering plane. The polarisation is given by

$$p = \frac{\sigma_n(\Theta, \mathbf{e}'_1) - \sigma_n(\Theta, \mathbf{e}'_2)}{\sigma_n(\Theta, \mathbf{e}'_1) + \sigma_n(\Theta, \mathbf{e}'_2)} = \frac{\sin^2\Theta}{\frac{\omega}{\omega'} + \frac{\omega'}{\omega} - \sin^2\Theta}.$$

---

[3] Note that the notation of the unit vectors in Fig. 3.7 is inverted, with respect to the one used here.

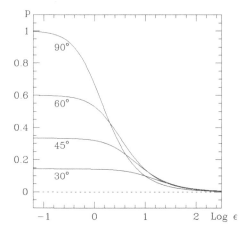

Fig. 15.5 Variation of the linear polarisation of the radiation scattered by Compton effect in the case of incident radiation that is not polarised. The *horizontal axis* shows the energy of the incident photons in terms of the rest mass of the electron. The *curves* refer to various values of the scattering angle Θ. For values of Θ between $90°$ and $180°$ the curves (not shown in figure) decrease more rapidly, with increasing ε, compared to the curves corresponding to the supplementary angle (with which they coincide for $\varepsilon \ll 1$)

The variation of p as a function of the parameter ε (for different values of the scattering angle) is shown in Fig. 15.5. The linear polarisation of the scattered radiation decreases monotonically with increasing energy of the incident photon and tends to zero for $\varepsilon \to \infty$.

To perform a more detailed analysis of the polarisation characteristics of the Compton effect we need to drop the assumption, introduced at the beginning of Sect. 15.6, that the polarisation unit vectors $\mathbf{e}$ and $\mathbf{e}'$ are real. In the more general case of complex unit vectors, it can be shown that the cross section at fixed polarisation (Eq. (15.36)) needs to be replaced by the following expression (which of course coincides with Eq. (15.36) for $\mathbf{e}$ and $\mathbf{e}'$ real)

$$\sigma\left(\Theta, \mathbf{e}, \mathbf{e}'\right) = \frac{r_c^2}{4} \frac{\omega'^2}{\omega^2} \left[\left(\frac{\omega}{\omega'} + \frac{\omega'}{\omega} - 2 \right)\left(1 + |\mathbf{e} \cdot \mathbf{e}'^*|^2 - |\mathbf{e} \cdot \mathbf{e}'|^2\right) + 4|\mathbf{e} \cdot \mathbf{e}'^*|^2 \right].$$

One can use this equation to show that the Stokes parameters of the scattered radiation are related to those of the incident radiation through a matrix that is a generalisation of the one we obtained for Thomson scattering (Eq. (3.27)). This matrix, in the usual geometry of Fig. 3.7 and with the usual conventions for the definition of the Stokes parameters, is given by

$$\begin{pmatrix} \frac{\omega}{\omega'} + \frac{\omega'}{\omega} - \sin^2\Theta & \sin^2\Theta & 0 & 0 \\ \sin^2\Theta & 1 + \cos^2\Theta & 0 & 0 \\ 0 & 0 & 2\cos\Theta & 0 \\ 0 & 0 & 0 & \left(\frac{\omega}{\omega'} + \frac{\omega'}{\omega}\right)\cos\Theta \end{pmatrix}.$$

One can easily verify that in the non-relativistic limit ($\omega' = \omega$) this matrix coincides with the one of the Thomson scattering. The structure of the matrix shows that the circular polarisation, which can exist in the scattered radiation only when it was already present in the incident radiation, still persists for large values of ε. This property is related to the conservation of angular momentum.[4]

15.9 Energy Exchange Between Photons and Electrons

In the process we have analysed in detail in the previous sections, the electron, initially at rest, undergoes a recoil and acquires kinetic energy. The energy acquired by the electron can be easily calculated taking into account the expression for the cross section $\sigma(\Theta)$ of Eq. (15.37) and considering that this energy is equal to $\hbar(\omega - \omega')$. Denoting by the symbol σ_E the cross section for the transfer of energy from the photon to the electron, this quantity is expressed by the integral

$$\sigma_E = \frac{1}{\hbar\omega} \oint_{4\pi} \sigma(\Theta)\hbar(\omega - \omega')\, d\Omega.$$

Recalling Eqs. (15.37) and (15.38), introducing the variable ε (Eq. (15.39)), and applying the change of variable $x = 1 + \varepsilon(1 - \cos\Theta)$, we obtain

$$\sigma_E = \pi r_c^2 \frac{1}{\varepsilon} \int_1^{1+2\varepsilon} \frac{1}{x^2}\left[\frac{1}{x} + x + \left(\frac{x-1}{\varepsilon}\right)^2 - 2\frac{x-1}{\varepsilon}\right]\left(1 - \frac{1}{x}\right) dx,$$

or, with simple algebra

$$\sigma_E = \pi r_c^2 \frac{1}{\varepsilon^3} \int_1^{1+2\varepsilon}\left[1 + (\varepsilon^2 - 2\varepsilon - 3)\frac{1}{x} - (\varepsilon^2 - 4\varepsilon - 3)\frac{1}{x^2}\right.$$
$$\left. + (\varepsilon^2 - 2\varepsilon - 1)\frac{1}{x^3} - \frac{\varepsilon^2}{x^4}\right] dx.$$

By executing the integrals, and normalising the cross section to the Thomson cross section, we obtain

$$\sigma_E = \sigma_T \mathcal{E}(\varepsilon),$$

where

$$\mathcal{E}(\varepsilon) = \frac{9 + 51\varepsilon + 93\varepsilon^2 + 51\varepsilon^3 - 10\varepsilon^4}{4\varepsilon^2(1 + 2\varepsilon)^3} - \frac{3}{8\varepsilon^3}(3 + 2\varepsilon - \varepsilon^2)\ln(1 + 2\varepsilon).$$

The behaviour of the function $\mathcal{E}(\varepsilon)$ is shown in Fig. 15.6. The asymptotic expansions are: for $\varepsilon \ll 1$

$$\mathcal{E}(\varepsilon) = \varepsilon - \frac{21}{5}\varepsilon^2 + \cdots,$$

---

[4] For a discussion see, for example, Fano (1949).

Fig. 15.6 Cross section for
the relative transfer of energy
from the photon to the
electron (initially at rest) in
Compton scattering. The
maximum of the curve
corresponds to $\varepsilon = 0.9821$

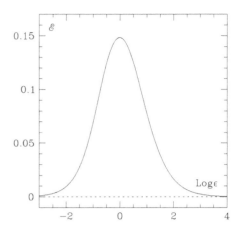

and, for $\varepsilon \gg 1$,

$$\mathcal{E}(\varepsilon) \simeq \frac{3}{8\varepsilon}\left(\ln\varepsilon + \ln 2 - \frac{5}{6}\right).$$

The variation of $\sigma_{\mathrm{E}}(\varepsilon)$ as a function of ε is easily interpreted by observing that, for small ε, in practice the photons bounce elastically on the electron without any energy transfer. Conversely, for large values of ε, the electron becomes practically "transparent" to the photon. From these considerations, we deduce that the function σ_{E} must have a maximum. This maximum occurs for $\varepsilon \simeq 1$, i.e. when the photon energy equals the energy of the electron at rest.

In the case of the electron at rest, the transfer of energy is always from the photon to the electron. However, when the electron is in motion, energy transfers can occur in both directions, depending on the kinematic parameters. In particular, when the kinetic energy of the electron is large compared to the energy of the photon, the energy is transferred from the electron to the photon. This phenomenon is commonly called inverse Compton scattering.

15.10 The Inverse Compton Scattering

We now determine in an approximate way the average energy that is transferred via inverse Compton scattering by an electron moving at velocity **v** to a set of photons that we assume being monochromatic and isotropic. We start by performing a Lorentz transformation from the laboratory reference system, where the photons have frequency ω_0 and direction distributed isotropically and the electron has velocity **v**, to the reference system where the electron is at rest. In this system the photons are not isotropic and monochromatic anymore. Their frequency distribution is obtained by recalling the formula of the Doppler effect

$$\omega = \omega_0 \gamma (1 - \beta \cos\alpha),$$

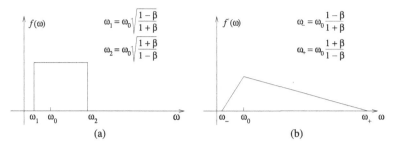

Fig. 15.7 Distribution function of the photons in the rest frame of the electron (**a**), and distribution function of the scattered photons in the laboratory system (**b**). Before the scattering, the distribution function in this system is a Dirac delta at the frequency ω_0. The distributions refer to the case $\beta = 0.7$

where $\beta = v/c$, $\gamma = (1 - \beta^2)^{-1/2}$, and α is the angle between the directions of the electron and the photon, in the laboratory system. On the other hand, the hypothesis of the isotropy of the photons in the laboratory system implies that

$$f(\omega)\, d\omega = \frac{1}{4\pi}\, d\Omega = \frac{1}{2} \sin\alpha\, d\alpha,$$

and since

$$d\omega = \omega_0 \gamma \beta \sin\alpha\, d\alpha,$$

defining the frequencies ω_1 and ω_2 by the equations

$$\omega_1 = \omega_0 \sqrt{\frac{1 - \beta}{1 + \beta}}, \qquad \omega_2 = \omega_0 \sqrt{\frac{1 + \beta}{1 - \beta}},$$

one obtains, for $\omega_1 \leq \omega \leq \omega_2$

$$f(\omega) = \frac{1}{2\omega_0 \gamma \beta}.$$

The result is an uniform distribution of ω values in a finite interval, as shown in Fig. 15.7, panel (a).

Suppose now that the frequency ω_0 is sufficiently low, in such a way that the whole interval of ω values in which the function $f(\omega)$ is different from zero falls within the regime of Thomson scattering. The $f(\omega)$ distribution is not altered during the scattering process because each photon maintains its frequency. If we also neglect the dependence of the Thomson cross section on the scattering angle (i.e. we assume that Thomson scattering is isotropic), we obtain the result that, after the scattering, the photons are distributed isotropically also in the rest frame of the electron. In this case we can then perform a new Lorentz transformation to return to the laboratory frame. The result for the frequency distribution of the scattered photons, $f'(\omega)$, is illustrated in Fig. 15.7, panel (b). The distribution is zero anywhere except in the interval

$$\omega_- \leq \omega \leq \omega_+,$$

where

$$\omega_- = \omega_0 \frac{1-\beta}{1+\beta}, \qquad \omega_+ = \omega_0 \frac{1+\beta}{1-\beta}.$$

The function increases linearly for values of ω that are between ω_- and ω_0, has a maximum for $\omega = \omega_0$, and then decreases linearly for ω between ω_0 and ω_+. The mean value of the distribution, $\langle \omega \rangle$, is obtained by evaluating the abscissa of the barycentre of the triangle of Fig. 15.7, panel (b). We have

$$\langle \omega \rangle = \frac{1}{3}(\omega_- + \omega_+ + \omega_0) = \omega_0 \frac{1 + \frac{1}{3}\beta^2}{1 - \beta^2}.$$

The fraction of the energy transferred on average from the electron to the photons for a single scattering event is therefore

$$\frac{\langle \omega \rangle - \omega_0}{\omega_0} = \frac{\frac{4}{3}\beta^2}{1 - \beta^2} = \frac{4}{3}\gamma^2 \beta^2.$$

An electron that propagates in a medium in which photons are present therefore behaves as if it were emitting electromagnetic energy. In reality, this is not a real emission but only the fact that the photons gain energy because of the impact with the electron. To calculate the power "emitted" in this process, we observe that, according to the previous formula, the relative gain of photon energy is independent of ω_0. If we denote then by u_{phot} the energy density of the photons (energy per unit volume) before the impact, and recall that in the limit that we have considered the cross section is σ_T, we can express the power emitted in the form

$$W = \frac{4}{3}c\sigma_T\gamma^2\beta^2 u_{\text{phot}}.$$

We can then use this formula to define the cross section for the inverse Compton scattering $\sigma_C^{(i)}$. Taking into account that the electron is moving with a velocity $v = c\beta$ and that its cross section sweeps per unit time the photon energy $c\beta u_{\text{phot}}$, we obtain

$$\sigma_C^{(i)} = \frac{4}{3}\gamma^2\beta\sigma_T.$$

Chapter 16
Appendix

16.1 Units of Measurement for Electromagnetic Phenomena

The units of measurement relative to electromagnetic phenomena have been introduced through a long and complex historical process. Without going into the details
of this process, we try here to summarise the fundamental points in light of a modern vision of the phenomena themselves. We stress that these notes are aimed at a
reader who is already familiar with the basic phenomenology of electromagnetism.

Within electrostatics, the fundamental law is Coulomb's law which is written, in
vacuum, in the general form

$$\mathbf{F} = k_C \frac{q_1 q_2}{r^2} \text{ vers } \mathbf{r},$$

where $\mathbf{F}$ is the force that a point charge q_1 exerts on the point charge q_2 placed at
the distance $\mathbf{r} = \mathbf{r}_2 - \mathbf{r}_1$, and where k_C is a constant that implicitly defines the unit
of measurement of the charge (assuming, of course, that the units of measurement
of the mechanical quantities have already been set). We note that k_C can be chosen
to be dimensional or dimensionless. The electric field vector $\mathbf{E}$ is defined in an
arbitrary point using the equation

$$\mathbf{E} = \frac{\mathbf{F}}{q_p},$$

where $\mathbf{F}$ is the electric force exerted on the "test" charge q_p placed at the same
location. From this definition and Coulomb's law we can deduce the expression of
the electric field due to a point charge q, which is

$$\mathbf{E} = k_C \frac{q}{r^2} \text{ vers } \mathbf{r},$$

from which the Gauss theorem (in its integral form) follows

$$\Phi(\mathbf{E}) = \int_\Sigma \mathbf{E} \cdot \mathbf{n} \, dS = 4\pi k_C Q,$$

E. Landi Degl'Innocenti, *Atomic Spectroscopy and Radiative Processes*,
UNITEXT for Physics, DOI 10.1007/978-88-470-2808-1_16, © Springer-Verlag Italia 2014

where Q is the charge contained within the surface Σ. In its differential form, we have

$$\operatorname{div} \mathbf{E} = 4\pi k_C \rho,$$

where ρ is the density of the electric charge (the charge contained within the unit volume).

Regarding the definition of the electric field vector, we note that it is not the only possible one, since we could have defined the electric field created by the charge q as

$$\mathbf{E} = k_C \delta \frac{q}{r^2} \text{ vers } \mathbf{r},$$

with δ an arbitrary constant (possibly dimensional), as long as the electric force that the field exerts on the test charge q_p is written in the form

$$\mathbf{F} = \frac{1}{\delta} q_p \mathbf{E}.$$

Fortunately, the constant δ has (historically) always been set to unity. The same is not true for magnetic phenomena.

Regarding magnetostatics, we have equations that are similar to those of electrostatics. In these equations, for historical reason, the fictitious concept of "magnetic mass" (or "magnetic pole") is introduced. The equations corresponding to those previously written are the following ones (the symbol m denoting the magnetic mass):

Gilbert's law[1] (analogous to Coulomb's law)

$$\mathbf{F} = k_G \frac{m_1 m_2}{r^2} \text{ vers } \mathbf{r}.$$

Definition of the vector of the magnetic induction generated from the magnetic mass m

$$\mathbf{B} = k_G \gamma \frac{m}{r^2} \text{ vers } \mathbf{r},$$

where γ is an arbitrary constant (possibly dimensional).
Force acting on the test magnetic mass m_p

$$\mathbf{F} = \frac{1}{\gamma} m_p \mathbf{B}. \tag{16.1}$$

Equivalent of the Gauss theorem (integral form)

$$\Phi(\mathbf{B}) = 0,$$

---

[1]This law is not universally attributed to Gilbert. In fact, this law was discovered experimentally by Coulomb himself and could therefore be rightly called the "second Coulomb's law". William Gilbert (1564–1603) was an English physician who lived well before Coulomb. He is remembered for his studies on the terrestrial magnetism and by the fact that he realised that the magnetic force should increase with decreasing distance.

since isolated magnetic masses (magnetic monopoles) do not exist.
Equivalent of the Gauss theorem (differential form)

$$\operatorname{div} \mathbf{B} = 0.$$

The first quantitative relations between electric and magnetic phenomena were established with experiments based on electric currents.[2] The intensity of the electric current i flowing in a conductor is defined by the simple equation

$$i = \frac{dq}{dt},$$

from which we can define the current density $\mathbf{j}$ as a vector directed along the direction of motion of the (positive) charges of magnitude

$$j = \frac{i}{\sigma},$$

where σ is the cross-sectional area of the conductor. The experiments performed during the first half of the nineteenth century especially by Ørsted, Ampère, and Faraday, led to the idea that electric currents create magnetic fields in their surroundings and that, at the same time, a magnetic field is able to exert a force on electric currents. During the same period, a new idea clearly emerged: that permanent magnets contain, at the microscopic level, a large number of elementary electric currents. These currents would be responsible, ultimately, for magnetostatic phenomena.

In modern terms, the magnetic properties of the currents can be summarised by a single law that is expressed by saying that, in stationary conditions, the current element of an elementary circuit (microscopic or macroscopic) $i_1 \, d\boldsymbol{\ell}_1$ acts on the current element of another elementary circuit, $i_2 \, d\boldsymbol{\ell}_2$, with an infinitesimal force $d\mathbf{F}$ given by

$$d\mathbf{F} = k_A i_2 \, d\boldsymbol{\ell}_2 \times \left(i_1 \, d\boldsymbol{\ell}_1 \times \frac{\operatorname{vers} \mathbf{r}}{r^2} \right),$$

where k_A is a new constant (which cannot be independent of those already introduced), and where $\mathbf{r}$ is the radius vector that goes from the current element $i_1 \, d\boldsymbol{\ell}_1$ to the current element $i_2 \, d\boldsymbol{\ell}_2$. This law allows the introduction of the magnetic induction vector. The definition of this vector is somewhat arbitrary and it is assumed, in general, that the current element $i \, d\boldsymbol{\ell}$ creates the elementary induction vector $d\mathbf{B}$ given by (first law of Laplace or Biot and Savart's law)

$$d\mathbf{B} = k_A \beta i \, d\boldsymbol{\ell} \times \frac{\operatorname{vers} \mathbf{r}}{r^2}, \tag{16.2}$$

---

[2]These experiments were made possible thanks to the discovery of the electric battery by Alessandro Volta.

and that a current element $i\,\mathrm{d}\boldsymbol{\ell}$ is subject, in the presence of an induction vector $\mathbf{B}$, to a force $\mathrm{d}\mathbf{F}$ given by (second law of Laplace)

$$\mathrm{d}\mathbf{F} = \frac{1}{\beta} i\,\mathrm{d}\boldsymbol{\ell} \times \mathbf{B}.$$

The quantity β introduced in these equations is arbitrary.

Let's see the mathematical consequences of Eq. (16.2) for a closed circuit. The magnetic induction vector is given by

$$\mathbf{B} = k_{\mathrm{A}}\beta \oint_{\mathrm{C}} i\,\mathrm{d}\boldsymbol{\ell} \times \frac{\mathrm{vers}\,\mathbf{r}}{r^2},$$

where C is the curve describing the closed circuit. Using standard mathematical methods, one finds the following equations

$$\mathrm{div}\,\mathbf{B} = 0,$$

which confirms the analogous equation for magnetostatics, and

$$\mathrm{rot}\,\mathbf{B} = 4\pi k_{\mathrm{A}}\beta\mathbf{j},$$

where $\mathbf{j}$ is the current density.

This equation, known as Ampère's law, applies only to stationary phenomena. As shown by Maxwell, it can be transformed into a more general equation that is also valid for phenomena that are variable in time. To do this we observe that, taking the divergence of both sides, we have

$$\mathrm{div}\,\mathbf{j} = 0,$$

while, in general, the continuity equation must hold

$$\mathrm{div}\,\mathbf{j} + \frac{\partial \rho}{\partial t} = 0,$$

ρ being the charge density. In order to rearrange things, we take the derivative (with respect to time) of the differential expression of Coulomb's law

$$\frac{\partial \rho}{\partial t} = \frac{1}{4\pi k_{\mathrm{C}}} \frac{\partial}{\partial t}(\mathrm{div}\,\mathbf{E}),$$

so that in general the following equation holds

$$\mathrm{div}\left(\mathbf{j} + \frac{1}{4\pi k_{\mathrm{C}}} \frac{\partial \mathbf{E}}{\partial t}\right) = 0.$$

The second term in parentheses is the so-called displacement current density. With its introduction, the equation for $\mathrm{rot}\,\mathbf{B}$, corrected to include non-stationary phenomena, is

$$\mathrm{rot}\,\mathbf{B} - \frac{k_{\mathrm{A}}\beta}{k_{\mathrm{C}}} \frac{\partial \mathbf{E}}{\partial t} = 4\pi k_{\mathrm{A}}\beta\mathbf{j}.$$

Finally, we need to consider the phenomena of magnetic induction. The law that describes them can be deduced, at least in a particular case, from the second law of Laplace. We have

$$\text{rot}\,\mathbf{E} = -\frac{1}{\beta}\frac{\partial \mathbf{B}}{\partial t},$$

so that, in summary, the laws governing the electromagnetic phenomena can all be enclosed in the following four Maxwell's equations

$$\text{div}\,\mathbf{E} = 4\pi k_C \rho, \qquad\qquad \text{div}\,\mathbf{B} = 0,$$

$$\text{rot}\,\mathbf{B} - \frac{k_A \beta}{k_C}\frac{\partial \mathbf{E}}{\partial t} = 4\pi k_A \beta \mathbf{j}, \qquad \text{rot}\,\mathbf{E} + \frac{1}{\beta}\frac{\partial \mathbf{B}}{\partial t} = 0.$$

We now consider Maxwell's equations in vacuum. Taking the curl of the third equation and substituting the fourth, we obtain the wave equation

$$\nabla^2 \mathbf{B} = \frac{k_A}{k_C}\frac{\partial^2 \mathbf{B}}{\partial t^2}.$$

On the other hand we know that electromagnetic waves propagate in vacuum with velocity c, so that we must have

$$\frac{k_A}{k_C} = \frac{1}{c^2},$$

or

$$k_A = \frac{k_C}{c^2},$$

that is, a relation between the quantities k_A and k_C that is independent of the unit system under consideration.

Let's see how we proceed in the two more common systems of units, the cgs system of Gauss (sometimes also called the Gauss-Hertz system) and the International System of Units (SI). In the cgs system, we assume $k_C = 1$, so that the unit of charge is defined as the charge that repels an equal charge, at a distance of one centimeter, with the force of one dyne. Such unit of charge is called Franklin or statcoulomb. Since $k_C = 1$, it follows that $k_A = 1/c^2$. Within this system we also assume that $\beta = c$, so that Maxwell's equations are written as

$$\text{div}\,\mathbf{E} = 4\pi\rho, \qquad\qquad \text{div}\,\mathbf{B} = 0,$$

$$\text{rot}\,\mathbf{B} - \frac{1}{c}\frac{\partial \mathbf{E}}{\partial t} = 4\pi\frac{\mathbf{j}}{c}, \qquad \text{rot}\,\mathbf{E} + \frac{1}{c}\frac{\partial \mathbf{B}}{\partial t} = 0.$$

Moreover, the first and the second law of Laplace, together with the law that summarises them, can be written in the form

$$d\mathbf{B} = \frac{i}{c} d\boldsymbol{\ell} \times \frac{\text{vers}\,\mathbf{r}}{r^2}, \qquad d\mathbf{F} = \frac{i}{c} d\boldsymbol{\ell} \times \mathbf{B},$$

$$d\mathbf{F} = \frac{i_2}{c} d\boldsymbol{\ell}_2 \times \left(\frac{i_1}{c} d\boldsymbol{\ell}_1 \times \frac{\text{vers}\,\mathbf{r}}{r^2} \right).$$

Within the International System, instead, two new constants are introduced. They are the vacuum permittivity (also called dielectric permittivity of the vacuum) ϵ_0 and the vacuum permeability (magnetic permeability of the vacuum) μ_0, such that

$$\epsilon_0 \mu_0 = \frac{1}{c^2}.$$

Using these quantities, we put

$$k_C = \frac{1}{4\pi \epsilon_0},$$

so that we have

$$k_A = \frac{k_C}{c^2} = \frac{1}{4\pi \epsilon_0 c^2} = \frac{\mu_0}{4\pi}.$$

Within this system we also put $\beta = 1$, so that Maxwell's equations are written as

$$\text{div}\,\mathbf{E} = \frac{\rho}{\epsilon_0}, \qquad\qquad \text{div}\,\mathbf{B} = 0,$$

$$\text{rot}\,\mathbf{B} - \frac{1}{c^2} \frac{\partial \mathbf{E}}{\partial t} = \mu_0 \mathbf{j}, \qquad \text{rot}\,\mathbf{E} + \frac{\partial \mathbf{B}}{\partial t} = 0.$$

Moreover, the first and the second law of Laplace and the law that summarises them are written, respectively, in the form

$$d\mathbf{B} = \frac{\mu_0}{4\pi} i \, d\boldsymbol{\ell} \times \frac{\text{vers}\,\mathbf{r}}{r^2}, \qquad d\mathbf{F} = i \, d\boldsymbol{\ell} \times \mathbf{B},$$

$$d\mathbf{F} = \frac{\mu_0}{4\pi} i_2 \, d\boldsymbol{\ell}_2 \times \left(i_1 \, d\boldsymbol{\ell}_1 \times \frac{\text{vers}\,\mathbf{r}}{r^2} \right). \tag{16.3}$$

With respect to the numerical values of ϵ_0 and μ_0, the Ampère (unit of measurement of the current) is defined as the current that, flowing along an infinite straight wire of negligible thickness in vacuum, attracts an equal wire, located at a distance of one meter, with a force per unit length equal to $2 \times 10^{-7}\,\text{N}\,\text{m}^{-1}$. Using Eq. (16.3) we deduce that in such a geometry the force per unit length that acts on one of the two conductors is attractive and has a magnitude given by the following expression

$$\frac{dF}{dl} = 2 \frac{\mu_0}{4\pi} \frac{i^2}{r},$$

so we must have[3]

$$\mu_0 = 4\pi \times 10^{-7} \, \mathrm{N\,A^{-2}} = 1.256637 \times 10^{-6} \, \mathrm{N\,A^{-2}},$$

and then, recalling that the Coulomb is the charge transported in one second by a current of one Ampère

$$\epsilon_0 = \frac{1}{\mu_0 c^2} = 8.854188 \times 10^{-12} \, \mathrm{C^2\,N^{-1}\,m^{-2}}.$$

Finally, it remains to analyse the relation between magnetic masses and currents. We can infer from Laplace's laws that a filiform planar circuit of area σ and current i behaves, at distances much larger than its size, as a magnetic dipole directed along the unit vector $\mathbf{n}$ perpendicular to the plane of the circuit. The direction of $\mathbf{n}$ is specified by the rule of the corkscrew (or the right screw). This is the so-called Ampère principle of equivalence, which is expressed by the formula

$$\boldsymbol{\mu} = k_\mathrm{P} i \sigma \mathbf{n},$$

where k_P is a new constant to be related to those previously introduced. To establish this relation, we evaluate, for example, the moment of the forces acting on an elementary dipole located at a point in space where the field $\mathbf{B}$ is present. Using Eq. (16.1), we have

$$\mathbf{M} = \frac{1}{\gamma} \boldsymbol{\mu} \times \mathbf{B} = \frac{1}{\gamma} k_\mathrm{P} i \sigma \mathbf{n} \times \mathbf{B}.$$

Instead, from the second law of Laplace we have

$$\mathbf{M} = \frac{1}{\beta} \oint i \mathbf{r} \times (\mathrm{d}\boldsymbol{\ell} \times \mathbf{B}),$$

which can be rewritten as

$$\mathbf{M} = \frac{1}{\beta} i \sigma \mathbf{n} \times \mathbf{B}.$$

Equating the two expressions for $\mathbf{M}$ we have

$$k_\mathrm{P} = \frac{\gamma}{\beta}.$$

Finally, considering the force exerted between two infinitesimal circuits, treated in the first instance as elementary dipoles and then as coils carrying a current, we obtain the relation

$$k_\mathrm{G} k_\mathrm{P}^2 = k_\mathrm{A},$$

---

[3]With the introduction of capacity and inductance, together with their units, the Farad (F) and the Henry (H), the units in which μ_0 and ϵ_0 are expressed are, respectively, $\mathrm{H\,m^{-1}}$ and $\mathrm{F\,m^{-1}}$.

which allows to write k_G in the form

$$k_G = \frac{k_A}{k_P^2} = \frac{k_C \beta^2}{c^2 \gamma^2}.$$

In the cgs system, since $k_C = 1$ and $\beta = c$, and assuming $\gamma = 1$, we obtain

$$k_P = \frac{1}{c}, \qquad k_G = 1.$$

Ampère's principle of equivalence is therefore

$$\boldsymbol{\mu} = \frac{i}{c}\sigma \mathbf{n},$$

and Gilbert's law

$$\mathbf{F} = \frac{m_1 m_2}{r^2} \text{ vers } \mathbf{r}.$$

In the International System, instead, since $k_C = 1/(4\pi\epsilon_0)$ and $\beta = 1$, assuming[4] $\gamma = \mu_0$ and recalling that $c^2 = 1/(\epsilon_0\mu_0)$, we obtain

$$k_P = \mu_0, \qquad k_G = \frac{1}{4\pi\mu_0}.$$

In this case Ampère's principle of equivalence is

$$\boldsymbol{\mu} = \mu_0 i\sigma \mathbf{n},$$

and Gilbert's law is

$$\mathbf{F} = \frac{1}{4\pi\mu_0} \frac{m_1 m_2}{r^2} \text{ vers } \mathbf{r}.$$

Finally, we note that, besides the two systems introduced here, there are other ones that have been used for the electromagnetic phenomena. In particular, it is worth mentioning the electrostatic cgs system, the electromagnetic cgs system and the cgs system of Heavyside.

16.2 Tensor Algebra

In this volume, we often need to deal with vectors and tensors, together with their differential expressions such as divergences, curls and gradients. It is therefore useful to give a brief introduction to this topic in order to make the reader familiar

---

[4]This convention is not universally accepted. Some authors prefer to assume $\gamma = 1$ also in the International System. In this case Ampère's principle of equivalence is written as $\boldsymbol{\mu} = i\sigma\mathbf{n}$ while in Gilbert's law the factor μ_0 is to appear in the numerator rather than in the denominator.

with a compact formalism that allows to easily deduce a series of vector and tensor identities, as well as various transformation formulae.

The traditional definition of a tensor that is commonly given in physics is based on the generalisation of the definition of a vector. In a Cartesian orthogonal reference system, the vector $\mathbf{v}$ is defined as an entity with three components (v_x, v_y, v_z) (or v_1, v_2, v_3) which, under an arbitrary rotation of the reference system, are modified according to the law

$$v_i' = \sum_j C_{ij} v_j,$$

where the coefficients C_{ij} are the direction cosines of the new axes with respect to the old ones. In close analogy, we define a tensor $\mathbf{T}$ of rank n as an entity with 3^n components ($T_{i \ldots j}$ with $i, \ldots, j = 1, 3$) which, under a rotation of the reference system, are transformed according to the law

$$T_{i \ldots j}' = \sum_{k, \ldots, l} C_{ik} \cdots C_{jl} T_{k \ldots l}.$$

The tensor most commonly known in physics is the stress tensor that characterises inside an elastic material the force $\mathbf{dF}$ that is exerted on a surface dS with normal $\mathbf{n}$. In components we have

$$dF_i = \sum_j T_{ij} n_j \, dS.$$

In addition to the stress tensor we can also mention, for their importance in various fields of physics, the deformation tensor, the inertia tensor, and the dielectric tensor.

A particular tensor of rank two is the so-called dyad that is obtained from two vectors $\mathbf{u}$ and $\mathbf{v}$ when the direct product of their components is considered. The dyad is indicated simply by the symbol $\mathbf{uv}$, and we have by definition

$$(\mathbf{uv})_{ij} = u_i v_j \quad (i, j = 1, 2, 3).$$

Obviously, in general

$$\mathbf{uv} \neq \mathbf{vu}.$$

A scalar quantity is, by definition, a tensor of rank zero, while a vector is, by definition, a tensor of rank one. Tensors of higher rank may be obtained by considering the direct product of tensors of lower rank. For example, by the direct product of two tensors of rank two a tensor of rank four is obtained.

The tensor algebra covers all operations that can be performed on tensors. We now provide some definitions

1. Given two tensors $\mathbf{T}$ and $\mathbf{V}$, the first of rank n ($n \geq 1$) and the second of rank n' ($n' \geq 1$), we define the scalar product (or internal product) of the two tensors a tensor of rank ($n + n' - 2$) obtained by a sum (or saturation) which operates over the last index of the first tensor and the first index of the second tensor. For

example, if n and n' are both equal to 2, defining $\mathbf{W}$ as the tensor obtained by the scalar product, we have that $\mathbf{W}$ is also a tensor of rank two defined by

$$W_{ij} = \sum_k T_{ik} V_{kj}.$$

2. Given a tensor of rank n (with $n \geq 1$), the divergence of such tensor is a tensor of rank $(n-1)$ obtained by saturating its first component with the formal vector ∇ (called "nabla" operator or "del" operator) defined by

$$\nabla \equiv \left(\frac{\partial}{\partial x}, \frac{\partial}{\partial y}, \frac{\partial}{\partial z} \right).$$

For example, for a tensor $\mathbf{T}$ of rank two, div $\mathbf{T}$ is a vector whose components are given by

$$(\text{div}\,\mathbf{T})_i = \sum_j \frac{\partial}{\partial x_j} T_{ji} = (\nabla \cdot \mathbf{T})_i.$$

3. Given a tensor of rank n (with $n \geq 0$), the gradient of such tensor is a tensor of rank $(n+1)$ obtained by applying to it the formal vector ∇ in such a way that the first index of the resulting tensor is the "derivation one". For example, for a tensor of rank 1, i.e. for a a vector $\mathbf{v}$, we have

$$(\text{grad}\,\mathbf{v})_{ij} = (\nabla\mathbf{v})_{ij} = \frac{\partial}{\partial x_i} v_j.$$

It should be noted that this convention is not universally adopted. Some authors prefer to indicate with the symbol grad $\mathbf{v}$ the quantity

$$(\text{grad}\,\mathbf{v})_{ij} = \frac{\partial}{\partial x_j} v_i.$$

The reader should therefore pay attention to the conventions used by each author before using the vector identities that are found in different books. For example, using our conventions, we have

$$\sum_i u_i \frac{\partial v_j}{\partial x_i} = (\mathbf{u} \cdot \text{grad}\,\mathbf{v})_j, \qquad \sum_i u_i \frac{\partial v_i}{\partial x_j} = \left[(\text{grad}\,\mathbf{v}) \cdot \mathbf{u} \right]_j.$$

Using the formal vector ∇, the quantities in the right-hand side can also be written, respectively, as

$$\left[(\mathbf{u} \cdot \nabla)\mathbf{v} \right]_j, \qquad \left[(\nabla\mathbf{v}) \cdot \mathbf{u} \right]_j.$$

4. Given a tensor of rank n ($n \geq 1$), the curl (also known as rotor) of such a tensor is a tensor of the same rank n with the first component being obtained by saturating the first component of the given tensor with the completely antisymmetric tensor

(known as the Ricci, or Ricci-Levi Civita tensor) and with the component of the formal vector ∇. For example, for a vector $\mathbf{v}$ we have

$$(\text{rot}\,\mathbf{v})_i = \sum_{jk} \epsilon_{ijk} \frac{\partial v_k}{\partial x_j} = (\nabla \times \mathbf{v})_i,$$

and for a tensor $\mathbf{T}$ of rank two

$$(\text{rot}\,\mathbf{T})_{ij} = \sum_{kl} \epsilon_{ikl} \frac{\partial T_{lj}}{\partial x_k} = (\nabla \times \mathbf{T})_{ij}.$$

The antisymmetric tensor of rank three ϵ_{ijk}, introduced in these expressions, is defined by the equation $\epsilon_{ijk} = 0$ if at least two of the three indices i, j, k are equal; by the equation $\epsilon_{ijk} = 1$ if the ordered triad (i, j, k) is an even permutation of the fundamental triad $(1, 2, 3)$; and by the equation $\epsilon_{ijk} = -1$ if the ordered triad (i, j, k) is an odd permutation of the fundamental triad $(1, 2, 3)$. Ultimately, only 6 of the 27 components of the tensor are different from zero. Note that the usual vector product between two vectors can be conveniently expressed through the antisymmetric tensor. If $\mathbf{w} = \mathbf{u} \times \mathbf{v}$, we have

$$w_i = \sum_{jk} \epsilon_{ijk} u_j v_k.$$

Note also that the vector product operation and the curl operator (which involve the antisymmetric tensor) imply a choice about the chirality of the Cartesian orthogonal system in which the components of the vectors (and tensors) are defined. The convention that is now almost universally accepted (and that we use) is to choose a right-handed triad, i.e. to suppose that, if the axes x and y are directed respectively along the thumb and index finger of the right hand, the z axis is directed along the middle finger.

The antisymmetric tensor has a number of properties. The first concerns the permutation of its indices. For an even permutation the tensor remains unchanged, while for an odd permutation the tensor changes sign. In formulae

$$\epsilon_{ijk} = \epsilon_{jki} = \epsilon_{kij} = -\epsilon_{jik} = -\epsilon_{ikj} = -\epsilon_{kji}.$$

In addition, the following saturation properties hold

$$\sum_k \epsilon_{ijk} \epsilon_{lmk} = \delta_{il}\delta_{jm} - \delta_{im}\delta_{jl},$$

$$\sum_{jk} \epsilon_{ijk} \epsilon_{ljk} = 2\delta_{il},$$

$$\sum_{ijk} \epsilon_{ijk} \epsilon_{ijk} = 6,$$

where δ_{ij} is the so-called Kronecker delta, i.e. the symbol defined by

$$\delta_{ij} = 1 \quad \text{if } i = j, \qquad \delta_{ij} = 0 \quad \text{if } i \neq j.$$

The above definitions and properties can be used to obtain a number of vector identities that are listed below. In these equations, the quantities f and g are scalars, **a** and **b** are vectors, and **T** is a tensor of rank 2.

- $\operatorname{div}(f\mathbf{a}) = \mathbf{a} \cdot \operatorname{grad} f + f \operatorname{div} \mathbf{a}.$ (16.4)

In fact we have

$$\operatorname{div}(f\mathbf{a}) = \sum_i \frac{\partial}{\partial x_i}(fa_i) = \sum_i a_i \frac{\partial f}{\partial x_i} + f \sum_i \frac{\partial a_i}{\partial x_i}.$$

- $\operatorname{grad}(fg) = g \operatorname{grad} f + f \operatorname{grad} g.$ (16.5)

In fact we have, for the i-th component

$$\left[\operatorname{grad}(fg)\right]_i = \frac{\partial}{\partial x_i}(fg) = g\frac{\partial f}{\partial x_i} + f\frac{\partial g}{\partial x_i}.$$

- $\operatorname{rot}(f\mathbf{a}) = \operatorname{grad} f \times \mathbf{a} + f \operatorname{rot} \mathbf{a}.$ (16.6)

In fact we have, for the i-th component

$$\left[\operatorname{rot}(f\mathbf{a})\right]_i = \sum_{jk} \epsilon_{ijk}\frac{\partial}{\partial x_j}(fa_k) = \sum_{jk}\epsilon_{ijk}\left[\left(\frac{\partial f}{\partial x_j}\right)a_k + f\frac{\partial a_k}{\partial x_j}\right] =$$

$$= \left[(\operatorname{grad} f) \times \mathbf{a}\right]_i + f[\operatorname{rot}\mathbf{a}]_i.$$

- $\operatorname{div}(\mathbf{a} \times \mathbf{b}) = \mathbf{b} \cdot \operatorname{rot}\mathbf{a} - \mathbf{a} \cdot \operatorname{rot}\mathbf{b}.$ (16.7)

In fact we have

$$\operatorname{div}(\mathbf{a} \times \mathbf{b}) = \sum_i \frac{\partial}{\partial x_i}\left(\sum_{jk}\epsilon_{ijk}a_jb_k\right) = \sum_{ijk}\epsilon_{ijk}\left[\left(\frac{\partial a_j}{\partial x_i}\right)b_k + a_j\left(\frac{\partial b_k}{\partial x_i}\right)\right]$$

$$= \sum_{ijk}b_k\epsilon_{kij}\frac{\partial a_j}{\partial x_i} - \sum_{ijk}a_j\epsilon_{jik}\frac{\partial b_k}{\partial x_i} = \sum_k b_k(\operatorname{rot}\mathbf{a})_k - \sum_j a_j(\operatorname{rot}\mathbf{b})_j.$$

- $\operatorname{grad}(\mathbf{a} \cdot \mathbf{b}) = (\operatorname{grad}\mathbf{a}) \cdot \mathbf{b} + (\operatorname{grad}\mathbf{b}) \cdot \mathbf{a}.$ (16.8)

In fact we have, for the i-th component

$$\left[\operatorname{grad}(\mathbf{a} \cdot \mathbf{b})\right]_i = \frac{\partial}{\partial x_i}\left(\sum_j a_jb_j\right) = \sum_j\left(\frac{\partial a_j}{\partial x_i}\right)b_j + \sum_j a_j\left(\frac{\partial b_j}{\partial x_i}\right)$$

$$= \left[(\operatorname{grad}\mathbf{a}) \cdot \mathbf{b}\right]_i + \left[(\operatorname{grad}\mathbf{b}) \cdot \mathbf{a}\right]_i.$$

- $\text{rot}(\mathbf{a} \times \mathbf{b}) = \mathbf{b} \cdot \text{grad}\,\mathbf{a} - \mathbf{a} \cdot \text{grad}\,\mathbf{b} + \mathbf{a}\,\text{div}\,\mathbf{b} - \mathbf{b}\,\text{div}\,\mathbf{a}.$ (16.9)

In fact we have, for the i-th component

$$
\begin{aligned}
\left[\text{rot}(\mathbf{a} \times \mathbf{b})\right]_i &= \sum_{jk} \epsilon_{ijk} \frac{\partial}{\partial x_j}(\mathbf{a} \times \mathbf{b})_k = \sum_{jklm} \epsilon_{ijk}\epsilon_{klm} \frac{\partial}{\partial x_j}(a_l b_m) \\
&= \sum_{jlm}(\delta_{il}\delta_{jm} - \delta_{im}\delta_{jl})\left[\left(\frac{\partial a_l}{\partial x_j}\right)b_m + a_l\frac{\partial b_m}{\partial x_j}\right] \\
&= \sum_{ij}\left(b_j\frac{\partial a_i}{\partial x_j} - b_i\frac{\partial a_j}{\partial x_j} + a_i\frac{\partial b_j}{\partial x_j} - a_j\frac{\partial b_i}{\partial x_j}\right) \\
&= [\mathbf{b} \cdot \text{grad}\,\mathbf{a}]_i - b_i\,\text{div}\,\mathbf{a} + a_i\,\text{div}\,\mathbf{b} - [\mathbf{a} \cdot \text{grad}\,\mathbf{b}]_i.
\end{aligned}
$$

- $\text{grad}(f\mathbf{a}) = (\text{grad}\,f)\mathbf{a} + f\,\text{grad}\,\mathbf{a}.$ (16.10)

In fact we have, for the ij-th component

$$
\left[\text{grad}(f\mathbf{a})\right]_{ij} = \frac{\partial}{\partial x_i}(fa_j) = \left(\frac{\partial f}{\partial x_i}\right)a_j + f\frac{\partial a_j}{\partial x_i} = (\text{grad}\,f)_i a_j + f(\text{grad}\,\mathbf{a})_{ij}.
$$

- $\text{div}(\mathbf{ab}) = \mathbf{b}\,\text{div}\,\mathbf{a} + \mathbf{a} \cdot \text{grad}\,\mathbf{b}.$ (16.11)

In fact we have, for the i-th component

$$
\begin{aligned}
\left[\text{div}(\mathbf{ab})\right]_i &= \sum_j \frac{\partial}{\partial x_j}(a_j b_i) = \sum_j\left[\left(\frac{\partial a_j}{\partial x_j}\right)b_i + a_j\frac{\partial b_i}{\partial x_j}\right] \\
&= b_i\,\text{div}\,\mathbf{a} + [\mathbf{a} \cdot \text{grad}\,\mathbf{b}]_i.
\end{aligned}
$$

- $\mathbf{a} \times \text{rot}\,\mathbf{b} = (\text{grad}\,\mathbf{b}) \cdot \mathbf{a} - \mathbf{a} \cdot \text{grad}\,\mathbf{b}.$ (16.12)

In fact we have, for the i-th component

$$
\begin{aligned}
[\mathbf{a} \times \text{rot}\,\mathbf{b}]_i &= \sum_{jk} \epsilon_{ijk} a_j (\text{rot}\,\mathbf{b})_k = \sum_{jklm} \epsilon_{ijk}\epsilon_{klm} a_j \frac{\partial b_m}{\partial x_l} \\
&= \sum_{jlm}(\delta_{il}\delta_{jm} - \delta_{im}\delta_{jl})a_j \frac{\partial b_m}{\partial x_l} \\
&= \sum_j\left(a_j\frac{\partial b_j}{\partial x_i} - a_j\frac{\partial b_i}{\partial x_j}\right) = \left[(\text{grad}\,\mathbf{b}) \cdot \mathbf{a}\right]_i - [\mathbf{a} \cdot \text{grad}\,\mathbf{b}]_i.
\end{aligned}
$$

- $\text{div}(f\mathbf{T}) = (\text{grad}\,f) \cdot \mathbf{T} + f\,\text{div}\,\mathbf{T}.$ (16.13)

In fact we have, for the i-th component

$$\left[\text{div}(f\mathbf{T})\right]_i = \sum_j \frac{\partial}{\partial x_j}(fT_{ji}) = \sum_j \left[\left(\frac{\partial f}{\partial x_j}\right)T_{ji} + f\frac{\partial T_{ji}}{\partial x_j}\right]$$

$$= \left[(\text{grad } f)\cdot\mathbf{T}\right]_i + f(\text{div }\mathbf{T})_i.$$

- $\quad \text{rot}(\text{rot }\mathbf{a}) = \text{grad div }\mathbf{a} - \nabla^2\mathbf{a}.$ (16.14)

In fact we have, for the i-th component

$$\left[\text{rot}(\text{rot }\mathbf{a})\right]_i = \sum_{jk}\epsilon_{ijk}\frac{\partial}{\partial x_j}(\text{rot }\mathbf{a})_k = \sum_{jklm}\epsilon_{ijk}\epsilon_{klm}\frac{\partial}{\partial x_j}\frac{\partial}{\partial x_l}a_m$$

$$= \sum_{jlm}(\delta_{il}\delta_{jm} - \delta_{im}\delta_{jl})\frac{\partial^2 a_m}{\partial x_j \partial x_l}$$

$$= \sum_j\left(\frac{\partial^2 a_j}{\partial x_j \partial x_i} - \frac{\partial^2 a_i}{\partial x_j \partial x_j}\right) = [\text{grad div }\mathbf{a}]_i - \left[\nabla^2\mathbf{a}\right]_i.$$

There are also other vector identities that apply only in integral form. They result from the theorems of Gauss and Stokes-Ampère, which we recall now.

Gauss theorem: If Σ is a closed surface enclosing the volume V and if $\mathbf{n}$ is the normal external to the surface, Gauss theorem is expressed by the equation

- $$\int_\Sigma \mathbf{a}\cdot\mathbf{n}\,\text{d}S = \int_V \text{div }\mathbf{a}\,\text{d}V,$$

where $\mathbf{a}$ is an arbitrary vector that is a function of the position.

Stokes-Ampère theorem: if ℓ is a closed circuit and if Σ is a surface that is leaning on this circuit, the Stokes-Ampère theorem is stated by the equation

- $$\oint_\ell \mathbf{a}\cdot\text{d}\boldsymbol{\ell} = \int_\Sigma \text{rot }\mathbf{a}\cdot\mathbf{n}\,\text{d}S,$$

where $\mathbf{n}$ is the normal external to the surface. We note that the validity of this equation implies a convention about the direction of integration along the circuit, which in turn depends on the implicit convention in the definition of the curl operator. When the (x, y, z) system used to define the vector components is a right-handed system, then the direction of integration along the circuit follows the corkscrew (or the right screw) rule, for which the direction of $\mathbf{n}$ coincides with the direction of advancement of the corkscrew.

Various identities can be obtained from the Gauss and Stokes-Ampère theorems. Some of them are collected below.

- $$\oint_\ell f\,\text{d}\boldsymbol{\ell} = \int_\Sigma \mathbf{n}\times\text{grad } f\,\text{d}S.$$

This identity can be proven by noting that, if $\mathbf{c}$ is an arbitrary constant vector, we have

$$\mathbf{c} \cdot \oint_\ell f \, d\boldsymbol{\ell} = \oint_\ell (f\mathbf{c}) \cdot d\boldsymbol{\ell},$$

and, applying the Stokes-Ampère theorem

$$\mathbf{c} \cdot \oint_\ell f \, d\boldsymbol{\ell} = \int_\Sigma \mathrm{rot}(f\mathbf{c}) \cdot \mathbf{n} \, dS.$$

Recalling the vector identity of Eq. (16.6), and taking into account that $\mathbf{c}$ is a constant vector, we have

$$\mathbf{c} \cdot \oint_\ell f \, d\boldsymbol{\ell} = \int_\Sigma \left[(\mathrm{grad}\, f) \times \mathbf{c} \right] \cdot \mathbf{n} \, dS = \mathbf{c} \cdot \int_\Sigma \mathbf{n} \times \mathrm{grad}\, f \, dS.$$

The identity therefore follows, because $\mathbf{c}$ is an arbitrary vector.

With entirely similar procedures and taking into account the vector identities demonstrated previously, we obtain the additional identities

- $$\oint_\ell \mathbf{a} \times d\boldsymbol{\ell} = \int_\Sigma \left[\mathbf{n} \, \mathrm{div}\, \mathbf{a} - (\mathrm{grad}\, \mathbf{a}) \cdot \mathbf{n} \right] dS.$$

- $$\int_\Sigma \mathbf{n} \times \mathbf{a} \, dS = \int_V \mathrm{rot}\, \mathbf{a} \, dV.$$

- $$\int_\Sigma f \mathbf{n} \, dS = \int_V \mathrm{grad}\, f \, dV.$$

In particular, if we put $f = 1$ in this last identity, we get

- $$\int_\Sigma \mathbf{n} \, dS = 0,$$

which is an important geometrical relation valid for an arbitrary closed surface.

16.3 The Dirac Delta Function

The Dirac delta function, traditionally indicated by the symbol $\delta(x)$, can be thought of as a function which is null for any value of x, except for an infinitesimal interval centered at the origin where the function has a very high peak which tends to infinity, but such that the integral of the function in dx is equal to 1. Obviously, it is not a function in strict mathematical sense, but can be thought of as the limit of a family of functions depending on a suitable parameter. For example, if we consider the family of functions $f(x, a)$

$$f(x, a) = \begin{cases} \frac{1}{a} & \text{for } |x| \le \frac{a}{2}, \\ 0 & \text{for } |x| > \frac{a}{2}, \end{cases}$$

we have that

$$\delta(x) = \lim_{a \to 0} f(x, a).$$

Similarly, if we consider the family

$$g(x, a) = \frac{1}{\sqrt{2\pi}a} e^{-(x/a)^2},$$

we also have

$$\delta(x) = \lim_{a \to 0} g(x, a).$$

There are endless possibilities to represent the Dirac delta as the limit of suitable families of functions. The most common representations in mathematical physics are the following ones

$$\delta(x) = \lim_{\Omega \to \infty} \frac{1}{\pi} \frac{\sin(\Omega x)}{x},$$

$$\delta(x) = \lim_{\Omega \to \infty} \frac{1}{\pi} \frac{\sin^2(\Omega x)}{\Omega x^2}.$$

The fundamental property of the Dirac delta is summarised in the following expression, which constitutes its formal definition

$$\int_{-\infty}^{\infty} F(x)\delta(x)\,dx = F(0),$$

and from which, by means of simple changes of variable, the following two relations are found

$$\int_{-\infty}^{\infty} F(x)\delta(x - x_0)\,dx = F(x_0),$$

$$\int_{-\infty}^{\infty} F(x)\delta(ax)\,dx = \frac{1}{|a|}F(0),$$

where a is any real number different from zero. From these equations we can get an important generalisation concerning the Dirac delta whose argument is an arbitrary real function $g(x)$. Denoting this quantity by the symbol $\delta[g(x)]$ and denoting by x_i the zeroes (if any) of the function $g(x)$, we have

$$\int_{-\infty}^{\infty} F(x)\delta[g(x)]\,dx = \sum_i \frac{1}{|g'(x_i)|} F(x_i),$$

where $g'(x)$ is the derivative of the function $g(x)$ with respect to its argument. Further generalisations to the case of the three-dimensional Dirac delta are described directly in the text (see Sect. 3.2).

Finally, we can give a meaning to the derivative of the Dirac delta function, $\delta'(x)$, defined by the usual relation

$$\delta'(x) = \lim_{\Delta x \to 0} \frac{\delta(x + \Delta x) - \delta(x)}{\Delta x}.$$

Using this definition we have, for an arbitrary function $F(x)$

$$\int_{-\infty}^{\infty} F(x)\delta'(x)\,dx = \lim_{\Delta x \to 0} \int_{-\infty}^{\infty} F(x)\frac{\delta(x + \Delta x) - \delta(x)}{\Delta x}\,dx,$$

from which we obtain

$$\int_{-\infty}^{\infty} F(x)\delta'(x)\,dx = \lim_{\Delta x \to 0} \frac{F(-\Delta x) - F(0)}{\Delta x} = -F'(0).$$

16.4 Recovering the Elementary Laws of Electromagnetism

In Chap. 3, starting from the Liénerd and Wiechart potentials, we calculated the expressions of the electric and magnetic field at an arbitrary point in space, due to a single moving charge. The results are contained in Eqs. (3.19) and (3.20). We are now going to show how the basic equations of electromagnetism valid for stationary phenomena can be derived from these equations in the non-relativistic limit. The purpose of this appendix is a simple consistency check, since it is obvious that the equations from which we start, being a consequence of Maxwell's equations, must already contain those results that, even historically, are the basis of Maxwell's equations themselves.

Consider a particle with electric charge e, moving within an electric conductor having a constant transverse section. Its velocity is much lower than the velocity of light. To fix ideas, we can think that the velocity is of the order of 10^{-2} cm s^{-1}, which represents the order of magnitude of the drift velocities of electrons inside a conductor in a typical macroscopic electric circuit. The corresponding value of β is of the order of 10^{-12}, so that the approximation $\beta^2 \ll 1$ is certainly verified. Furthermore, the effects of the curvature of the conductor (causing very small accelerations) can certainly be neglected so that we can assume that the electric field is given only by the Coulomb term of Eq. (3.19). Neglecting terms of the order of β^2, such field is written in the form

$$\mathbf{E}(\mathbf{r}, t) = \frac{e}{\kappa^3 R^2}(\mathbf{n} - \boldsymbol{\beta}),$$

where κ, R, $\mathbf{n}$ are the quantities introduced in Chap. 3 and that need to be calculated at the retarded time t'. The magnetic field is then given by Eq. (3.20), i.e.

$$B(\mathbf{r}, t) = \mathbf{n} \times \mathbf{E}(\mathbf{r}, t).$$

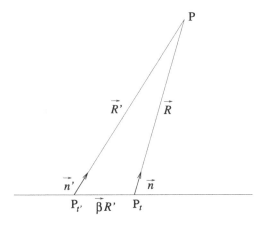

Fig. 16.1 We want to evaluate the electric field in the point P at time t. P_t is the position of the particle at the same time, while $P_{t'}$ is the position of the particle at the retarded time

We can immediately notice that, if we put $\beta = 0$ (i.e. we consider an electric charge at rest), obviously we do not need to consider the difference between real time and retarded time, so we obtain, being $\kappa = 1$

$$\mathbf{E}(\mathbf{r}) = \frac{e\mathbf{n}}{R^2}, \qquad \mathbf{B}(\mathbf{r}) = 0.$$

These are the ordinary equations of electrostatics which represent, in terms of fields, Coulomb's law.

We are now going to see what we get at first order in β. With simple considerations it can be shown that the electric field $E(\mathbf{r}, t)$ is exactly equal to what one would calculate using Coulomb's law and assuming, hypothetically, that the velocity of light were infinite (i.e. neglecting the difference between real and retarded time). In fact, referring to Fig. 16.1 and denoting by a single quote the quantities measured at the retarded time t' and without superscript the same quantities at time t, we have

$$t' = t - \frac{R'}{c}, \qquad \mathbf{R}' = \mathbf{R} + (t - t')\mathbf{v} = \mathbf{R} + R'\boldsymbol{\beta},$$

from which it follows, dividing by R'

$$\mathbf{n}' - \boldsymbol{\beta} = \frac{\mathbf{R}}{R'}. \tag{16.15}$$

Introducing the new notations in the expression for the electric field and recalling that $\boldsymbol{\beta}$ is constant we obtain

$$\mathbf{E}(\mathbf{r}, t) = \frac{e}{\kappa'^3 R'^2}(\mathbf{n}' - \beta) = \frac{e\mathbf{R}}{\kappa'^3 R'^3}.$$

On the other hand we have by definition that

$$\kappa' = 1 - \boldsymbol{\beta} \cdot \mathbf{n}',$$

and applying Carnot's theorem to the triangle $PP_{t'}P_t$

$$R = R'\sqrt{1 - 2\boldsymbol{\beta} \cdot \mathbf{n}' + \beta^2}. \tag{16.16}$$

Substituting in the expression of the electric field, we obtain the result we anticipated. In fact we obtain, apart from terms of the order of β^2

$$\mathbf{E}(\mathbf{r}, t) = \frac{e\mathbf{R}(1 - 2\boldsymbol{\beta} \cdot \mathbf{n}' + \beta^2)^{3/2}}{R^3(1 - \boldsymbol{\beta} \cdot \mathbf{n}')^3} \simeq \frac{e\mathbf{n}}{R^2}.$$

It remains to evaluate the contribution of the magnetic field. We have

$$\mathbf{B}(\mathbf{r}, t) = \mathbf{n}' \times \mathbf{E}(\mathbf{r}, t) = \frac{e\mathbf{n}' \times \mathbf{n}}{R^2}.$$

On the other hand, also apart from terms of the order of β^2, we have, from Eqs. (16.15) and (16.16)

$$\mathbf{n}' = \boldsymbol{\beta} + \mathbf{n}(1 - \boldsymbol{\beta} \cdot \mathbf{n}),$$

so that

$$\mathbf{B}(\mathbf{r}, t) = \frac{e\boldsymbol{\beta} \times \mathbf{n}}{R^2}.$$

Now we apply this equation to the case of an element of a conductor, of length $d\ell$. Denoting by N the number density of the moving charged particles and with S the transverse section of the conductor, the element contains a number of particles given by $NS\,d\ell$ with velocity $\mathbf{v} = c\boldsymbol{\beta}$ parallel to $d\ell$. There is an equal number of fixed particles of opposite charge, so the resulting electric field is null for the property previously demonstrated. For the magnetic field we have instead

$$B(\mathbf{r}, t) = e\frac{NSv}{c}d\ell \times \frac{\mathbf{n}}{R^2}.$$

On the other hand, if we denote by i the intensity of the current flowing in the conductor

$$i = eNSv,$$

so that the equation for the magnetic field is written

$$B(\mathbf{r}, t) = \frac{i}{c}d\ell \times \frac{\mathbf{n}}{R^2}.$$

This is just the Biot and Savart law expressing the magnetic field generated by a current element. As is clear from our deduction, although electric charges move within the conductor at very low speed, they are nevertheless able to create a relativistic effect which is manifested by the presence of the magnetic field.

16.5 The Relativistic Larmor Equation

Within the radiation zone, Eqs. (3.18) and (3.20) provide the expressions for the
electric and magnetic field due to a moving charge

$$\mathbf{E}(\mathbf{r}, t) = \frac{e}{c^2 \kappa^3 R} \mathbf{n} \times \left[(\mathbf{n} - \boldsymbol{\beta}) \times \mathbf{a} \right], \qquad \mathbf{B}(\mathbf{r}, t) = \mathbf{n} \times \mathbf{E}(\mathbf{r}, t),$$

where e is the value of the electric charge, c is the speed of light, $\mathbf{n}$ is the unit
vector along the direction of $\mathbf{R}$, the vector that goes from the charge to the point of
coordinates $\mathbf{r}$, $\boldsymbol{\beta} = \mathbf{v}/c$ is the velocity of the charge in units of the speed of light, $\mathbf{a}$
is the acceleration, and κ is defined by the equation

$$\kappa = 1 - \mathbf{n} \cdot \boldsymbol{\beta}.$$

We recall that the quantities R, κ, $\mathbf{n}$, $\boldsymbol{\beta}$, and $\mathbf{a}$ that appear in the previous equations
must be evaluated at the retarded time t', related to the time t by the equation

$$t' = t - \frac{R}{c}.$$

Expanding the double vector product, we obtain

$$\mathbf{E}(\mathbf{r}, t) = \frac{e}{c^2 \kappa^3 R} \left[(\mathbf{n} \cdot \mathbf{a})(\mathbf{n} - \boldsymbol{\beta}) - \kappa \mathbf{a} \right].$$

On the other hand, we know that the Poynting vector is given by

$$\mathbf{S}(\mathbf{r}, t) = \frac{c}{4\pi} E^2(\mathbf{r}, t) \mathbf{n},$$

and expanding the square of the electric field we obtain with simple algebra

$$\mathbf{S}(\mathbf{r}, t) = \frac{e^2}{4\pi c^3 R^2} \left[\frac{a^2}{\kappa^4} + 2\frac{(\mathbf{n} \cdot \mathbf{a})(\boldsymbol{\beta} \cdot \mathbf{a})}{\kappa^5} - \frac{(1 - \beta^2)(\mathbf{n} \cdot \mathbf{a})^2}{\kappa^6} \right] \mathbf{n}.$$

This expression shows that, in the general case, the angular distribution of the emitted radiation (i.e. the radiation diagram) is quite complex. The special cases where
the acceleration is either parallel or perpendicular to the velocity have been discussed in the text. Here, it is sufficient to emphasize the fact that, for any velocity
and acceleration, there are always two directions where the Poynting vector is zero.
This can be shown simply from the expression of the electric field. The electric field
is obviously zero along the directions characterised by those unit vectors $\mathbf{n}_0$ such
that the vector $\mathbf{n}_0 - \boldsymbol{\beta}$ is parallel to the vector $\mathbf{a}$. The same holds for the Poynting
vector. The directions $\mathbf{n}_0$ are then contained in the plane defined by the vectors $\boldsymbol{\beta}$
and $\mathbf{a}$, and are given by the solutions of the equation

$$(\mathbf{n}_0 - \boldsymbol{\beta}) \times \mathbf{a} = 0.$$

Denoting by α the angle between the velocity and the acceleration vectors, the unit vectors $\mathbf{n}_0$ are defined by the angles $\theta_\pm$ (which start from the acceleration vector and increase in the same direction as α) given by

$$\theta_+ = \arcsin(\beta \sin\alpha), \qquad \theta_- = \pi - \arcsin(\beta \sin\alpha).$$

For example, if $\alpha = 45°$ and $\beta = 0.8$, we have $\theta_+ = 34°.45$ and $\theta_- = 145°.55$.

Let us now move on to the calculation of the power. We note that if the integral

$$\mathcal{I} = \oint \mathbf{S}(\mathbf{r}, t) \cdot \mathbf{n} R^2 \, d\Omega, \tag{16.17}$$

were simply executed over a sphere of radius R centred on the position of the charge at the retarded time $(t - R/c)$, we would obtain the ratio between the energy that flows across the sphere in a time interval dt and the dt itself. This quantity is however of not much interest. It is more interesting to obtain the power emitted by the charged particle. In order to do this, we need to take into account the fact that the energy that flows across the sphere in a time dt was emitted by the particle in the time dt' which depends on the direction and is related to dt by

$$dt = \kappa \, dt'.$$

To find the power W emitted by the charged particle we therefore need to calculate the integral

$$W = \oint \mathbf{S}(\mathbf{r}, t) \cdot \mathbf{n} \frac{dt}{dt'} R^2 \, d\Omega = \oint \mathbf{S}(\mathbf{r}, t) \cdot \mathbf{n} \kappa R^2 \, d\Omega.$$

Substituting the above expression of the Poynting vector, we find

$$W = \frac{e^2}{4\pi c^3} \oint \left[\frac{a^2}{\kappa^3} + 2\frac{(\mathbf{n} \cdot \mathbf{a})(\boldsymbol{\beta} \cdot \mathbf{a})}{\kappa^4} - \frac{(1 - \beta^2)(\mathbf{n} \cdot \mathbf{a})^2}{\kappa^5} \right] d\Omega.$$

To calculate this integral, we introduce a system of polar coordinates (ψ, χ) with the polar axis directed along the velocity vector and the azimuth χ measured from the plane containing the velocity and the acceleration. With obvious notations, the three vectors $\boldsymbol{\beta}$, $\mathbf{a}$, and $\mathbf{n}$ in this system of coordinates are given by

$$\boldsymbol{\beta} = \beta\mathbf{k}, \qquad \mathbf{a} = a_\perp\mathbf{i} + a_\parallel\mathbf{k}, \qquad \mathbf{n} = \sin\psi\cos\chi\,\mathbf{i} + \sin\psi\sin\chi\,\mathbf{j} + \cos\psi\,\mathbf{k},$$

so that the integrand can be written in the form

$$\frac{a_\parallel^2 + a_\perp^2}{(1 - \beta\cos\psi)^3} + 2\beta a_\parallel \frac{\sin\psi\cos\chi\,a_\perp + \cos\psi\,a_\parallel}{(1 - \beta\cos\psi)^4}$$
$$- (1 - \beta^2)\frac{\sin^2\psi\cos^2\chi\,a_\perp^2 + 2\sin\psi\cos\psi\cos\chi\,a_\perp a_\parallel + \cos^2\psi\,a_\parallel^2}{(1 - \beta\cos\psi)^5},$$

and $d\Omega$ is given by $\sin\psi\,d\psi\,d\chi$. By integrating in $d\chi$ within the interval $(0, 2\pi)$, the factors that do not contain any function of χ result in 2π, those containing $\cos\chi$ produce zero, while the factor containing $\cos^2\chi$ gives π. In summary we have

$$W = \frac{e^2}{2c^3}\int_0^\pi \left[\frac{a_\parallel^2 + a_\perp^2}{(1-\beta\cos\psi)^3} + \frac{2\beta\cos\psi\,a_\parallel^2}{(1-\beta\cos\psi)^4}\right.$$
$$\left. - (1-\beta^2)\frac{\frac{1}{2}\sin^2\psi\,a_\perp^2 + \cos^2\psi\,a_\parallel^2}{(1-\beta\cos\psi)^5}\right]\sin\psi\,d\psi.$$

The integrals in $d\psi$ appearing in this expression are simple and can be evaluated either by integrating by parts or by changing the integration variable from ψ to $x = 1 - \beta\cos\psi$. We obtain

$$\frac{1}{2}\int_0^\pi \frac{1}{(1-\beta\cos\psi)^3}\sin\psi\,d\psi = \frac{1}{(1-\beta^2)^2},$$

$$\frac{1}{2}\int_0^\pi \frac{\cos\psi}{(1-\beta\cos\psi)^4}\sin\psi\,d\psi = \frac{4}{3}\frac{\beta}{(1-\beta^2)^3},$$

$$\frac{1}{2}\int_0^\pi \frac{\sin^2\psi}{(1-\beta\cos\psi)^5}\sin\psi\,d\psi = \frac{2}{3}\frac{1}{(1-\beta^2)^3},$$

$$\frac{1}{2}\int_0^\pi \frac{\cos^2\psi}{(1-\beta\cos\psi)^5}\sin\psi\,d\psi = \frac{1}{3}\frac{1+5\beta^2}{(1-\beta^2)^4}.$$

Substituting these expressions and grouping separately the terms in $a_\parallel^2$ and in $a_\perp^2$, we get

$$W = \frac{e^2}{2c^3}\left\{a_\parallel^2\left[\frac{1}{(1-\beta^2)^2} + \frac{8}{3}\frac{\beta^2}{(1-\beta^2)^3} - \frac{1}{3}\frac{1+5\beta^2}{(1-\beta^2)^3}\right]\right.$$
$$\left. + a_\perp^2\left[\frac{1}{(1-\beta^2)^2} - \frac{1}{3}\frac{1}{(1-\beta^2)^2}\right]\right\},$$

or, expanding,

$$W = \frac{2e^2}{3c^3}\left[\frac{a_\parallel^2}{(1-\beta^2)^3} + \frac{a_\perp^2}{(1-\beta^2)^2}\right].$$

Recalling the definition of the relativistic factor γ

$$\gamma = \frac{1}{\sqrt{1-\beta^2}},$$

the expression for the power emitted by a relativistic charge in accelerated motion can also be written in the more representative form

$$W = \frac{2e^2}{3c^3}\left(\gamma^6 a_\parallel^2 + \gamma^4 a_\perp^2\right).$$

This formula is a generalisation of the Larmor equation (3.23) to the relativistic case. Obviously, for $\gamma = 1$ we find Larmor equation since $a_\parallel^2 + a_\perp^2 = a^2$.

To conclude, we note that, if we had executed the integral of the Poynting vector on the sphere without taking into account the difference between the dt and the dt' (i.e. the integral $\mathcal{I}$ of Eq. (16.17)), we would have obviously obtained a different expression. Taking into account that

$$\frac{1}{2}\int_0^\pi \frac{1}{(1-\beta\cos\psi)^4}\sin\psi\,d\psi = \frac{1}{3}\frac{3+\beta^2}{(1-\beta^2)^3},$$

$$\frac{1}{2}\int_0^\pi \frac{\cos\psi}{(1-\beta\cos\psi)^5}\sin\psi\,d\psi = \frac{1}{3}\frac{\beta(5+\beta^2)}{(1-\beta^2)^4},$$

$$\frac{1}{2}\int_0^\pi \frac{\sin^2\psi}{(1-\beta\cos\psi)^6}\sin\psi\,d\psi = \frac{2}{15}\frac{5+\beta^2}{(1-\beta^2)^4},$$

$$\frac{1}{2}\int_0^\pi \frac{\cos^2\psi}{(1-\beta\cos\psi)^6}\sin\psi\,d\psi = \frac{1}{15}\frac{5+38\beta^2+5\beta^4}{(1-\beta^2)^5},$$

we have, in fact, that

$$\mathcal{I} = \frac{2e^2}{3c^3}\left[\gamma^8\left(1+\frac{1}{5}\beta^2\right)a_\parallel^2 + \gamma^6\left(1+\frac{2}{5}\beta^2\right)a_\perp^2\right].$$

This difference between the power emitted by the particle (W) and the power received on the sphere ($\mathcal{I}$) is a simple kinematic effect and has nothing to do with relativity. A similar effect occurs in the case of acoustic waves emitted, for example, by an airplane travelling at a speed close to the velocity of sound. While the power emitted by the plane into acoustic waves is fixed, the received power can be very large and, at the limit, almost infinite if the plane travels for a long time at exactly the speed of sound (the so-called sonic bang is precisely due to this phenomenon).

16.6 Gravitational Waves

The equations that we have obtained for the radiation of electromagnetic waves can also be applied, with some slight modifications, to treat gravitational radiation. Obviously, this is not rigorous, since the laws of gravitational radiation should be derived from the general theory of relativity. The approach followed here is however sufficient to describe the fundamental properties of the mechanisms for the generation of gravitational waves and leads to formulae that are substantially correct (as can be verified a posteriori).

We start by performing a formal transformation to the equations for electromagnetic radiation described in Sect. 3.10

$$e_i \to m_i \quad (i = 1, \dots, N),$$

i.e. we replace, for each particle, the charge with the mass. Furthermore, in the equations that express the Poynting vector (i.e. in those which express the radiated power), we multiply the right-hand side by the universal gravitational constant G. We note, incidentally, that in these equations the dimensional factor $[e^2]$ is replaced by the dimensional factor $[Gm^2]$ having the same dimensions. The various quantities introduced in Sect. 3.10, i.e. the electric dipole moment **D** (Eq. (3.31)), the magnetic dipole moment **M** (Eq. (3.32)), and the symmetric tensor of order two (related to the electric quadrupole moment) $\mathcal{Q}$ (Eq. (3.33)) are transformed in as many quantities for which we use, respectively, the symbols $\mathbf{D}_G$, $\mathbf{M}_G$, and $\mathcal{Q}_G$, i.e.

$$\mathbf{D} = \sum_{i=1}^{N} e_i \mathbf{s}_i \rightarrow \mathbf{D}_G = \sum_{i=1}^{N} m_i \mathbf{s}_i,$$

$$\mathbf{M} = \frac{1}{2c} \sum_{i=1}^{N} e_i \mathbf{s}_i \times \mathbf{v}_i \rightarrow \mathbf{M}_G = \frac{1}{2c} \sum_{i=1}^{N} m_i \mathbf{s}_i \times \mathbf{v}_i,$$

$$\mathcal{Q} = \sum_{i=1}^{N} e_i \mathbf{s}_i \mathbf{s}_i \rightarrow \mathcal{Q}_G = \sum_{i=1}^{N} m_i \mathbf{s}_i \mathbf{s}_i.$$

We now note that the quantity $\mathbf{D}_G$, the analogous of the electric dipole, is, by definition, the coordinate of the centre of mass of the system of N particles $\mathbf{r}_G$ multiplied by the total mass. We therefore have

$$\mathbf{D}_G = \sum_{i=1}^{N} m_i \mathbf{s}_i = \mathcal{M} \mathbf{r}_G,$$

where

$$\mathcal{M} = \sum_{i=1}^{N} m_i.$$

We then have, for an isolated system,

$$\ddot{\mathbf{D}}_G = \mathcal{M} \frac{\mathrm{d}^2}{\mathrm{d}t^2} \mathbf{r}_G = 0.$$

Furthermore, the analogous of the magnetic dipole, the quantity $\mathbf{M}_G$, is proportional to the total angular momentum of the system **J**, since

$$2c\mathbf{M}_G = \sum_{i=1}^{N} m_i \mathbf{s}_i \times \mathbf{v}_i = \mathbf{J}.$$

We then obtain, for an isolated system,

$$\dot{\mathbf{M}}_G = \frac{\mathrm{d}}{\mathrm{d}t} \mathbf{J} = 0,$$

and therefore also

$$\ddot{\mathbf{M}}_G = 0.$$

According to our analogy, we can then conclude that for gravitational waves there is neither the analogue of electric dipole radiation, nor the analogue of magnetic dipole radiation. It therefore only remains the analogue of the electric quadrupole radiation (in addition, obviously, to the radiation due to higher multipoles). The tensor $\mathcal{Q}_G$ is traditionally denoted by the symbol $\mathcal{I}$, because it is essentially an inertia tensor. It is however not to be confused with the ordinary inertia tensor $\mathcal{I}$ that is used to describe the dynamics of a rigid body, and that is defined as

$$\mathcal{I} = \sum_{i=1}^{N} m_i \left(s_i^2 \mathbf{U} - \mathbf{s}_i \mathbf{s}_i \right),$$

where $\mathbf{U}$ is the unitary tensor. We have, obviously,

$$\mathcal{I} = -\mathcal{I} + \frac{1}{2} (\mathrm{Tr}\,\mathcal{I}) \mathbf{U},$$

since, recalling the definition of the trace of a tensor

$$\mathrm{Tr}\,\mathcal{I} = \sum_i m_i \left(3s_i^2 - x_i^2 - y_i^2 - z_i^2 \right) = 2 \sum_i m_i s_i^2.$$

The two tensors $\mathcal{I}$ and $\mathcal{I}$ differ by a quantity which is proportional to the unitary tensor. This property is strictly analogous to the one that exists between the tensors $\mathcal{Q}$ and $\mathcal{Q}$ in electrodynamics. Therefore, within our analogy, the power emitted in gravitational waves at the lowest order (of the multipolar expansion) can be obtained from Eq. (3.34) and is given by

$$W_G = \frac{G}{20c^5} \sum_{jk} (\dddot{\mathcal{I}}_{jk})^2.$$

This formula is correct in all respects, aside from the numerical factor. The calculations based on general relativity produce a similar result, where the factor $\frac{1}{20}$ is replaced by the factor $\frac{1}{5}$. Intuitively, one can justify this multiplication by a factor of four noting that an electromagnetic wave is described by two vectors $\mathbf{E}$ and $\mathbf{B}$ that are not independent and are perpendicular to the direction of propagation, say z. Only two components of one of the two fields, such as E_x and E_y, are sufficient to describe the wave. A gravitational wave is instead described by two independent tensors also perpendicular to the direction of propagation. If we denote these tensors by the traditional symbols $\mathbf{e}^+$ and $\mathbf{e}^\times$, the wave is described by the eight components $(e_{xx}^+, e_{xy}^+, e_{yx}^+, e_{yy}^+, e_{xx}^\times, e_{xy}^\times, e_{yx}^\times, e_{yy}^\times)$. The factor of four is therefore associated with, say, the degrees of freedom of the polarisation. The correct formula for the power emitted in gravitational waves is then

$$W_G = \frac{G}{5c^5} \sum_{jk} (\dddot{\mathcal{I}}_{jk})^2.$$

Finally, we note that if we change the origin of coordinates putting

$$\mathbf{s}'_i = \mathbf{b} + \mathbf{s}_i,$$

with $\mathbf{b}$ a constant vector, we obtain that the new inertia tensor $\mathcal{I}'$ is

$$\mathcal{I}' = \mathcal{I} + \mathcal{M}[(2\mathbf{b} \cdot \mathbf{r}_G + b^2)\mathbf{U} - \mathbf{bb} - \mathbf{br}_G - \mathbf{r}_G\mathbf{b}],$$

so that, for an isolated system,

$$\ddot{\mathcal{I}}' = \ddot{\mathcal{I}},$$

and, all the more so, $\dddot{\mathcal{I}}' = \dddot{\mathcal{I}}$. This equation allows to calculate the inertia tensor in a coordinate system having an arbitrary origin in order to determine the power emitted in gravitational waves.

16.7 Calculation of the Thomas-Fermi Integral

Some applications of atomic physics based on the Thomas-Fermi model require the calculation of the following integral

$$\mathcal{I} = \int_0^\infty \frac{(1+\chi)\chi^{3/2}}{x^{1/2}} \, dx,$$

where $\chi(x)$ is the solution of the Thomas-Fermi equation

$$x^{1/2}\chi'' = \chi^{3/2},$$

which satisfies the boundary condition

$$\chi(0) = 1, \qquad \lim_{x \to \infty} \chi(x) = 0.$$

The integral is split into the sum of two integrals

$$\mathcal{I} = \mathcal{I}_1 + \mathcal{I}_2, \tag{16.18}$$

where

$$\mathcal{I}_1 = \int_0^\infty \frac{\chi^{3/2}}{x^{1/2}} \, dx, \qquad \mathcal{I}_2 = \int_0^\infty \frac{\chi^{5/2}}{x^{1/2}} \, dx.$$

The first integral is trivial since, taking into account the Thomas-Fermi equation and the boundary conditions of the function χ, we have

$$\mathcal{I}_1 = \int_0^\infty \chi'' \, dx = -\chi'(0). \tag{16.19}$$

The calculation of the second integral is more complex. It can be done in the following way. On one hand, we have

$$\mathcal{I}_2 = \int_0^\infty \frac{\chi^{5/2}}{x^{1/2}} \, dx = \int_0^\infty \chi \chi'' \, dx,$$

and, integrating by parts and taking into account that $\chi(0) = 1$

$$\mathcal{I}_2 = -\chi'(0) - \int_0^\infty \chi'^2 \, dx. \tag{16.20}$$

On the other hand, considering the quantity $x^{-1/2} \, dx$ as a differential factor, by integrating again by parts and recalling the Thomas-Fermi equation, we obtain

$$\mathcal{I}_2 = \int_0^\infty \frac{\chi^{5/2}}{x^{1/2}} \, dx = -5 \int_0^\infty x^{1/2} \chi^{3/2} \chi' \, dx = -5 \int_0^\infty x \chi' \chi'' \, dx.$$

Now we note that the product $\chi' \chi''$ can be expressed in the form

$$\chi' \chi'' = \frac{1}{2} \frac{d\chi'^2}{dx},$$

and integrating again by parts we get

$$\mathcal{I}_2 = \frac{5}{2} \int_0^\infty \chi'^2 \, dx.$$

Comparing this expression with Eq. (16.20), we obtain

$$\int_0^\infty \chi'^2 \, dx = -\frac{2}{7} \chi'(0), \quad \text{or,} \quad \mathcal{I}_2 = -\frac{5}{7} \chi'(0).$$

Finally, recalling Eqs. (16.18) and (16.19) we get

$$\mathcal{I} = \int_0^\infty \frac{(1+\chi)\chi^{3/2}}{x^{1/2}} \, dx = -\frac{12}{7} \chi'(0). \tag{16.21}$$

16.8 Energy of the Ground Configuration of the Silicon Atom

As an application of the results obtained in Chap. 8, we evaluate the energy of the ground configuration of the silicon atom, i.e. of the $1s^2 2s^2 2p^6 3s^2 3p^2$ configuration.

Before performing these calculations, we need to evaluate some 3-j symbols. By means of the analytical formula given in Eq. (7.16), we have

$$\begin{pmatrix} 0 & 0 & 0 \\ 0 & 0 & 0 \end{pmatrix}^2 = 1, \qquad \begin{pmatrix} 0 & 1 & 1 \\ 0 & 0 & 0 \end{pmatrix}^2 = \frac{1}{3}, \qquad \begin{pmatrix} 1 & 1 & 2 \\ 0 & 0 & 0 \end{pmatrix}^2 = \frac{2}{15}.$$

The configuration contains four closed subshells and one open subshell. We start by evaluating the degenerate contribution to the energy. The Hamiltonian H_0 (defined in Eq. (7.3)) and the $\mathcal{F}$ part of the Hamiltonian $\mathcal{H}_1$ (defined in Eqs. (8.2) and (8.3)) produce five terms, one for each subshell (open or closed). The corresponding energy (that we denote by $\mathcal{E}_1$) is obtained from Eq. (8.7) and is given by

$$\mathcal{E}_1 = 2W_0(1s) + 2W_0(2s) + 6W_0(2p) + 2W_0(3s) + 2W_0(3p)$$
$$+ 2I(1s) + 2I(2s) + 6I(2p) + 2I(3s) + 2I(3p),$$

where W_0 is defined by Eq. (7.11) and $I(n, l)$ is the integral defined in Eq. (8.6). The energy of the Coulomb interaction (i.e. the part $\mathcal{G}$ of the Hamiltonian $\mathcal{H}_1$) resulting from closed subshells contributes four terms. Denoting by $\mathcal{E}_2$ the corresponding energy, we have, using Eq. (8.17),

$$\mathcal{E}_2 = F^0(1s, 1s) + F^0(2s, 2s) + 15F^0(2p, 2p) - \frac{6}{5}F^2(2p, 2p) + F^0(3s, 3s),$$

where the quantities $F^k(n_a l_a, n_n l_b)$ are defined in Eq. (8.9). Considering the energy of the Coulomb interaction between different closed subshells, we have six contributions, as many as the number of the distinct pairs that can be formed with the four closed subshells. Denoting by $\mathcal{E}_3$ the corresponding energy, we have, using Eqs. (8.14) and (8.16)

$$\mathcal{E}_3 = 4F^0(1s, 2s) - 2G^0(1s, 2s) + 12F^0(1s, 2p) - 2G^1(1s, 2p) + 4F^0(1s, 3s)$$
$$- 2G^0(1s, 3s) + 12F^0(2s, 2p) - 2G^1(2s, 2p) + 4F^0(2s, 3s) - 2G^0(2s, 3s)$$
$$+ 12F^0(2p, 3s) - 2G^1(2p, 3s),$$

where the quantities $G^k(n_a l_a, n_n l_b)$ are defined in Eq. (8.10). Finally, we need to evaluate the contribution of the Coulomb interaction between the open subshell $3p$ and the four closed subshells. Denoting by $\mathcal{E}_4$ the corresponding energy, we have, using Eqs. (8.13) and (8.15),

$$\mathcal{E}_4 = 4F^0(1s, 3p) - \frac{2}{3}G^1(1s, 3p) + 4F^0(2s, 3p) - \frac{2}{3}G^1(2s, 3p) + 12F^0(2p, 3p)$$
$$- 2G^0(2p, 3p) - \frac{4}{5}G^2(2p, 3p) + 4F^0(3s, 3p) - \frac{2}{3}G^1(3s, 3p).$$

The four contributions to the energy that we have calculated are degenerate with respect to all the states of the configuration. For the degenerate part of the energy of the ground configuration of the silicon atom, $\mathcal{E}$, we then have

$$\mathcal{E} = \mathcal{E}_1 + \mathcal{E}_2 + \mathcal{E}_3 + \mathcal{E}_4.$$

What remains to calculate is given by Eq. (8.11), with the sum extended to the only pair of electrons belonging to the open subshell $3p$. The explicit computation is done in Sect. 8.6. The two $3p$ electrons give rise to three terms that, in order of increasing energy, are ^3P, ^1D, and ^1S. The ratio between the intervals $(^1S - {}^1D)$ and $(^1D - {}^3P)$ is equal to $3/2$.

16.9 Calculation of the Fine-Structure Constant of a Term

The calculation of the constant $\zeta(\alpha, LS)$, which characterises the fine-structure intervals of the terms belonging to a given configuration, can be carried out with a process based on the diagonal sum rule. A similar process was followed in Sect. 8.1 to determine the energy of the terms. The starting point is Eq. (9.8) which, in the case of diagonal matrix elements, is

$$\sum_i \langle \alpha L S M_L M_S | \xi(r_i) \boldsymbol{\ell}_i \cdot \mathbf{s}_i | \alpha L S M_L M_S \rangle = \zeta(\alpha, LS) M_L M_S.$$

On the other hand, for any eigenstate of the configuration of the form $\psi^A(a_1, a_2, \ldots, a_N)$ of Eq. (7.1), the diagonal matrix element of the same operator is given by

$$\sum_i \langle \psi^A(a_1, a_2, \ldots, a_N) | \xi(r_i) \boldsymbol{\ell}_i \cdot \mathbf{s}_i | \psi^A(a_1, a_2, \ldots, a_N) \rangle = \sum_i \zeta_{n_i l_i} m_i m_{si},$$

where $\zeta_{n_i l_i}$ is the quantity defined in Eq. (9.10).

We now consider the particular case of the pf configuration which, as shown in Table 7.3, gives rise to the six terms ^1D, ^1F, ^1G, ^3D, ^3F, and ^3G. We start from a state having the highest values of the quantum numbers M_L and M_S, i.e. $M_L = 4$, $M_S = 1$. This state can only originate from the ^3G term. Considering instead single particle states, this state is of the type $m_1 = 1$, $m_{s1} = \frac{1}{2}$, $m_2 = 3$, $m_{s2} = \frac{1}{2}$, where the indices 1 and 2 refer, respectively, to the p and f electron. Using the same notations as in Sect. 8.1 we can write the equality[5]

$$[4, 1] = \left(1^+, 3^+\right),$$

---

[5] The symbol $[M_L, M_S]$ means the sum of the diagonal matrix elements of the Hamiltonian of the spin-orbit interaction over all the states ψ^A for which M_L and M_S are the eigenvalues of L_z and S_z, respectively. Similarly, the notation $(m_1^\pm, m_2^\pm)$ is used to denote the diagonal matrix element of the same Hamiltonian on the state where the electron 1 has the magnetic quantum number m_1 and spin quantum number $+1/2$ or $-1/2$ and, similarly, the electron 2 has the magnetic quantum number m_2 and spin quantum number $+1/2$ or $-1/2$.

that, according to the previous equation, is[6]

$$4\zeta\left(^3G\right) = \frac{1}{2}\zeta_{np} + \frac{3}{2}\zeta_{nf}.$$

We therefore obtain the result

$$\zeta\left(^3G\right) = \frac{1}{8}\zeta_{np} + \frac{3}{8}\zeta_{nf}.$$

We then proceed by lowering the value of M_L (maintaining $M_S = 1$). We obtain the equations

$$[3, 1] = \left(0^+, 3^+\right) + \left(1^+, 2^+\right), \qquad [2, 1] = \left(-1^+, 3^+\right) + \left(0^+, 2^+\right) + \left(1^+, 1^+\right),$$

from which we have, noting that the combination $[M_L = 3, M_S = 1]$ can originate from the ^3G and ^3F terms, and that the combination $[M_L = 2, M_S = 1]$ can originate from the ^3G, ^3F, and ^3D terms,

$$3\left[\zeta\left(^3G\right) + \zeta\left(^3F\right)\right] = \frac{3}{2}\zeta_{nf} + \frac{1}{2}\zeta_{np} + \zeta_{nf},$$

$$2\left[\zeta\left(^3G\right) + \zeta\left(^3F\right) + \zeta\left(^3D\right)\right] = -\frac{1}{2}\zeta_{np} + \frac{3}{2}\zeta_{nf} + \zeta_{nf} + \frac{1}{2}\zeta_{np} + \frac{1}{2}\zeta_{nf}.$$

By solving the system, we arrive at the following expressions (which can also be obtained from Eq. (9.11))

$$\zeta\left(^3F\right) = \frac{1}{24}\zeta_{np} + \frac{11}{24}\zeta_{nf}, \qquad \zeta\left(^3D\right) = -\frac{1}{6}\zeta_{np} + \frac{2}{3}\zeta_{nf}.$$

In principle, we could also consider the values of $M_S = 0$. For example,

$$[4, 0] = \left(1^+, 3^-\right) + \left(1^-, 3^+\right).$$

However, in so doing we obtain equations of the form $0 = 0$ and the value of $\zeta\left(^1G\right)$ is undetermined. This is entirely consistent, since the singlet states do not have fine structure and the constant ζ is not defined.

The cases of the configurations of equivalent electrons are also interesting, because, by repeating the same arguments, we obtain directly the third Hund's rule. For example, consider the configuration p^2 which produces, as shown in Table 7.4, the three terms ^1S, ^1D, and ^3P. For the singlet terms, the fine structure constant remains undetermined, as usual. For the triplet term we have instead

$$[1, 1] = \left(0^+, 1^+\right),$$

---

[6]We note that even if there are electrons in the closed subshells they do not produce any contribution to the equation.

from which we obtain

$$\zeta\left(^3P\right) = \frac{1}{2}\zeta_{np}.$$

If we consider the complementary configuration p^4, we get the same structure of terms. This time, to find the fine structure constant of the ^3P, term, we need to consider the equation[7]

$$[1,1] = \left(1^+, 1^-, 0^+, -1^+\right),$$

and we obtain

$$\zeta\left(^3P\right) = -\frac{1}{2}\zeta_{np},$$

i.e. a value that is exactly the same (but opposite in sign) to the one of the configuration p^2. These arguments can be repeated for any configuration of equivalent electrons and for the corresponding complementary configuration and lead to the third Hund's rule. In the particular case of configurations that fill half of a subshell (such as p^3, d^5, and f^7), the configuration coincides with the complementary one, and the fine structure constant is zero for all the terms.

16.10 The Fundamental Principle of Statistical Thermodynamics

Consider, in all generality, a macroscopic physical system. We suppose that the system is in thermal equilibrium with an ideal heat reservoir having temperature T (canonical ensemble). We also suppose to identify with the index i all possible microscopic states of the system and we denote by E_i the energy of the i-th state. Macroscopically, the system is in a steady state. On the other hand, from the microscopical point of view, we can think that it constantly evolves from one microscopic state to another. We can then introduce a statistical description denoting by p_i the probability that the system is in the i-th microscopic state. The following normalisation property should obviously be valid

$$\sum_i p_i = 1.$$

We now need to relate the probability p_i with the energy E_i. To do so, we give a definition of the entropy by putting, according to an hypothesis originally due to Boltzmann

$$S = -k_B \sum_i p_i \ln p_i,$$

---

[7]The quantity $(1^+, 1^-, 0^+, -1^+)$, relative to the configuration p^4, is obtained from the corresponding quantity $(0^+, 1^+)$, relative to the configuration p^2, taking the "complementary" of the latter, i.e. $(-1^+, -1^-, 0^-, 1^-)$, and then changing sign to all the values of m and m_s.

where k_B is the Boltzmann constant.

This definition can be justified by considering that the entropy of a system measures the amount of "disorder" contained in the system itself and noting that the function defined above has the mathematical property of assuming its maximum value when all the probabilities p_i are equal to one another, and take the minimum value (which is equal to 0) when a single p_i is equal to 1 and all the others are equal to 0. The proof of the second property is trivial. To prove the first property we note that giving an arbitrary variation δp_i to the probabilities, the corresponding change δS of the entropy is

$$\delta S = -k_B \sum_i (\ln p_i + 1)\delta p_i.$$

On the other hand, being

$$\sum_i \delta p_i = 0,$$

it follows that if $\ln p_i$ is constant (i.e. independent of i), δS is null and therefore the entropy presents an extreme. It is then easy to verify that such an extreme is actually a maximum, since

$$\frac{d^2 S}{dp_i^2} = -k_B \frac{1}{p_i} < 0.$$

Having justified the definition of the entropy, we now take into account that the internal energy of the system is given by the expression

$$U = \sum_i p_i E_i.$$

If we consider an infinitesimal thermodynamical transformation of the system, the internal energy will vary, in general, because both the probabilities p_i and the energies E_i change. We then have

$$\delta U = \sum_i (\delta p_i) E_i + \sum_i p_i (\delta E_i).$$

If, however, the external conditions of the system are not varied, the quantities E_i remain fixed to the initial value and the second term of the right hand side is null. On the other hand, to keep constant the external conditions of the system means that the system does not accomplish mechanical work on the ambient medium, so we can write, according to the first principle of thermodynamics

$$\delta U = \sum_i (\delta p_i) E_i = \delta Q = T \delta S,$$

where δQ is the heat exchanged with the reservoir, and taking into account that

$$\delta S = -k_B \sum_i (\delta p_i) \ln p_i,$$

we obtain the equation

$$\sum_i (\delta p_i) E_i = -k_B T \sum_i (\delta p_i) \ln p_i.$$

This equation must be satisfied for an arbitrary thermodynamic transformation (as long as no work is done). The following relation therefore must hold

$$E_i = -k_B T \ln p_i + \text{const.},$$

which leads to the relation

$$p_i = A e^{-\beta E_i},$$

where A is a constant and where we have put

$$\beta = \frac{1}{k_B T}.$$

The constant A is determined by imposing the normalisation condition. Since we must have

$$\sum_i p_i = \sum_i A e^{-\beta E_i} = 1,$$

it follows that

$$A = \frac{1}{\mathcal{Z}},$$

where the quantity $\mathcal{Z}$, known as the sum over states, is given by

$$\mathcal{Z} = \sum_i e^{-\beta E_i}.$$

The expression of p_i can therefore be written in its final form

$$p_i = \frac{1}{\mathcal{Z}} e^{-\beta E_i} = \frac{e^{-\beta E_i}}{\sum_j e^{-\beta E_j}}. \tag{16.22}$$

This expression, often referred to as Gibbs principle, is of extreme generality and can rightly be considered the basis of all statistical thermodynamics. It can be written in an alternative form by assuming that the microscopic states of the system are not discrete (and therefore countable) but are identified by the representative point in the phase space of the system having dimension $2\mathcal{N}$, where

$\mathcal{N}$ is the number of degrees of freedom of the system itself. In this case, denoting by dP the probability that the representative point of the system is in the cell $d\Gamma = dq_1\, dq_2 \cdots dq_\mathcal{N}\, dp_1\, dp_2 \cdots dp_\mathcal{N}$ of the phase space centered around the values (q_i, p_i), and denoting by $H(q_i, p_i)$ the Hamiltonian of the system, we have

$$dP = \frac{1}{\mathcal{Z}} e^{-\beta H(q_i, p_i)}\, d\Gamma = \frac{e^{-\beta H(q_i, p_i)}\, d\Gamma}{\int e^{-\beta H(q_i, p_i)}\, d\Gamma}, \tag{16.23}$$

where the integral is over the entire volume of the phase space available to the system. Equations (16.22) and (16.23) coincide, respectively, with Eqs. (10.2) and (10.1) which we have assumed as the basic principles for the deduction of the various laws of thermodynamical equilibrium in Chap. 10.

16.11 Transition Probability for the Coherences

In Chap. 11, we have introduced the so-called random phase approximation and we have determined the kinetic equations for the diagonal matrix elements ρ_α of the density matrix operator of the physical system. The result that we found is the kinetic equation (9.11), which is interpreted by introducing the transition probability per unit time between different states of the system. This probability is given by Fermi's golden rule, expressed by Eq. (11.10). We now want to generalise these results by determining the kinetic equations for the so-called coherences, i.e. for the non-diagonal matrix elements of the density matrix operator.

We start again from Eq. (8.11) and introduce the hypothesis, less restrictive than that of the random phases, that in the physical system there might exist coherences, even if only within pairs of states, $|\alpha\rangle$ and $|\alpha'\rangle$, having the same energy eigenvalue (degenerate states) and such that the matrix element of the interaction Hamiltonian between them, $\mathcal{H}^I_{\alpha\alpha'}$, is zero. Taking into account this approximation, when evaluating the product $c_\alpha(t)c^*_{\alpha'}(t)$ we obtain, considering only the terms that are at most quadratic in the matrix elements of $\mathcal{H}^I$,

$$c_\alpha(t)c^*_{\alpha'}(t) = c_\alpha(0)c^*_{\alpha'}(0)$$

$$+ \frac{1}{\hbar^2} \sum_{\beta\beta'} \mathcal{H}^I_{\alpha\beta} \mathcal{H}^I_{\beta'\alpha'} c_\beta(0)c^*_{\beta'}(0) \frac{e^{i\omega_{\alpha\beta}t} - 1}{\omega_{\alpha\beta}} \frac{e^{-i\omega_{\alpha'\beta'}t} - 1}{\omega_{\alpha'\beta'}}$$

$$+ \frac{1}{\hbar^2} \left[\sum_{\beta\gamma} \mathcal{H}^I_{\alpha\beta} \mathcal{H}^I_{\beta\gamma} c_\gamma(0)c^*_{\alpha'}(0) \left(\frac{e^{i\omega_{\alpha\gamma}t} - 1}{\omega_{\alpha\gamma}\omega_{\beta\gamma}} - \frac{e^{i\omega_{\alpha\beta}t} - 1}{\omega_{\alpha\beta}\omega_{\beta\gamma}} \right) \right.$$

$$\left. + C.C.(\alpha \leftrightarrow \alpha') \right],$$

where the symbol $[\cdots + C.C.(\alpha \leftrightarrow \alpha')]$ means that we need to add to the term in brackets its complex conjugate (with the exchange of the indices α and α').

We now need to recall the approximation we have introduced in that the states between which coherences exist are iso-energetic. As regards the second line of the previous equation, this implies that $\omega_{\alpha\beta} = \omega_{\alpha'\beta'}$ and so the two temporal factors are one the complex conjugate of the other. Regarding the third line, we can consider the limit $(\omega_{\alpha\gamma} \to 0)$ and the temporal function between round brackets, which we indicate with $\mathcal{F}(t)$, is equal to

$$\mathcal{F}(t) = \frac{e^{i\omega_{\alpha\gamma}t} - 1}{\omega_{\alpha\gamma}\omega_{\beta\gamma}} - \frac{e^{i\omega_{\alpha\beta}t} - 1}{\omega_{\alpha\beta}\omega_{\beta\gamma}} = -\frac{it}{\omega_{\alpha\beta}} + \frac{e^{i\omega_{\alpha\beta}t} - 1}{\omega_{\alpha\beta}^2}$$

$$= -\frac{2\sin^2(\omega_{\alpha\beta}t/2)}{\omega_{\alpha\beta}^2} + i\frac{\sin(\omega_{\alpha\beta}t) - \omega_{\alpha\beta}t}{\omega_{\alpha\beta}^2}.$$

We now proceed by evaluating the statistical average over the physical system. We introduce the notation of the density matrix by putting[8] $\rho_{\alpha\alpha'} = \langle c_\alpha(t)c_{\alpha'}^*(t)\rangle$. Changing the index of the sum γ in α'', the kinetic equation for the coherences becomes

$$\rho_{\alpha\alpha'}(t) = \rho_{\alpha\alpha'}(0) + \frac{1}{\hbar^2}\sum_{\beta\beta'}\mathcal{H}_{\alpha\beta}^I\mathcal{H}_{\beta'\alpha'}^I\rho_{\beta\beta'}(0)\frac{4\sin^2(\omega_{\alpha\beta}t/2)}{\omega_{\alpha\beta}^2}$$

$$+ \frac{1}{\hbar^2}\left[\sum_{\beta\alpha''}\mathcal{H}_{\alpha\beta}^I\mathcal{H}_{\beta\alpha''}^I\rho_{\alpha''\alpha'}(0)\mathcal{F}(t) + C.C.(\alpha \leftrightarrow \alpha')\right]. \quad (16.24)$$

We consider the limit of this equation for $t \to \infty$. As we have seen on various occasions within the text (cf. Fig. 11.1)

$$\lim_{t\to\infty}\frac{4\sin^2(\omega_{\alpha\beta}t/2)}{\omega_{\alpha\beta}^2} = 2\pi t\delta(\omega_{\alpha\beta}) = 2\pi\hbar t\delta(E_\alpha - E_\beta),$$

where we have used the definition of the Bohr frequencies in terms of the energies of the states of the physical system. Regarding the function $\mathcal{F}(t)$, while its real part produces again a Dirac delta over the energy, the imaginary part behaves, at the limit of $t \to \infty$, as shown in Fig. 16.2. It can rigorously be shown within the distribution theory that we have

$$\lim_{t\to\infty}\mathcal{F}(t) = -\pi t\delta(\omega_{\alpha\beta}) - itPP\frac{1}{\omega_{\alpha\beta}} = -\pi\hbar t\left[\delta(E_\alpha - E_\beta) + \frac{i}{\pi}PP\frac{1}{E_\alpha - E_\beta}\right],$$

where the symbol PP means the Cauchy principal value.

We are now able to write the kinetic equation that generalises Eq. (11.9), valid for the diagonal elements of the density matrix, to the case of coherences. Starting

---

[8]We note that in Chap. 11 we only introduced the diagonal elements of the density matrix, denoted for simplicity by the symbol ρ_α instead of $\rho_{\alpha\alpha}$.

Fig. 16.2 The imaginary part of $\mathcal{F}(t)$ is plotted as a function of $\omega_{\alpha\beta}$ for a fixed time t. As t increases, the behaviour of the function becomes more and more similar to that one of the function $-t/\omega_{\alpha\beta}$ (*dotted line*), except in the origin where it is zero

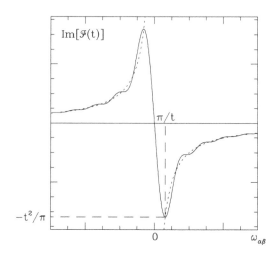

with Eq. (16.24) and noting that all the terms in the right hand side behave linearly with t, we can write, for $t \to \infty$

$$\frac{\mathrm{d}}{\mathrm{d}t}\rho_{\alpha\alpha'} = \sum_{\beta\beta'} T_{\alpha\alpha'\beta\beta'}\rho_{\beta\beta'} - \sum_{\alpha''}\left(R_{\alpha\alpha''}\rho_{\alpha''\alpha'} + R^*_{\alpha'\alpha''}\rho_{\alpha\alpha''}\right), \qquad (16.25)$$

where $T_{\alpha\alpha'\beta\beta'}$, the rate of transfer from the coherence $\rho_{\beta\beta'}$ to the coherence $\rho_{\alpha\alpha'}$, is given by

$$T_{\alpha\alpha'\beta\beta'} = \frac{2\pi}{\hbar}\mathcal{H}^I_{\alpha\beta}\mathcal{H}^I_{\beta'\alpha'}\delta(E_\alpha - E_\beta),$$

and where $R_{\alpha\alpha''}$, the relaxation rate that relates the coherence $\rho_{\alpha\alpha'}$ to the coherence $\rho_{\alpha''\alpha'}$, is given by

$$R_{\alpha\alpha''} = \frac{\pi}{\hbar}\sum_\beta \mathcal{H}^I_{\alpha\beta}\mathcal{H}^I_{\beta\alpha''}\left[\delta(E_\alpha - E_\beta) + \frac{i}{\pi}\mathrm{PP}\frac{1}{E_\alpha - E_\beta}\right].$$

It is easy to show that Eq. (16.25) coincides with Eq. (11.9) in the case of the random phase approximation, i.e. when we consider only the diagonal elements of the density matrix. We have, in fact,

$$T_{\alpha\alpha\beta\beta} = \frac{2\pi}{\hbar}\left|\mathcal{H}^I_{\alpha\beta}\right|^2\delta(E_\alpha - E_\beta) = P_{\alpha\beta},$$

where $P_{\alpha\beta}$ is the transition probability per unit time between the states $|\alpha\rangle$ and $|\beta\rangle$ (or between the states $|\beta\rangle$ and $|\alpha\rangle$) given by Eq. (11.10) (Fermi's golden rule). Similarly,

$$R_{\alpha\alpha} + R^*_{\alpha\alpha} = \frac{2\pi}{\hbar}\sum_\beta\left|\mathcal{H}^I_{\alpha\beta}\right|^2\delta(E_\alpha - E_\beta) = \sum_\beta P_{\alpha\beta}.$$

Equation (16.25) can therefore be rightly considered the generalisation of Fermi's golden rule to the case of coherences. It is important to stress the presence of the imaginary factor in the relaxation rates. Such factor is responsible for some phenomena typical of the interaction between material systems and the radiation field such as, in particular, the anomalous dispersion phenomena that occur in the propagation of polarised radiation in an anisotropic medium (Faraday effect, Macaluso-Corbino effect, etc.).

16.12 Sums over the Magnetic Quantum Numbers

Here, we want to prove Eq. (11.17). That is, we want to show that, for any polarisation unit vector $\mathbf{e}$, having defined the averages of the square moduli of the dipole matrix elements $\mathcal{A}$ and $\mathcal{A}'$ over the magnetic quantum numbers by the equations

$$\mathcal{A} = \frac{1}{g_a g_b} \sum_{\alpha, \beta} |\mathbf{r}_{b\beta, a\alpha} \cdot \mathbf{e}|^2,$$

$$\mathcal{A}' = |\mathbf{r}_{ba}|^2 = \frac{1}{g_a g_b} \sum_{\alpha, \beta} |\mathbf{r}_{b\beta, a\alpha}|^2 = \frac{1}{g_a g_b} \sum_{\alpha, \beta} \langle u_{b\beta}|\mathbf{r}|u_{a\alpha}\rangle \cdot \langle u_{a\alpha}|\mathbf{r}|u_{b\beta}\rangle,$$

we have

$$\mathcal{A} = \frac{1}{3}\mathcal{A}'.$$

The indices a and b in the previous equations denote any two energy levels of the atomic system while the indices α and β denote the respective magnetic sublevels, which are degenerate with respect to the energy. To demonstrate the equation, we need to introduce a more detailed notation which takes into account the fact that the atomic levels are normally characterised not only by a set of internal quantum numbers γ (which specify the configuration and the term), but also by the quantum number for the angular momentum J and the magnetic quantum number M. Applying the formal substitutions

$$|u_{a\alpha}\rangle \rightarrow |\gamma_a J_a M_a\rangle, \qquad |u_{b\beta}\rangle \rightarrow |\gamma_b J_b M_b\rangle,$$

$$g_a \rightarrow 2J_a + 1, \qquad g_b \rightarrow 2J_b + 1,$$

we obtain

$$\mathcal{A} = \sum_{M_a M_b} \frac{|\langle \gamma_b J_b M_b|\mathbf{r} \cdot \mathbf{e}|\gamma_a J_a M_a\rangle|^2}{(2J_a + 1)(2J_b + 1)},$$

$$\mathcal{A}' = \sum_{M_a M_b} \frac{\langle \gamma_b J_b M_b|\mathbf{r}|\gamma_a J_a M_a\rangle \cdot \langle \gamma_a J_a M_a|\mathbf{r}|\gamma_b J_b M_b\rangle}{(2J_a + 1)(2J_b + 1)}.$$

To calculate $\mathcal{A}$ we apply the Wigner-Eckart theorem, noting that the scalar product $\mathbf{r} \cdot \mathbf{e}$ can be expressed in terms of the spherical components of the two vectors. We have in fact (cf. Eq. (9.5))

$$\mathbf{r} \cdot \mathbf{e} = \sum_q (-1)^q r_q e_{-q}.$$

Using Eq. (9.4) we have

$$\langle \gamma_b J_b M_b | r_q | \gamma_a J_a M_a \rangle$$

$$= (-1)^{J_a + M_b + 1} \sqrt{2J_b + 1} \begin{pmatrix} J_b & J_a & 1 \\ -M_b & M_a & q \end{pmatrix} \langle \gamma_b J_b \| \mathbf{r} \| \gamma_a J_a \rangle,$$

and we obtain

$$\mathcal{A} = \sum_{qq'} (-1)^{q+q'} e_{-q} (e_{-q'})^*$$

$$\times \sum_{M_a M_b} \begin{pmatrix} J_b & J_a & 1 \\ -M_b & M_a & q \end{pmatrix} \begin{pmatrix} J_b & J_a & 1 \\ -M_b & M_a & q' \end{pmatrix} \frac{|\langle \gamma_b J_b \| \mathbf{r} \| \gamma_a J_a \rangle|^2}{2J_a + 1}.$$

The sum over M_a and M_b of the product of the two 3-j symbols can be calculated using the property of the 3-j symbols of Eq. (7.18). We have

$$\sum_{M_a M_b} \begin{pmatrix} J_b & J_a & 1 \\ -M_b & M_a & q \end{pmatrix} \begin{pmatrix} J_b & J_a & 1 \\ -M_b & M_a & q' \end{pmatrix} = \frac{1}{3} \delta_{qq'},$$

and we obtain, being $\sum_q e_q (e_q)^* = 1$

$$\mathcal{A} = \frac{1}{3} \frac{|\langle \gamma_b J_b \| \mathbf{r} \| \gamma_a J_a \rangle|^2}{2J_a + 1}.$$

To calculate the quantity $\mathcal{A}'$ we proceed in a similar way first noting that $\langle \gamma_a J_a M_a | \mathbf{r} | \gamma_b J_b M_b \rangle = \langle \gamma_b J_b M_b | \mathbf{r} | \gamma_a J_a M_a \rangle^*$. We obtain

$$\mathcal{A}' = \sum_q \sum_{M_a M_b} \begin{pmatrix} J_b & J_a & 1 \\ -M_b & M_a & q \end{pmatrix} \begin{pmatrix} J_b & J_a & 1 \\ -M_b & M_a & q' \end{pmatrix} \frac{|\langle \gamma_b J_b \| \mathbf{r} \| \gamma_a J_a \rangle|^2}{2J_a + 1},$$

and using the same property of the 3-j symbols and summing over q we arrive at the result that we wanted to prove, i.e.

$$\mathcal{A}' = \frac{|\langle \gamma_b J_b \| \mathbf{r} \| \gamma_a J_a \rangle|^2}{2J_a + 1} = 3\mathcal{A}.$$

The above results can be used to express the quantity $|\mathbf{r}_{ba}|^2$ that we introduced within the text in terms of the reduced matrix elements of the spherical tensor $\mathbf{r}$.

Since $|\mathbf{r}_{ba}|^2 = \mathcal{A}'$, we have

$$|\mathbf{r}_{ba}|^2 = \frac{|\langle \gamma_b J_b \| \mathbf{r} \| \gamma_a J_a \rangle|^2}{2J_a + 1}.$$

On the other hand, being $|\mathbf{r}_{ba}|^2 = |\mathbf{r}_{ab}|^2$, we obtain by symmetry

$$|\mathbf{r}_{ba}|^2 = |\mathbf{r}_{ab}|^2 = \frac{|\langle \gamma_b J_b \| \mathbf{r} \| \gamma_a J_a \rangle|^2}{2J_a + 1} = \frac{|\langle \gamma_a J_a \| \mathbf{r} \| \gamma_b J_b \rangle|^2}{2J_b + 1},$$

an equation that relates the reduced matrix elements under the exchange of the bra with the ket.

In spectroscopy, the concept of line (or transition) strength is commonly used. Such quantity is invariant with respect to the exchange of the lower and upper level, and is defined by

$$S = g_b \left| \langle \gamma_b J_b \| \mathbf{d} \| \gamma_a J_a \rangle \right|^2 = g_a \left| \langle \gamma_a J_a \| \mathbf{d} \| \gamma_b J_b \rangle \right|^2,$$

where $\mathbf{d} = -e_0 \mathbf{r}$ is the electric dipole operator. The quantities introduced in the text are therefore related to the line strength via the relation

$$|\mathbf{r}_{ba}|^2 = |\mathbf{r}_{ab}|^2 = \frac{1}{e_0^2} \frac{S}{g_a g_b}.$$

These relations can then be used to express the Einstein coefficients in terms of the line strength instead of in terms of the dipole matrix elements. For example, recalling Eq. (11.20), the Einstein coefficient A_{ab} can be written in the form

$$g_a A_{ab} = \frac{64\pi^4 \nu_{ab}^3}{3hc^3} S.$$

An alternative quantity that is also used to characterise the strength of a line (or a transition) is the so-called oscillator strength. This quantity is introduced in the following way. The absorption coefficient of a plasma of "classical" atoms, described by the Lorentz atomic model and integrated in frequency is given by

$$\left[k_R^{(a)} \right]_{class} = \mathcal{N} \frac{\pi e_0^2}{mc},$$

where $\mathcal{N}$ is the number density of atoms. Comparing this expression with that one for $k_R^{(a)}$ obtained in Sect. 11.9 (Eq. (11.33)), we see that the two quantities coincide if we identify $\mathcal{N}$ with $\mathcal{N}_b$ and multiply the classical expression for the dimensionless quantity f_{ba}, known as the oscillator strength of the transition, given by

$$f_{ba} = \frac{8\pi^2 m \nu_{ab}}{3h} g_a |\mathbf{r}_{ba}|^2.$$

The oscillator strength may be considered as a parameter measuring the efficiency of the transition, since it represents a sort of "equivalent number" of classical oscillators. Typically, it is a relatively small number that can reach values of the order of unity only for the strongest spectral lines. The relations between oscillator strength, line strength, and Einstein coefficients are easily obtained using the previous relations. For example, we have

$$
g_b f_{ba} = \frac{8\pi^2 m v_{ab}}{3he_0^2}\mathcal{S}, \qquad g_a A_{ab} = \frac{8\pi^2 e_0^2 v_{ab}^2}{mc^3} g_b f_{ba}.
$$

16.13 Calculation of a Matrix Element

We wish to calculate the probability per unit time that the following elementary process occurs: a non-relativistic free electron of momentum $\mathbf{q}$ undergoes a transition to a free state of momentum $\mathbf{q}'$ due to absorption of a photon with wave vector $\mathbf{k}$. According to Fermi's golden rule, repeating the arguments presented in Sect. 11.4 but without introducing the dipole approximation,[9] such probability is proportional to the squared modulus of the matrix element $\mathcal{M}$ given by

$$
\mathcal{M} = \langle u_f | e^{i\mathbf{k}\cdot\mathbf{r}} \mathbf{p} \cdot \mathbf{e} | u_i \rangle,
$$

where $|u_i\rangle$ and $|u_f\rangle$ are the eigenvectors of the atomic system (in our case of the free electron) in the initial and final state, respectively, $\mathbf{e}$ is the polarisation unit vector of the absorbed photon, and $\mathbf{p}$ is the momentum operator of the electron. Within the representation of the wavefunctions, where the operator $\mathbf{p}$ is given by $-i\hbar\,\mathrm{grad}$, the matrix element $\mathcal{M}$ is

$$
\mathcal{M} = -i\hbar\mathbf{e} \cdot \int \psi_f^*(\mathbf{r})e^{i\mathbf{k}\cdot\mathbf{r}}\,\mathrm{grad}\big[\psi_i(\mathbf{r})\big]\,d^3\mathbf{r}.
$$

On the other hand, the eigenfunctions ψ_f and ψ_i are of the type of a plane wave, i.e.

$$
\psi_f(\mathbf{r}) = \frac{1}{\sqrt{\mathcal{V}}}e^{i\mathbf{q}'\cdot\mathbf{r}/\hbar}, \qquad \psi_i(\mathbf{r}) = \frac{1}{\sqrt{\mathcal{V}}}e^{i\mathbf{q}\cdot\mathbf{r}/\hbar},
$$

where $\mathcal{V}$ is the normalisation volume. Substituting in the integral we have

$$
\mathcal{M} = \frac{\mathbf{e}\cdot\mathbf{q}}{\mathcal{V}}\int e^{-i(\mathbf{q}'-\hbar\mathbf{k}-\mathbf{q})\cdot\mathbf{r}/\hbar}\,d^3\mathbf{r}.
$$

---

[9]The dipole approximation is appropriate when considering the interaction between radiation and electrons that are bound in an atom. For free electrons, described by eigenfunctions of the type of a plane wave, the approximation cannot be applied.

The integral is null unless the argument of the exponential is zero. This leads to the equality $\mathbf{q}' = (\hbar\mathbf{k} + \mathbf{q})$ which represents the conservation of momentum. In such case, the integral is simply equal to $\mathcal{V}$, so we obtain

$$\mathcal{M} = \mathbf{e} \cdot \mathbf{q} = \mathbf{e} \cdot (\mathbf{q}' - \hbar\mathbf{k}).$$

16.14   Gauge Invariance in Quantum Electrodynamics

Consider the quantity $\mathcal{R}_{\mathrm{fi}}$ defined in Eq. (15.22) of the text that we rewrite here in the form

$$\mathcal{R}_{\mathrm{fi}} = P + Q,$$

where

$$P = W_s^{\dagger}(\mathbf{p}')(\alpha \cdot \mathbf{e}'^*)\frac{H_{\mathbf{g}} + \epsilon + \hbar\omega}{(\epsilon + \hbar\omega)^2 - \epsilon_{\mathbf{g}}^2}(\alpha \cdot \mathbf{e})W_r(\mathbf{p}),$$

$$Q = W_s^{\dagger}(\mathbf{p}')(\alpha \cdot \mathbf{e})\frac{H_{\mathbf{h}} + \epsilon - \hbar\omega'}{(\epsilon - \hbar\omega')^2 - \epsilon_{\mathbf{h}}^2}(\alpha \cdot \mathbf{e}'^*)W_r(\mathbf{p}).$$

We want to demonstrate that $\mathcal{R}_{\mathrm{fi}}$ is invariant with respect to the transformation

$$\alpha \cdot \mathbf{e} \rightarrow \alpha \cdot \mathbf{e} + C(\alpha \cdot \mathbf{u} - 1), \tag{16.26}$$

where C is an arbitrary constant and where $\mathbf{u}$ is the unit vector of the direction of the initial photon ($\mathbf{u} = \mathbf{k}/k$). Performing such transformation, the quantities P and Q are transformed according to the equations

$$P \rightarrow P + CP', \qquad Q \rightarrow Q + CQ',$$

where

$$P' = W_s^{\dagger}(\mathbf{p}')(\alpha \cdot \mathbf{e}'^*)\frac{H_{\mathbf{g}} + \epsilon + \hbar\omega}{(\epsilon + \hbar\omega)^2 - \epsilon_{\mathbf{g}}^2}[(\alpha \cdot \mathbf{u}) - 1]W_r(\mathbf{p}),$$

$$Q' = W_s^{\dagger}(\mathbf{p}')[(\alpha \cdot \mathbf{u}) - 1]\frac{H_{\mathbf{h}} + \epsilon - \hbar\omega'}{(\epsilon - \hbar\omega')^2 - \epsilon_{\mathbf{h}}^2}(\alpha \cdot \mathbf{e}'^*)W_r(\mathbf{p}).$$

We multiply the two quantities P' and Q' by the product $c\hbar k$ and note that

$$c\hbar k[(\alpha \cdot \mathbf{u}) - 1] = c\hbar(\alpha \cdot \mathbf{k}) - \hbar\omega.$$

Recalling the kinematic relations of the Compton effect and noting that the quantities $\mathbf{g}$ and $\mathbf{h}$, contained respectively in P' and Q', are given by (see Eq. (15.15))

$$\mathbf{g} = \mathbf{p} + \hbar\mathbf{k}, \qquad \mathbf{h} = \mathbf{p} - \hbar\mathbf{k}' = \mathbf{p}' - \hbar\mathbf{k},$$

we can perform the following substitution in the expression of P'

$$\hbar\mathbf{k} = \mathbf{g} - \mathbf{p},$$

and in the expression of Q'

$$\hbar\mathbf{k} = \mathbf{p}' - \mathbf{h}.$$

Substituting, and recalling also that $\epsilon - \hbar\omega' = \epsilon' - \hbar\omega$, we obtain

$$c\hbar k\, P' = W_s^\dagger(\mathbf{p}')(\boldsymbol{\alpha}\cdot\mathbf{e}'^*)\frac{H_\mathbf{g} + \epsilon + \hbar\omega}{(\epsilon + \hbar\omega)^2 - \epsilon_\mathbf{g}^2}[c\boldsymbol{\alpha}\cdot\mathbf{g} - c\boldsymbol{\alpha}\cdot\mathbf{p} - \hbar\omega]W_r(\mathbf{p}),$$

$$c\hbar k\, Q' = W_s^\dagger(\mathbf{p}')[c\boldsymbol{\alpha}\cdot\mathbf{p}' - c\boldsymbol{\alpha}\cdot\mathbf{h} - \hbar\omega]\frac{H_\mathbf{h} + \epsilon' - \hbar\omega}{(\epsilon' - \hbar\omega)^2 - \epsilon_\mathbf{h}^2}(\boldsymbol{\alpha}\cdot\mathbf{e}'^*)W_r(\mathbf{p}).$$

Within the square brackets, we add and subtract the factor βmc^2 and recall that an expression of the type $(c\boldsymbol{\alpha}\cdot\mathbf{q} + \beta mc^2)$, with $\mathbf{q}$ arbitrary, is the Dirac Hamiltonian $H_\mathbf{q}$. We obtain

$$c\hbar k\, P' = W_s^\dagger(\mathbf{p}')(\boldsymbol{\alpha}\cdot\mathbf{e}'^*)\frac{H_\mathbf{g} + \epsilon + \hbar\omega}{(\epsilon + \hbar\omega)^2 - \epsilon_\mathbf{g}^2}[H_\mathbf{g} - H_\mathbf{p} - \hbar\omega]W_r(\mathbf{p}),$$

$$c\hbar k\, Q' = W_s^\dagger(\mathbf{p}')[H_{\mathbf{p}'} - H_\mathbf{h} - \hbar\omega]\frac{H_\mathbf{h} + \epsilon' - \hbar\omega}{(\epsilon' - \hbar\omega)^2 - \epsilon_\mathbf{h}^2}(\boldsymbol{\alpha}\cdot\mathbf{e}'^*)W_r(\mathbf{p}).$$

Now, noting that

$$H_\mathbf{p} W_r(\mathbf{p}) = \epsilon\, W_r(\mathbf{p}), \qquad W_s^\dagger(\mathbf{p}')H_{\mathbf{p}'} = \epsilon'\, W_s^\dagger(\mathbf{p}'),$$

we have

$$c\hbar k\, P' = W_s^\dagger(\mathbf{p}')(\boldsymbol{\alpha}\cdot\mathbf{e}'^*)\frac{H_\mathbf{g} + \epsilon + \hbar\omega}{(\epsilon + \hbar\omega)^2 - \epsilon_\mathbf{g}^2}[H_\mathbf{g} - \epsilon - \hbar\omega]W_r(\mathbf{p}),$$

$$c\hbar k\, Q' = W_s^\dagger(\mathbf{p}')[\epsilon' - H_\mathbf{h} - \hbar\omega]\frac{H_\mathbf{h} + \epsilon' - \hbar\omega}{(\epsilon' - \hbar\omega)^2 - \epsilon_\mathbf{h}^2}(\boldsymbol{\alpha}\cdot\mathbf{e}'^*)W_r(\mathbf{p}).$$

Finally, taking into account that

$$[H_\mathbf{g} + \epsilon + \hbar\omega][H_\mathbf{g} - \epsilon - \hbar\omega] = \epsilon_\mathbf{g}^2 - (\epsilon + \hbar\omega)^2,$$

$$[\epsilon' - H_\mathbf{h} - \hbar\omega][H_\mathbf{h} + \epsilon' - \hbar\omega] = (\epsilon' - \hbar\omega)^2 - \epsilon_\mathbf{h}^2,$$

we obtain

$$c\hbar k\, P' = -W_s^\dagger(\mathbf{p}')(\boldsymbol{\alpha}\cdot\mathbf{e}'^*)W_r(\mathbf{p}), \qquad c\hbar k\, Q' = W_s^\dagger(\mathbf{p}')(\boldsymbol{\alpha}\cdot\mathbf{e}'^*)W_r(\mathbf{p}),$$

from which it follows that $(P' + Q') = 0$. This shows that the quantity $\mathcal{R}_{\mathrm{fi}}$ is invariant with respect to the transformation (16.26). In all similarity, it can be shown that $\mathcal{R}_{\mathrm{fi}}$ is also invariant with respect to the transformation

$$\boldsymbol{\alpha} \cdot \mathbf{e}'^* \to \boldsymbol{\alpha} \cdot \mathbf{e}'^* + C'(\boldsymbol{\alpha} \cdot \mathbf{u}' - 1),$$

where C' is an arbitrary constant and where $\mathbf{u}'$ is the unit vector of the direction of the final photon $(\mathbf{u}' = \mathbf{k}'/k')$.

An alternative way to express these invariant properties is to formally consider the quantity $\mathcal{R}_{\mathrm{fi}}$ as a function of the matrix $(\boldsymbol{\alpha} \cdot \mathbf{e})$, or, alternatively, of the matrix $(\boldsymbol{\alpha} \cdot \mathbf{e}'^*)$. From the above proof it follows that

$$\mathcal{R}_{\mathrm{fi}}\{\boldsymbol{\alpha} \cdot \mathbf{e} \to \boldsymbol{\alpha} \cdot \mathbf{u}\} = \mathcal{R}_{\mathrm{fi}}\{\boldsymbol{\alpha} \cdot \mathbf{e} \to 1\},$$
$$\mathcal{R}_{\mathrm{fi}}\{\boldsymbol{\alpha} \cdot \mathbf{e}'^* \to \boldsymbol{\alpha} \cdot \mathbf{u}'\} = \mathcal{R}_{\mathrm{fi}}\{\boldsymbol{\alpha} \cdot \mathbf{e}'^* \to 1\}. \tag{16.27}$$

16.15 The Gamma Matrices and the Relativistic Invariants

The relation between energy $\epsilon_{\mathbf{p}}$ and momentum $\mathbf{p}$ of a relativistic particle of mass m is

$$\epsilon_{\mathbf{p}}^2 = c^2 p^2 + m^2 c^4.$$

In particular, for a photon $(m = 0)$ we have

$$\epsilon_{\mathbf{p}} = cp, \quad \text{with } p = |\mathbf{p}|,$$

or, in terms of frequency and wavenumber

$$\hbar\omega = c\hbar k, \quad \text{with } k = |\mathbf{k}|.$$

Such relations may be formally simplified if we adopt a unit system in which $\hbar = c = 1$. The introduction of this convention is equivalent to define the unit time interval as the time needed by light to travel the unit of length. With this definition, the energy, the momentum, and the mass (and similarly for a photon, the angular frequency and the wavenumber) all assume the dimensions of the reciprocal of a length (or a time). The relation between momentum and energy is written, in this system of units, in the form

$$\epsilon_{\mathbf{p}}^2 = p^2 + m^2, \quad \text{or} \quad \epsilon_{\mathbf{p}}^2 - p^2 = m^2,$$

and for a photon

$$\epsilon_{\mathbf{p}} = p, \quad \text{or} \quad \omega = k.$$

We now introduce with the symbol $\mathcal{P}_\mu$ $(\mu = 0, 1, 2, 3)$ the quadrivector momentum-energy of the particle. It is an entity with four components that are defined in this way

$$\mathcal{P}_0 = \epsilon_{\mathbf{p}}, \qquad \mathcal{P}_1 = p_1 = p_x, \qquad \mathcal{P}_2 = p_2 = p_y, \qquad \mathcal{P}_3 = p_3 = p_z,$$

or, in a more compact form

$$\mathcal{P} = (\epsilon_{\mathbf{p}}, \mathbf{p}).$$

Defining the metric tensor $g_{\mu\nu}$ as

$$g_{00} = 1, \qquad g_{0i} = g_{i0} = 0 \quad (i = 1, 2, 3), \qquad g_{jk} = -\delta_{ik} \quad (j, k = 1, 2, 3),$$

the scalar product of two quadrivectors $\mathcal{P}$ and $\mathcal{Q}$ is

$$(\mathcal{PQ}) = \sum_{\mu\nu} g_{\mu\nu} \mathcal{P}_\mu \mathcal{Q}_\nu = \epsilon_{\mathbf{p}} \epsilon_{\mathbf{q}} - \mathbf{p} \cdot \mathbf{q}.$$

In particular we have

$$\mathcal{P}^2 = (\mathcal{PP}) = \sum_{\mu\nu} g_{\mu\nu} \mathcal{P}_\mu \mathcal{P}_\nu = \epsilon_{\mathbf{p}}^2 - p^2 = m^2.$$

These quantities (the scalar product of two quadrivectors defined by the above metric tensor and, in particular, the square of a quadrivector) are relativistic invariants, i.e. do not change under Lorentz transformations. We are now going to show how the probability amplitudes of Compton scattering can be expressed in terms of these invariants.

Consider the quantity $\mathcal{R}_{\mathrm{fi}}$ defined in Eq. (15.22). This quantity is composed of two terms that we denote by P and Q. For the first one we have, taking into account the system of units we have introduced ($c = \hbar = 1$)

$$P = W_s^\dagger (\mathbf{p}') (\boldsymbol{\alpha} \cdot \mathbf{e}'^*) \frac{H_{\mathbf{g}} + \epsilon + \omega}{(\epsilon + \omega)^2 - \epsilon_{\mathbf{g}}^2} (\boldsymbol{\alpha} \cdot \mathbf{e}) W_r (\mathbf{p}),$$

where

$$H_{\mathbf{g}} = \boldsymbol{\alpha} \cdot \mathbf{g} + \beta m,$$

with

$$\mathbf{g} = \mathbf{p} + \mathbf{k}, \qquad \epsilon_{\mathbf{g}} = \sqrt{g^2 + m^2}.$$

Recalling that the square of the Dirac matrix β is unity, we can write

$$P = W_s^\dagger (\mathbf{p}') (\boldsymbol{\alpha} \cdot \mathbf{e}'^*) \beta^2 \frac{H_{\mathbf{g}} + \epsilon + \omega}{(\epsilon + \omega)^2 - \epsilon_{\mathbf{g}}^2} \beta^2 (\boldsymbol{\alpha} \cdot \mathbf{e}) W_r (\mathbf{p}).$$

If we now also recall that the Dirac matrix β anticommutes with any of the $\boldsymbol{\alpha}$ matrices, we obtain

$$P = W_s^\dagger (\mathbf{p}') \beta (\boldsymbol{\alpha} \cdot \mathbf{e}'^*) \frac{\beta (H_{\mathbf{g}} + \epsilon + \omega)}{(\epsilon + \omega)^2 - \epsilon_{\mathbf{g}}^2} \beta (\boldsymbol{\alpha} \cdot \mathbf{e}) \beta W_r (\mathbf{p}).$$

We now define the matrices γ_μ $(\mu = 0, 1, 2, 3)$

$$\gamma_0 = \beta, \qquad \gamma_1 = \beta\alpha_1, \qquad \gamma_2 = \beta\alpha_2, \qquad \gamma_3 = \beta\alpha_3.$$

The fundamental property of these matrices regards their anticommutator which is (as easily derived from the properties of the $\boldsymbol{\alpha}$ and β matrices)

$$\{\gamma_\mu, \gamma_\nu\} = \gamma_\mu\gamma_\nu + \gamma_\nu\gamma_\mu = 2g_{\mu\nu}, \tag{16.28}$$

where $g_{\mu\nu}$ is the metric tensor that we have previously defined. Moreover, given an arbitrary quadrivector $\mathcal{V}$, we define by the symbol $\rlap{/}{\mathcal{V}}$ the matrix

$$\rlap{/}{\mathcal{V}} = \sum_\mu \gamma_\mu \mathcal{V}_\mu.$$

With these definitions, the quantity P can be written in the form

$$P = W_s^\dagger(\mathbf{p}')\rlap{/}{\mathcal{E}}'^* \frac{\rlap{/}{\mathcal{G}} + m}{\mathcal{G}^2 - m^2}\rlap{/}{\mathcal{E}}\gamma_0 W_r(\mathbf{p}), \tag{16.29}$$

where the quadrivectors $\mathcal{G}$, $\mathcal{E}$ and $\mathcal{E}'$ are given by

$$\mathcal{G} = (\omega + \epsilon, \mathbf{g}), \qquad \mathcal{E} = (0, \mathbf{e}), \qquad \mathcal{E}' = (0, \mathbf{e}').$$

We note that the quadrivector $\mathcal{G}$ can also be written in the form

$$\mathcal{G} = \mathcal{P} + \mathcal{K},$$

where

$$\mathcal{K} = (\omega, \mathbf{k}).$$

If we now consider the quantity P^*, complex conjugate of P, we need to proceed carefully because the γ matrices (except γ_0) are not Hermitian. We have in fact

$$\gamma_0^\dagger = \beta^\dagger = \beta = \gamma_0,$$

$$\gamma_i^\dagger = (\beta\alpha_i)^\dagger = \alpha_i^\dagger\beta^\dagger = \alpha_i\beta = -\beta\alpha_i = -\gamma_i \quad (i = 1, 2, 3).$$

These properties can be summarised in only one relation

$$\gamma_\mu^\dagger = \gamma_0\gamma_\mu\gamma_0,$$

which implies, for an arbitrary quadrivector

$$\rlap{/}{\mathcal{V}}^\dagger = \gamma_0\rlap{/}{\mathcal{V}}^*\gamma_0.$$

We therefore obtain

$$P^* = W_r^\dagger(\mathbf{p})\gamma_0\gamma_0\rlap{/}{\mathcal{E}}^*\gamma_0\frac{\gamma_0\rlap{/}{\mathcal{G}}\gamma_0 + m}{\mathcal{G}^2 - m^2}\gamma_0\rlap{/}{\mathcal{E}}'\gamma_0 W_s(\mathbf{p}'),$$

or

$$P^* = W_r^\dagger(\mathbf{p})\not{\mathcal{E}}^* \frac{\not{\mathcal{G}}+m}{\mathcal{G}^2-m^2}\not{\mathcal{E}}' \gamma_0 W_s(\mathbf{p}').$$ (16.30)

Similar considerations can be repeated for the other term Q of Eq. (15.22) defined by

$$Q = W_s^\dagger(\mathbf{p}')(\alpha \cdot \mathbf{e})\frac{H_\mathbf{h}+\epsilon-\omega'}{(\epsilon-\omega')^2-\epsilon_\mathbf{h}^2}(\alpha \cdot \mathbf{e}'^*)W_r(\mathbf{p}),$$

where

$$H_\mathbf{h} = \alpha \cdot \mathbf{h} + \beta m,$$

with

$$\mathbf{h} = \mathbf{p} - \mathbf{k}', \qquad \epsilon_\mathbf{h} = \sqrt{h^2+m^2}.$$

We have

$$Q = W_s^\dagger(\mathbf{p}')\not{\mathcal{E}}\frac{\not{\mathcal{H}}+m}{\mathcal{H}^2-m^2}\not{\mathcal{E}}'^* \gamma_0 W_r(\mathbf{p}),$$ (16.31)

where the quadrivector $\mathcal{H}$ is defined by

$$\mathcal{H} = (\epsilon-\omega', \mathbf{h}) = \mathcal{P} - \mathcal{K}',$$

being

$$\mathcal{K}' = (\omega', \mathbf{k}').$$

Still in analogy to what discussed before, we also have

$$Q^* = W_r^\dagger(\mathbf{p})\not{\mathcal{E}}'\frac{\not{\mathcal{H}}+m}{\mathcal{H}^2-m^2}\not{\mathcal{E}}^* \gamma_0 W_s(\mathbf{p}').$$ (16.32)

We can now evaluate the square of the modulus of the quantity $\mathcal{R}_{\mathrm{fi}}$. It is

$$|\mathcal{R}_{\mathrm{fi}}|^2 = (P+Q)(P^*+Q^*) = PP^* + PQ^* + QP^* + QQ^*,$$

where, using Eqs. (16.29)–(16.32), the four terms are given by

$$PP^* = W_s^\dagger(\mathbf{p}')\not{\mathcal{E}}'^* \frac{\not{\mathcal{G}}+m}{\mathcal{G}^2-m^2}\not{\mathcal{E}}\gamma_0 W_r(\mathbf{p}) W_r^\dagger(\mathbf{p})\not{\mathcal{E}}^*\frac{\not{\mathcal{G}}+m}{\mathcal{G}^2-m^2}\not{\mathcal{E}}'\gamma_0 W_s(\mathbf{p}'),$$

$$PQ^* = W_s^\dagger(\mathbf{p}')\not{\mathcal{E}}'^* \frac{\not{\mathcal{G}}+m}{\mathcal{G}^2-m^2}\not{\mathcal{E}}\gamma_0 W_r(\mathbf{p}) W_r^\dagger(\mathbf{p})\not{\mathcal{E}}'\frac{\not{\mathcal{H}}+m}{\mathcal{H}^2-m^2}\not{\mathcal{E}}^*\gamma_0 W_s(\mathbf{p}'),$$

$$QP^* = W_s^\dagger(\mathbf{p}')\not{\mathcal{E}}\frac{\not{\mathcal{H}}+m}{\mathcal{H}^2-m^2}\not{\mathcal{E}}'^*\gamma_0 W_r(\mathbf{p}) W_r^\dagger(\mathbf{p})\not{\mathcal{E}}^*\frac{\not{\mathcal{G}}+m}{\mathcal{G}^2-m^2}\not{\mathcal{E}}'\gamma_0 W_s(\mathbf{p}'),$$

$$QQ^* = W_s^\dagger(\mathbf{p}')\not{\mathcal{E}}\frac{\not{\mathcal{H}}+m}{\mathcal{H}^2-m^2}\not{\mathcal{E}}'^*\gamma_0 W_r(\mathbf{p}) W_r^\dagger(\mathbf{p})\not{\mathcal{E}}'\frac{\not{\mathcal{H}}+m}{\mathcal{H}^2-m^2}\not{\mathcal{E}}^*\gamma_0 W_s(\mathbf{p}').$$

These expressions can be simplified when one considers the average over the initial spin states of the electron and the sum over the final spin states of the electron. Taking into account the results we have obtained in Sect. 15.5 (Eqs. (15.18) and (15.20)) we have

$$\sum_{r=1,2} W_r(\mathbf{p}) W_r^\dagger(\mathbf{p}) = \frac{\epsilon_\mathbf{p} + H_\mathbf{p}}{2\epsilon_\mathbf{p}}, \qquad \sum_{s=1,2} W_s(\mathbf{p}') W_s^\dagger(\mathbf{p}') = \frac{\epsilon_{\mathbf{p}'} + H_{\mathbf{p}'}}{2\epsilon_{\mathbf{p}'}},$$

and defining the quadrivector

$$\mathcal{P} = (\epsilon_\mathbf{p}, \mathbf{p}), \qquad \mathcal{P}' = (\epsilon_{\mathbf{p}'}, \mathbf{p}'),$$

we obtain

$$\sum_{r=1,2} \gamma_0 W_r(\mathbf{p}) W_r^\dagger(\mathbf{p}) = \frac{\cancel{P} + m}{2\epsilon_\mathbf{p}}, \qquad \sum_{s=1,2} \gamma_0 W_s(\mathbf{p}') W_s^\dagger(\mathbf{p}') = \frac{\cancel{P}' + m}{2\epsilon_{\mathbf{p}'}}.$$

By denoting with the symbol $\langle \cdots \rangle$ the average over the spin states and using the definition of the trace of a matrix, according to which a scalar product of the form $W_1 \mathcal{X} W_2^\dagger$, with W_1 and W_2 arbitrary spinors and $\mathcal{X}$ an arbitrary matrix, can be written in the form $\mathrm{Tr}(W_2 W_1^\dagger \mathcal{X})$, we have

$$\langle PP^* \rangle = \frac{1}{2} \sum_{r=1,2} \sum_{s=1,2} PP^*$$

$$= \frac{1}{8\epsilon_\mathbf{p}\epsilon_{\mathbf{p}'}} \mathrm{Tr}\left\{ (\cancel{P}' + m)\cancel{\mathscr{E}}'^* \frac{\cancel{G} + m}{G^2 - m^2} \cancel{\mathscr{E}} (\cancel{P} + m)\cancel{\mathscr{E}}^* \frac{\cancel{G} + m}{G^2 - m^2} \cancel{\mathscr{E}}' \right\},$$

with similar expressions for the other three terms $\langle PQ^* \rangle$, $\langle QP^* \rangle$, and $\langle QQ^* \rangle$.

This last result can be greatly simplified if we sum over the polarisation states of the final photon and we average over the polarisation states of the initial photon. The average over the polarisation states of the initial photon, for example, is obtained by applying to the previous formula the formal substitution

$$\cancel{\mathscr{E}} (\cancel{P} + m) \cancel{\mathscr{E}}^* \rightarrow \frac{1}{2} \sum_{i=1,2} \cancel{\mathscr{E}}^{(i)} (\cancel{P} + m) \cancel{\mathscr{E}}^{(i)},$$

where

$$\mathcal{E}^{(i)} = (0, \mathbf{e}^{(i)}),$$

$\mathbf{e}^{(i)}$ ($i = 1, 2$) being two unit vectors that we can assume real, perpendicular to each other and perpendicular to the direction of the initial photon. Taking into account the invariance under gauge transformations described in Sect. 16.14 and, in particular, recalling Eq. (16.27), the sum can be modified by extending it to a third "unit quadrivector" (that we denote by $\mathcal{E}^{(3)}$) and subtracting then the contribution from another unit quadrivector. According to special relativity, this unit quadrivector, which we denote by $\mathcal{E}^{(0)}$, is of the purely temporal type. Defining

$$\mathcal{E}^{(3)} = \left(0, \mathbf{e}^{(3)}\right), \qquad \mathcal{E}^{(0)} = (1, \mathbf{0}),$$

where $\mathbf{e}^{(3)}$ is an unit vector directed along the direction of the incoming photon ($\mathbf{e}^{(3)} = \mathbf{k}/k$), and recalling the definition of the metric tensor, we apply the following transformation

$$\not{\mathcal{E}}(\not{P} + m)\not{\mathcal{E}}^* \rightarrow \frac{1}{2}\mathcal{S}, \quad \text{where } \mathcal{S} = -\sum_{i,j=0}^{3} g_{ij}\not{\mathcal{E}}^{(i)}(\not{P} + m)\not{\mathcal{E}}^{(j)}.$$

We note that the sum $\mathcal{S}$ can also be written in the form

$$\mathcal{S} = -\gamma_0(\not{P} + m)\gamma_0 + \left(\boldsymbol{\gamma} \cdot \mathbf{e}^{(1)}\right)(\not{P} + m)\left(\boldsymbol{\gamma} \cdot \mathbf{e}^{(1)}\right)$$
$$+ \left(\boldsymbol{\gamma} \cdot \mathbf{e}^{(2)}\right)(\not{P} + m)\left(\boldsymbol{\gamma} \cdot \mathbf{e}^{(2)}\right) + \left(\boldsymbol{\gamma} \cdot \mathbf{e}^{(3)}\right)(\not{P} + m)\left(\boldsymbol{\gamma} \cdot \mathbf{e}^{(3)}\right),$$

where $\boldsymbol{\gamma}$ is the formal vector defined by $\boldsymbol{\gamma} = (\gamma_1, \gamma_2, \gamma_3)$. The right-hand side can then be transformed to get

$$\mathcal{S} = -\gamma_0(\not{P} + m)\gamma_0 + \sum_{i,j=1}^{3} \gamma_i(\not{P} + m)\gamma_j\left[e_i^{(1)}e_j^{(1)} + e_i^{(2)}e_j^{(2)} + e_i^{(3)}e_j^{(3)}\right].$$

On the other hand, taking into account Eq. (15.7), the quantity in square brackets is equal to the Kronecker delta δ_{ij}, so we obtain

$$\mathcal{S} = -\gamma_0(\not{P} + m)\gamma_0 + \sum_{i=1}^{3} \gamma_i(\not{P} + m)\gamma_i = -\sum_{\mu,\nu} g_{\mu\nu}\gamma_\mu(\not{P} + m)\gamma_\nu.$$

Finally, we take into account the properties of the γ matrices. From Eq. (16.28) we have

$$\gamma_\mu\gamma_\nu = -\gamma_\nu\gamma_\mu + 2g_{\mu\nu}.$$

Moreover, it is easy to verify that the following relation holds

$$\sum_{\mu\nu} g_{\mu\nu}\gamma_\mu\gamma_\nu = 4,$$

and that, given the properties of the metric tensor,

$$\sum_{\mu} g_{\mu\nu}g_{\mu\rho} = \delta_{\nu\rho},$$

so that

$$\sum_{\mu\nu} g_{\mu\nu}g_{\mu\rho}\gamma_\nu = \gamma_\rho.$$

Taking advantage of these properties, we get after some algebra

$$\mathcal{S} = 2(\rlap{/}P - 2m).$$

Summarising the foregoing, the average on the states of polarisation of the initial photon is obtained by performing the formal transformation

$$\rlap{/}e(\rlap{/}P + m)\rlap{/}e^* \rightarrow \rlap{/}P - 2m.$$

Similarly, the sum over the polarisation states the final photon is obtained by performing the formal transformation

$$\rlap{/}e'(\rlap{/}P' + m)\rlap{/}e'^* \rightarrow 2(\rlap{/}P' - 2m).$$

Now we denote by the symbol $\langle\langle P P^* \rangle\rangle$ the quantity obtained by taking the average of $\langle P P^* \rangle$ over the states of initial polarisation and the sum of the same quantity over the states of final polarisation.[10] We have

$$\langle\langle P P^* \rangle\rangle = \frac{1}{4\epsilon_{\mathbf{p}}\epsilon_{\mathbf{p}'}} \mathrm{Tr}\left\{ (\rlap{/}P' - 2m)\frac{\rlap{/}G + m}{G^2 - m^2}(\rlap{/}P - 2m)\frac{\rlap{/}G + m}{G^2 - m^2} \right\}.$$

At this point it is necessary to briefly discuss the traces of the products of the γ matrices. It is easy to verify that the trace of the product of an odd number of γ matrices is null. When instead the number of γ matrices is zero or even, the result is different from zero. Denoting by a an arbitrary constant, with $\mathcal{A}$, $\mathcal{B}$, $\mathcal{C}$, and $\mathcal{D}$ four arbitrary quadrivectors, and recalling the definition of the scalar product of quadrivectors, we have

$$\mathrm{Tr}\{a\} = 4a, \qquad \mathrm{Tr}\{\rlap{/}A\rlap{/}B\} = 4(\mathcal{A}\mathcal{B}),$$

$$\mathrm{Tr}\{\rlap{/}A\rlap{/}B\rlap{/}C\rlap{/}D\} = 4\big[(\mathcal{A}\mathcal{B})(\mathcal{C}\mathcal{D}) - (\mathcal{A}\mathcal{C})(\mathcal{B}\mathcal{D}) + (\mathcal{A}\mathcal{D})(\mathcal{B}\mathcal{C})\big].$$

The first relation is obvious. For the second one we have

$$\mathrm{Tr}\{\rlap{/}A\rlap{/}B\} = \sum_{\mu\nu} \mathrm{Tr}\{\gamma_\mu\gamma_\nu\}A_\mu B_\nu,$$

and using the anticommutation property of the γ matrices

$$\mathrm{Tr}\{\gamma_\mu\gamma_\nu\} = 8g_{\mu\nu} - \mathrm{Tr}\{\gamma_\nu\gamma_\mu\}.$$

From the cyclic property of the trace it then follows that

$$\mathrm{Tr}\{\gamma_\mu\gamma_\nu\} = 4g_{\mu\nu},$$

---

[10]Recall that the first average, $\langle P P^* \rangle$, has a similar meaning with respect to the spin states of the electron.

which proves by simple substitution the second relation. For the third relation we have

$$\text{Tr}\{A\!\!\!/B\!\!\!/C\!\!\!/D\!\!\!/\} = \sum_{\mu\nu\rho\sigma} \text{Tr}\{\gamma_\mu\gamma_\nu\gamma_\rho\gamma_\sigma\}A_\mu B_\nu C_\rho D_\sigma,$$

and, for the anticommutation property of the γ matrices,

$$\begin{aligned}
\text{Tr}\{\gamma_\mu\gamma_\nu\gamma_\rho\gamma_\sigma\} &= 2g_{\mu\nu}\,\text{Tr}\{\gamma_\rho\gamma_\sigma\} - \text{Tr}\{\gamma_\nu\gamma_\mu\gamma_\rho\gamma_\sigma\} \\
&= 2g_{\mu\nu}\,\text{Tr}\{\gamma_\rho\gamma_\sigma\} - 2g_{\mu\rho}\,\text{Tr}\{\gamma_\nu\gamma_\sigma\} + \text{Tr}\{\gamma_\nu\gamma_\rho\gamma_\mu\gamma_\sigma\} \\
&= 2g_{\mu\nu}\,\text{Tr}\{\gamma_\rho\gamma_\sigma\} - 2g_{\mu\rho}\,\text{Tr}\{\gamma_\nu\gamma_\sigma\} + 2g_{\mu\sigma}\,\text{Tr}\{\gamma_\nu\gamma_\rho\} \\
&\quad - \text{Tr}\{\gamma_\nu\gamma_\rho\gamma_\sigma\gamma_\mu\}.
\end{aligned}$$

Using the cyclic property of the trace we then have

$$\text{Tr}\{\gamma_\mu\gamma_\nu\gamma_\rho\gamma_\sigma\} = g_{\mu\nu}\text{Tr}\{\gamma_\rho\gamma_\sigma\} - g_{\mu\rho}\,\text{Tr}\{\gamma_\nu\gamma_\sigma\} + g_{\mu\sigma}\,\text{Tr}\{\gamma_\nu\gamma_\rho\},$$

and using of the result previously obtained

$$\text{Tr}\{\gamma_\mu\gamma_\nu\gamma_\rho\gamma_\sigma\} = 4g_{\mu\nu}g_{\rho\sigma} - 4g_{\mu\rho}g_{\nu\sigma} + 4g_{\mu\sigma}g_{\nu\rho}.$$

The third relation is then finally obtained by simple substitution of this identity.

The result obtained for $\langle\langle PP^*\rangle\rangle$ shows that the trace contained in this quantity can be expressed exclusively in terms of scalar products of quadrivectors, i.e. in terms of relativistic invariants. Similar considerations can then be repeated for the other quantities $\langle\langle PQ^*\rangle\rangle$, $\langle\langle QP^*\rangle\rangle$, and $\langle\langle QQ^*\rangle\rangle$, which, once calculated, allow one to obtain the transition probability per unit time and the cross section. Obviously, in the particular case in which the electron is initially at rest, one finds again for the cross section the Klein-Nishina equation in the form of Eq. (15.37), which refers to the average over the polarisation states of the initial photon and the sum over the polarisation states of the final photon.

The formalism of the γ matrices presented in this chapter is very powerful and elegant. It allows one to deal with relative ease even with the most complex problems in quantum electrodynamics. In any case, we emphasize that the formalism that we have used in the text to deduce the Klein-Nishina equation, which does not make use of the γ matrices, was the first to be used in the applications.

16.16 Physical Constants

The constants are expressed in the cgs system of units with at most six significant digits.

Constant of gravitation: $G = 6.67428 \times 10^{-8}\ \text{cm}^3\,\text{g}^{-1}\,\text{s}^{-2}$
Velocity of light in vacuum: $c = 2.99792 \times 10^{10}\ \text{cm}\,\text{s}^{-1}$
Planck constant: $h = 6.62607 \times 10^{-27}\ \text{erg}\,\text{s}$

Reduced Planck constant: $\hbar = h/(2\pi) = 1.05457 \times 10^{-27}$ erg s
Boltzmann constant: $k_B = 1.38065 \times 10^{-16}$ erg K^{-1}
Charge of the electron (absolute value): $e_0 = 4.80320 \times 10^{-10}$ esu
Electron mass: $m = 9.10938 \times 10^{-28}$ g
Reduced electron mass: $m_r = m M_p/(m + M_p) = 9.10442 \times 10^{-28}$ g
Atomic mass unit (amu): $m_H = 1.66054 \times 10^{-24}$ g
Proton mass: $M_p = 1.67262 \times 10^{-24}$ g
Proton/electron mass ratio: $M_p/m = 1.83615 \times 10^3$
Avogadro constant: $N_A = 6.02214 \times 10^{23}$ mol^{-1}
Fine-structure constant: $\alpha = e_0^2/(\hbar c) = 7.29735 \times 10^{-3}$
Reciprocal of the fine-structure constant: $1/\alpha = \hbar c/e_0^2 = 137.036$
Classical radius of the electron: $r_c = e_0^2/(mc^2) = 2.81794 \times 10^{-13}$ cm
Compton wavelength of the electron: $\lambda_C = h/(mc) = 2.42631 \times 10^{-10}$ cm
Radius of the first Bohr orbit: $a_0 = \hbar^2/(me_0^2) = 5.29177 \times 10^{-9}$ cm
Rydberg constant: $R = me_0^4/(4\pi c\hbar^3) = 1.09737 \times 10^5$ cm^{-1}
Rydberg constant (hydrogen atom): $R_H = m_r e_0^4/(4\pi c\hbar^3) = 1.09677 \times 10^5$ cm^{-1}
Bohr magneton: $\mu_0 = e_0\hbar/(2mc) = 9.27401 \times 10^{-21}$ erg G^{-1}
Thomson cross section: $\sigma_T = 8\pi r_c^2/3 = 6.65246 \times 10^{-25}$ cm^2
Stefan-Boltzmann constant:[11] $\sigma = 5.67040 \times 10^{-5}$ erg cm^{-2} s^{-1} K^{-4}
Radiation density constant:[11] $a = 7.56577 \times 10^{-15}$ erg cm^{-3} K^{-4}
First radiation constant: $c_1 = 2\pi hc^2 = 3.74177 \times 10^{-5}$ erg cm^2 s^{-1}
Second radiation constant: $c_2 = hc/k_B = 1.43877$ cm K

---

[11] $\sigma = \dfrac{2\pi^5 k_B^4}{15h^3c^2}$, $a = \dfrac{4\sigma}{c} = \dfrac{8\pi^5 k_B^4}{15h^3c^3}$.

References

Abramowitz, M., Stegun, I.A.: Handbook of Mathematical Functions. Dover, New York (1971)
Bowen, I.S.: Astrophys. J. **67**, 1 (1928)
Bethe, H.A.: Z. Phys. **57**, 815 (1929)
Bethe, H.A., Salpeter, E.E.: Quantum Mechanics of One- and Two-Electron Atoms. Springer, Berlin (1957)
Bjorken, J.D., Drell, S.D.: Relativistic Quantum Fields. McGraw-Hill, New York (1965)
Brink, D.M., Satchler, G.R.: Angular Momentum. Clarendon Press, Oxford (1968)
Condon, E.U., Shortley, G.H.: The Theory of Atomic Spectra. Cambridge University Press, Cambridge (1935)
Dirac, P.A.M.: The Principles of Quantum Mechanics, 4th. edn. Clarendon Press, Oxford (1958)
Edlén, B.: Z. Astrophys. **22**, 30 (1942)
Fano, U.: J. Opt. Soc. Am. **39**, 859 (1949)
Giachetti, R., Sorace, E.: J. Phys. A **39**, 15207 (2006)
Landau, L., Lifchitz, E.: Théorie du Champ. Éditions Mir, Moscow (1966)
Landi Degl'Innocenti, E.: Fisica Solare. Springer, Milano (2008)
Landi Degl'Innocenti, E., Landolfi, M.: Polarization in Spectral Lines. Kluwer Academic, Dordrecht (2004)
McLennan, J.C.: Proc. R. Soc. Lond. A **120**, 327 (1928)
Mihalas, D.: Stellar Atmospheres, 2nd edn. Freeman, San Francisco (1978)
Racah, G.: Phys. Rev. **62**, 438 (1942)
Rybicki, G.B., Lightman, A.P.: Radiative Processes in Astrophysics. Wiley, New York (1979)
Schrödinger, E.: Statistical Thermodynamics. Heinemann, London (1961)
Toraldo di Francia, G.: La Diffrazione della Luce. Edizioni Scientifiche Einaudi. Boringhieri, Torino (1958)

Index

Printed in the United States
By Bookmasters